Ultrasonic Modelling for Non-destructive Testing

Ultrasonic Modelling for Non-destructive Testing

Editor

Michel DARMON

MDPI

Basel • Beijing • Wuhan • Barcelona • Belgrade • Novi Sad • Cluj • Manchester

Editor
Michel DARMON
Paris-Saclay University
Palaiseau
France

Editorial Office
MDPI
St. Alban-Anlage 66
4052 Basel, Switzerland

This is a reprint of articles from the Special Issue published online in the open access journal *Applied Sciences* (ISSN 2076-3417) (available at: https://www.mdpi.com/journal/applsci/special_issues/Ultrasonic_Modelling_Nondestructive_Testing).

For citation purposes, cite each article independently as indicated on the article page online and as indicated below:

Lastname, A.A.; Lastname, B.B. Article Title. *Journal Name* **Year**, *Volume Number*, Page Range.

ISBN 978-3-7258-0757-4 (Hbk)
ISBN 978-3-7258-0758-1 (PDF)
doi.org/10.3390/books978-3-7258-0758-1

Cover image courtesy of Michel DARMON

Contents

About the Editor

Michel DARMON

Michel DARMON is Research Director at The French Alternative Energies and Atomic Energy Commission (CEA). In 2015, he obtained the habilitation to supervise research (Paris-Saclay University). He previously received a diploma of engineering at ESPCI PSL, Paris, France, in 1999 and a Ph.D. degree in physical acoustics from the Paris-Cité University in 2002 for modeling the ultrasonic response of solid inclusions. He is a co-author of 70 scientific publications and belongs to the Editorial Board or is editor for several international journals. His research activities are mainly in ultrasonic modeling with a special interest in flaws scattering, wave propagation in complex media, multiple scattering, laser ultrasonics, non-linear acoustics, imaging, and artificial intelligence.

applied sciences

MDPI

Editorial

Special Issue on Ultrasonic Modeling for Non-Destructive Testing

Michel Darmon

Université Paris-Saclay, CEA, List, F-91120 Palaiseau, France; michel.darmon@cea.fr

1. Introduction

This Special Issue of Applied Sciences focuses on advancing modeling methods for the ultrasonic Non-destructive Testing (NDT) of materials. Ultrasonic techniques are employed for non-destructive purposes to evaluate the properties and damage states of structures devised for numerous applications (engineering, building materials, medicine, etc.). The advantages and inconveniences of ultrasonic testing compared to other NDT methods have been widely reviewed [1]. The coexistence of all the different NDT methods can allow the inspection of different kinds of structures and damages [2]. The scope of this Special Issue ranges from ultrasonic wave techniques for classical non-destructive evaluation, the structural health and condition monitoring of structures, existing or novel methods for imaging, ultrasonic characterization, non-linear acoustics, acoustic emission, laser ultrasonics, additive manufacturing, medical applications, and sensors, to, signal and noise analysis.

The current Special Issue notably aims to explore advances in ultrasonic modeling methods for understanding or predicting NDT inspections. Different authors present novel achievements in the understanding and modeling of ultrasonic waves for NDT applications. High-quality research and review papers on theoretical, practical, and validation aspects were accepted, leading to a collection of 20 published papers. These are briefly reviewed here, classified into different topics. A brief overview of the main developments in the field is recalled in each topic.

2. The Feasibility of Using Innovative NDT Methods on Complex and Various Materials

Ultrasonic NDT methods can investigate different kinds of material properties (mechanical, chemical, physical, biological, etc.) with various physical states/compositions (e.g., solid, liquid, heterogeneous [3], inhomogeneous, complex, and moving media). Ultrasonic NDT is broadly used on metals, plastics, composites and ceramics. Ultrasonic testing has even been used on wooden elements using tomography or acoustic emission [4] and fabrics [5].

In this Special Issue, the first paper from Malla, Mehrabi et al. reviews the efficiency of two classical NDT techniques in detecting embedded FRP Reinforcements: Ground-Penetrating Radar (GPR) and Phased Array Ultrasonic (PAU). GPR can detect GFRP (Glass–Fiber Reinforced Plastic) bars/strands and CFRP (Carbon–FRP) strands to some extent, with a detectability potential increasing with the emission center frequency, whereas PAU can only identify GFRP and CFRP strands.

Liu and Lacidogna proposed a non-destructive method based on the Southwell procedure that considers the temperature effect to evaluate the critical load, critical thickness, and service life of internally corroded shells under external pressure. The technique appears more practical than other methods, and its accuracy suits engineering applications.

It is also important to efficiently evaluate the repair of industrial structures. Feng and Ferri Aliabadi demonstrated the utility of using embedded PZT transducers to detect damage along the bond line and on the surface of a composite repair patch. The fabrication of the PZT sensors was validated using electro-mechanical impedance results.

Citation: Darmon, M. Special Issue on Ultrasonic Modeling for Non-Destructive Testing. *Appl. Sci.* **2024**, *14*, 2008. https://doi.org/10.3390/app14052008

Received: 25 January 2024
Accepted: 17 February 2024
Published: 29 February 2024

Noise reduction in experimental data acquired with an ultrasonic system is also challenging. Burrascano et al. used the Hammerstein model identification technique based on swept sine excitation signals to model nonlinear systems. The robustness of the model parameter estimation in the presence of measured noise was evaluated, and a technical solution to moderate the noise effects and improve the model parameters estimation was proposed.

3. Acoustic/Elastic Bulk Waves Propagation and Scattering

The main principles of elastic or ultrasonic wave propagation and scattering in solids can be found in several classical textbooks [6–9]. Simulating an NDT measurement generally requires modeling the propagation and scattering of ultrasonic waves from targets/flaws/damage/interfaces [10]. Developed simulation tools may rely on different mathematical/physical theories or assumptions; for instance, semi-analytical, numerical, and hybrid models [11] may be used for direct simulation and model benchmarking to ensure their confidence level [12,13].

Van der Neut et al. developed a theory to obtain the Green function in a lossless layered isotropic elastic medium from two-sided data. To that aim, they introduced an alternative Marchenko equation and two other equations using both reflection and transmission data.

Since the ultrasonic wave propagation field can be seen as spatio-temporal data, Gantala and Balasubramaniam developed two different spatio-temporal deep learning (SDL) models that simulate forward and reflected ultrasonic wave propagation in 2D. The training was ensured by using data obtained from finite element simulations.

In order to simulate the ultrasonic inspection of embedded cracks in 3D configurations, an experimental and theoretical comparison of 3D models was performed by Darmon, Toullelan and Dorval: they notably compared the analytical Physical Theory of Diffraction to a hybrid method in terms of flaw scattering. Based on the spectral finite element method, the hybrid method can provide a finer prediction in configurations involving small flaws or head waves (first arrival waves [14] of a complex nature that are complicated to simulate analytically [15,16]).

For NDT applications, ultrasonic wave generation can involve physics other than piezoelectricity. In the framework of photoacoustics, Hou, Glorieux, Marsh et al. numerically investigated the efficiency of laser-excited ultrasounds (surface and bulk elastic waves) using phased arrays by comparing a finite element method to k-wave simulations and by varying the array characteristics.

The simulation of the scattering of waves from obstacles in fluids also remains a significant issue [17]; it requires first- modeling ultrasonic wave propagation in fluids [18]. Schmelt and Twiefel extended the spec-radiation method based on acoustic holography to consider propagation in multiple fluid layers. This technique was experimentally validated on a wooden particleboard assumed to behave as a fluid.

Nagaso, Moysan et al. developed a high-performance ultrasonic simulation tool for inhomogeneous moving fluids. The CFD simulation of a real experiment of liquid metal jet mixing was performed, and the moving fluid's temperature (varying in time) served as the input toa spectral finite element model for wave propagation. The authors showed that an ultrasonic transmission measurement system can monitor the principal flow fluctuation.

4. Lamb Waves in Plates

Guided waves are increasingly employed for structural health monitoring [19] due to their ability to propagate over long distances. It has, notably, a wide spectrum of applications for different materials such as rails or composites [20] for aviation aircraft.

Davey, Assier, and Abrahams proposed a semi-analytical model to asymptotically derive the reflection of Lamb waves in a semi-infinite elastic waveguide. Their method deals with singularity at traction-free elastic corners with an internal angle greater than π by adding into the Lamb mode expansion newly considered corner modes that satisfy boundary and radiation conditions.

Kazys and Žukauskas proposed to employ ultrasonic linear air-coupled arrays that are electronically readjusted to optimally excite and receive A0 and S0 guided wave modes in thin plastic films. The feasibility of such a measurement was evaluated through 2D and 3D guided waves simulations with a single transducer and a linear phased array.

Singh, Bentahar, El Guerjouma et al. modelled the scattering of an incident A0 guided wave mode from an impacted damaged zone (conical-shaped geometry with decayed elastic stiffness properties) in a quasi-isotropic composite plate. The simulation was experimentally validated by comparing the scattering directivity, and it enabled the geometrical characterization of the impact.

Capineri, Bulletti, and Marino-Merlo analyzed errors in the impact location by varying the number of piezoelectric sensors from four to eight and using the S0 mode in an aluminum plate. A good compromise between the number of sensors and the error in impact localization was obtained with six sensors.

5. Imaging for Medical and Engineering Applications

Inversion theory and artificial intelligence (AI) are growing research domains for imaging/localizing damage. Imaging techniques can classically employ array transducers [21] for applications in the NDT of materials.

Ultrasonic imaging is also crucial for medical applications [22]. Zhang et al. proposed a new technique to image the human thorax's acoustic velocity distribution using ultrasound travel time tomography. A forward model employed shortest-path ray tracing, and the supervised descent method was applied to a training set, leading to successful numerical experiments.

Vallee, Chaix et al. investigated the application of the topological energy method to leaky Lamb waves. To validate the method, measurements were carried out on a single immersed plate, successively considering—the plate edge and a notch as defects. The two kinds of flaws were precisely localized using the proposed imaging method.

To design automatic tools for crack characterization, Fradkin, Darmon et al. proposed a code including a signal-processing module based on a modified total focusing method and AI modules, leading to a crack characterization report (size, localization, and orientation). It has been promisingly tested on two similar datasets, including planar notches, both embedded and surface-breaking.

Xu et al. designed an ice-coupled ultrasonic tomography to image flaws in complex-shaped parts. Ice-coupling allows for a reduction in the acoustic impedance mismatch, as compared to the immersion method. Applying the k-space pseudo spectral method, full matrix capture, and a frequency-domain full waveform inversion (FWI) based on the L-BFGS method, the ice-coupled ultrasonic FWI technique demonstrated a promising ability to detect several kinds of flaws in complex parts.

Funding: This research received no external funding.

Acknowledgments: I want to thank all the authors, anonymous reviewers, academic editors, and Editorial Office of Applied Sciences members who have participated in this Special Issue. It also benefited from the coordination and intensive, remarkable work of the managing editor of this Special Issue.

Conflicts of Interest: The author declares no conflicts of interest.

References

1. Gupta, M.; Khan, M.; Butola, R.; Singari, R. Advances in Applications of Non-Destructive Testing (NDT): A Review. *Adv. Mater. Process. Technol.* **2021**, *8*, 2286–2307. [CrossRef]
2. Corneloup, G.; Gueudré, C.; Ploix, M.-A. *Non-Destructive Testing and Testability of Materials and Structures*; EPFL Press: Lausanne, Switzerland, 2021.
3. Jodhani, J.; Handa, A.; Gautam, A.; Ashwni; Rana, R. Ultrasonic Non-Destructive Evaluation of Composites: A Review. *Mater. Today Proc.* **2023**, *78*, 627–632. [CrossRef]

4. Zielińska, M.; Rucka, M. Non-Destructive Testing of Wooden Elements. *IOP Conf. Ser. Mater. Sci. Eng.* **2021**, *1203*, 032058. [CrossRef]
5. Kumpati, R.; Skarka, W.; Ontipuli, S.K. Current Trends in Integration of Nondestructive Testing Methods for Engineered Materials Testing. *Sensors* **2021**, *21*, 6175. [CrossRef] [PubMed]
6. Auld, B.A. *Acoustic Fields and Waves in Solids*, 2nd ed.; Krieger Publishing Company: Malabar, FL, USA, 1990; ISBN 978-0-89874-782-9.
7. Rose, J.L. *Ultrasonic Waves in Solid Media*; Cambridge University Press: Cambridge, UK, 2004; ISBN 978-0-521-54889-2.
8. Royer, D.; Valier-Brasier, T. *Elastic Waves in Solids: Propagation*; ISTE Ltd. and John Wiley & Sons Inc.: London, UK, 2022; ISBN 978-1-78630-814-6.
9. Valier-Brasier, T.; Royer, D. *Elastic Waves in Solids, Volume 2: Radiation, Scattering, Generation*, 1st ed.; Wiley-ISTE: Hoboken, NJ, USA, 2022; ISBN 978-1-78630-815-3.
10. Schmerr, L.W.; Song, S.-J. *Ultrasonic Nondestructive Evaluation Systems: Models and Measurements*; Springer: New York, NY, USA, 2007.
11. Imperiale, A.; Leymarie, N.; Fortuna, T.; Demaldent, E. Coupling Strategies Between Asymptotic and Numerical Models with Application to Ultrasonic Non-Destructive Testing of Surface Flaws. *J. Theor. Comp. Acout.* **2019**, *27*, 1850052. [CrossRef]
12. Dib, G.; Crawford, S.; Jacob, R. *Validation of Ultrasonic Nondestructive Examination (NDE) Computational Models–Phase 1*; PNNL Report; Pacific Northwest National Laboratory: Richland, WA, USA, 2017.
13. Raillon, R.; Bey, S.; Dubois, A.; Mahaut, S.; Darmon, M. Results of the 2009 Ut Modeling Benchmark Obtained with Civa: Responses of Notches, Side-drilled Holes and Flat-bottom Holes of Various Sizes. *AIP Conf. Proc.* **2010**, *1211*, 2157–2164. [CrossRef]
14. Ferrand, A.; Darmon, M.; Chatillon, S.; Deschamps, M. Modeling of Ray Paths of Head Waves on Irregular Interfaces in TOFD Inspection for NDE. *Ultrasonics* **2014**, *54*, 1851–1860. [CrossRef] [PubMed]
15. Gridin, D. High-Frequency Asymptotic Description of Head Waves and Boundary Layers Surrounding the Critical Rays in an Elastic Half-Space. *J. Acoust. Soc. Am.* **1998**, *104*, 1188–1197. [CrossRef]
16. Gridin, D.; Fradkin, L.J. The High-Frequency Asymptotic Description of Pulses Radiated by a Circular Normal Transducer into an Elastic Half-Space. *J. Acoust. Soc. Am.* **1998**, *104*, 3190–3198. [CrossRef]
17. Lu, B.; Darmon, M.; Potel, C.; Zernov, V. Models Comparison for the Scattering of an Acoustic Wave on Immersed Targets. *J. Phys. Conf. Ser.* **2012**, *353*, 012009. [CrossRef]
18. Bruneau, M.; Potel, C. *Materials and Acoustics Handbook*; John Wiley & Sons: Hoboken, NJ, USA, 2013; ISBN 978-1-118-62286-5.
19. Mitra, M.; Gopalakrishnan, S. Guided Wave Based Structural Health Monitoring: A Review. *Smart Mater. Struct.* **2016**, *25*, 053001. [CrossRef]
20. Ricci, F.; Monaco, E.; Boffa, N.D.; Maio, L.; Memmolo, V. Guided Waves for Structural Health Monitoring in Composites: A Review and Implementation Strategies. *Prog. Aerosp. Sci.* **2022**, *129*, 100790. [CrossRef]
21. Drinkwater, B.W.; Wilcox, P.D. Ultrasonic Arrays for Non-Destructive Evaluation: A Review. *NDT E Int.* **2006**, *39*, 525–541. [CrossRef]
22. Moran, C.M.; Thomson, A.J.W. Preclinical Ultrasound Imaging—A Review of Techniques and Imaging Applications. *Front. Phys.* **2020**, *8*, 124. [CrossRef]

applied sciences

MDPI

Article

Feasibility of Conventional Non-Destructive Testing Methods in Detecting Embedded FRP Reinforcements

Pranit Malla [1,*], Seyed Saman Khedmatgozar Dolati [1], Jesus D. Ortiz [2], Armin B. Mehrabi [1,*], Antonio Nanni [2] and Kien Dinh [3]

1 Department of Civil and Environmental Engineering, Florida International University, Miami, FL 33174, USA; skhed004@fiu.edu
2 Department of Civil and Architectural Engineering, University of Miami, Coral Gables, FL 33146, USA; jdo72@miami.edu (J.D.O.); nanni@miami.edu (A.N.)
3 NDT Concrete LLC, Deltona, FL 32725, USA; kien@ndt-concrete.com
* Correspondence: pmall011@fiu.edu (P.M.); amehrabi@fiu.edu (A.B.M.)

Abstract: Fiber-Reinforced Polymer (FRP) bars/strands are the most promising alternative to their steel counterparts for reinforcing concrete elements due to their resistance to corrosion, lighter weight, higher strength and better durability. However, very limited research has been conducted in relation to non-destructive testing (NDT) methods that are applicable to damage detection in FRP bars or the detection of FRP reinforcements embedded in concrete. The ability to assess the condition of the relatively new and unique FRP reinforcements will increase the confidence of the construction industry in their use as a reliable substitute for steel reinforcements. This paper investigates the ability of two of the most commonly used NDT methods, Ground Penetrating Radar (GPR) and Phased Array Ultrasonic (PAU), in detecting FRP bars/strands embedded in concrete elements. GPR and PAU tests were performed on two slab specimens reinforced with GFRP (Glass-FRP) bars, the most commonly used FRP bar, with variations in their depth, size and configuration, and a slab specimen with different types of available FRP reinforcements. The results show that GPR devices can detect GFRP bars/strands and CFRP (Carbon-FRP) strands to some extent, and their detectability increases with the increase in their antenna center frequency. On the contrary, PAU is only capable of detecting GFRP and CFRP strands. The results of this paper also emphasize the need for further research and developments related to NDT applications to embedded FRP bars.

Keywords: fiber-reinforced polymer (FRP); ground penetrating radar (GPR); ultrasonic testing (UT); phased array ultrasonic (PAU); non-destructive testing (NDT); reinforced concrete

Citation: Malla, P.; Khedmatgozar Dolati, S.S.; Ortiz, J.D.; Mehrabi, A.B.; Nanni, A.; Dinh, K. Feasibility of Conventional Non-Destructive Testing Methods in Detecting Embedded FRP Reinforcements. *Appl. Sci.* **2023**, *13*, 4399. https://doi.org/10.3390/app13074399

Academic Editor: Michel Darmon

Received: 2 March 2023
Revised: 28 March 2023
Accepted: 29 March 2023
Published: 30 March 2023

1. Introduction

Corrosion of steel reinforcement is one of the main problems in traditional concrete structures, which severely affect its safety and serviceability. Several corrosion protection measures, including but not limited to cathodic protection, epoxy-coated bars and galvanized steel reinforcements have been implemented in the past; however, they have only managed to delay corrosion rather than eradicating it [1]. The most promising alternatives available that could entirely stop the process of corrosion within concrete is the use of fiber-reinforced polymer (FRP) bars/strands [2–6]. Depending upon the types of reinforcing fibers used, the FRP bars/strands can be classified as GFRP (Glass-FRP), CFPR (Carbon-FRP), BFRP (Basalt-FRP) and AFRP (Aramid-FRP) bars, respectively for glass, carbon, basalt and aramid fibers. The FRP bars are resistant to all the elements that leads to corrosion in steel reinforced concrete (RC) structures, such as reduction of pH of the concrete from carbonation, chloride contamination of the concrete and diffusion of halides and chemicals [7,8]. The use of FRP bars for reinforcing new structures is even more preferable because they also have 1.5–2 times higher tensile strength than their steel counterparts [9–11]. In addition, although the initial cost of FRP bars is higher than that of the conventional steel bars, their life cycle cost analysis indicates that it can be quite economical in the long term [12]. Given these circumstances, the embedded FRP bar is gradually becoming a trusted material in civil engineering.

To ensure construction quality and continued structural reliability, the necessity for innovative non-destructive testing (NDT) methods or research on the feasibility of existing NDTs becomes more apparent with the rise in use of unique structural materials such as FRP bars. Although higher durability and performance are associated with the FRP bars in some respects when compared to steel, concerns still remain regarding damages and defects in this material, many of them rather unique, such as debonding, delamination and aging from UV exposure. It is equally important to understand the damages and defects associated with the use of FRP bars as it is for other structural materials. Further, it is even more crucial to identify the signs of possible failure of structures reinforced with FRP bars at the earliest because they are not as ductile as conventional constructions [13] and do not display any exterior warnings of damage until they break [14].

However, there has only been minimal or non-existing work carried out for assessment of FRP embedded within concrete. The only available comprehensive guide for the inspection of FRP application in civil engineering is the NCHRP report 564 [15], which is limited to the inspection of FRP bridge decks. Currently, there is no guide or manual for the inspection of FRP-reinforced concrete elements [16,17]. This is in most part because of the lack of research on non-destructive testing and inspection of such elements. Most FRP bars are undetectable or have low detectability and therefore cannot be effectively located in the inspection, which makes detection of their damages difficult. The use of FRP material in highway construction has increased consistently [18–21], but the lack of methods for condition assessment of FRP-RC has noticeably dampened the proliferation of FRP use. Bridge owners are not comfortable with products that cannot be detected properly and whose condition cannot be effectively assessed for maintenance purposes. Hence, there is a strong need for research on the means and methods of condition assessment of FRP-RC, the availability of which will have exponential effects in increasing the use of FRP in future constructions. The main objective of this paper is to determine the feasibility of the most common NDT methods in detecting embedded FRP bars/strands, which would act as an initial step for propelling future studies on the nondestructive testing of FRP bars/strands.

Many NDT techniques, such as visual inspection (VT) [9,22], tap testing (TT) [23–27], impact echo testing (IE) [28–32], microwave testing (MW) [33–38], ground penetrating radar (GPR) [39–47], ultrasonic testing (UT) or phased array ultrasonic testing (PAU) [48–55], infrared thermography testing (IR) [56–64], acoustic emission testing (AE) [65–68], laser testing (LT) [69–75], radiographic testing (RT) [76–81], global structural response testing [82–87], etc., have been studied for detecting damages in the externally applied FRP composites. However, the inspection of FRP-reinforced concrete members are limited to detection of debonding between internal FRP bars and concrete [88,89] or initiation of breakage in FRP [14,90] rather than the detection of the bars themselves. Hence, among several available NDTs, this paper aims to determine the feasibility of using commercially available GPR and PAU for detecting embedded FRP bars in concrete as they are the most common methods in the NDT practice for steel RC elements [16,91,92]. In 2019, Drobiec et al. confirmed that GPR and PAU have limited detectability of non-metallic reinforcement (FRP reinforcements), but they had only conducted tests on the fiber meshes of FRPs [93]. However, this paper will further explore the potential of these devices in detecting the FRP bars and strands used in FRP-reinforced concrete. The results of this study show that lower frequency GPR devices are able to detect larger FRP bars/strands at shallower depths with improvements in their detectability with the use of higher antenna center frequency devices. The results also indicate that PAU devices are only effective for detecting GFRP and CFRP prestressing strands.

2. Materials and Methods

In order to determine whether GPR and PAU can detect internal FRP reinforcement, small-scale concrete slabs were fabricated. Two slab specimens (i.e., labeled as J and L) were 30 in. wide by 30 in. long and 7 in. deep, and the third specimen (viz. Slab C) was 36 in. wide by 36 in. long and 5 in. deep. The concrete mix used to cast the slab specimens was the 'Class II 4500 Bridgedeck' concrete, as per the Florida Department of Transportation (FDOT). This class specified a guaranteed compressive strength of 4600 psi (31 MPa). Type II Cement was used with a water to cementitious material ratio (w/m) of 0.44, and #57 stone and silica sand was used as coarse and fine aggregate, respectively. To obtain the actual strength value; concrete cylinders were tested at 28 days according to ASTM C39 [94,95]. An average compressive strength of 31.70 MPa was obtained with a standard deviation of 0.69 MPa (coefficient of variation of 2.2%).

The slab specimens were fabricated, targeting different parameters such as FRP bars/strands type (GFRP, CFRP, BFRP), bar diameter, bar direction and bar depths. Table 1 shows the identification of the slab specimens by group, highlighting their main parameters. Because GFRP bars are the most commonly used FRP reinforcement in concrete elements, the first two slabs constructed (Slabs C and J) were only reinforced with GFRP bars of different sizes, at different depths and different configuration (bars in one direction and bars in two orthogonal directions, i.e., mesh). The concrete cover specified by ACI CODE-440.11-22 [96] for the concrete members reinforced with GFPR bars ranges from 0.75 in. to 3 in., hence the depth variations in the slab specimens were included to represent the layers of FRP reinforcement, which could be anywhere within the concrete cover range specified. Further, the third slab (Slab L) has different types of internal reinforcement (bars and strands) embedded into it, including one steel bar whose detectability acts as a control for this research. Having GFRP, CFRP, BFRP and steel bars/strands on the same slab specimen allows comparison of the detectability of different FRP bar/stands with the steel bar under the same test conditions.

Table 1. Identification of small-scale concrete slab specimens.

Specimen Group	Slab ID	Bar Diameter	No. of Bars
Slabs with GFRP bars in one direction	Slab C	#4 and #6	6
Slabs with GFRP bars in two directions (mesh)	Slab J	#6	10
Slab with different FRP bars in one direction	Slab L	#3, #5, #8 GFRP bars, #5 CFRP strand, #3 steel bar, #6 GFRP strand, #3, #5 BFRP bars	9

All specimens were fabricated and labeled following the same layout as shown in Figure 1. Every slab is identified with a letter (C, J and L), and every side has an identification number (from 1 to 4). The direction of the measurement will be determined by the number where the measurement is started to the end of the measurement.

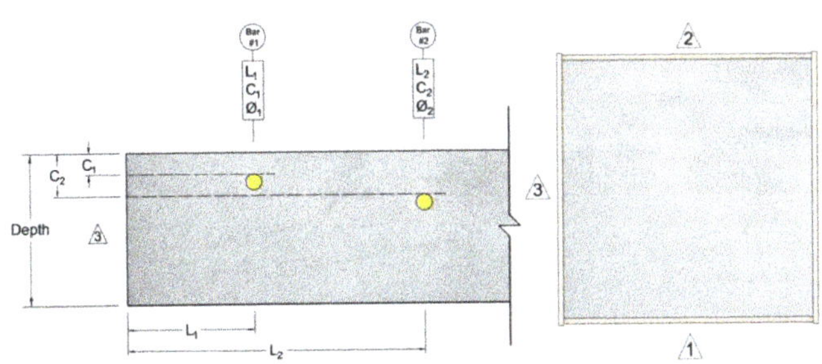

Figure 1. Labeling of slab specimens.

Table 2 shows the details and dimensions of the specimens. For each specimen, the distance to edge, depth to surface, bar diameter, bar material and slab depth are presented based on the convention shown in Figure 1. The formwork for each slab specimen constructed is shown in Figure 2.

Table 2. Reinforcement/Dimension details of detectability slab specimens.

Slab ID	Parameter [Symbol/Units]	Reference Side	Bar 1	Bar 2	Bar 3	Bar 4	Bar 5	Bar 6	Bar 7	Bar 8	Bar 9	Bar 10
C	Distance to edge reference (L/in.)	3	3	9	15	21	27	33				
	Depth to surface (C/in.)		0.75	0.75	1.5	1.5	3	3				
	Diameter of bar (*F*/in.)		#4	#6	#4	#6	#4	#6				
	Material of bar (T)		Glass									
	Depth of slab (h/in.)		5.0									
J	Distance to edge reference (L/in.)	3	3.3	8.6	14.5	20.2	25.4	(Bottom mesh)				
	Depth to surface (C/in.)		3.3	3.3	3.6	3.5	3.8					
	Diameter of bar (*F*/in.)		#6									
	Material of bar (T)		Glass									
	Depth of slab (h/in.)		7.0									
	Distance to edge reference (L/in.)	1	(Top Mesh)					3.3	8.6	14.5	20.2	25.4
	Depth to surface (C/in.)							2.6	2.6	2.9	2.8	3.1
	Diameter of bar (*F*/in.)							#6				
	Material of bar (T)							Glass				
	Depth of slab (h/in.]							7.0				
L	Distance to edge reference (L/in.)	3	4.3	7.1	10.3	13.1	16.4	18.8	21.6	24.8	27.6	
	Depth to surface (C/in.)		4.2	3.3	3.6	3.9	3.9	4.0	3.8	4.1	4.0	
	Diameter of bar (*F*/in.)		#3	#8	#5	#5	#5	#3	#6	#3	#5	
	Material of bar (T)		Glass				C-Std *	Steel	G-Std *	Basalt		
	Depth of slab (h/in.)		7.0									

* C-Std. (i.e., CFRP strands) and G-Std (i.e., GFRP strands) are labelled as Bar 5 and Bar 7 in Slab L.

2.1. GPR Tests

Ground penetrating radar (GPR) is a non-destructive testing (NDT) method that is used to the analyze internal characteristics of structures or elements in real-time. It works on the principle that electromagnetic waves reflect back when they encounter an interface between two materials with different dielectric constants [31,39]. GPR functions by emitting electromagnetic waves (in the form of radio waves) through the test material and then detecting the waves that bounce back off any discontinuities within the material. These discontinuities can take various forms, including interfaces between different materials, such as the concrete-bar interface shown in Figure 3 or the concrete-air/water interfaces (subsurface defects like voids, cracks, debonding and delamination) [97]. Due to the limited research on the use of GPR for FRP-reinforced structures, this method has not yet been able to establish itself as a reliable NDT method. This paper will explore the possibility of using GPR for FRP bar detection.

In this study, the GPR tests were conducted both as individual line scans along a straight line and as grid scans over a test area. The line scans were conducted as a reconnaissance survey or preliminary inspection prior to the detailed grid/area scan to form a quick idea of what to expect inside the structure, the orientation of reinforcement or other subsurface features, and the depth of exploration. The GPR line scans conducted along a straight line on the surface of the test specimen were used to obtain its cross-section image through a plane normal to the surface along the direction of the scan. However, going back and forth between several line scans and interpretating each line scan was time consuming and labor intensive. Thus, in order to simplify the interpretation of results, grid scans were also conducted over the surface of the test specimens. Grid scans were performed simply by taking GPR data along the straight lines of a grid over an area covering the test specimen, as shown in Figure 4. The grid required for a grid scan was arranged using a grid mat secured to the surface so that it did not move while taking the GPR data.

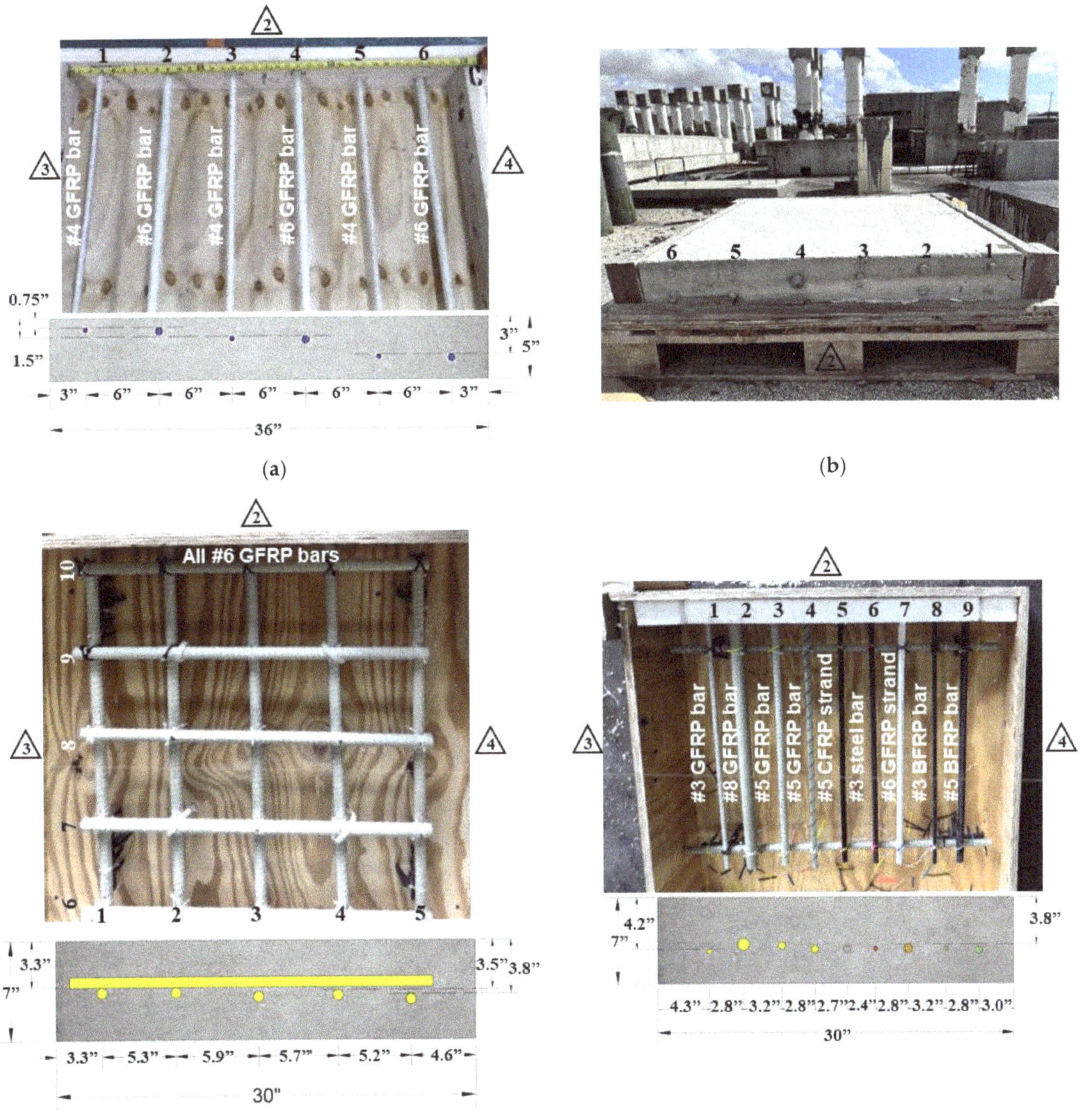

Figure 2. FRP bars in different slabs: (**a**,**b**) Slab C, (**c**) Slab J, (**d**) Slab L.

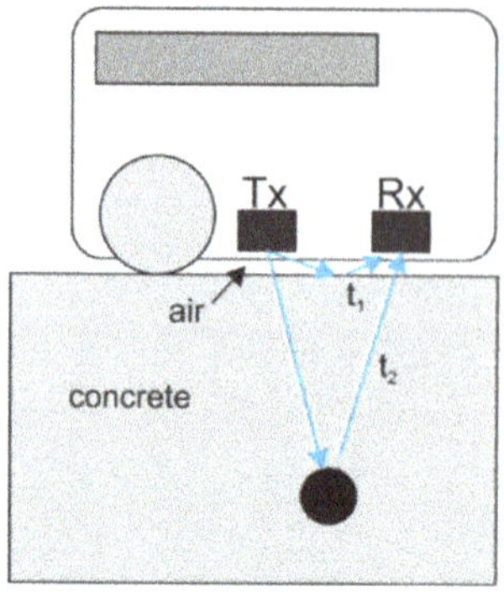

Figure 3. Reflection of incident wave at concrete–bar interface [98].

Figure 4. GPR grid scan test setup for Slab L.

The spacing between each scan in the grid were set as 2 in. for this study. The spacing between the individual lines in the grid determines the resolution of the data collected and it is termed as grid resolution. The lower the spacing between each line scans in a grid, the higher the resolution of the data collected. Hence, the images will be clearer, and it would be easier to interpret the data. Apart from using closely spaced grid lines, a higher GPR frequency can also be used to increase the resolution. The depth of penetration and the resolution of a GPR depends upon the frequency of the pulse transmitted into the material. Lower frequencies allow deeper penetration with lower resolution, whereas higher frequencies allow detection of small defects (higher resolution) but localized within a shallower depth [31,93,99,100]. Four different GPR systems with different center frequency ranges were used in this experiment to determine the effect of GPR frequency on the bar detectability, as shown in Table 3 and Figure 5. The data acquired from grid scans were used to give the cross-section image of the test specimen through the plane parallel to the surface of the specimen along its depth. The cross-sectional image along the depth of the specimen is termed as a depth slice or time slice image.

Table 3. Technical specifications of the GPR systems used [101–104].

GPR Systems	Center Frequency	Radar Technology	Depth Range	Manufacturer
Conquest 100 Enhanced	1000 MHz	Monostatic GPR antenna	24 in. (60 cm)	Sensors and Software, Canada
C-Thrue radar	2000 MHz	Dual polarization antenna for multi-level detection	31.5 in. (80 cm)	IDS GeoRadar, Italy

Table 3. *Cont.*

GPR Systems	Center Frequency	Radar Technology	Depth Range	Manufacturer
Proceq GP8800	400–6000 MHz	Stepped-frequency continuous-wave (SFCW) GPR	25.6 in. (65 cm)	Screening Eagle Technologies, Switzerland
Proceq GP8000	200–4000 MHz	Stepped-frequency continuous-wave (SFCW) GPR	31.5 in. (80 cm)	Screening Eagle Technologies, Switzerland

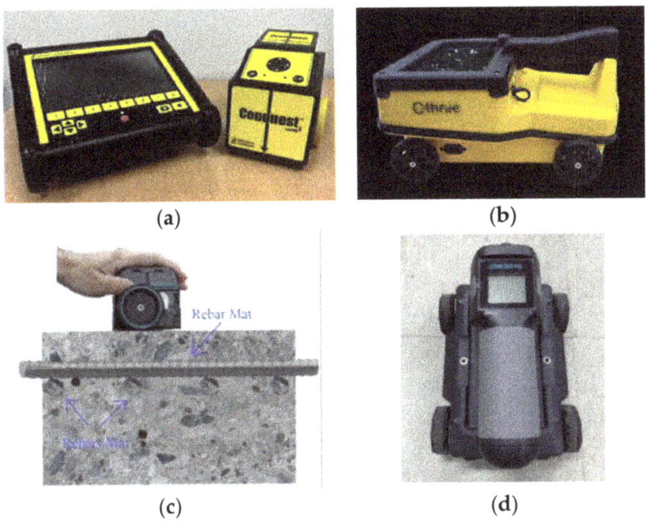

Figure 5. GPR systems used: (**a**) Conquest 100 Enhanced, (**b**) C-Thrue radar [105], (**c**) Proceq GP8800 [106], (**d**) Proceq GP8000 [107].

2.2. PAU Tests

Ultrasonic testing (UT) operates on the principle that the incident ultrasonic waves generated by ultrasonic transducers (which transform electrical or optical signals into ultrasonic waves and vice versa) will be reflected back when they encounter an interface between two materials with different acoustic impedances. A phased array ultrasonic (PAU) is an advancement over the UT technology that can be achieved when several of these transducers are arranged together into an array and operated at slightly different times (either electrically or physically) so that the individual waves interact both positively and negatively, allowing beam focusing and beam steering, as shown in Figure 6 [108]. The advantages of PAU in comparison to conventional UT include a 5 to 10 times faster scanning rate, better images due to multiple angles and frequencies, which require less interpretation, higher resolution, capability of beam focusing, reliability, portability and mobility [109,110]. Unlike conventional UT, PAU testing allows signal focusing at desired locations and angles, which is advantageous for the testing of composite materials that have an anisotropic structure, creating challenges in signal evaluation [53,111]. However, PAU is a relatively new technique compared to the traditional NDTs, and hence the limitations of portable units of PAU include the uncertainty in its application as it has yet to be completely proven.

Line scan using PAU was performed by moving the array of ultrasonic transducers in a sideways direction along the desired line of inspection, as shown in Figure 7. Each scan at individual positions is stitched together to give one continuous line scan, which is the cross-sectional image through a plane normal to the surface along the direction of the scan. Similarly, the area scan using PAU was conducted as a stripe scan. The stripe scan was performed by positioning the array of ultrasonic transducers perpendicular to the desired line of inspection and then taking scans while moving forward with desired spacing between the consecutive scans, as shown in Figure 7. Each individual scan collected during stripe scan is simply a line scan having a width of measurement equal to that of the PAU device or the array of transducers. These individual scans along the line of inspection are stitched together to give a time slice or depth slice view, which can be further processed into a full 3D iso-surface representation. The depth slice view is simply the cross-section image parallel to the surface scanned.

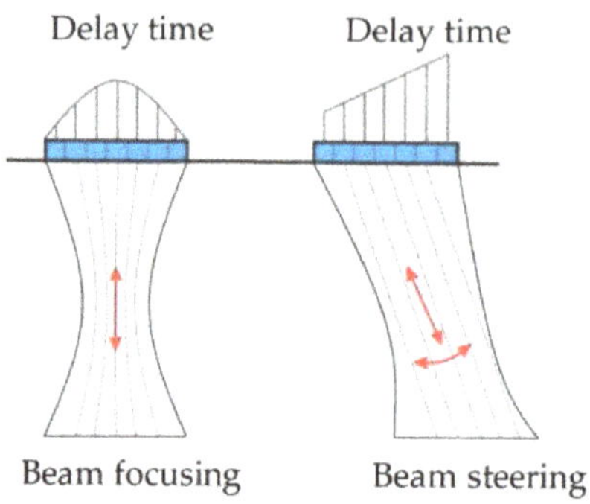

Figure 6. Phased array with angled wavefront [112].

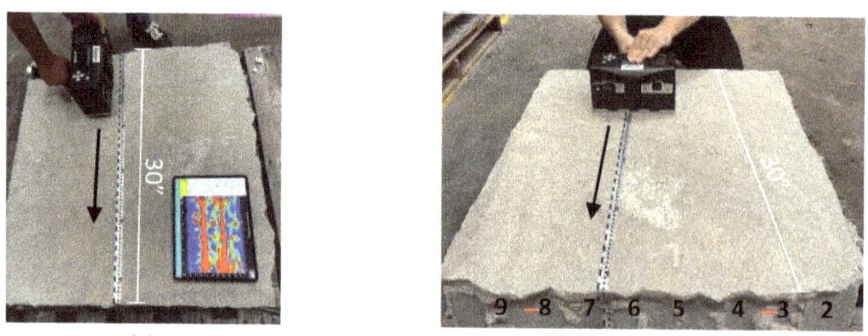

(**a**) (**b**)

Figure 7. PAU test setup: (**a**) Line scan, (**b**) Stripe scan.

Two different PAU systems were used in this experiment to determine their FRP bar detection capability, as shown in Table 4 and Figure 8. The MIRA 3D device had 64 ultrasonic transducers located in a 16 × 4 grid at 3 cm spacing (extended 16 rows of 4 transducers each) where the Pundit Live Array Pro had 24 ultrasonic transducers located in an 8 × 3 grid at 3 cm spacing (8 rows of 3 transducers each).

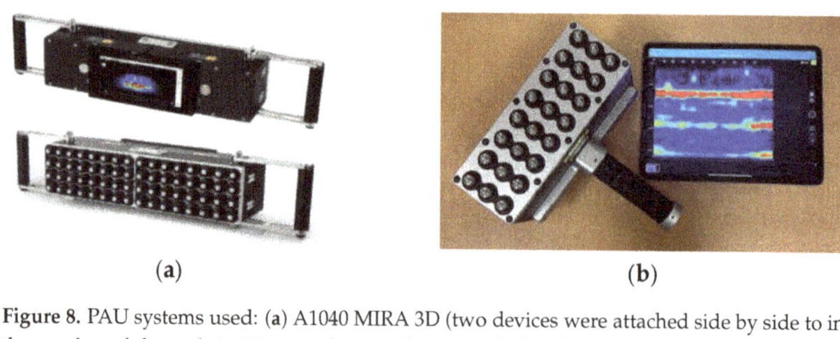

(**a**) (**b**)

Figure 8. PAU systems used: (**a**) A1040 MIRA 3D (two devices were attached side by side to increase the number of channels to 16 rows of 4 transducers each) [115], (**b**) Pundit live array pro.

Table 4. Technical specifications of the PAU systems used [113–115].

PAU Systems	Number of Channels	Technology	Transducer	Bandwidth	Depth Range	Manufacturer
A1040 MIRA 3D	8 × 4 (extendable to 16 × 4, 24 × 4, etc.)	Multi-channel ultrasonic pulse echo tomograph	Active dry point contact (A-DPC) transducers	10–100 KHz	6.5 ft (2 m)	ACS-Solutions GmbH, Germany
Pundit live array pro	8 × 3 (with upgrade option to 16 × 3)	Phased array ultrasonic pulse echo	Dry-contact Pundit Array transducer	15–100 KHz	6.6 ft (2 m)	Screening Eagle Technologies, Switzerland

3. Results

3.1. GPR Tests Results

GPR response or line views traversing perpendicular to bars embedded in each slab specimens obtained using a 1 GHz GPR device are shown in Figure 9, where the top of the hyperbolic shape, i.e., inverted U shape, indicates the location of the bar and the shape of the tails gives a measure of velocity and depth [116]. The background subtraction filter was used to enhance the hyperbolas from embedded FRP bars. Without the application of this filter, the top of the hyperbola of the shallow targets (FRP bars close to the top surface) would have been obscured by the direct wave band that appears at the top surface. However, using the filter also removed the horizontal band indicating the bottom surface of the slab, but because the main objective of this research was to detect the FRP bars, the detection of the bottom surface was deemed not important. Only the line scans conducted at the centerline of the slabs using the lowest frequency GPR device has been presented in this paper for the brevity of the results as it would represent the lowest resolution among the GPR devices used in this study. Further, it should be noted that in order to confirm that the hyperbolas detected were indeed the FRP bars, but not other internal features (such as voids), several line scans were performed and checked for repetition of the same pattern of hyperbolas in each line scan (which indicates the presence of a continuous internal target such as a reinforcement bar). For Slab C with GFRP bars in one direction, bars at up to a depth of 1.5 inch were detectable. For Slab J with GFRP bars in two directions (mesh), the hyperbolas for the bars on both the top and bottom mesh were detectable. The bars near the edge were shadowed by the hyperbola due to the edge of the slab, which could have been avoided by leaving an offset from the edge for taking the measurements. The GPR line scan of Slab L was able to detect carbon strand (Bar 5), steel bar (Bar 6) and GFRP strand (Bar 7), along with the larger diameter #8 GFRP bar (Bar 2). However, it was not able to detect other glass (Bars 1, 3, 4) and basalt (Bars 8, 9) bars. The bar detectability of the line scans conducted using lower frequency device is further summarized in Table 5.

Table 5. Bar detectability in line scans performed using Conquest 100 Enhanced.

Slab ID	Bar 1	Bar 2	Bar 3	Bar 4	Bar 5	Bar 6	Bar 7	Bar 8	Bar 9	Bar 10
C	✓	✓	✓	✓	X	X	-	-	-	-
J	✓	✓	✓	✓	✓	X	✓	✓	✓	✓
L	X	✓	X	X	✓	✓	✓	X	X	-

Note: ✓ = detectable, X = not detectable.

The depth slices obtained for each slab specimen using each GPR device are further illustrated in Figure 10, where it can be seen that the bar detectability improves as the device central frequency increases. Although the line scan in Figure 9a shows four hyperbolas for Bars 1–4, the depth slice in Figure 10a only shows the presence of Bar 2, which is the larger GFRP bar (#6 GFRP bar) at shallower depth in Slab C. This could be because of the limitations in vertical resolution due to a lower frequency of the GPR device. Larger GFRP bars are detectable using lower frequency GPR device at shallower depth, but they return weaker hyperbolas in comparison to metallic objects. Moreover, the relatively higher amplitude of shallower GFRP bars may have dominated and masked the lower amplitude returns from the GFRP bar that are deeper and smaller in diameter, which could be the reason only Bar 2 was visible in Figure 10a. However, the depth slices in Figure 10b,c, obtained using higher frequency GPR devices, are capable of detecting a smaller diameter bar, even at deeper depths.

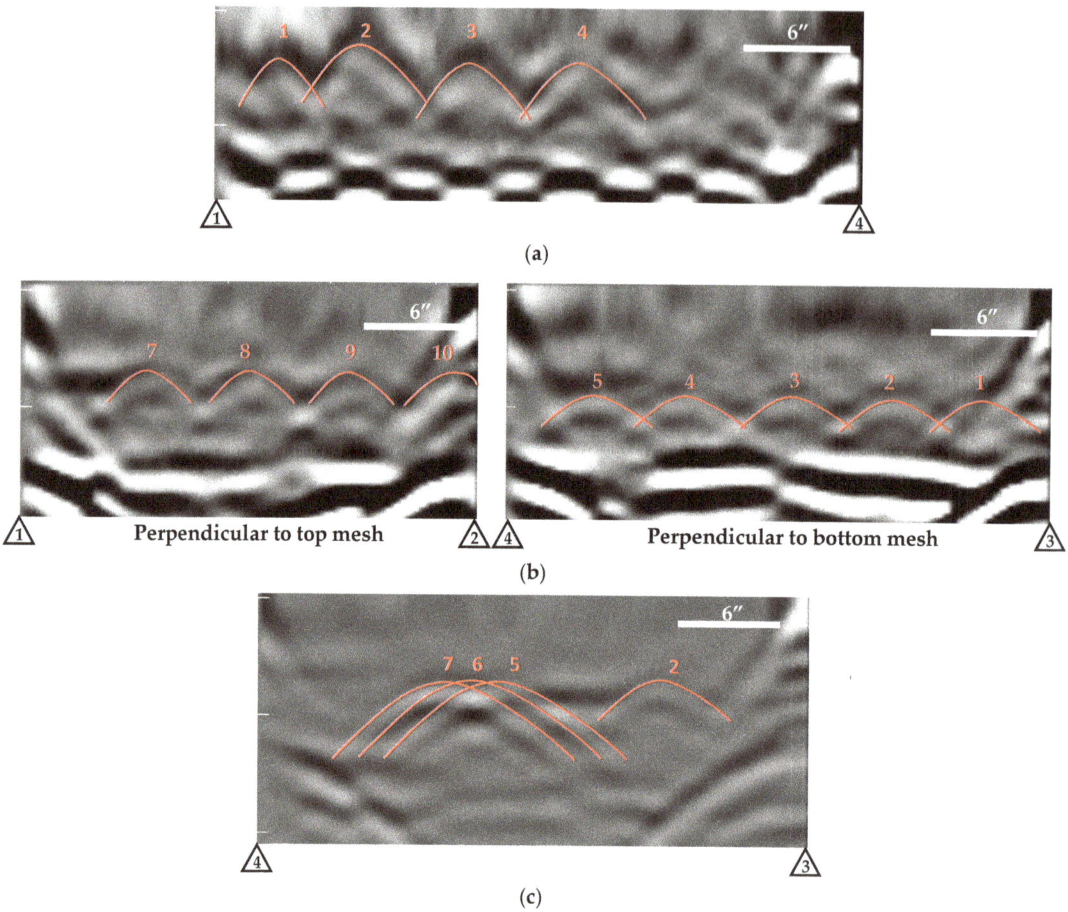

Figure 9. Line scan results of Conquest 100 Enhanced for different slab specimens: (**a**) Slab C, (**b**) Slab J, (**c**) Slab L.

In the case of depth slices for Slab J with a top and bottom mesh of GFRP bars, all the GPR devices could clearly detect the top mesh of #6 GFRP bars at a shallower depth, while the bottom mesh was only visible for the higher frequency GPRs, as shown in Figure 10d–f. The inability of lower frequency GPR to show the bottom mesh could be due to the same reason for not showing deeper GFRP bars in Slab C, as explained previously. In addition to the higher center frequency of the 2 GHz GPR device, its dual polarization feature further permits detection on both first and second levels of bars, whereas for the GPR device with maximum center frequency of 4 GHz and 6 GHz, the ability to detect the bottom mesh is solely due to the higher resolution, which it can afford because of its higher frequency.

The line scan of Slab L shows that Bar 2 (#8 GFRP bar) is visible as a faded hyperbola in addition to distinctive hyperbolas of Bars 5, 6 and 7. But again, Bar 2 is not visible in the depth slice in Figure 10g because GFRP bars return weaker hyperbolas in comparison to steel bars (Bar 6) and the high amplitude steel bar dominates and hides the lower amplitude returns from the GFRP bar. In Figure 10g–i, it can be seen that the strongest detection is that of the steel bar (Bar 6), followed by the carbon strand (Bar 5) and GFRP strand (Bar 7). Carbon strands, in addition to having cavities from the twisting shape of exterior wires that are filled (or not filled) with concrete, they are also electrically conductive, which could be the reason they are distinctively visible, and the GFRP strands are visible only because of the cavities within the GFRP twisting wires. Apart from Bars 2, 5, 6 and 7, the higher frequency GPR was also able to capture other bars, as shown in Figure 10i.

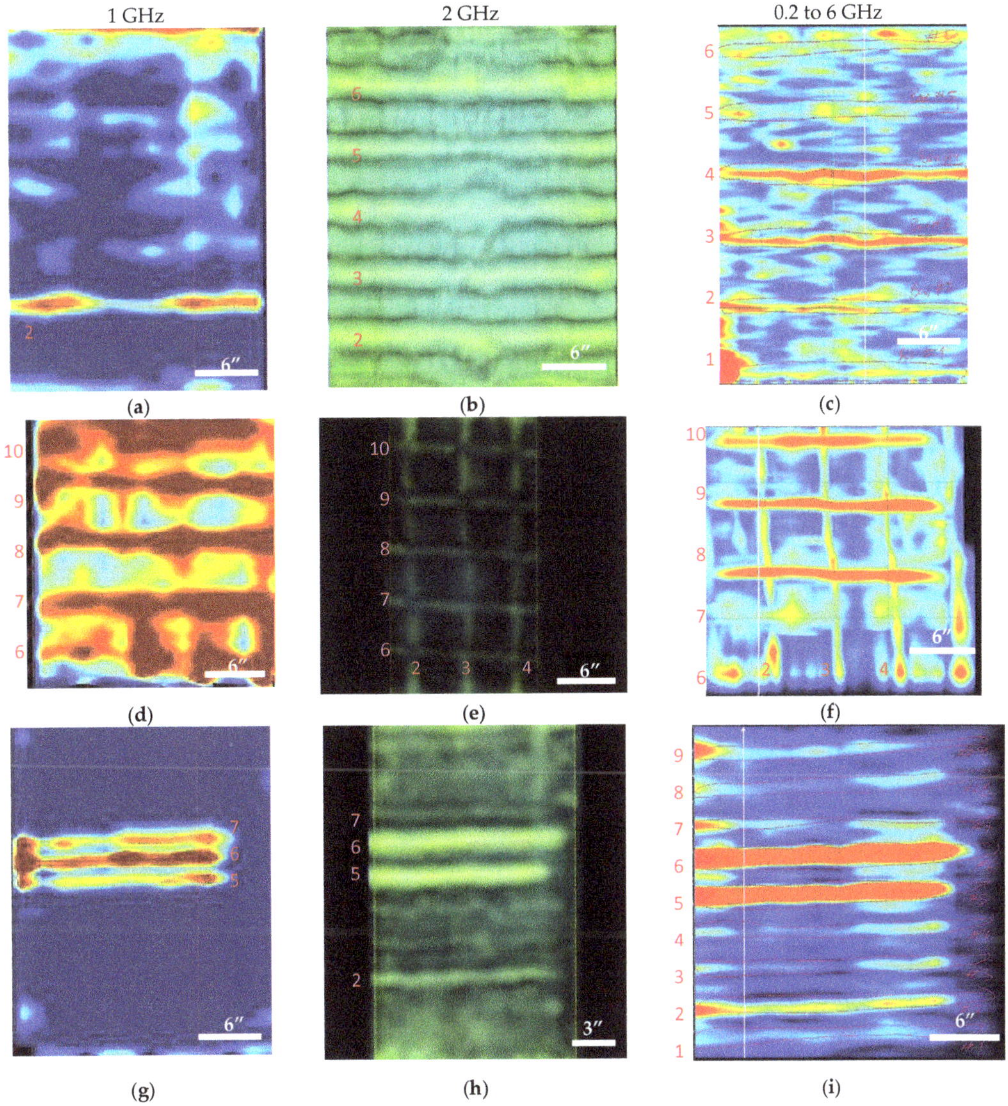

Figure 10. Depth slices: (**a**) Slab C using Conquest 100 Enhanced, (**b**) Slab C using C-Thrue, (**c**) Slab C using Proceq GP8800, (**d**) Slab J using Conquest 100 Enhanced, (**e**) Slab J using C-Thrue, (**f**) Slab J using Proceq GP8800, (**g**) Slab L using Conquest 100 Enhanced, (**h**) Slab L using C-Thrue, (**i**) Slab L using Proceq GP8000.

3.2. PAU Tests Results

The results of line scans and the area scans obtained from the PAU testing on each slab specimen using MIRA 3D and Pundit Live Array Pro devices are shown in Figure 11. It can be seen that PAU could not detect bars in Slab C and Slab J. However, it was able to detect carbon strands, steel bars and GFRP strands in Slab L. As ultrasonic testing and PAU is very sensitive in detecting the presence of (air) voids, their ability for detecting CFRP and GFRP strands could be attributed to the presence of (air) voids within the twisted FRP cables of these strands.

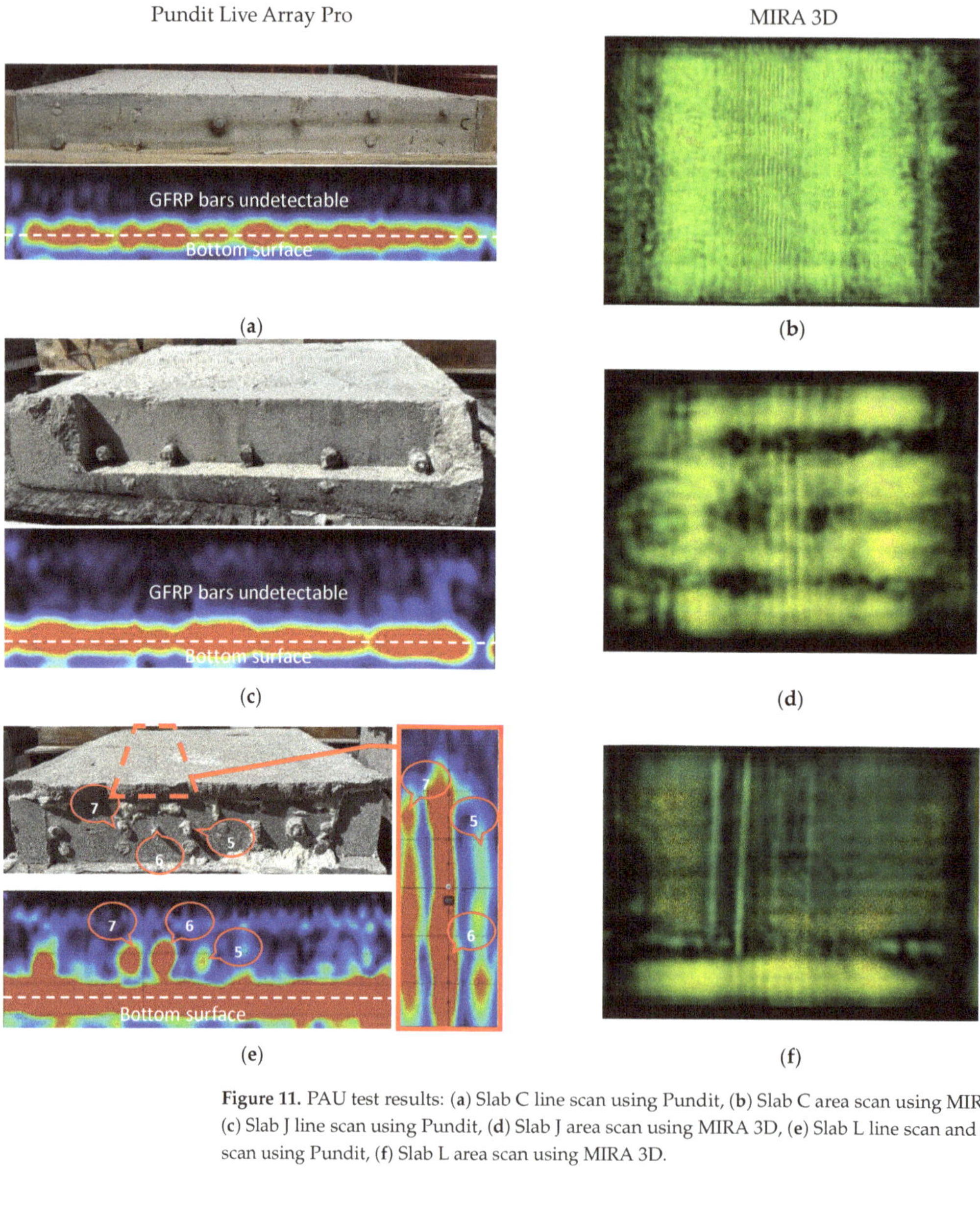

Figure 11. PAU test results: (**a**) Slab C line scan using Pundit, (**b**) Slab C area scan using MIRA 3D, (**c**) Slab J line scan using Pundit, (**d**) Slab J area scan using MIRA 3D, (**e**) Slab L line scan and stripe scan using Pundit, (**f**) Slab L area scan using MIRA 3D.

4. Discussion

The GFRP bars that dominate the embedded concrete reinforcement application are not conductive and have a density similar to concrete. Therefore, the conventional NDT methods that use electromagnetic and stress waves for the detection of steel bars fail to detect GFRP bars with the same clarity as steel. As explained earlier in Sections 2.1 and 2.2, the GPR and PAU devices work on the basis of differences in dielectric constant and acoustic impedance encountered at an interface. The reflection coefficient of the electromagnetic wave (for GPR tests) and ultrasonic waves (for PAU tests) passing through concrete when they encounter air (bottom surface), steel bar and GFRP bars are shown in Table 6. It can be seen that the concrete–GFRP interface reflection coefficient, i.e., the amount of reflected energy, is less than 8% for GPR tests and has a very minimal value of 0.04% for PAU tests, which is the reason why GPR had limited detectability of GFRP bars and PAU was not able to detect them at all. The dielectric constants and acoustic impedances for average concrete, air and steel used in Table 6 were obtained from the literature [116–118]. From the specifications chart provided by the manufacturer of the GFRP bar, the dielectric constant was determined to be a value of less than 5, and the acoustic impedance was estimated to be 10.7×10^6 kg/m^2 s (from the equation $Z = \sqrt{E.\rho}$, where E = 54.5 GPa is the modulus of elasticity and ρ = 2.1 g/cm^3 is the density of the GFRP bar).

Table 6. Reflection coefficients for GPR and PAU tests.

Interface	Relative Dielectric Constant or Permittivity, ε		GPR Reflection Coefficient, R	Acoustic Impedance (10^6 kg/m^2 s), Z		PAU Reflection Coefficient, R
	ε_1	ε_2	$\frac{\sqrt{\varepsilon_1}-\sqrt{\varepsilon_2}}{\sqrt{\varepsilon_1}+\sqrt{\varepsilon_2}}$	Z_1	Z_2	$\frac{(Z_2-Z_1)^2}{(Z_2+Z_1)^2}$
Concrete-Air	7	1	45%	9.6	0.000429	99%
Concrete-Steel	7	∞	100%	9.6	46.5	43%
Concrete-GFRP	7	<5	<8%	9.6	10.7	0.30%

However, the depth slices of GPR tests obtained using four different devices with different antenna center frequency demonstrated that the detection of FRP bars becomes better with the increase in frequency, as shown in Table 7. Apart from the frequencies playing the major role in the detection of FRP bars, different detectability levels could also be due to several other factors related to the radar technologies being used in each of the GPR devices. For example, the 2 GHz GPR device also had a dual polarization feature that enhances the detection of multilevel bars. The maximum frequency GPR devices used in this study also had a Stepped Frequency Continuous Wave (SFCW) system. This provides both the benefits of higher resolution in the detection of shallow targets and the increase in penetration depth. Moreover, there were some disparities in the results of the line scans and the grid scans for the lower frequency GPR device as the bars visible in the line scan were not visible in the depth slices. This could be due to the fact that the depth slices are produced through the interpolation of a series of line scans, and it would show the bars with stronger signals, i.e., stronger hyperbola more dominantly in comparison to the bars with relatively weaker hyperbola. Thus, although it is time consuming, it is recommended to go through each line scan of a grid while analyzing the results for GPR tests. On the other hand, it was found that PAU can only detect FRP strands, which could be due to the presence of (air) voids within the twisted wires of FRP or irregularity of the shape of their surface in comparison with smooth bars.

Table 7. Bar detectability in depth slices performed using devices with different center frequency.

Slab ID	Center Frequency	Bar 1	Bar 2	Bar 3	Bar 4	Bar 5	Bar 6	Bar 7	Bar 8	Bar 9	Bar 10
C	1 GHz	X	✓	X	X	X	X	-	-	-	-
	2 GHz	✓	✓	✓	✓	✓	✓	-	-	-	-
	0.2 to 6 GHz	✓	✓	✓	✓	✓	✓	-	-	-	-
J	1 GHz	✓	✓	✓	✓	✓	X	X	X	X	X
	2 GHz	✓	✓	✓	✓	✓	✓	✓	✓	✓	✓
	0.2 to 6 GHz	✓	✓	✓	✓	✓	✓	✓	✓	✓	✓
L	1 GHz	X	X	X	X	✓	✓	✓	X	X	-
	2 GHz	X	✓	X	X	✓	✓	✓	X	X	-
	0.2 to 6 GHz	✓	✓	✓	✓	✓	✓	✓	✓	✓	-

Note: ✓ = detectable, X = not detectable.

5. Summary and Conclusions

The objective of this research was to determine the feasibility of using commercially available NDT methods such as GPR and PAU in detecting FRP bars embedded in concrete. These two NDT methods were tested over several parameters, such as FRP type, bar diameter, bar direction, and bar depths to determine their limitations and detection capabilities. Detecting the FRP bars embedded in concrete is the first obstacle to be overcome before being able to detect damages for ensuring their structural safety. Thus, the long-term objective of this study is to promote the research on NDT methods applicable to embedded FRP bars, which could help in proliferating their use in the construction industry.

In conclusion, the results of this study show that NDT methods that rely on electromagnetic waves such as Ground Penetrating Radar (GPR) become less effective, if not obsolete for non-metallic/non-conductive embedded bars. However, with the increase in center frequency of the GPR device, the detectability of FRP bars can be fairly improved. Similarly, other NDT methods such as Ultrasonic Testing (UT) or Phased Array Ultrasonic (PAU) that are based on stress waves have some capability for detecting steel reinforcement, but they perform poorly for the detection of the most commonly used FRP embedded bars (GFRP bars) but are good for detecting FRP strands.

However, there are some limitations related to the research conducted in this paper, which can be further explored and incorporated in the future studies related to NDT methods for the inspection of FRP-reinforced concrete elements. The scope of this study was limited to the real time test results obtained from post processing tools embedded within the devices using GPR and PAU, which were originally tuned for the detection of steel bars in traditional reinforced concrete elements. The detectability of the FRP bars embedded in concrete can be further verified by the use of advanced post processing algorithms such as SAFT (synthetic aperture focusing technique) or FMC/TFM (full matrix capture/total focusing method) in a follow up future study. Similarly, this study is limited to only three specimens with limited variation in depth of bars up to 3 in., which can be overcome in future studies by either conducting experimental verification on several specimens or by conducting numerical simulation analysis with adequate test parameters to collect more data for a statistically sound validation.

Author Contributions: Conceptualization, P.M., S.S.K.D. and A.B.M.; methodology, P.M., S.S.K.D. and J.D.O.; software, P.M., S.S.K.D. and K.D.; validation, A.B.M., A.N. and K.D.; formal analysis, A.B.M. and A.N.; investigation, P.M., S.S.K.D. and K.D.; resources, A.B.M., A.N. and K.D.; data curation, P.M., S.S.K.D. and J.D.O.; writing—original draft preparation, P.M.; writing—review and editing, A.B.M., A.N., S.S.K.D. and J.D.O.; visualization, P.M., S.S.K.D. and J.D.O.; supervision, A.B.M. and A.N.; project administration, A.B.M. and A.N.; funding acquisition, A.B.M. and A.N. All authors have read and agreed to the published version of the manuscript.

Funding: This research received no external funding.

Institutional Review Board Statement: Not applicable.

Informed Consent Statement: Not applicable.

Data Availability Statement: The data presented in this study are available by request from the corresponding author.

Acknowledgments: The authors greatly acknowledge the support by the Department of Civil and Environmental Engineering at Florida International University, the Department of Civil and Architectural Engineering at the University of Miami, especially Ana De Diego Castro, and Screening Eagle Technologies, Switzerland. The contents of this paper reflect the views of the authors, who are responsible for the facts and the accuracy of the information presented herein.

Conflicts of Interest: The authors declare no conflict of interest.

References

1. Wang, W. *Durability Behavior of Fiber Reinforced Polymer and Steel Reinforced Polymer for Infrastructure Applications*; Missouri University of Science and Technology: Rolla, MO, USA, 2017.
2. Benmokrane, B.; El-Salakawy, E.; El-Ragaby, A.; Lackey, T. Designing and Testing of a Concrete Bridge Deck Reinforced with Glass FRP Bars. *J. Bridg. Eng.* **2006**, *11*, 217–229. [CrossRef]
3. Khedmatgozar Dolati, S.S.; Mehrabi, A. NSM FRP Pile-Splice System for Prestressed Precast Concrete Piles. *Pract. Period. Struct. Des. Constr.* **2022**, *27*, 4022046. [CrossRef]
4. Dolati, S.S.K.; Mehrabi, A. FRP Sheet/Jacket System as an Alternative Method for Splicing Prestressed-Precast Concrete Piles. *Case Stud. Constr. Mater.* **2022**, *16*, e00912.
5. Mehrabi, A.; Khedmatgozar Dolati, S.S. NSMB Pile Splice System for Precast Concrete Piles. U.S. Patent 11,319,689, 3 May 2022.
6. Khedmatgozar Dolati, S.S.; Mehrabi, A. FRP Splice System for Joining Structural Elements. U.S. Patent 11,319,706, 3 May 2022.
7. Mohamed, H.M.; Benmokrane, B. Design and Performance of Reinforced Concrete Water Chlorination Tank Totally Reinforced with GFRP Bars: Case Study. *J. Compos. Constr.* **2014**, *18*, 1–11. [CrossRef]
8. Benmokrane, B.; Mohamed, H.M.; Mousa, S.; Elsafty, A.; Nolan, S. Design, Construction, Testing, and Behavior of Driven Precast Concrete Piles Reinforced with GFRP Bars and Spirals. *J. Bridg. Eng.* **2021**, *26*, 1–13. [CrossRef]
9. Yazdani, N.; Garcia, E.C.; Riad, M. *Field Assessment of Concrete Structures Rehabilitated with FRP*; Elsevier Ltd.: Amsterdam, The Netherlands, 2018; ISBN 9780081021811.
10. Dolati, S.S.K.; Matamoros, A.; Ghannoum, W. Evaluating the Effects of Loading Protocol on the Strength and Deformation Capacity of Flexure-Shear Critical Concrete Columns. *Eng. Struct.* **2023**, *279*, 115592. [CrossRef]
11. Javed, A.; Krishna, C.; Ali, K.; Afzal, M.F.U.D.; Mehrabi, A.; Meguro, K. Micro-Scale Experimental Approach for the Seismic Performance Evaluation of RC Frames with Improper Lap Splices. *Infrastructures* **2023**, *8*, 56. [CrossRef]
12. Malnati, P. A Hidden Revolution: FRP Rebar Gains New Strength. *Compos. Technol.* **2011**, *17*, 6.
13. Mirmiran, A.; Wei, Y. Damage Assessment of FRP-Encased Concrete Using Ultrasonic Pulse Velocity. *J. Eng. Mech.* **2001**, *127*, 126–135. [CrossRef]
14. Ghaib, M. Detection and Localization of Damage in Fiber Reinforced Polymer Bars Using Acoustic Emission, Micro Computed Tomography, and Scanning Electron Microscopy Techniques. Ph.D. Thesis, University of Manitoba, Winnipeg, MB, Canada, 2018.
15. Telang, N.M.; Dumlao, C.; Mehrabi, A.B.; Ciolko, A.T.; Gutierrez, J. NCHRP Report 564: Field Inspection of In-Service FRP Bridge Decks. *Transp. Res. Board* **2006**, *564*, 175. [CrossRef]
16. Khedmatgozar Dolati, S.S.; Malla, P.; Ortiz, J.D.; Mehrabi, A.; Nanni, A. Non-Destructive Testing Applications for in-Service FRP Reinforced/Strengthened Concrete Bridge Elements. In *Nondestructive Characterization and Monitoring of Advanced Materials, Aerospace, Civil Infrastructure, and Transportation XVI*; SPIE: Bellingham, WA, USA, 2022; Volume 12047, pp. 59–74.
17. Ortiz, J.D.; Khedmatgozar Dolati, S.S.; Malla, P.; Nanni, A.; Mehrabi, A. FRP-Reinforced/Strengthened Concrete: State-of-the-Art Review on Durability and Mechanical Effects. *Materials* **2023**, *16*, 1990. [CrossRef]
18. Kim, Y.J. *Use of Fiber-Reinforced Polymers in Highway Infrastructure*; National Academics: Washington, DC, USA, 2017; ISBN 9780309390040.
19. Karbhari, V.M.; Chin, J.W.; Hunston, D.; Benmokrane, B.; Juska, T.; Morgan, R.; Lesko, J.J.; Sorathia, U.; Reynaud, D. Durability Gap Analysis for Fiber-Reinforced Polymer Composites in Civil Infrastructure. *J. Compos. Constr.* **2003**, *7*, 238–247. [CrossRef]
20. Mohammedameen, A.; Gülşan, M.E.; Alzeebaree, R.; Çevik, A.; Niş, A. Mechanical and Durability Performance of FRP Confined and Unconfined Strain Hardening Cementitious Composites Exposed to Sulfate Attack. *Constr. Build. Mater.* **2019**, *207*, 158–173. [CrossRef]
21. Benmokrane, B.; El-Salakawy, E.; El-Ragaby, A.; El-Gamal, S. Performance Evaluation of Innovative Concrete Bridge Deck Slabs Reinforced with Fibre-Reinforced-Polymer Bars. *Can. J. Civ. Eng.* **2007**, *34*, 298–310. [CrossRef]
22. Ettouney, S.; Alampalli, M. *Infrastructure Health in Civil Engineering*; National Academics: Washington, DC, USA, 2011; ISBN 9781439866542.
23. Khanal, S. *Review of Modern Nondestructive Testing Techniques for Civil Infrastructure*; West Virginia University: Morgantown, WV, USA, 2020.

24. Wheeler, A.S. *Nondestructive Evaluation of Concrete Bridge Columns Rehabilitiated with Fiber Reinforced Polymers Using Digital Tap Hammer and Infrared Thermography Nondestructive Evaluation of Concrete Bridge Columns*; West Virginia University: Morgantown, WV, USA, 2018.
25. Halabe, U.B.; Joshi, R.M.; Gangarao, H.V.S. Nondestructive Testing of FRP Composite Structural Components and FRP Rehabilitated Bridge Using Digital Tap Testing. *J. Multidiscip. Eng. Sci. Technol.* **2020**, *7*, 11477–11482.
26. Taillade, F.; Quiertant, M.; Benzarti, K.; Dumoulin, J.; Aubagnac, C. Nondestructive Evaluation of FRP Strengthening Systems Bonded on RC Structures Using Pulsed Stimulated Infrared Thermography. In *Infrared Thermography*; IntechOpen: London, UK, 2012; Chapter 9; pp. 193–208.
27. Ekenel, M.; Myers, J.J. Nondestructive Evaluation of RC Structures Strengthened with FRP Laminates Containing Near-Surface Defects in the Form of Delaminations. *Sci. Eng. Compos. Mater.* **2007**, *14*, 299–315. [CrossRef]
28. Hsieh, C.T.; Lin, Y. Detecting Debonding Flaws at the Epoxy-Concrete Interfaces in near-Surface Mounted CFRP Strengthening Beams Using the Impact-Echo Method. *NDT E Int.* **2016**, *83*, 1–13. [CrossRef]
29. Crawford, K.C. Non-Destructive Testing of FRP-Structural Systems Applied to Concrete Bridges. In *Nondestructive Testing of Materials and Structures*; Springer: Dordrecht, The Netherlands, 2013; pp. 835–840.
30. Crawford, K.C. NDT Evaluation of Long-Term Bond Durability of CFRP-Structural Systems Applied to RC Highway Bridges. *Int. J. Adv. Struct. Eng.* **2016**, *8*, 161–168. [CrossRef]
31. *ACI Committee 228*; ACI 228.2R-13: Report on Nondestructive Test Methods for Evaluation of Concrete in Structures. ACI: Farmington Hills, MI, USA, 2013.
32. Dong, Y.; Ansari, F. Non-Destructive Testing and Evaluation (NDT/NDE) of Civil Structures Rehabilitated Using Fiber Reinforced Polymer (FRP) Composites. *Serv. Life Estim. Ext. Civ. Eng. Struct.* **2011**, *1*, 193–222. [CrossRef]
33. Gower, M.; Lodeiro, M.; Aktas, A.; Shaw, R.; Maierhofer, C.; Krankenhagen, R.; Augustin, S.; Rollig, M.; Knazovicka, L.; Blahut, A.; et al. Design and Manufacture of Reference and Natural Defect Artefacts for the Evaluation of NDE Techniques for Fibre Reinforced Plastic (FRP) Composites in Energy Applications. In Proceedings of the 19th World Conference on Non-Destructive Testing, Munich, Germany, 13–17 June 2016.
34. Aboukhousa, M.; Qaddoumi, N. Near-Field Microwave Imaging of Subsurface Inclusions in Laminated Composite Structures. In Proceedings of the 16th World Conference on Nondestructive Testing, Montreal, QC, Canada, 30 August–3 September 2004.
35. Kharkovsky, S.; Ryley, A.C.; Stephen, V.; Zoughi, R. Dual-Polarized near-Field Microwave Reflectometer for Noninvasive Inspection of Carbon Fiber Reinforced Polymer-Strengthened Structures. *IEEE Trans. Instrum. Meas.* **2008**, *57*, 168–175. [CrossRef]
36. Navagato, M.D.; Narayanan, R.M. Microwave Imaging of Multilayered Structures Using Ultrawideband Noise Signals. *NDT E Int.* **2019**, *104*, 19–33. [CrossRef]
37. Ekenel, M.; Stephen, V.; Myers, J.J.; Zoughi, R. Microwave NDE of Reinforced Concrete Beams Strengthened with CFRP Laminates Containing Surface Defects and Tested under Cyclic Loading. In Proceedings of the 16th World Conference on Nondestructive Testing, Montreal, QC, Canada, 30 August–3 September 2004.
38. Akuthota, B.; Hughes, D.; Zoughi, R.; Myers, J.; Nanni, A. Near-Field Microwave Detection of Disbond in Carbon Fiber Reinforced Polymer Composites Used for Strengthening Cement-Based Structures and Disbond Repair Verification. *J. Mater. Civ. Eng.* **2004**, *16*, 540–546. [CrossRef]
39. Dutta, S.S. Nondestructive Evaluation of FRP Wrapped Concrete Cylinders Using Infrared Thermography and Groud Penetrating Radar. Master of Science Thesis, West Virginia University, Morgantown, WV, USA, 2006.
40. Jackson, D.; Islam, M.; Alampalli, S. Feasibility of Evaluating the Performance of Fiber Reinforced Plastic (FRP) Wrapped Reinforced Concrete Columns Using Ground Penetrating RADAR (GPR) and Infrared (IR) Thermography Techniques. In *Structural Materials Technology IV—An NDT Conference*; CRC Press: Boca Raton, FL, USA, 2000; pp. 390–395.
41. Hing, C.L.C.; Halabe, U.B. Nondestructive Testing of GFRP Bridge Decks Using Ground Penetrating Radar and Infrared Thermography. *J. Bridg. Eng.* **2010**, *15*, 391–398. [CrossRef]
42. Yazdani, N.; Beneberu, E.; Riad, M. Nondestructive Evaluation of FRP-Concrete Interface Bond Due to Surface Defects. *Adv. Civ. Eng.* **2019**, *2019*, 2563079. [CrossRef]
43. Büyüköztürk, O.; Yu, T.Y. Far-Field Radar NDT Technique for Detecting GFRP Debonding from Concrete. *Constr. Build. Mater.* **2009**, *23*, 1678–1689. [CrossRef]
44. Sen, R. Developments in the Durability of FRP-Concrete Bond. *Constr. Build. Mater.* **2015**, *78*, 112–125. [CrossRef]
45. Riad, M.; Yazdani, N.; Almomani, Y.; Gonzalez, E. Utilization of Ground Penetrating Radar (GPR) in the Non-Destructive Assessment of FRP Laminate-Concrete Bond Strength. In *Advances and Challenges in Structural Engineering, Proceedings of the International Congress and Exhibition "Sustainable Civil Infrastructures", Cairo, Egypt, 10–19 November 2019*; Springer Science and Business Media: Berlin, Germany, 2019; pp. 244–267.
46. Jaishankar, P.; Kanchidurai, S.; Thomas, A.C.; Mohan, K.S.R. Experimental Investigation on Non-Destructive Behaviour of Repaired FRP Concrete Beams. *Mater. Today Proc.* **2022**, *64*, 990–994. [CrossRef]
47. Liu, Z.; Gu, X.; Chen, J.; Wang, D.; Chen, Y.; Wang, L. Automatic Recognition of Pavement Cracks from Combined GPR B-Scan and C-Scan Images Using Multiscale Feature Fusion Deep Neural Networks. *Autom. Constr.* **2023**, *146*, 104698. [CrossRef]
48. La Malfa Ribolla, E.; Hajidehi, M.R.; Rizzo, P.; Scimemi, G.F.; Spada, A.; Giambanco, G. Ultrasonic Inspection for the Detection of Debonding in CFRP-Reinforced Concrete. *Struct. Infrastruct. Eng.* **2017**, *14*, 807–816. [CrossRef]

49. Concu, G.; Trulli, N. Direct and Semi-Direct Ultrasonic Testing for Quality Control of FRC-Concrete Adhesion. *Structures* **2021**, *32*, 54–64. [CrossRef]
50. Gunes, O. Failure Modes in Structural Applications of Fiber-Reinforced Polymer (FRP) Composites and Their Prevention. In *Developments in Fiber-Reinforced Polymer (FRP) Composites for Civil Engineering*; Woodhead Publishing Series in Civil and Structural Engineering; Woodhead Publishing: Sawston, UK, 2013; ISBN 9781845691455.
51. Ribolla, E.L.M.; Hajidehi, M.R.; Scimemi, G.F.; Spada, A.; Giambanco, G. Assessment of Bonding Defects in FRP Reinforced Structures via Ultrasonic Technique. *Chall. J. Struct. Mech.* **2016**, *2*, 3. [CrossRef]
52. Ray, B.C.; Hasan, S.T.; Clegg, D.W. Evaluation of Defects in FRP Composites by NDT Techniques. *J. Reinf. Plast. Compos.* **2007**, *26*, 1187–1192. [CrossRef]
53. Taheri, H.; Hassen, A.A. Nondestructive Ultrasonic Inspection of Composite Materials: A Comparative Advantage of Phased Array Ultrasonic. *Appl. Sci.* **2019**, *9*, 1628. [CrossRef]
54. Boychuk, A.S.; Generalov, A.S.; Stepanov, A.V. Nondestructive Testing of FRP by Using Phased Array Ultrasonic Technology. In Proceedings of the ICNDT 2013: 12th International NDT Conference—Application of Contemporary Non-Destructive Testing in Engineering, Portorož, Slovenia, 4–6 September 2013; pp. 51–55.
55. Meola, C.; Boccardi, S.; Carlomagno, G.M.; Boffa, N.D.; Monaco, E.; Ricci, F. Nondestructive Evaluation of Carbon Fibre Reinforced Composites with Infrared Thermography and Ultrasonics. *Compos. Struct.* **2015**, *134*, 845–853. [CrossRef]
56. Wen, B.; Zhou, Z.; Zeng, B.; Yang, C.; Fang, D.; Xu, Q.; Shao, Y.; Wan, C. Pulse-Heating Infrared Thermography Inspection of Bonding Defects on Carbon Fiber Reinforced Polymer Composites. *Sci. Prog.* **2020**, *103*, 36850420950131. [CrossRef]
57. Caldeira, M.M.; Padaratz, I.J. Potentialities of Infrared Thermography to Assess Damage in Bonding between Concrete and GFRP. *Rev. IBRACON Estrut. Mater.* **2015**, *8*, 296–322. [CrossRef]
58. Brown, J.R.; Chittineni, S.H. Comparison of Lock-in and Pulse-Phase Thermography for Defect Characterization in FRP Composites Applied to Concrete. *Therm. Infrared Appl. XXXVII* **2015**, *9485*, 94850B. [CrossRef]
59. Milovanović, B.; Banjad Pečur, I. Review of Active IR Thermography for Detection and Characterization of Defects in Reinforced Concrete. *J. Imaging* **2016**, *2*, 11. [CrossRef]
60. Riad, M. Quantitative Non-Destructive Evaluation (NDE) of FRP Laminate-Concrete Bond Strength. Ph.D. Thesis, The University of Texas at Arlington, Arlington, TX, USA, 2017.
61. Tashan, J.; Al-Mahaidi, R.; Mamkak, A. Defect Size Measurement and Far Distance Infrared Detection in CFRP-Concrete and CFRP-Steel Systems. *Aust. J. Struct. Eng.* **2016**, *17*, 2–13. [CrossRef]
62. Mabry, N.J.; Peters, K.J.; Seracino, R. Depth Detection of Bond Defects in Multilayered Externally Bonded CFRP-to-Concrete Using Pulse Phase Thermography. *J. Compos. Constr.* **2015**, *19*, 04015002. [CrossRef]
63. Gu, J.C.; Unjoh, S.; Naito, H. Detectability of Delamination Regions Using Infrared Thermography in Concrete Members Strengthened by CFRP Jacketing. *Compos. Struct.* **2020**, *245*, 112328. [CrossRef]
64. Yumnam, M.; Gupta, H.; Ghosh, D.; Jaganathan, J. Inspection of Concrete Structures Externally Reinforced with FRP Composites Using Active Infrared Thermography: A Review. *Constr. Build. Mater.* **2021**, *310*, 125265. [CrossRef]
65. Wang, B.; Zhong, S.; Lee, T.-L.; Fancey, K.S.; Mi, J. Non-Destructive Testing and Evaluation of Composite Materials/Structures: A State-of-the-Art Review. *Adv. Mech. Eng.* **2020**, *12*, 1687814020913761. [CrossRef]
66. Carpinteri, A.; Lacidogna, G.; Paggi, M. Acoustic Emission Monitoring and Numerical Modeling of FRP Delamination in RC Beams with Non-Rectangular Cross-Section. *Mater. Struct. Constr.* **2007**, *40*, 553–566. [CrossRef]
67. Degala, S.; Rizzo, P.; Ramanathan, K.; Harries, K.A. Acoustic Emission Monitoring of CFRP Reinforced Concrete Slabs. *Constr. Build. Mater.* **2009**, *23*, 2016–2026. [CrossRef]
68. Saeedifar, M.; Fotouhi, M.; Ahmadi Najafabadi, M.; Hosseini Toudeshky, H.; Minak, G. Prediction of Quasi-Static Delamination Onset and Growth in Laminated Composites by Acoustic Emission. *Compos. Part B Eng.* **2016**, *85*, 113–122. [CrossRef]
69. Taillade, F.; Quiertant, M.; Benzarti, K.; Aubagnac, C.; Moser, E. Non-Destructive Evaluation (NDE) of Composites: Using Shearography to Detect Bond Defects. In *Non-Destructive Evaluation (NDE) of Polymer Matrix Composites*; Woodhead Publishing Series in Composites Science and Engineering; Woodhead Publishing: Sawston, UK, 2013; pp. 542–556. [CrossRef]
70. Yu, T.; Cheng, T.K.; Zhou, A.; Lau, D. Remote Defect Detection of FRP-Bonded Concrete System Using Acoustic-Laser and Imaging Radar Techniques. *Constr. Build. Mater.* **2016**, *109*, 146–155. [CrossRef]
71. Qiu, Q.; Lau, D. Defect Detection of FRP-Bonded Civil Structures under Vehicle-Induced Airborne Noise. *Mech. Syst. Signal Process.* **2021**, *146*, 106992. [CrossRef]
72. Qiu, Q.; Lau, D. A Novel Approach for Near-Surface Defect Detection in FRP-Bonded Concrete Systems Using Laser Reflection and Acoustic-Laser Techniques. *Constr. Build. Mater.* **2017**, *141*, 553–564. [CrossRef]
73. Qiu, Q.; Lau, D. Experimental Evaluation on the Effectiveness of Acoustic-Laser Technique towards the FRP-Bonded Concrete System. In *Structural Health Monitoring and Inspection of Advanced Materials, Aerospace, and Civil Infrastructure 2015, Proceedings of the SPIE Smart Structures and Materials + Nondestructive Evaluation and Health Monitoring, San Diego, CA, USA, 8–12 March 2015*; SPIE: San Diego, CA, USA; Volume 9437, p. 943705. [CrossRef]
74. Yang, L. Recent Developments in Digital Shearography for Nondestructive Testing. *Mater. Eval.* **2006**, *64*, 704–709.
75. Choi, S.W.; Lee, J.H. Nondestructive Evaluation of Internal Defects for Composite Materials by Using Shearography. *Key Eng. Mater.* **2004**, *270–273*, 781–786. [CrossRef]

76. Wan, B. *Using Fiber-Reinforced Polymer (FRP) Composites in Bridge Construction and Monitoring Their Performance: An Overview*; Woodhead Publishing: Sawston, UK, 2014; ISBN 9780857097019.
77. Garney, G. *Defects Found Through Non-Destructive Testing Methods of Fiber Reinforced Polymeric Composites*; California State University: Fullerton, CA, USA; Ann Arbor, MI, USA, 2006.
78. Weissenböck, J.; Reh, A.; Salaberger, D.; Heinzl, C.; Kastner, J. Advanced Visualization and Exploration Techniques for Fiber Reinforced Polymers. In Proceedings of the 11th European Conference on Non-Destructive Testing (ECNDT 2014), Prague, Czech Republic, 6–11 October 2014; pp. 1–10.
79. Rodríguez-Hortalá, M.; Hatzmann, J.; Degischer, H.P. Structural Characterisation of Defects in CFRP Laminates Produced with Different Epoxy Resin Systems and Their Influence on the Mechanical Properties. In Proceedings of the ECCM 2012—Composites at Venice, 15th European Conference on Composite Materials, Tarrytown, NY, USA, 6–8 June 2012; pp. 24–28.
80. Khosravani, M.R.; Reinicke, T. On the Use of X-Ray Computed Tomography in Assessment of 3D-Printed Components. *J. Nondestruct. Eval.* **2020**, *39*, 75. [CrossRef]
81. Kim, S.Y.; Sesso, M.L.; Franks, G.V. In-Situ 4-Point Flexural Testing and Synchrotron Micro X-Ray Computed Tomography of 3D Printed Hierarchical-Porous Ultra-High Temperature Ceramic. *Addit. Manuf.* **2022**, *54*, 102728. [CrossRef]
82. Karbhari, V.M.; Kaiser, H.; Navada, R.; Ghosh, K.; Lee, L. *Methods for Detecting Defects in Composite Rehabilitated Concrete Structures*; Oregon Department of Transportation: Salem, OR, USA; Federal Highway Administration: Washington, DC, USA, 2005.
83. Alampalli, S. Field Performance of an FRP Slab Bridge. *Compos. Struct.* **2006**, *72*, 494–502. [CrossRef]
84. Guan, H.; Karbhari, V.M.; Sikorsky, C.S. Long-Term Structural Health Monitoring System for a FRP Composite Highway Bridge Structure. *J. Intell. Mater. Syst. Struct.* **2007**, *18*, 809–823. [CrossRef]
85. Hag-elsafi, O.; Kunin, J.; Alampalli, S.; Conway, T. *Strengthening of Route 378 Bridge Over Wynantskill Creek In New York Using FRP Laminates*; National Academies of Sciences, Engineering, and Medicine: Washington, DC, USA, 2001.
86. Hag-elsafi, O.; Lund, R.; Alampalli, S. *Strengthening of Church Street Bridge Pier Capbeam Using Bonded FRP Composite Plates: Strengthening and Load Testing*; Transportation Research and Development Bureau: New York, NY, USA, 2002.
87. Afzal, M.F.U.D.; Matsumoto, Y.; Nohmi, H.; Sakai, S.; Su, D.; Nagayama, T. Comparison of Radar Based Displacement Measurement Systems with Conventional Systems in Vibration Measurements at a Cable Stayed Bridge. In Proceedings of the 11th German-Japan Bridge Symposium, Osaka, Japan, 30–31 August 2016.
88. Li, W.; Ho, S.C.M.; Patil, D.; Song, G. Acoustic Emission Monitoring and Finite Element Analysis of Debonding in Fiber-Reinforced Polymer Rebar Reinforced Concrete. *Struct. Health Monit.* **2017**, *16*, 674–681. [CrossRef]
89. Xu, K.; Ren, C.; Deng, Q.; Jin, Q.; Chen, X. Real-Time Monitoring of Bond Slip between GFRP Bar and Concrete Structure Using Piezoceramic Transducer-Enabled Active Sensing. *Sensors* **2018**, *18*, 2653. [CrossRef]
90. Sharma, G.; Sharma, S.; Sharma, S.K. Fracture Monitoring of Steel and GFRP Reinforced Concrete Beams Using Acoustic Emission and Digital Image Correlation Techniques. *Struct. Concr.* **2021**, *22*, 1962–1976. [CrossRef]
91. Dolati, S.S.K.; Mehrabi, A.; Dolati, S.S.K.; Caluk, N. NDT Methods for Damage Detection in Steel Bridges. *Proc. SPIE* **2022**, *12048*, 385–394.
92. Khedmatgozar Dolati, S.S.; Caluk, N.; Mehrabi, A.; Khedmatgozar Dolati, S.S. Non-Destructive Testing Applications for Steel Bridges. *Appl. Sci.* **2021**, *11*, 9757. [CrossRef]
93. Drobiec, Ł.; Jasiński, R.; Mazur, W. The Use of Non-Destructive Methods to Detect Non-Metallic Reinforcement in Concrete and Masonry. *Preprints* **2019**, 2019010021, *Preprints*. [CrossRef]
94. *ASTM C39*; Standard Test Method for Compressive Strength of Cylindrical Concrete Specimens. ASTM: West Conshohocken, PA, USA, 2010.
95. Hameed, A.; Rasool, A.M.; Ibrahim, Y.E.; Afzal, M.F.U.D.; Qazi, A.U.; Hameed, I. Utilization of Fly Ash as a Viscosity-Modifying Agent to Produce Cost-Effective, Self-Compacting Concrete: A Sustainable Solution. *Sustainability* **2022**, *14*, 1559. [CrossRef]
96. *ACI CODE-440.11-22*; Building Code Requirements for Structural Concrete Reinforced with Glass Fiber-Reinforced Polymer (GFRP) Bars—Code and Commentary. ACI: Farmington Hills, MI, USA, 2022.
97. Javed, A.; Sadeghnejad, A.; Rehmat, S.; Yakel, A.; Azizinamini, A. *Magnetic Flux Leakage Method for Damage Detection in Internal Post-Tensioning Tendons*; Florida Department of Transportation: Tallahassee, FL, USA, 2021; 166p.
98. Tosti, F.; Ferrante, C. *Using Ground Penetrating Radar Methods to Investigate Reinforced Concrete Structures*; Springer: Dordrecht, The Netherlands, 2020; Volume 41, ISBN 0123456789.
99. Liu, Z.; Yeoh, J.K.W.; Gu, X.; Dong, Q.; Chen, Y.; Wu, W.; Wang, L.; Wang, D. Automatic Pixel-Level Detection of Vertical Cracks in Asphalt Pavement Based on GPR Investigation and Improved Mask R-CNN. *Autom. Constr.* **2023**, *146*, 104689. [CrossRef]
100. Capozzoli, L.; Rizzo, E. Combined NDT Techniques in Civil Engineering Applications: Laboratory and Real Test. *Constr. Build. Mater.* **2017**, *154*, 1139–1150. [CrossRef]
101. Sensors and Software. *Conquest 100 User's Guide*; Sensors and Software: Mississauga, ON, Canada, 2015.
102. IDS Georadar. *C-Thrue User Manual: All-in-One Ground Penetrating Radar (GPR) for Non-Destructive Testing of Concrete Structures User*; IDS Georadar: Pisa, Italy, 2017.
103. Screening Eagle. *Sales Flyer: Proceq Ground Penetrating Radars GP8000*; Screening Eagle: Austin, TX, USA, 2022.
104. Screening Eagle. *Sales Flyer: Proceq Ground Penetrating Radars GP8800*; Screening Eagle: Austin, TX, USA, 2022.
105. Chen, R.; Tran, K.T.; Dinh, K.; Ferraro, C.C. Evaluation of Ultrasonic SH-Waveform Tomography for Determining Cover Thickness and Rebar Size in Concrete Structures. *J. Nondestruct. Eval.* **2022**, *41*, 1–16. [CrossRef]

106. Torres-González, M.; Mantero, J.; Hurtado, S.; Flores-Alés, V.; Alejandre, F.J.; Alducín-Ochoa, J.M. Characterization and Radioactive Evaluation of the Concrete from a Radiotherapy Bunker. *Struct. Concr.* **2022**, *23*, 3102–3113. [CrossRef]
107. Park, S.; Kim, J.; Jeon, K.; Kim, J.; Park, S. Improvement of Gpr-Based Rebar Diameter Estimation Using Yolo-V3. *Remote Sens.* **2021**, *13*, 2011. [CrossRef]
108. Shull, P.J. *Nondestructive Evaluation Theory, Techniques, and Applications*; Marcel Dekker, Inc.: New York, NY, USA, 2001; ISBN 0824788729.
109. Boychuk, A.S.; Generalov, A.S.; Stepanov, A.V. CFRP Structural Health Monitoring by Ultrasonic Phased Array Technique. In Proceedings of the 7th European Workshop on Structural Health Monitoring. EWSHM 2014—2nd European Conference of the Prognostics and Health Management Society, Nantes, France, 8–11 July 2014; pp. 2206–2211.
110. Ryan, T.W.; EricMann, J.; Chill, Z.M.; Ott, B.T. Bridge Inspector's Reference Manual. *Fhwa* **2012**, *BIRM 1*, 1020.
111. Taheri, H.; Delfanian, F.; Du, J. Acoustic Emission and Ultrasound Phased Array Technique For Composite Material Evaluation. In Proceedings of the ASME 2013 International Mechanical Engineering Congress and Exposition, San Diego, CA, USA, 15–21 November 2013; pp. 1–8.
112. Prabhakara, P.; Mielentz, F.; Stolpe, H.; Behrens, M.; Lay, V.; Niederleithinger, E. Validation of Novel Ultrasonic Phased Array Borehole Probe by Using Simulation and Measurement. *Sensors* **2022**, *22*, 9823. [CrossRef]
113. Screening Eagle. *Sales Flyer: Pundit Live Array Pro Pulse Echo Testing with A.I.*; Screening Eagle: Austin, TX, USA, 2018.
114. Acoustic Control Systems. *Product Flyer: A1040 MIRA 3D*; Acoustic Control Systems: North Yorkshire, UK, 2022.
115. Dinh, K.; Tran, K.; Gucunski, N.; Ferraro, C.C.; Nguyen, T. Imaging Concrete Structures with Ultrasonic Shear Waves—Technology Development and Demonstration of Capabilities. *Infrastructures* **2023**, *8*, 53. [CrossRef]
116. Annan, A.P. Ground Penetrating Radar Principles, Procedures & Applications. In *Ground Penetrating Radar Theory and Application*; Elsevier Science: Amsterdam, The Netherlands, 2009; p. iv. [CrossRef]
117. Gehrig, M.D.; Morris, D.V.; Bryant, J.T. Ground Penetrating Radar for Concrete Evaluation Studies. In Proceedings of the Foundation Performance Association Meeting, London, UK, 24 March 2004; pp. 1–17.
118. Tsalicoglou, I. The Three (Four) Musketeers of Multi-Technology Concrete NDT for Accuracy and Productivity. In Proceedings of the 3rd Japanese-Swiss Workshop—"Durability Testing of Concrete, on Site and in the Lab", Wildegg, Switzerland, 16 October 2018.

applied sciences

MDPI

Article

A Non-Destructive Method for Predicting Critical Load, Critical Thickness and Service Life for Corroded Spherical Shells under Uniform External Pressure Based on NDT Data

Cheng Huijuan Liu [1] and Giuseppe Lacidogna [2,*]

[1] Department of Mathematics, Aberystwyth University, Aberystwyth, Ceredigion, Wales SY23 3BZ, UK; hul18@aber.ac.uk

[2] Department of Structural, Geotechnical and Building Engineering, Politecnico di Torino, 10129 Turin, Italy

* Correspondence: giuseppe.lacidogna@polito.it

Featured Application: This work can be potentially valuable to be used as a reference for existing estimating methods based on NDT.

Abstract: A pressurized spherical shell that is continuously corroded will likely buckle and lose its stability. There are many analytical and numerical methods to study this problem (critical load, critical thickness, and service life), but the friendliness (operability) in engineering test applications is still not ideal. Therefore, in this paper, we propose a new non-destructive method by combining the Southwell non-destructive procedure with the stable analysis method of corroded spherical thin shells. When used carefully, it can estimate the critical load (critical thickness) and service life of these thin shells. Furthermore, its procedure proved to be more practical than existing methods; it can be easily mastered, applied, and generalized in most engineering tests. When used properly, its accuracy is acceptable in the field of engineering estimations. In the context of the high demand for non-destructive analysis in industry, it may be of sufficient potential value to be used as a reference for existing estimating methods based on NDT data.

Keywords: pressurized spherical shell; corrosion; non-destructive method; critical load; critical thickness; service life; NDT data

Citation: Liu, C.H.; Lacidogna, G. A Non-Destructive Method for Predicting Critical Load, Critical Thickness and Service Life for Corroded Spherical Shells under Uniform External Pressure Based on NDT Data. *Appl. Sci.* **2023**, *13*, 4172. https://doi.org/10.3390/app13074172

Academic Editor: Michel Darmon

Received: 6 February 2023
Revised: 3 March 2023
Accepted: 13 March 2023
Published: 24 March 2023

1. Introduction

Due to the increasing use of shell-type structures in spacecrafts, submarines, buildings, and storage tanks, there has been a corresponding increase in the interest of researchers and practical engineers in the stability of shells. Hemispherical shells are the most important structural element in engineering applications because they can resist higher pure internal pressure loads than any other geometric vessel with the same wall thickness and radius.

In practice, most pressure vessels experience external loads due to hydrostatic pressure or external shocks-. Therefore, they should be designed to withstand the worst load combinations without failure. Loads transmitted by cylindrical rigid actuators applied on top of the sphere are considered common external loads. Therefore, it is important to study its effect on the initial buckling behaviour of such shells. Meanwhile, corrosion is defined as the gradual destruction of a material due to chemical reactions within the environment. The most common type of corrosion is uniform corrosion or general corrosion, which is distributed almost uniformly over the entire exposed surface. General wear can occur both with the formation of a fully protective ultra-thin coating of corrosion products and without an oxide layer. The formation of a blocking passivation film, as well as changes in the concentration of one or the other reactants, may inhibit when the corrosion rate should decay exponentially (decline) with time [1–3].

On the other hand, as with other types of damage (e.g., [4]), corrosion can be enhanced by the applied load [5]. Experimental data suggests that there is a stress corrosion threshold, after which mechanical stress accelerates corrosion [5–7]. In this case, the stress changes due to the reduction in shell thickness, and the changed stress in turn enhances the corrosion process. In general, for the strength analysis of structural elements under mechano-chemical corrosion conditions, an initial boundary value problem with unknown evolutionary boundaries must be solved.

In addition to stress, there are many other effects that can affect the corrosion rate, such as temperature; it has a great influence on the rate of galvanic corrosion of metals. In the case of neutral-solution corrosion (oxygen depolarization), elevated temperature has a favourable effect on the overpotential and oxygen diffusion rate for oxygen depolarization but leads to a decrease in oxygen solubility. When corrosion (hydrogen depolarization) occurs in acidic media (such as sea water), the corrosion rate increases exponentially with increasing temperature due to the reduced hydrogen evolution overpotential. An Arrhenius-type experimental dependence was observed between corrosion rate and temperature [8]. The effect of temperature on acid corrosion, most commonly in hydrochloric and sulfuric acid, has been the subject of extensive research [9–22]. In hydrochloric acid, the effective activation energies for corrosion processes vary from 57.7 to 87.8 kJ mol/L, where most are concentrated around 60.7 kJ mol/L. In some cases, studies were performed at only three temperature values using a single experimental method, which increased the likelihood of erroneous determination of the corrosion activation energy. In this regard, further research is advisable, as it may provide a reliable comparative basis for discussing the obtained results.

In summary, we know that temperature has a great influence on corrosion rate, corrosion and stress can interact with each other, and finally, they can jointly affect the stability of the shell. In this regard, we should explore the relationships of temperature, corrosion, and stress to shell stability. Experimental [23], analytical [24–28], and numerical methods [29–31] were used to study the buckling of (uniformly compressed hemispherical of moderate thickness) metal shells using corrosion and temperature. A related study [28] also demonstrated the high accuracy of these methods. However, in practical engineering applications, they often lack operability and are less friendly to workers.

Non-destructive estimation methods, such as in the field of pressure vessels, have been a research hotspot due to their advantages of simple operation, widespread use, and low cost. As one of the non-destructive methods, the non-destructive approach of Southwell's column analysis is now extended to spherical shells, which are subjected to uniform external pressure [32]. However, there are few non-destructive methods for estimating the buckling of pressure spherical shells of this type. The paper [33] established a non-destructive estimation method for a spherical shell under external pressure that can predict its critical load. It is based on an exact first-order solution of the critical stress and requires two assumptions: one is that the thickness of the shell does not change; and the second is that the temperature does not affect the critical buckling of the shell. In any case, when the shell is in a corrosive environment, the thickness of the shell varies and the effect of temperature on both the stress and the corrosion rate is unavoidable. No non-destructive method has been found as of yet to predict the critical loads (stresses) of the shell under this condition, including the service life. Therefore, in the industry, people still hope to obtain a new lossless method. This method can predict the critical load (or stress) and critical thickness of the shell under external pressure from corrosion on the basis of a higher-order (second-order) exact solution. At the same time, we are also oriented to predict its useful (remaining) life. For this reason, in this paper, we will analytically extend the Southwell process method for the non-destructive prediction of critical loads, critical stresses, and service life of hemispherical shells subjected to uniform external pressure considering corrosion and ambient temperature through rigorous mathematical derivation.

2. Problem Description

A model of a spherical shell is considered. It is affected by the external pressure p_0, and its inner surface undergoes mechano-chemical corrosion, as shown in Figure 1. Assuming that the rate of internal corrosion is v_0, the corrosion process causes the thickness h of the shell to change with time t. We adopt the effective stress definition to characterize stress here, which is commonly used when corrosion is present.

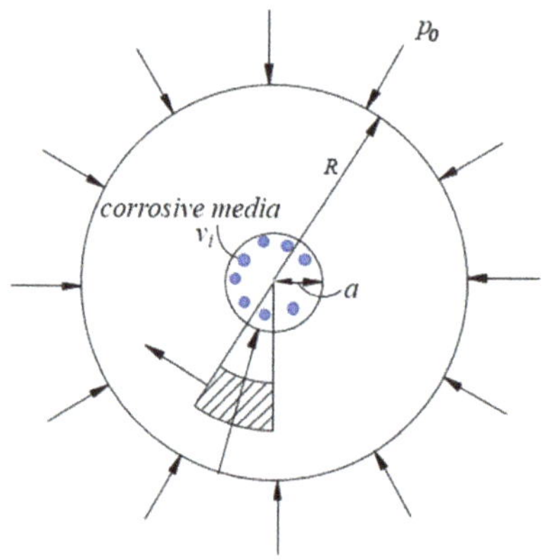

Figure 1. A spherical shell subjected to both external pressure and internal corrosion.

The problem in this paper is to predict (estimate) the critical thickness, critical stress, and service life of the spherical shell based on non-destructive testing data in a temperature and corrosive environment. This is essentially a non-destructive estimation problem involving variable boundary conditions. For clarity, we describe this complex and combined problem in two steps.

2.1. Before All, It Is Necessary to Solve the Estimation of the Critical Thickness, Critical Stress, and Service Life of the Shell Based on NDT Data in a Non-Corrosion and Temperature-Independent Environment

It involves two sub-problems. First, theoretically, we need to obtain the second-order buckling critical stress for the spherical shell model (see Appendix A), which is the accurate solution. Then, based on the format of this second-order solution, a test data-based estimation method for the shell critical stress will be established by introducing an existing non-destructive method (see [33]).

The mathematical problem involved in obtaining the second-order buckling critical stress is related to the solution of the following Equations (1)–(4):

$$\frac{dN_x}{d\theta} + (N_x - N_y)\cot\theta - Q_x + N_y\left(\frac{u}{R} + \frac{dw}{Rd\theta}\right) - Q_x\left(\frac{d^2w}{Rd\theta^2} + \frac{w}{R}\right) = 0, \quad (1)$$

$$\frac{dQ_x}{d\theta} + Q_x\cot\theta + N_x + N_y + pR + N_x\left(\frac{d^2w}{Rd\theta^2} + \frac{du}{Rd\theta}\right) + N_y\left(\frac{u}{R} + \frac{dw}{Rd\theta}\right)\cot\theta = 0, \quad (2)$$

$$\frac{dM_x}{d\theta} + (M_x - M_y)\cot\theta - Q_xR + M_y\left(\frac{u}{R} + \frac{dw}{Rd\theta}\right) = 0, \quad (3)$$

$$\sigma = \frac{pR}{2h_0} \quad (4)$$

where u is the displacement of the shell element in x direction, v is the displacement in y direction, w is the displacement in z direction, t_0 is the shell thickness, p_{cr} is the classical buckling pressure, N_x, N_y are the resultant forces, Q_x, Q_y are the shear forces, M_x, M_y are the bending moments, and θ, ψ are the angles of the shell element.

When introducing the non-destructive method, the corresponding mathematical problem involved is how to express the relationship between non-destructively measurable quantities (such as: w and w/p at any point on the outer surface of the shell) as an equation of a straight line. The physical quantity to be estimated (e.g., the critical load p_{cr} of the shell) needs to appear exactly in the expression of the line slope. Assuming that we can obtain this equation line by non-destructive testing in the elastic stage, the quantity to be estimated can be obtained immediately. For example, let us take Equation (12) in which the equation slope is $\frac{p_{cr}^2}{Em^3\Delta^2}$. Then suppose we get the slope by testing and drawing; therefore, the only unknown quality in it, p_{cr}, can be expressed (obtained) by the slope easily.

Moreover, the assumptions within this non-destructive method are that the shell deformation is axisymmetric and the compression in the shell is uniform. At the same time, it does not assume the specific location of buckling and the number of buckling waves, as shown in Figure 2.

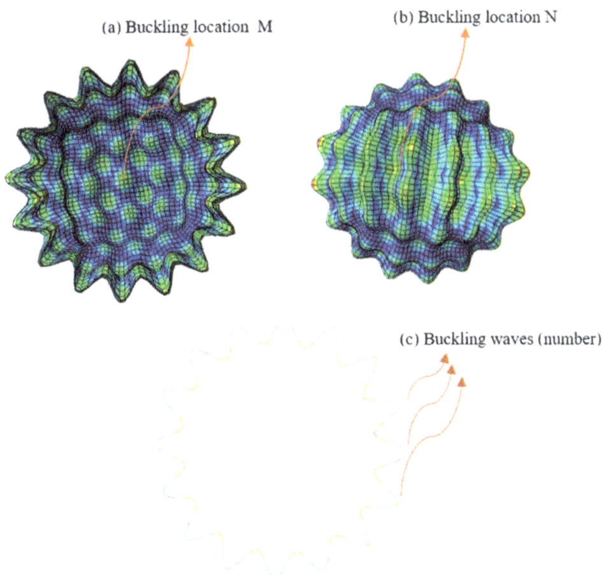

Figure 2. The non-destructive method does not assume the exact location of spherical shell buckling (e.g., the point M or N) and the number of buckling waves.

2.2. Then, on the Basis of the Previous Step 2.1, We Can Further Solve the Problem of Critical Load (Stress) and Service Life in an Environment Where Corrosion and Temperature Coexist

The mathematical problem that needs to be solved is similar to that in Section 2.1, but an extra equation (corrosion kinetic equation) needs to be taken into consideration—the relationships between the corrosion rate (that is, the derivative of thickness), temperature, and stress:

$$\frac{dh}{dt} = -f(\sigma, T)\frac{dh}{dt} = -f(\sigma, T) \tag{5}$$

here, for f (σ, T) we adopt:

$$f(\sigma, t) = v_0 \exp(\frac{V\sigma}{RT_g}) \tag{6}$$

where h is the thickness of the shell, t is the corrosion time, v_0 is the initial corrosion rate, σ is the stress, T is the temperature, R_g is the molar gas constant, and V is the material molar volume.

It is noted that the derivation process of Equation (6) is shown in Appendix B, where we found that the relationship between corrosion rate and temperature conforms to the Arrhenius type [5]. Meanwhile, it should be observed that according to the physical definition of shell buckling, the critical thickness in the case described in this section will be identical to its expression obtained in Section 2.1.

3. Problem Solving

In order to solve the problems highlighted in this paper (see Section 2), we run the following two steps to explain our methods.

3.1. First Step: Establish a Non-Destructive Method for Predicting Spherical Shell Life Regardless of Corrosion and Temperature

From (A50) in Appendix A.1, we rewrite the it as:

$$\frac{2Em}{1-v^2} = \frac{p}{\varphi} \tag{7}$$

Here, m is the ratio of the thickness to the radius of the sphere.

We rewrite Equation (A51) in Appendix A.1 as:

$$p_{cr}^2 = Em.Em.m^2\left(\frac{2}{\sqrt{3(1-v^2)}} - \frac{mv}{1-v^2}\right)^2, \tag{8}$$

here, $m = \frac{h_0}{R}$.

Let $\Delta = \frac{2}{\sqrt{3(1-v^2)}} - \frac{mv}{1-v^2}$, from Equations (7) and (8), we can obtain:

$$\frac{p}{\phi} = \frac{2Em}{1-v^2} = \frac{2p_{cr}^2}{Em^3(1-v^2)\Delta^2} \tag{9}$$

and then

$$\frac{1}{\phi} = \frac{2Em}{1-v^2} = \frac{2p_{cr}^2}{pEm^3(1-v^2)\Delta^2} \tag{10}$$

By substituting Equation (10) into Equation (A69) in Appendix A.2, we get:

$$w \cong \frac{B_0'(v-1)}{1 + \frac{1-v^2}{2}.\frac{2p_{cr}^2}{pEm^3(1-v^2)\Delta^2}} \tag{11}$$

As can be seen from the equations above, we have successfully established the relationship between the displacement and the pressure of the shell by appropriately rewriting the form of the second-order critical load and combining the relationship between the displacement w and the rotation angle φ shown in Equation (5).

By cross-multiplying in Equation (11), we can achieve:

$$w + \frac{p_{cr}^2}{Em^3\Delta^2}\frac{w}{p} = B_0'(v-1). \tag{12}$$

Formally, Equation (12) is the equation of a straight line following the Southwell procedure described in Section 2.1. It has w as one axis and w/p as another. Furthermore, the expression of the slope of this line contains the unknown critical load p_{cr}. Therefore, the slope of this line can be gained experimentally, and the critical load can then be obtained.

Let $S = \frac{p_{cr}^2}{Em^3\Delta^2}$; Equation (12) can be rewritten as:

$$w + S\frac{w}{p} = B_0'(v-1) \tag{13}$$

one can get:

$$p_{cr} = m\Delta\sqrt{SEm} \tag{14}$$

By substituting Equation (A53) in Appendix A.1 into Equation (7), we get:

$$\phi = \frac{(h^*)^2\sqrt{\frac{1-v^2}{3}}}{Rh}. \tag{15}$$

Substituting Equation (15) into (A69) in Appendix A, we get:

$$w + J(wh) = B_0'(v-1), \tag{16}$$

where:

$$J = \frac{R\sqrt{3(1-v^2)}}{2(h^*)^2}, \tag{17}$$

then we obtain:

$$h^* = \frac{\sqrt{R}\sqrt[4]{3(1-v^2)}}{\sqrt{2J}}. \tag{18}$$

With the previous steps, we have obtained a non-destructive estimation of the critical thickness (see Equation (18)) in the same way that we have used to solve the critical load above (see Equation (14)).

3.2. Second Step: Establish a Non-Destructive Method for Predicting Critical Load, Critical Thickness, and Service Life of Spherical Shells in the Presence of Corrosion and Temperature

According to the description in Section 2.2, this case requires an additional corrosion-rate equation than in Section 2.1—see (A77) in Appendix B.

Considering Equation (10), from Equation (A77) we obtain:

$$\frac{d\sigma}{dt} = \frac{2\sigma^2}{pR}v_0 \exp\left(\overline{E}_{c0}\left(1-\frac{\overline{E}_c}{\overline{T}}\right)\right)\exp\left(\frac{V\sigma}{R_gT}\right). \tag{19}$$

Through performing variable separation on Equation (19) in $[t_0, t^*]$, $[\sigma_0, \sigma^*]$, we get:

$$\sigma^{-2}\exp\left(-\frac{V\sigma}{R_gT}\right)d\sigma = \frac{2}{pR}v_0\exp\left(\overline{E}_{c0}\left(1-\frac{\overline{E}_c}{\overline{T}}\right)\right)dt, \tag{20}$$

where, t_0 refers to the initial time, t^* to the service life, σ_0 to the stress at time t_0, and σ^* to the critical stress.

Integrating Equation (20), we get:

$$\int_{\sigma_0}^{\sigma^*}\exp\left(-\frac{V\sigma}{R_gT}\right)\sigma^{-2}d\sigma = \frac{2}{pR}v_0\exp\left(\overline{E}_{c0}\left(1-\frac{\overline{E}_c}{\overline{T}}\right)\right)\int_{t_0}^{t^*}dt. \tag{21}$$

Considering that Equation (10) establishes consistency, through $\sigma \equiv \frac{pR}{2h}$, the service life t^* can be obtained:

$$t^* = \frac{h_0\sigma_0}{v_0}\left[-\overline{E}_{co}(1-\overline{E}_c/\overline{T}\right]\int_{\sigma^0}^{\sigma^*}\sigma^{-2}\exp\left(\frac{-V\sigma}{R_gT}\right)d\sigma. \tag{22}$$

If we still consider Equation (10), Equation (A77) can be transformed into:

$$\frac{dh}{dt} = v_0\exp\left(\overline{E}_{c0}\left(1-\frac{\overline{E}_c}{\overline{T}}\right)\right)\exp\left(\frac{VpR}{2R_gTh}\right). \tag{23}$$

In order to separate the variables t and h in $[t_0, t^*]$ and $[h_0, h^*]$, first we get:

$$\exp\left(-\frac{VpR}{2R_gTh}\right)dh = v_0 \exp\left(\overline{E}_{c0}\left(1-\frac{\overline{E}_c}{\overline{T}}\right)\right)dt, \tag{24}$$

then:

$$\int_{h_0}^{h^*} \exp\left(-\frac{VpR}{2R_gTh}\right)dh = v_0 \exp\left(\overline{E}_{c0}\left(1-\frac{\overline{E}_c}{\overline{T}}\right)\right)\int_{t_0}^{t^*} dt, \tag{25}$$

and finally:

$$t^* = \frac{1}{v_0 \exp\left(\overline{E}_{c0}\left(1-\frac{\overline{E}_c}{\overline{T}}\right)\right)}\int_{h_0}^{h^*} \exp\left(-\frac{VpR}{2R_gTh}\right)dh + t_0. \tag{26}$$

Note that in this section we still have $p_{cr} = m\Delta\sqrt{SEm}$, $\sigma^* = \frac{mR\Delta\sqrt{SEm}}{2h_0}$, and $h^* = \frac{\sqrt{R}\sqrt[4]{3(1-v^2)}}{\sqrt{2J}}$.

4. Example Analysis

The whole process described above (including those in Appendices A and B), showed that the non-destructive method of this paper used mathematical logic rigorously. To further validate it, we compare its results with those of other methods on this section. Due to the limitation of NDT data found in literature on these kinds of shells, we compare our method within a special case in Section 2.1 (without corrosion and temperature) with another existing method.

The NDT data adopted here is taken from an experiment described in [33]. To maintain consistency with the experimental model of [33], we also apply internal suction to our FE model to simulate equivalently the external pressure. The geometric parameters for the FE model (meeting thin wall hypothesis) are: the radius of shell r = 0.05 m and the wall thickness h_0 = 0.0005 m, yielding a h_0/r ratio of 0.01. The mechanical properties are: the Young modulus E = 650 MPa, Poisson ratio ν = 0.4, and density ρ = 1150 kg/m^3.

A 1/8 symmetrical spherical shell model built by the ANSYS software is shown in Figure 3. The first two buckling modes obtained from the eigenvalue buckling analysis are shown in Figure 4.

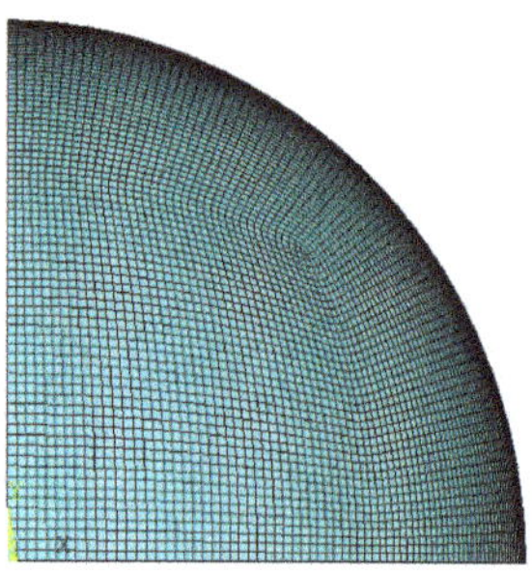

Figure 3. 1/8 symmetrical spherical shell FE model.

First, we calculate the value of p^* using the described method. The results for w and w/p are plotted in Figure 5.

Within the plot, it is shown that the fitting line equation for the w and w/p points is:

$$w + 7.27385\, w/p = 16.59268, \tag{27}$$

therefore, we get S = 7.27385 by Equation (13), and then obtain p^* = 0.08593 through Equation (14).

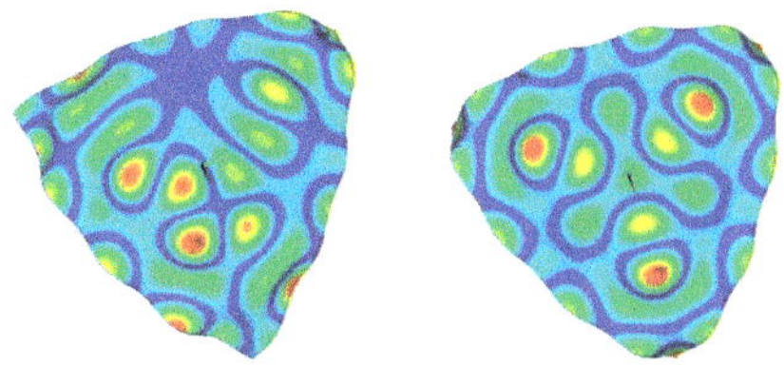

(a) First mode (b) Second mode

Figure 4. Buckling modes obtained from the eigenvalue buckling analysis.

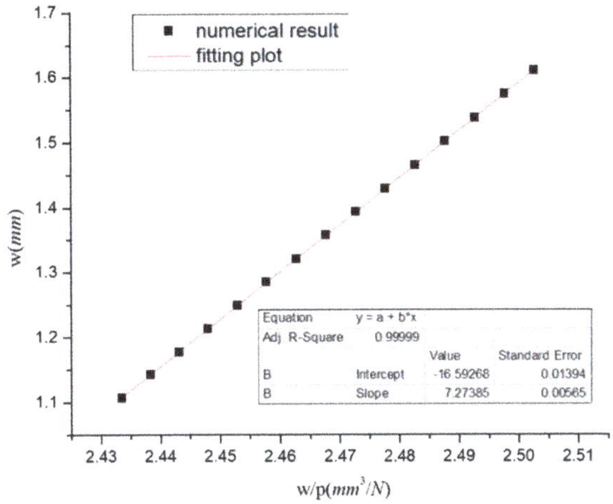

Figure 5. Plot of w/p vs. w (R = 50 mm, h = 0.5 mm).

In order to verify the practicality and accuracy of this method, we also calculate the p^* value adopting the other five existing methods, including one-order analytical value, numerical value, two-order analytical value, etc., which are shown in Table 1. The highest error is produced from the numerical method of [33], while the lowest error is produced from the lowest eigenvalue method.

Table 1. Critical load of our method comparing with other methods.

Method No.	1st	2nd	3rd	4th	5th	6th
Method name	The lowest Eigenvalue method	Traditional one-order analytical method	Non-destruction Our method	Analytical two-order method [33]	Numerical method [33]	Experimental method [33]
Critical load value p^*	0.081574	0.08127	0.08593	0.08189	0.08790	0.07800
Formulas	$[K]+\lambda_i[S])\{\psi_i\}=0$. Here $[K]$,$[S]$ are the constants, λ_i is the buckling load multiplier, $\{\psi_i\}$ is the buckling mode.	$p_{cr}=\frac{2Em^2}{\sqrt{3(1-v^2)}}$	$p_{cr}=m\Delta\sqrt{SEm}$	$p_{cr}=\frac{2Em^2}{\sqrt{3(1-v^2)}}-\frac{Em^3v}{(1-v^2)}$	Ansys FEM software.	Non-destructive Testing
Relative Error	0.03718	0.04193	0.10167	0.04988	0.12692	Reference

Because the error values from our method, when used correctly, are within the range of the other approaches, its accuracy can be considered acceptable in the field of engineering. Besides that, compared with other methods, there is no doubt that our method is the most practical and easiest one to operate in engineering.

From Table 1, we also note that the experimental values are the smallest relative to any other method (numerical methods of various orders, analytical methods, etc.). There may be several reasons to explain this behavior: for example, the roundness of the actual shell (not ideal), but with geometric imperfections; also, the realistic boundary support is not a 100%-hinged support or a 100%-fixed support either. However, we use the 100%-hinged support in the numerical model for simplification.

5. Practical Implementation of This Method

This method is friendly to inspectors working for practical projects and easy to implement. Its general steps are as follows: basically, we should apply an external uniform pressure (or equivalent uniform pressure) p by n times to the shell. Here, $p = p_0 + \Delta p_i$, $i = 1, 2, 3, \ldots, n$. At the same time, we record the deformation w_i after each loading. Here, the parameter S can be obtained through Equation (13). Then the p_{cr} is obtained by Equation (14). Then by the same way, we can obtain parameter J by Equation (17), and accordingly obtain h^* by Equation (18), and finally obtain t^* by Equation (26).

6. Conclusions

The highlight of this study is that it proposes a non-destructive method based on the Southwell procedure to estimate the critical load, critical thickness, and service life of internally corroded shells under external pressure while considering the temperature's effect. Of course, it is also applicable for the shell in special cases, such as industrial environments without corrosion or temperature.

Based on rigorous theories (shell stability theory, the Southwell plot method, and corrosion dynamic theory), and following a scientific route from simple to complex, we derived and acquired our new method step-by-step. We first derived the critical load and critical thickness of the spherical shell in the absence of corrosion and temperature using the Southwell procedure method. Second, we derived non-destructive methods for critical load, critical thickness, and service life prediction under corrosion and temperature conditions. Third, we compared our method with other methods.

The results show that, if used properly, engineering precision requirements can be met. Furthermore, when the results are carefully interpreted, this technique provides useful estimates of elastic buckling loads (critical thickness and service life). The utility of this approach lies in the fact that it is versatile, simple, and non-destructive. Furthermore, it does not require any assumptions about the buckling wave number or the precise location of buckling.

It has to be pointed out that since the service life of a spherical shell is usually very long (and so is its corrosion process), NDT data for the shells that are regularly detected, especially those collected regularly during the shell corrosion process, and fully recorded and published data, has not been found. A very small part of it was found in some data from experiments, which is employed in this paper in Section 4. This is why the method in this paper cannot be successfully verified in more working conditions currently. However, the authors will try to obtain (retrieve data from peer industries) NDT data under more operating conditions to fully evaluate the non-destructive prediction ability of this method for the shell service life (such as serving in corrosive and temperature environments) in the near future.

In summary, the highlight of this method in evaluating critical loads (thickness, lifetime) for shells is that it is non-destructive. Additionally, it is the first time introducing the Southwell plot method into the stabilization analysis of shells of time-varying thickness. Its precision meets engineering requirements, and more importantly, it is practice- and implementation-friendly.

Author Contributions: Conceptualization, methodology, software, validation, formal analysis, C.H.L. Investigation; resources; data curation; writing—review and editing, visualization, supervision, funding acquisition, G.L. All authors have read and agreed to the published version of the manuscript.

Funding: This work was supported by the National Natural Science Foundation of China [Grant number, 52068003] and by the Opening Project of Guangxi Laboratory on the Study of Coral Reefs in the South China Sea, China [Grant number, GXLSCRSCS201900*]. Moreover, the sponsorship guaranteed with basic research funds provided by Politecnico di Torino (Italy) is acknowledged. Opinions, findings, and conclusions expressed in this paper are those of the authors and not necessarily those of the sponsors. The author C.H.L. also acknowledges funding for the Training Program of 1000 Young and Middle-aged Outstanding Teachers in Colleges and Universities from Guangxi Education Department (CED) of China (No.: t2010097921).

Institutional Review Board Statement: Not applicable.

Informed Consent Statement: Not applicable.

Data Availability Statement: Available when asked for academic use.

Acknowledgments: The author (C.H.L.) is thankful to the Department of Mathematics for hospitality during her stay at Aberystwyth University. The authors (C.H.L. & G.L.) thank H. Yuan in USA and H. Nunn in Aberystwyth University for fruitful discussions. The author (C.H.L.) also thanks S. Shi in the Department of Mathematics of Hefei University of Technology, S. Huang and I. Burgess in the Department of Civil and Structural Engineering of the University of Sheffield for discussing in its earlier stage.

Conflicts of Interest: The authors declare no conflict of interest.

Notation List

S	slope of w vs. w/p line
u	displacement of the shell element in x direction
v	displacement of the shell element in y direction
w	displacement of the shell element in z direction
U_0	effect of initial imperfections
V	shearing force in straight members in y direction (buckling coefficient to be determined experimentally)
ε_x	unit elongation or strain in x-direction
ε_y	unit elongation or strain in y-direction
ε_y	unit elongation of middle surface in x-direction
ε_2	unit elongation of middle surface in y-direction
μ	Poisson's ratio
χ_y	change of curvature in x-direction
χ_x	change of curvature in y-direction
t_0	thickness of the shell
p_{cr}	buckling pressure
$H(\,)$	mathematical operator
N_x, N_y	resultant forces
Q_x, Q_y	shear forces
M_x	My bending moments
p_o	outer pressure
v_i	inner mechano-chemical corrosion rate
t	time
h	thickness
σ	principal stress
σ_e	effective stress
r	distance between a point in the shell material and the origin of the coordinate system/radius of two concentric spheres
r_o	distance between the point in the outer shell surface and the origin of the coordinate system
b	corrosion inhibition effect
$\mathrm{d}r$	radius of two concentric spheres
$\mathrm{d}\theta, \mathrm{d}\varphi$	top angles of four wedge-shaped sections
r_c	midsurface radius

x	thickness to midsurface radius ratio
t^*	time required for a corroded pressure shell to fail for the first time due to buckling or yielding
h	corresponding thickness of the shell under the critical failure state

Appendix A. Buckling Formulas for Spherical Shells in the Case of Section 2.1 (without Corrosion and Temperature)

Appendix A.1. Derivation of the Second-Order Critical Buckling Load (Stress)

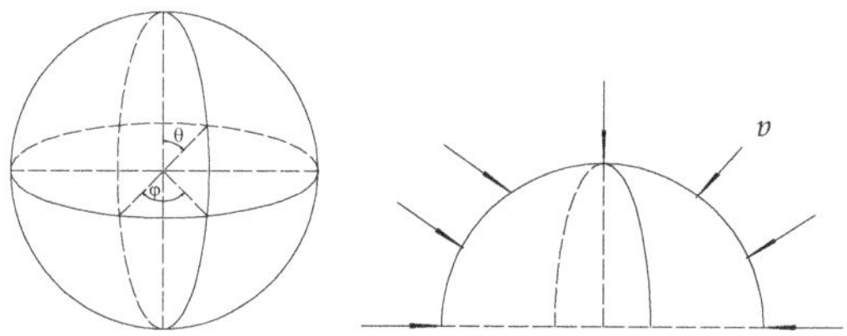

Figure A1. Hemispherical shells and one-half analysis model.

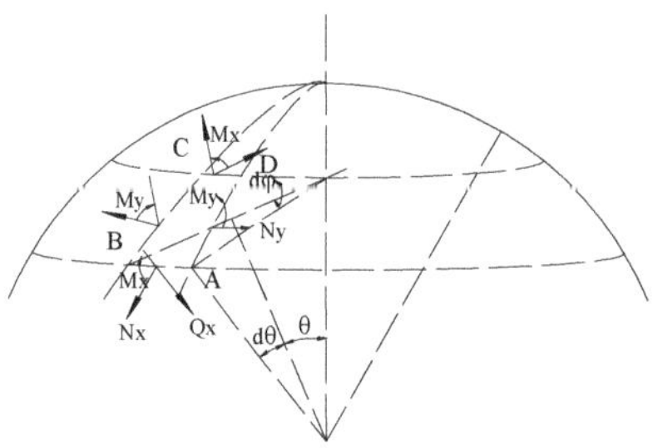

Figure A2. Hemispherical shell element and corresponding forces.

Using the above derived angles instead of the initial ones, $d\theta$, $\sin\theta d\psi$, and $\cos\theta d\psi$, the equations of equilibrium of the element $ABCD$ become [33]:

$$w + 7.27385\, w/p = 16.59268, \tag{A1}$$

$$\frac{dQ_x}{d\theta} + Q_x\cot\theta + N_x + N_y + pR + N_x\left(\frac{d^2w}{Rd\theta^2} + \frac{du}{Rd\theta}\right) + N_y\left(\frac{u}{R} + \frac{dw}{Rd\theta}\right)\cot\theta = 0, \tag{A2}$$

$$\frac{dM_x}{d\theta} + \left(M_x - M_y\right)\cot\theta - Q_xR + M_y\left(\frac{u}{R} + \frac{dw}{Rd\theta}\right) = 0. \tag{A3}$$

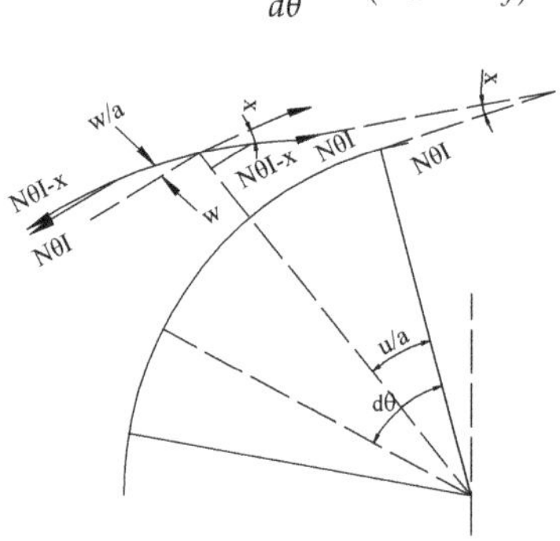

Figure A3. Meridian of a spherical shell before and after buckling.

Here, u is the displacement of the shell element in x direction, v is the displacement of the shell element in y direction, w is the displacement of the shell element in z direction, t_0 is the thickness of the shell, p_{cr} is the classical buckling pressure, N_x, N_y are the resultant forces, Q_x, Q_y are the shear forces, M_x, M_y are the bending moments, and θ, ψ are the angles of the shell element.

If a spherical shell is submitted to a uniform external pressure, there will be a uniform compression whose magnitude is:

$$\sigma = \frac{pR}{2h_0}. \tag{A4}$$

Let u,v, and w represent the components of small displacements during buckling from the compressed spherical form, then N_x and N_y differ little from the uniform compressive force $\frac{pR}{2}$ and become:

$$N_x = -\frac{pR}{2} + N'_x, \tag{A5}$$

$$N_y = -\frac{pR}{2} + N'_y, \tag{A6}$$

where N'_x and N'_y are the resultant forces due to small displacements u, v, and w.

Due to the stretching of the surface, p becomes $p(1 + \varepsilon_1 + \varepsilon_2)$. Therefore, substituting Equations (A5) and (A6) back into the differential equations of equilibrium (A1), (A2), and (3), and simplifying and neglecting the small terms, such as the products of N'_x, N'_y, and Q_x with the derivations of u, v, and w, we obtain:

$$\frac{dN'_x}{d\theta} + \left(N'_x - N'_y\right)\cot\theta - Q_x - 0.5pR\left(\frac{u}{R} + \frac{dw}{Rd\theta}\right) = 0, \tag{A7}$$

$$\begin{aligned} &\tfrac{dQ_x}{d\theta} + Q_x\cot\theta + N'_x + N'_y \\ &\quad + pR\left(\tfrac{du}{Rd\theta} + \tfrac{u}{R}\cot\theta - \tfrac{2w}{R}\right) \\ &\quad - 0.5pR\left(\tfrac{du}{d\theta} + \tfrac{d^2w}{ad\theta^2}\right) + \\ &-0.5pR\cot\theta\left(\tfrac{u}{R} + \tfrac{dw}{Rd\theta}\right) = 0, \end{aligned} \tag{A8}$$

$$\frac{dM_x}{d\theta} + (M_x - M_y)\cot\theta - Q_xR = 0. \tag{A9}$$

From Equation (A9) we get:

$$Q_x = \frac{dM_x}{Rd\theta} + (M_x - M_y)\frac{\cot\theta}{R}. \tag{A10}$$

Substituting Q_x into the Equations (A7) and (A8), and considering the following equations:

$$\sigma_x = \frac{E}{1-\mu^2}(\varepsilon_x + \mu\varepsilon_y) = \frac{N'_x}{h}, \tag{A11}$$

$$\sigma_y = \frac{E}{1-\mu^2}(\varepsilon_y + \mu\varepsilon_x) = \frac{N'_y}{h}, \tag{A12}$$

$$M_x = D(\chi_x + \mu\chi_y), \tag{A13}$$

$$M_y = D(\chi_y + \mu\chi_x), \tag{A14}$$

together with:

$$\varepsilon_x = \frac{\partial u}{\partial x} + \frac{1}{2}\left(\frac{\partial w}{\partial x}\right)^2, \tag{A15}$$

$$\varepsilon_y = \frac{\partial v}{\partial y} + \frac{1}{2}\left(\frac{\partial w}{\partial y}\right)^2, \tag{A16}$$

$$\chi_x = -\frac{\partial^2 w}{\partial x^2}, \tag{A17}$$

$$\chi_y = -\frac{\partial^2 w}{\partial y^2}, \tag{A18}$$

Here, ε_x is the unit elongation or strain in the x-direction, ε_y is the unit elongation or strain in the y-direction, ε_1 is the unit elongation of middle surface in the x-direction, ε_2 is the unit elongation of middle surface in the y-direction, μ is Poisson's ratio, χ_y is the change of curvature in the x-direction, and χ_x is the change of curvature in the y-direction. Then they can be written as:

$$\varepsilon_1 = \frac{du}{Rd\theta} - \frac{w}{R}, \tag{A19}$$

$$\varepsilon_2 = \frac{u}{R}\cot\theta - \frac{w}{R}, \tag{A20}$$

$$\chi_x = \frac{u}{R}\cot\theta - \frac{w}{R}, \tag{A21}$$

$$\chi_y = \frac{d^2 w}{R^2 d\theta^2} + \frac{du}{R^2 d\theta}, \tag{A22}$$

$$\chi_y = \left(\frac{u}{R^2} + \frac{d^2 w}{R^2 d\theta}\right)\cot\theta, \tag{A23}$$

And then we get:

$$N_x' = \frac{Eh_0}{1-v^2}\left[\frac{du}{Rd\theta} - \frac{w}{R} + v\left(u\frac{\cot\theta}{R} - \frac{w}{R}\right)\right], \tag{A24}$$

$$N_y' = \frac{Et_0}{(1-v^2)R}\left[u\cot\theta - w + v\left(\frac{du}{d\theta} - w\right)\right], \tag{A25}$$

$$M_x = -\frac{D}{R^2}\left[\frac{du}{d\theta} + \frac{d^2 w}{d\theta^2} + v\left(u + \frac{dw}{d\theta}\right)\cot\theta\right], \tag{A26}$$

$$M_y = -\frac{D}{R^2}\left[\left(u + \frac{dw}{d\theta}\right)\cot\theta + v\left(\frac{du}{d\theta} + \frac{d^2 w}{d\theta^2}\right)\right]. \tag{A27}$$

Now, introducing two dimensionless parameters, α and φ—which are defined as $\alpha = \frac{D(1-v^2)}{R^2 Eh_0} = \frac{h_0}{12R^2}$ and $\varphi = \frac{pR(1-v^2)}{2Eh_0}$—and using the elastic law to express the forces and moments in terms of u and w, one obtains the differential equations of equilibrium ((A7) and (A8)):

$$\begin{gathered}(1+\alpha)\left[\tfrac{d^2u}{d\theta^2} + \cot\theta\tfrac{du}{d\theta} - \left(v+\cot^2\theta\right)u\right] - (1+v)\tfrac{dw}{d\theta} + \alpha\left[\tfrac{d^3w}{d\theta^3} + \cot\theta\tfrac{d^2w}{d\theta^2} - \left(v+\cot^2\theta\right)\tfrac{dw}{d\theta}\right] \\ \pm\,\varphi\left(u - \tfrac{dw}{d\theta}\right) = 0,\end{gathered} \tag{A28}$$

$$\begin{gathered}(1+v)\left[\tfrac{du}{d\theta} + u\cot\theta - 2w\right] + \\ +\alpha\left[\tfrac{d^3u}{d\theta^3} - 2\cot\theta\tfrac{d^2u}{d\theta^2} + \left(1+v+\cot^2\theta\right)\left(\tfrac{du}{d\theta} + \tfrac{d^2w}{d\theta^2}\right) - \cot\theta\left(2-v+\cot^2\theta\right)(u+1) - \tfrac{d^4w}{d\theta^4}\right. \\ \left. + - 2\cot\theta\tfrac{d^3w}{d\theta^3}\right] - \varphi\left[-ucot\theta - \tfrac{du}{d\theta} + 4w + cot\tfrac{dw}{d\theta} + \tfrac{d^2w}{d\theta^2}\right] = 0.\end{gathered} \tag{A29}$$

These two equations may be simplified by neglecting, in comparison with unity, the first term, since the shell is thin, and therefore, the $\frac{h_0}{R}$ ratio is very small. Moreover, due largely to angular displacement χ, we make good use of this situation by introducing an auxiliary variable u, such that $u = -\frac{d\psi}{d\theta}$.

Thus, the expressions in the brackets in Equation (A29) become identical. Then, using the mathematical operator H, it turns to:

$$H(\ ) = \frac{d^2(\ldots)}{d\theta^2} + \cot\theta \frac{d(\ldots)}{d\theta} + 2(\ldots), \quad \text{(A30)}$$

$$\frac{d}{d\theta}[H(\psi) + \alpha H(w) - (1+v)(\psi + w) - \alpha(1+v)w - \varphi(\psi + w)] = 0. \quad \text{(A31)}$$

The fourth term, containing the factor, may be neglected in comparison with the third in Equation (A31). Integrating the Equation (A31) with respect to θ and assuming that the constant of integration is equal to zero, we obtain:

$$H(\psi) + \alpha H(w) - (1+v)(\psi + w) - \varphi(\psi + w) = 0, \quad \text{(A32)}$$

$$\begin{aligned} &\alpha HH(\psi + w) - (1+v)H(\psi) - (3+v)\alpha H(w) \\ &+2(1+v)(\psi + w) + \\ &+\varphi[-H(\psi) + H(w) + 2(\psi + w)] = 0. \end{aligned} \quad \text{(A33)}$$

Now, any regular function of cosθ in the interval $-1 \leq \cos\theta \leq 1$ may be expanded in a series of Legendre functions:

$$\begin{aligned} &P_0(\cos\theta) = 1, \\ &P_1(\cos\theta) = \cos\theta, \\ &P_2(\cos\theta) = 0.25(\cos 2\theta + 1), \\ &P_n(\cos\theta) = 2\tfrac{1\times3\times5\times\ldots\times(2n-1)}{2^n n!} \times \\ &\times \Big[\cos n\theta + \tfrac{1}{1} \times \tfrac{n}{2n-1}\cos(n-2)\theta \\ &+ \tfrac{1\times3}{1\times2} \times \tfrac{n(n-1)}{(2n-1)(2n-3)}\cos(n-4)\theta + \ldots\Big] \\ &\tfrac{d^2 p_n}{d\theta^2} + \cot\theta \tfrac{dp_n}{d\theta} + n(n+1)p_n = 0, \end{aligned} \quad \text{(A34)}$$

$$H(P_n) = -\lambda_n p_n, \quad \text{(A35)}$$

$$HH(P_n) = \lambda_n^2 p_n, \quad \text{(A36)}$$

in which $\lambda_n = n(n+1) - 2$, and n is an integer.

Assuming the general expressions of ψ and w for any symmetrical buckling of spherical shells, we have:

$$\psi = \sum_{n=0}^{\infty} A_n P_n \quad \text{(A37)}$$

$$w = \sum_{n=0}^{\infty} B_n P_n, \quad \text{(A38)}$$

Substituting them back to Equations (A35) and (A36), we can have:

$$\sum_{n=0}^{\infty} \{A_n[\lambda_n + (1+v) + \varphi] + B_n[\alpha\lambda_n + (1+v) + \varphi]\}P_n = 0, \quad \text{(A39)}$$

$$\sum_{n=0}^{\infty} \left\{ \begin{array}{c} A_n\left[\lambda_n^2 + (1+v)(\lambda_n + 2) + \varphi(\lambda_n + 2)\right] \\ +B_n\left[\alpha\lambda_n^2 + (3+v)\alpha\lambda_n + 2(1+v) - \varphi(\lambda_n - 2)\right] \end{array} \right\} P_n = 0. \quad \text{(A40)}$$

The Legendre functions form a complete set of functions. Therefore, the two series cannot vanish identically unless each coefficient vanishes.

From Equation (A29) we can get:

$$B_n = -\frac{A_n[\lambda_n + (1+v) + \varphi]}{\alpha\lambda_n + (1+v) + \varphi}, \tag{A41}$$

and eliminating B_n from the above set of equations, Equation (A41) can then be written as:

$$\begin{array}{l} A_n\left[\alpha\lambda_n^2 + (1+v)(\lambda_n+2)+\right] \\ -\frac{A_n[\lambda_n+(1+v)+\varphi]}{[\alpha\lambda_n+(1+v)+\varphi]\left[\alpha\lambda_n^2+(3+v)\alpha\lambda_n+2(1+v)-\varphi(\lambda_n-2)\right]}, \end{array} \tag{A42}$$

$$A_n\left(1-v^2\right)\lambda_n + \alpha\lambda_n\left[\lambda_n^2 + 2\lambda_n + (1+v)^2\right] - \varphi\lambda_n[\lambda_n + (1+3v)] = 0. \tag{A43}$$

Buckling of the shells becomes possible if these equations, for some value of n, yield for A_n and B_n a solution different than zero, which means a trivial solution; in other words, having a zero determinant of the system of equations. Thus:

$$\left(1-v^2\right)\lambda_n + \alpha\lambda_n\left[\lambda_n^2 + 2\lambda_n + (1+v)^2\right] - \varphi\lambda_n[\lambda_n + (1+3v)] = 0, \tag{A44}$$

a solution of which is $\lambda_n = 0$, which corresponds to a value of n equals to unity. Substituting this value of λ_n into Equation (A40), one obtains:

$$A_1 = -B_1. \tag{A45}$$

Now, for $\lambda_n \neq 0$, other than zero:

$$\varphi = \frac{(1-v^2) + \alpha\left[\lambda_n^2 + 2\lambda_n + (1+v)^2\right]}{\lambda_n + (1+3v)}, \tag{A46}$$

which yields for its minimum, or for $\frac{d\varphi}{d\lambda_n}$ after simplification:

$$\lambda_n^2 + 2(1+3v)\lambda_n - \frac{1+v^2}{\alpha} = 0, \tag{A47}$$

$$\lambda_n = -(1-3v)\lambda_n + \sqrt{\frac{1-v^2}{\alpha}}, \tag{A48}$$

$$\varphi_{min} = 2\sqrt{(1-v^2)\alpha} - 6v\alpha, \tag{A49}$$

$$\varphi = \frac{pR(1-v^2)}{2Eh_0}, \tag{A50}$$

and φ_{min} yields the first p_{cr} critical load [33]:

$$p_{cr} = \frac{2Eh_0^2\varphi_{min}}{R(1-v^2)} = \frac{2Eh_0}{R(1-v^2)}\left[\sqrt{\frac{1-v^2}{3}} \times \frac{h_0}{R} - \frac{vh_0}{2R^2}\right], \tag{A51}$$

$$\sigma_{cr} = \frac{p_{cr}R}{2h} = \frac{E}{1-v^2}\left[\sqrt{\frac{1-v^2}{3}} \times \frac{h_0}{R} - \frac{vh_0}{2R^2}\right]. \tag{A52}$$

Equation (A51) was derived with the assumption that the shell wall thickness is specified, whereas the critical value of external pressure Q is the unknown quantity.

On the contrary, if we assume that the external pressure Q value is specified, then the shell wall thickness h^*, corresponding to the stability loss, will be equal to:

$$h^* = \frac{4\sqrt{3(1-v^2)}\sqrt{p}R}{\sqrt{2E}}. \tag{A53}$$

Appendix A.2. Derive the Relationship between w and φ of the Spherical Shell

In 1934, Southwell began using his approach on columns. In this part of the section, an attempt is made to demonstrate that uniformly compressed spherical shells can also be analyzed by the Southwell procedure. In the derivation of the formula, as it was done for the classical theory of buckling shells (see previous part), it is assumed that the displacements u and w may be expressed as [33]:

$$\psi = \frac{du}{d\theta} = \sum_{n=0}^{\infty} A_n P_n, \tag{A54}$$

$$w = \sum_{n=0}^{\infty} B_n P_n, \tag{A55}$$

where P_n is the Legendre functions of the orders n, while A_n and B_n are the real constants from before:

$$\psi_0 = \sum_{n=0}^{\infty} A'_n P_n, \tag{A56}$$

$$w_0 = \sum_{n=0}^{\infty} B'_n P_n. \tag{A57}$$

Additionally, it is assumed that the manufacturing imperfections of ψ_0 is equal to zero. Thus, it is tried only with the direction w.

When the compressive load p is applied to the shell, each point of the middle surface undergoes elastic displacements u and w, and its normal distance from the reference sphere then becomes $w + w_0$. It is assumed that w_0 is from the order of the elastic deformation, and then the element of the shell looks like the deformed elements, which are used to establish the differential equations of the buckling problem. Again, going through the same procedure, one finds that the terms of those equations belong to two groups (see the proceeding section). In those terms, which contain the factor φ, the quantities u and w describe the difference in shape between the deformed element and an element of a true sphere. In these terms, w must now be replaced by $w + w_0$. On the other hand, all terms that do not have the factor φ, can be traced back to terms of the elastic law, and represent the stress resultants acting on the shell element. Before the application of the load, the shell is free of stress, and the stress resultants depend only on the elastic displacements u and w. Consequently, in all these terms, w does not need to be replaced by $w + w_0$, but stays w.

Thus, one arrives at the following set of differential equations:

$$H(\psi + w) + \alpha H(w) - (1-v)(\psi + w) - \varphi(\psi + w + w_0) = 0, \tag{A58}$$

$$\begin{aligned} \alpha HH(\psi + w) - (1+v)H(\psi) - (3+v)\alpha H(w) \\ +2(1+v)(\psi + w) + \\ \varphi[-H(\psi) + H(w + w_0) + H(w + w_0) + 2(\psi + w + w_0)] = 0, \end{aligned} \tag{A59}$$

in which HH denotes the same operator as before:

$$H(\ ) = \frac{d^2(\ldots)}{d\theta^2} + \text{cotg}\frac{d(\ldots)}{d\theta} + 2(\ldots), \tag{A60}$$

$$A_n(\lambda_n + 1 + v + \varphi) + B_n(\alpha\lambda_n + 1 + v + \varphi) = -B_n'\varphi, \tag{A61}$$

$$\begin{aligned} A_n(\alpha\lambda_n^2 + \lambda_n + 2 \quad & +v\lambda_n + 2v + \varphi\lambda_n + 2\varphi) \\ & +B_n(\alpha\lambda_n^2 + 3\alpha\lambda_n + v\alpha\lambda_n + 2 + 2v \\ & -\varphi\lambda_n + 2\varphi) \\ & = B_n'\varphi(\lambda_n - 2). \end{aligned} \tag{A62}$$

Thus, the problem is reduced to solving this set of equations. Eliminating A_n from the above set of equations:

$$\begin{aligned} [\alpha(\alpha-1)\lambda_n^3 + (\quad & \varphi\alpha - 2\alpha + \varphi)\lambda_n^2 \\ & +(v\alpha + v^2 + 2\varphi v + \varphi + 2\varphi^2 - 1 - \alpha - 3v\alpha - v^2\alpha + v\varphi \\ & -\alpha\varphi - v\alpha\varphi)\lambda_n]B_n = -B_n'\varphi[(\alpha+1)\lambda_n^2 + (2v + 2\varphi)\lambda_n], \end{aligned} \tag{A63}$$

$$B_n = -\frac{B_n'\varphi\lambda_n[(\alpha+1)\lambda_n + 2(v+\varphi)]}{[-\alpha\lambda_n^2(1-\alpha) + (\varphi - 2\alpha + \varphi\alpha)\lambda_n + v(3\varphi - 2\alpha) + \varphi + v^2 - \alpha - 1]\lambda_n}, \tag{A64}$$

After canceling λ_n, and neglecting the small quantities as α, φ, and their products in comparison with unity, we obtain:

$$B_n = -\frac{B_n'\varphi[\lambda_n + 2(v+\varphi)]}{-\alpha\lambda_n^2 + (\varphi - 2\alpha + \varphi\alpha)\lambda_n + v^2 - 1}. \tag{A65}$$

Coming back to the definition of the displacement w, one may write the equation $w = \sum_{n=0}^{\infty} B_n P_n$—or writing it in detail:

$$\begin{aligned} & w = B_0P_0 + B_1P_1 + B_2P_2 + \cdots\cdots \\ & w = [B_0 + 0.25B_2 + \tfrac{9}{64}B_4 + \cdots] + [B_1 + \tfrac{3}{8}B_3 + \cdots]\cos\theta + [B_2 + \cdots]\cos 2\theta + \end{aligned} \tag{A66}$$

here $\lambda_n = n(n+1) - 2$, which is the minimum for $n = \frac{1}{2}$; therefore, it has the same values for n equals to minus one and zero. Since n must be an integer, it is chosen as zero, which yields $\lambda = -2$, and corresponds to B_0, which is a function of λ_n, and gets smaller when λ_n becomes greater. Thus, it is possible to neglect all the terms and simply write $w \cong B_0$ since the terms which contain $\cos\theta$, $\cos 2\theta \ldots$ are much smaller—so, buckling is usually expected at the places where θ is large [33]:

$$w \cong \frac{B_0'\varphi[-2 + 2(v+\varphi)]}{4\alpha + (\varphi - 2\alpha + \varphi\alpha)2 + 1 - v^2}, \tag{A67}$$

$$w \cong \frac{2B_0'\varphi(v + \varphi - 1)}{2\varphi(1+\varphi) + 1 - v^2}, \tag{A68}$$

$$w \cong \frac{B_0'(v-1)}{1 + \frac{1-v^2}{2\varphi}}. \tag{A69}$$

Appendix B. Deriving the Corrosion Rate of the Spherical Shell as a Function of Temperature and Stress, Based on the Arrhenius Type

It is assumed that the shell is subjected to the simultaneous action of a constant external pressure Q and uniform internal corrosion. The rate of the thickness decrease at each moment of time t is equal to the corrosion rate [5]:

$$\frac{dh}{dt} = -f(\sigma, T)\frac{dh}{dt} = -f(\sigma, T), \tag{A70}$$

where $f(\sigma, T)$ is a sufficiently smooth function of compressive stress σ, defined by Equation (A2), and the temperature T. The corrosion rate dependence on the stress value, according to experimental data, can be approximated by exponential function suggested in [1], while the temperature effect is usually described by the Arrhenius-type law [11]. Combination of both functions yields the following relation Equation (A73).

Temperature has a great effect on the rate of metal electrochemical corrosion. In the case of corrosion in a neutral solution (oxygen depolarization), the increase of the corrosion rate increases exponentially with temperature increase because the hydrogen evolution overpotential decreases. Experimental dependence of the Arrhenius type is observed between the corrosion rate and temperature. Using the current density j_{corr}, we express the corrosion rate [1]:

$$j_{corr} = \lambda \exp(E_a / R_g T). \tag{A71}$$

where E_a is the effective activation energy of the corrosion process in kJ mol^{-1}, R_g *is* the molar gas constant in J mol^{-1} K^{-1}, T is the absolute temperature in K, λ *is* the pre-exponential factor, and j_{corr} *is* the corrosion current density, A cm^{-2}. Equation (A71) provides the determination of the effective activation energy of the corrosion process.

When $\sigma = 0$, $E_a = E_{co}$, we have $T = T_0$, and $j_{corr} = \lambda \exp(E_{c0}/R_g T)$.

Moreover, when $\sigma \neq 0$, $T = T$, and $E_a = E_c$, we get:

$$j_{corr} = \lambda \exp(E_c / R_g T). \tag{A72}$$

Considering $f(\sigma, t) = v_0 \exp\left(\frac{V\sigma}{R_g T}\right)$ [1], and then using linear scale method, we get:

$$v = v_0 \frac{\exp\left(\frac{E_C}{R_g T}\right)}{\exp\left(\frac{E_0}{R_g T_0}\right)} \exp\left(\frac{V\sigma}{R_g T}\right) = v_0 \exp(\overline{E}_{c0}\left(1 - \frac{\overline{E}_c}{\overline{T}}\right)) \exp\left(\frac{V\sigma}{R_g T}\right). \tag{A73}$$

From the above equation, we know:

$$\frac{dh}{dt} = f(\sigma)\,,\ \sigma \equiv \frac{pR}{2h}, \tag{A74}$$

therefore, we can easily get:

$$\frac{dh}{dt} = \frac{d\left(\frac{pR}{2\sigma}\right)}{d\sigma}\frac{d\sigma}{dt} = -\frac{1}{2\sigma^2} pR \frac{d\sigma}{dt}, \tag{A75}$$

and thus:

$$\frac{d\sigma}{dt} = \frac{2\sigma^2}{pR}\frac{dh}{dt}. \tag{A76}$$

Finally, we have [5]:

$$\frac{dh}{dt} = v_0 \exp\left(\overline{E}_{c0}\left(1 - \frac{\overline{E}_c}{\overline{T}}\right)\right) \exp\left(\frac{V\sigma}{R_g T}\right). \tag{A77}$$

in which R_g is the molar gas constant, V is the material molar volume, $\overline{E}_{co} = \frac{E_{co}}{R_g T_0}$, and $\overline{E}_c = \frac{E_c}{E_{co}}$.

References

1. Li, L.; Gao, J.; Wang, Y. Evaluation of cyto-toxicity and corrosion behavior of alkali-heat-treated magnesium in simulated body fluid. *Surf. Coat. Technol.* **2004**, *185*, 92–98. [CrossRef]
2. Ben Seghier, M.E.A.; Höche, D.; Zheludkevich, M. Prediction of the internal corrosion rate for oil and gas pipeline: Implementation of ensemble learning techniques. *J. Nat. Gas Sci. Eng.* **2022**, *99*, 104425. [CrossRef]

3. Islam, M.A. Corrosion behaviours of high strength TMT steel bars for reinforcing cement concrete structures. *Procedia Eng.* **2015**, *125*, 623–630. [CrossRef]
4. Páczelt, I.; Kucharski, S.; Mróz, Z. The experimental and numerical analysis of quasi-steady wear processes for a sliding spherical indenter. *Wear* **2012**, *274–275*, 127–148. [CrossRef]
5. Gutman, E.M. *Mechanochemistry of Solid Surfaces*; World Scientific Publishing Company: Singapore, 1994. [CrossRef]
6. Kashani, M.M.; Crewe, A.J.; Alexander, N.A. Nonlinear stress-strain behaviour of corrosion-damaged reinforcing bars including inelastic buckling. *Eng. Struct.* **2013**, *48*, 417–429. [CrossRef]
7. Möller, B.; Albrecht, S.; Wagener, R.; Melz, T. Fatigue strength of laser beam welded steel-aluminium joints considering variable amplitude loading and corrosive environment. *Procedia Struct. Integr.* **2019**, *18*, 556–559. [CrossRef]
8. Popova, A.; Sokolova, E.; Raicheva, S.; Christov, M. AC and DC study of the temperature effect on mild steel corrosion in acid media in the presence of benzimidazole derivatives. *Corros Sci.* **2003**, *45*, 33–58. [CrossRef]
9. Tanveer, N.; Mobin, M. Corrosion Protection of Carbon Steel by Poly (aniline-co-o-toluidine) and Poly (pyrrole-co-o-toluidine) Copolymer Coatings. *J. Miner Mater. Charact. Eng.* **2011**, *10*, 735–753. [CrossRef]
10. Noor, E.A. Temperature effects on the corrosion inhibition of mild steel in acidic solutions by aqueous extract of fenugreek leaves. *Int. J. Electrochem. Sci.* **2007**, *2*, 996–1017.
11. Al-Majedy, Y.K.; Ibraheem, H.H.; Falih, M.S.; Al-Amiery, A.A. New coumain derivatives as corrosion inhibitor. *IOP Conf. Ser. Mater. Sci. Eng.* **2019**, *579*, 012051. [CrossRef]
12. Abdel-Azim, A.A.; Milad, R.; El-Ghazawy, R.; Kamal, R. Corrosion inhibition efficiency of water soluble ethoxylated trimethylol propane by gravimetric analysis. *Egypt J. Pet.* **2014**, *23*, 15–20. [CrossRef]
13. AL-Saadie, K.; Abdul Karime, N.; Al-Mousawi, I.M. Corrosion Inhibition of Zinc in Hydrochloric Acid Medium Using Urea Inhibitor. *J. Al-Nahrain Univ. Sci.* **2007**, *10*, 31–38. [CrossRef]
14. Patil, D.B.; Sharma, A.R. Study on the corrosion kinetics of iron in acid and base medium. *E-J. Chem.* **2011**, *8* (Suppl. 1), 358–362. [CrossRef]
15. Zhang, H.H.; Qin, C.K.; Chen, Y.; Zhang, Z. Inhibition behaviour of mild steel by three new benzaldehyde thiosemicarbazone derivatives in 0.5 M H2SO4. Experimental and computational study. *R. Soc. Open Sci.* **2019**, *6*, 190192. [CrossRef]
16. Nicol, M.J. A comparative study of the kinetics of the oxidation of iron(II) by oxygen in acidic media - mechanistic and practical implications. *Hydrometallurgy* **2020**, *192*, 105246. [CrossRef]
17. Hashim, N.Z.N.; Kassim, K. The effect of temperature on mild steel corrosion in 1 M HCL by Schiff bases. *Malaysian J. Anal. Sci.* **2014**, *18*, 28–36.
18. Solutions, I.A. Electrodissolution Kinetics of Iron and Its Alloys Containing Titanium in Chloride Solutions. 1975. Available online: https://core.ac.uk/download/pdf/147423716.pdf (accessed on 31 October 2022).
19. Gao, M.; Wang, H.; Song, Y.; Han, E.H. Corrosion behavior on carbon steel in a simulated soil solution under the interaction effect of chloride and bicarbonate ions. *J. Mater. Res. Technol.* **2022**, *21*, 3014–3024. [CrossRef]
20. Piatti, R.C.V.; Arvía, A.J.; Podestá, J.J. The electrochemical kinetic behaviour of nickel in acid aqueous solutions containing chloride and perchlorate ions. *Electrochim. Acta* **1969**, *14*, 541–560. [CrossRef]
21. Fassina, P.; Bolzoni, F.; Fumagalli, G.; Lazzari, L.; Vergani, L.; Sciuccati, A. Influence of hydrogen and low temperature on mechanical behaviour of two pipeline steels. *Eng. Fract. Mech.* **2012**, *81*, 43–55. [CrossRef]
22. Garrigues, L.; Pebere, N.; Dabosi, F. An investigation of the corrosion inhibition of pure aluminum in neutral and acidic chloride solutions. *Electrochim. Acta* **1996**, *41*, 1209–1215. [CrossRef]
23. Colacino, E.; Carta, M.; Pia, G.; Porcheddu, A.; Ricci, P.C.; Delogu, F. Processing and Investigation Methods in Mechanochemical Kinetics. *ACS Omega* **2018**, *3*, 9196–9209. [CrossRef] [PubMed]
24. Bergman, R.M.; Levitsky, S.P.; Haddad, J.; Gutman, E.M. Stability loss of thin-walled cylindrical tubes, subjected to longitudinal compressive forces and external corrosion. *Thin-Walled Struct.* **2006**, *44*, 726–729. [CrossRef]
25. Gutman, E.M.; Bergman, R.M.; Levitsky, S.P. Influence of internal uniform corrosion on stability loss of a thin-walled spherical shell subjected to external pressure. *Corros. Sci.* **2016**, *111*, 212–215. [CrossRef]
26. Gutman, E.; Haddad, J.; Bergman, R. Stability of thin-walled high-pressure vessels subjected to uniform corrosion. *Thin-Walled Struct.* **2000**, *38*, 43–52. [CrossRef]
27. Elso, M.I. Finite Element Method studies on the stability behavior of cylindrical shells under axial and radial uniform and non-uniform loads. Master's Thesis, University of Applied Sciences, Krefeld, Germany, May 2012.
28. Wang, H.; Zhu, Y.; He, X.; Guan, W.; Zhan, M.; Zhang, J. Strength Prediction of Spherical Electronic Cabins with Pitting Corrosion. *Metals* **2022**, *12*, 1120. [CrossRef]
29. Naseri Ghalghachi, R.; Showkati, H.; Eyvazinejad Firouzsalari, S. Buckling behaviour of GFRP cylindrical shells subjected to axial compression load. *Compos. Struct.* **2021**, *260*, 113269. [CrossRef]
30. Amaya-Gómez, R.; Sánchez-Silva, M.; Bastidas-Arteaga, E.; Schoefs, F.; Muñoz, F. Reliability assessments of corroded pipelines based on internal pressure—A review. *Eng. Fail. Anal.* **2019**, *98*, 190–214. [CrossRef]
31. De Meo, D.; Oterkus, E. Finite element implementation of a peridynamic pitting corrosion damage model. *Ocean Eng.* **2017**, *135*, 76–83. [CrossRef]

32. Do, V.D.; Le Grognec, P.; Rohart, P. Closed-form solutions for the elastic—Plastic buckling design of shell structures under external pressure. *Eur. J. Mech. A Solids* **2023**, *98*, 104861. [CrossRef]
33. Nayyeri Amiri, S.; Rasheed, H.A. Nondestructive method to predict the buckling load in elastic spherical shells. *Eng. Struct.* **2017**, *150*, 300–317. [CrossRef]

applied sciences

MDPI

Article

Smart Patch for Structural Health Monitoring of Composite Repair

Tianyi Feng * and M. H. Ferri Aliabadi

Structural Integrity and Health Monitoring, Department of Aeronautics, Imperial College London, South Kensington, London SW7 2AZ, UK; m.h.aliabadi@imperial.ac.uk
* Correspondence: t.feng17@imperial.ac.uk

Abstract: The bondline integrity of a repair patch to the parent composite laminate is considered the most important factor in the repair design. A smart repair patch is proposed here to allow for real-time ultrasonic guided wave monitoring of repaired composites. A diagnostic film with lead zirconate titanate (PZT) transducers and inkjet-printed wires is embedded into the repair patch using a cut-out method. The electro-mechanical impedance (EMI) method is used to verify the integrity of the embedded PZT transducers. The performance of the smart repair patch is assessed on the external panel with artificial bondline delamination and surface-mounted artificial damage. The damage index correlation coefficient and delay-and-sum (DAS) algorithm are used for damage detection and localization. The results show that the developed repair patch can successfully detect and locate damages.

Keywords: smart repair patch; structural health monitoring (SHM); embedded PZT transducers; damage detection and localization; delay-and-sum (DAS) algorithm

Citation: Feng, T.; Aliabadi, M.H.F. Smart Patch for Structural Health Monitoring of Composite Repair. *Appl. Sci.* **2022**, *12*, 4916. https://doi.org/10.3390/app12104916

Academic Editor: Yun-Kyu An

Received: 21 March 2022
Accepted: 10 May 2022
Published: 12 May 2022

1. Introduction

Repair of in-service composite parts plays an important role in the sustainability of composite airframes due to the recent increase in the utilization of composite structures in modern aircraft. In certain practical cases, the repair of composite structures according to the Structural Repair Manual (SRM) can reduce the cost of replacement without compromising the mechanical of composite structures [1–3]. However, the strength and durability of the repaired area still need to be considered since the stress transfer at the interface is under service load with different environmental conditions [4]. The mechanical properties of the exposed thermosetting adhesives and resin of bonded area will absorb the moisture, which will affect the durability [5–8]. The defects will be also generated stress concentrations and initiate cracks when the repaired structure subjected to a cyclic loading [9–11]. In addition, failures may occur due to inconsistent processing methods which will degrade the mechanical properties between the repair patch and composite bondline [4,12].

To meet the requirements of airworthiness certification, repair based on Structural Health Monitoring (SHM) system can be an effective way to monitor the integrity of the bonded repair patch for aircrafts [3,8,12,13]. Generally, the smart patch consists of an array of actuators/sensors. Compared to traditional non-destructive inspection (NDI) techniques, SHM enables bondline inspection possible without destroying the repaired patch and can be used to monitor in real-time the bonding area of the repaired part [12,14]. It has been shown that the smart patch can detect debonding, delamination of composite layers and damage growth in the repaired area [8].

Piezoelectric lead zirconate titanate (PZT) transducers and fibre Bragg gratings (FBG) are two main types of sensors commonly utilized for SHM applications. The advantages of PZT transducers include wide frequency range, low price and small size and good coupling capacity which are particularly suitable for embedding [15,16]. In addition, they can simultaneously exhibit actuator/sensor behaviours, which allow for both passive and active detections [17–19]. While FBG sensors are super-light, small size, sensitive, lower power consumption, immune to electromagnetic interferences/corrosion and have high bandwidth and multiplexing sensors [15,16,20–23]. By comparing PZT and FBG sensors, PZT transducers can capture the 'integrated' signals in the entire covered area, while FBG sensors are directional dependent [16]. In addition, PZT transducers are sensitive to both symmetric (S_0) and anti-symmetric (A_0) modes, while FBG sensors are less sensitive to the A_0 mode [16]. For damage localization, RAPID (reconstruction algorithm for probabilistic inspection of defects) [24,25] and delay-and-sum (DAS) [26–28] algorithms are imaging methods and have been widely used. The RAPID algorithm does not need to acquire the information of wave modes and group velocity for ultrasonic guided waves (UGW) while the DAS algorithm only assumes one wave mode exists and little mode conversion exists when the wave interacts with damage.

Many works were conducted on the SHM system for the smart repair patch. Rito R.L. et al. [29] numerically and experimentally studied bondline monitoring using a composite repair patch with embedded FBG sensors. Both experimental and modelling results showed that the smart repair patch embedded with chirped FBG sensor can be used to monitor the initial disbond. Lambinet F. et al. used surface-mounted PZT transducers [13] and hybrid system (surface-mounted PZT transducers + embedded FBG sensors) [12] to conduct bending fatigue and impact tests for the step-sanded composite repair structures and used scaling subtraction method and RAPID to detect and locate these damages. In addition, they [3] also proposed a so-called Minimal Intersection Score (MIS) algorithm to detect and locate the damage to the smart repair patch under different environmental conditions. Roth W. et al. [30] numerically and experimentally used a smart patch with phased-array PZT transducers to monitor the artificial disbond. The Teflon tape was inserted in the edge between the patch and host composite structure during manufacturing. The results showed that disbond can be detected by using the electro-mechanical impedance (EMI) method. Qing X. et al. [31] used the smart patch combined with the SMART layer (Stanford Multi-Actuator-Receiver Transduction Layer) system to monitor the curing progress and disbond by using the active sensing SHM technique. Their results showed that the combined SMART layer system for the smart repair patch can be used to monitor the curing progress and integrity of the bonding quality of the composite repair structure. Later, Bekas D. G. et al. [32] used inkjet-printed interdigital sensors to monitor the bondline integrity of bonded composite joints by using the EMI method. Sánchez-Romate X.F. et al. [33] used carbon nanotube (CNT) based adhesive films to investigate the crack sensing capabilities of adhesive film for composite bonded repair. They demonstrated that the use of CNT adhesive films for the bonded repair composite structures did not affect their mechanical performance and the electrometrical results showed it can be used for SHM purposes.

Furthermore, Bekas D. G. et al. developed the SHM layer and renamed it as diagnostic film [34]. This layer had been shown to reduce the thickness of integrated layers (25.4 μm) [35] by 50% compared to that of the SMART layer (50.8 μm) [36]. They found this diagnostic film was effective for thin composites and can be used under extreme environmental and operational conditions. In the authors' previous work, a novel embedding technique using this diagnostic film with PZT transducers based on the edge cut-out method for the composite structure was developed [35]. This technique allowed edge trimming possible without damaging the printed circuits, which met industrial requirements for the next higher assembly. In addition, the EMI properties and sensing performance of embedded PZT transducers remain stable up to 1 million loading cycles under fatigue tests, and reductions of tensile and compressive modulus for the composite coupons remain acceptable for the worst-case scenario [37]. Earlier, Salmanpour M. S. et al. [38] reported that the SMART layer did not meet the operational and environmental conditions of regional aircraft as related to low/high temperatures changes under cyclic loading, while diagnostic film utilizing inkjet-printed technology performed well in the tests [39]. In addition, the SMART layer was the most fragile before and during the bonding procedures [38].

This paper is the first application and assessment of the embedded PZT transducers for detection of the damage along the bondline and on the surface of a composite repair patch. In previous work, PZT transducers were installed on the surface of the host structure to detect damage to the repair patch. This approach as shown in [3,13] required the placement of many sensors surrounding the entire patch and the guided waves had to travel the entire length of the patch to detect possible damage. There was also scattering from the edge to the patch since the guided wave was generated from the host structure and not internally to the patch as with the approach proposed in this paper. Furthermore, the use of a hybrid PZT-FBG acquisition system has already been reported in [12]. Hence, a smart repair patch with developed embedded diagnostic film and PZT transducers is the most innovative because it reduces the number of sensors.

In this paper, the *edge cut-out method* is applied to manufacture a smart repair patch combined with an embedded diagnostic layer and PZT transducers. Then this patch will bond to the composite host structure together to simulate composite repair. The active sensing method will be used for SHM purposes. The aim of this paper is to investigate the monitoring ability of the smart repair patch using the diagnostic film for bondline inspection of the composite repair structure. First, the EMI method will be used to verify the bonding properties between PZT transducers and the repair patch. Second, the damage index (DI) correlation coefficient [40] and the DAS algorithm will be used to detect and locate the artificial delamination and surface-mounted artificial damage.

2. Experimental Setup

Unidirectional carbon fibre prepregs Hexply® IM7/8552 were used in this experiment. For fabrication of the composite repair patch, the quasi-isotropic stacking sequence $[(0°/+45°/-45°/+90°)_2]_s$ was used for the lay-up and the thickness of the repair patch was about 2 mm after curing. The KAPTON® film (DuPontLM HM, melting temperature 400 °C) with a thickness of 25.4 μm was used as a diagnostic film and DuraActTM PZT transducers (P-876.K025 – PIC255, Curie temperature: 350 °C) were used during the manufacturing. Dimatix printer (DMP-2580) was used to print the circuits on the KAPTON® film. In addition, thermoplastic film (TPU-Pontocal AG) was used to pre-bond the PZT transducer to KAPTON® film for diagnostic film preparation. Furthermore, resin film Hexply® M56 was used during embedding to bond the surface and increase the bonding properties between the prepreg, KAPTON® films and PZT transducers.

The preparation of the diagnostic film, sensor installation and details of the novel cut-out method for the embedding were reported in the author's previous work [35]. PZT transducers needed to be pre-bonded to the printed diagnostic film using two layers of thermoplastic films for preparing the printed diagnostic film to ensure there is no delamination between the printed diagnostic film and transducers. Figure 1a shows the general drawing of the smart repair patch with the configuration of PZT transducers and Figure 1b shows the schematic of the embedding procedure during lay-up.

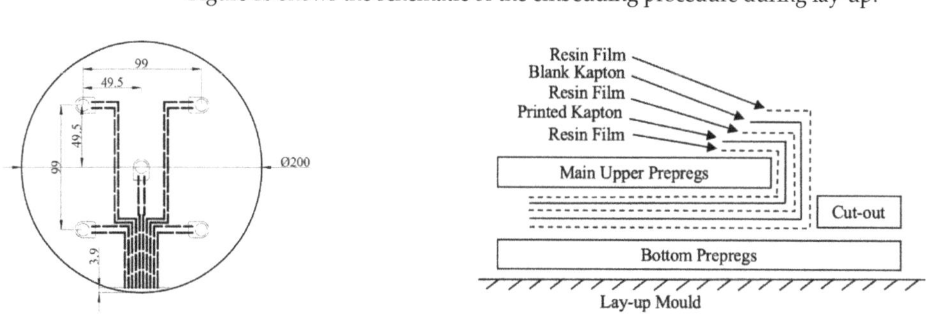

(a) (b)

Figure 1. Schematics of (**a**) general drawing of repair patch and (**b**) embedding procedure during layup.

For the embedding, a release film was placed in the middle of the cut-out area shown in Figure 1b to prevent the bonding of the diagnostic film to the prepregs within the cut-out area. Then the resin film and a blank diagnostic film were applied on the surface of the bottom prepregs respectively to prevent the short-circuited of printed circuits due to the conductivity of carbon fibre prepregs, followed by the resin film, the prepared KAPTON® film, resin film and main upper prepregs. After that, the exposed area of prepared KAPTON® films was applied on the surface of the main upper prepregs and the release film placed on the cut-out area was removed. Finally, cut-out prepregs were applied to the designated position.

After lay-up and bagging, the patch was cured at 180 °C in the autoclave and the curing cycle was set up according to the prepregs' datasheet. After curing and trimming, a connector (RS 514-4408, operating temperature range: −40 °C to +85 °C) was mounted on the surface of the repair patch and bonded by a super glue (RS 473-445, operating temperature range: −50 °C to +80 °C) to connect circuits and embedded PZT transducers by conductive epoxy adhesive resin/hardener (RS 186-3616). After the fabrication, the trimmed repair patch is shown in Figure 2b. For manufacturing the host composite structure, a quasi-isotropic stacking sequence $(0°/+45°/-45°/+90°)_{4s}$ was used for the layup and the size was 300 mm × 260 mm, and the thickness of the host structure was about 4 mm after curing.

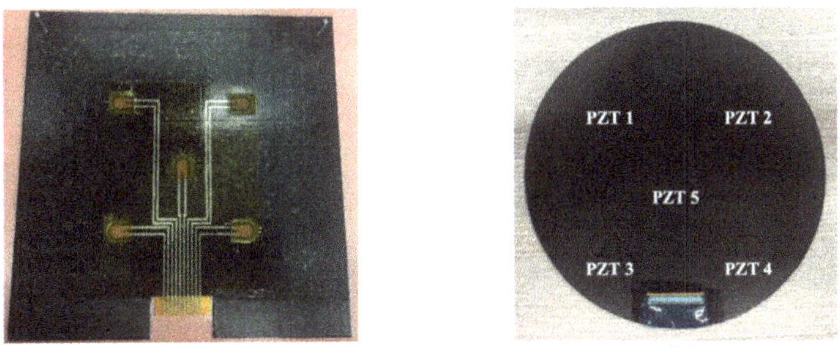

(a) (b)

Figure 2. Schematics of (**a**) embedding procedure and (**b**) trimmed smart repair patch.

After manufacturing, the EMI method was used to evaluate the bonding properties between the PZT transducer and the repair patch. The EMI method can be used to evaluate local damage severities, transducers' fractures, mechanical/electrical properties for the degradation and integrity of bonding properties [35]. According to the EMI method, the imaginary part of admittance at a low-frequency range will make the mechanical impedance of the composite structure close to zero and only the mechanical impedance of the PZT transducer is considered [41]. Therefore, any slope change in the imaginary part of admittance at the low-frequency range will determine the integrity of bonding properties between the host structure and the PZT transducers [35].

In the experiments, a SinePhase Impedance Analyzer (Model 16777K) was used for measuring the imaginary and real parts of admittance at room temperature. Figure 3 presents EMI results for the imaginary and real parts of admittances for different embedded PZT transducers. As is shown in Figure 3a, slopes of imaginary parts of the admittance for all embedded PZT transducers at the low-frequency range do not show an obvious difference. Hence, the bonding qualities were good for these embedded PZT transducers, and they can be further used for active sensing SHM purposes.

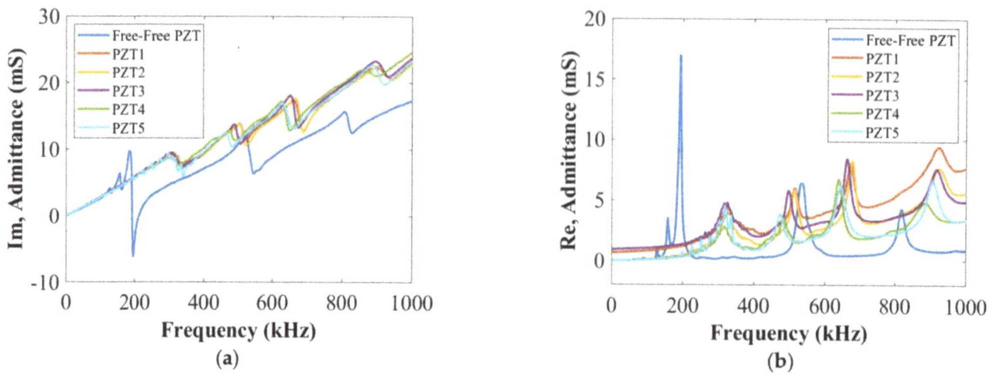

Figure 3. The EMI results of (**a**) imaginary and (**b**) real part of admittance for different PZT transducers.

3. Damage Detection and Localization

To verify if the smart repair patch embedded with diagnostic film and PZT transducers can monitor the integrity of the bondline quality, the damage index (DI) correlation coefficient and delay-and-sum (DAS) algorithm based on active sensing were used to detect and locate the bondline defects. In this section, two types of defects were studied, which were artificial delamination and surface-mounted artificial damage. In addition, blue contact gel was fully applied to the repair patch to bond it with the composite host structure as baseline signals (as is shown in Figure 4a). Here, the contact gel was used in DolphiCam C-Scan to improve the coupling on rougher surfaces. The reason for using this contact gel was that the repair patch could be easily removed at room temperature and used for further measurements. For simulating delamination and measuring the current signals, the repair patch was removed and two layers of KAPTON® films were inserted in the designated positions of the repair patch (shown in Figure 4b). The repair patch was then put back to the original position as close as possible. To detect the surface-mounted artificial damage, a weighted blue-tack was placed in designated positions on the surface of the repair patch and bottom of the host structure as the current signals.

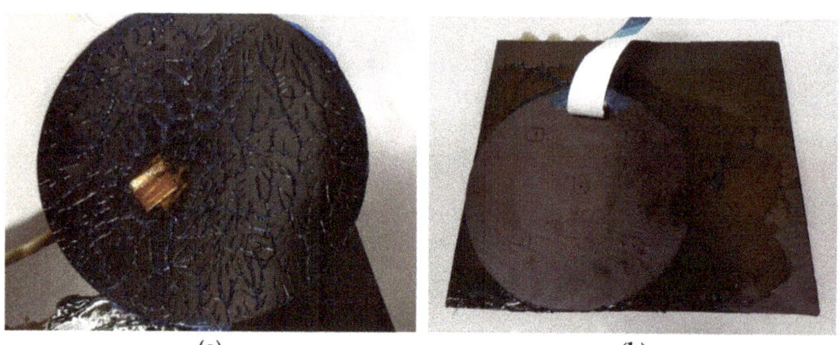

(**a**) (**b**)

Figure 4. Schematics of (**a**) artificial delamination and (**b**) bonding with blue contact gel for the bonded repair structures.

For measuring guided waves, a National Instrument (NI) PXIe-1073 and an arbitrary signal generator (NI PXI-5412) were used for signal generation, and a digitizer (NI PXI-5105) was used to record UGW signals. During the measurements, the time-of-arrival (ToA) of the A_0 mode could not be distinguished at 50 kHz due to the overlap between the crosstalk and the first wave packet of each measured signal. Therefore, a five-cycle Hanning-windowed toneburst signal at 250 kHz [42] was used as the actuation signal. The actuation amplitude and sampling frequency were 6 V and 100 MHz, respectively. Both baseline and current signals were measured at room temperature.

3.1. Delamination

Figures 5 and 6 compare the results of damage detections and localizations for two artificial delamination positions at 250 kHz by using three and four PZT transducers, respectively. The results show that the DI results using both three and four PZT transducers can detect the damage. The DAS results show that the artificial delamination can be located accurately using three PZT transducers but cannot when using four PZT transducers. Two factors affected the locating results during the measurements. First, the repair patch cannot be placed in the same position as the baseline status after inserting the KAPTON® films into bondline positions. In addition, the thickness of the contact gel used to bond the repair patch and the host structure cannot be kept at the same value since the contact gel will come back again every time after hardly pushing the repair patch. Both factors would affect the measured signals. Therefore, using four or more PZT transducers will cause the inaccuracy of damage localization due to the inconsistency caused by the contact gel.

However, the smart repair patch will be bonded to the host structure permanently by resin film for the actual situation. First, the repaired structure bonded by the resin film will make the patch maintain its original position when delamination happened and retain the same thickness of the bonding area. Furthermore, the selection of the contact glue to bond the patch and host structure is a compromising way under laboratory conditions and no alternative material can be found to bond the repaired structure together and peel them after measuring. Hence, the above two factors would not affect the measuring results when using four PZT transducers, and it can be confirmed that the DAS results would be accurate when using four PZT transducers for the actual repair.

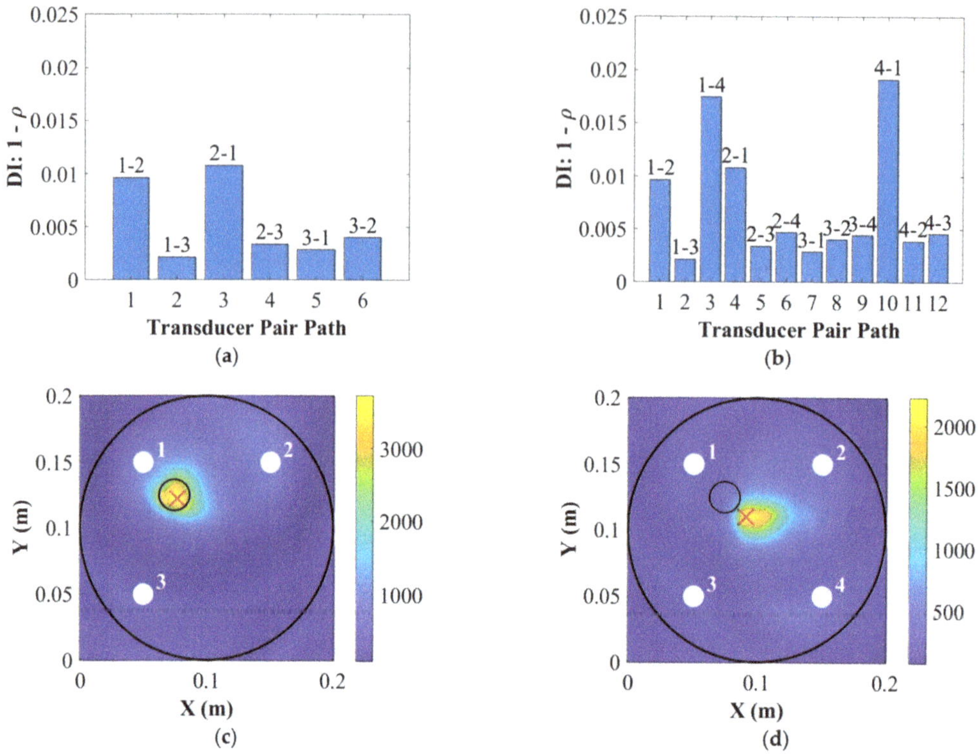

Figure 5. Delamination detection using (**a**) three PZT transducers and (**b**) four PZT transducers, and localization using (**c**) three PZT transducers and (**d**) four PZT transducers at 250 kHz for the position 1 (where the "○" is the position for real damage and the "×" is the position for predicted damage).

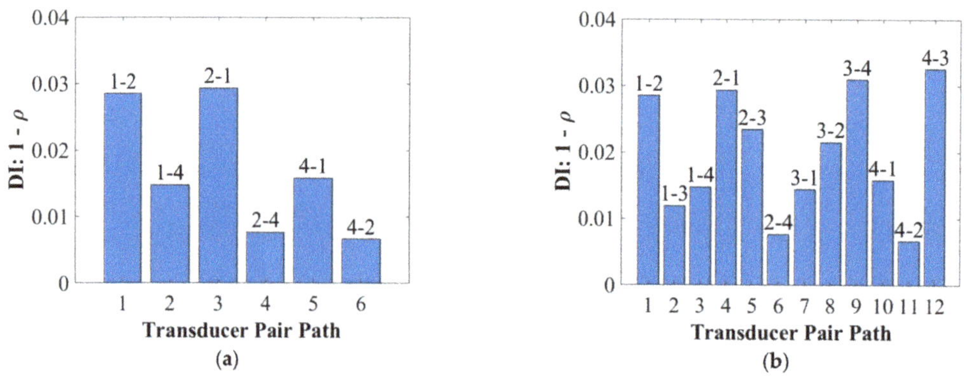

Figure 6. *Cont.*

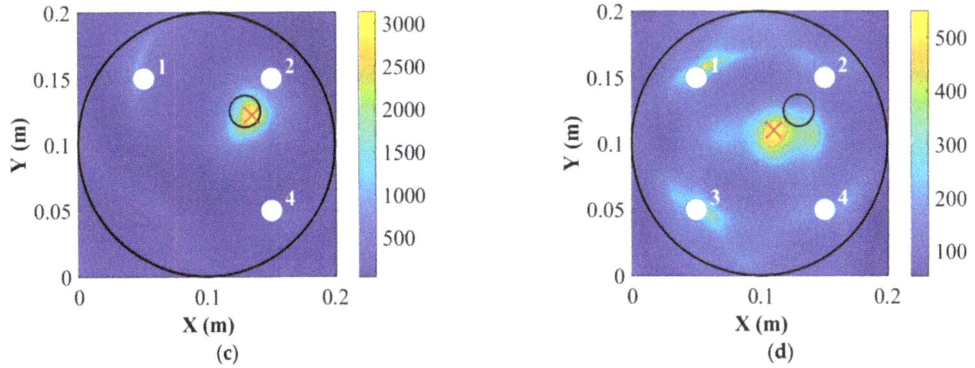

Figure 6. Delamination detection using (**a**) three PZT transducers and (**b**) four PZT transducers, and localization using (**c**) three PZT transducers and (**d**) four PZT transducers at 250 kHz for the position 2 (where the "○" is the position for real damage and the "×" is the position for predicted damage).

3.2. Surface-Mounted Artificial Damage—Surface of the Repair Patch

To simulate the repair patch suffering surface damage, a weighted blue tack was attached on the surface of the repair patch. Figures 7 and 8 show the results of damage detection and localization for the surface-mounted artificial damage attached on the surface of the repair patch. As can be seen in Figures 7 and 8, using both three and four PZT transducers can detect and locate the damage accurately. In this situation, the blue tack was removed from the surface of the repair patch only, so the using of contact gel would not become the factor affecting the accuracy of the localization results.

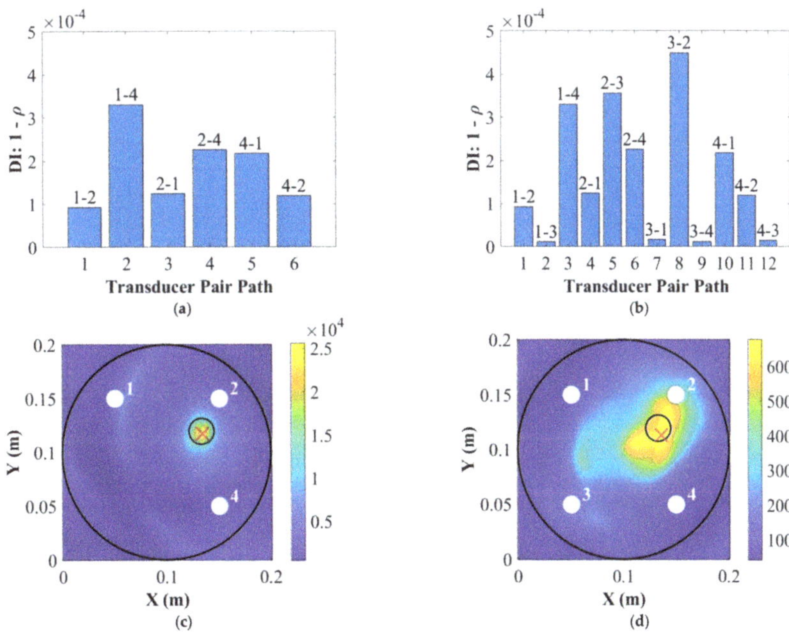

Figure 7. Detection of surface-mounted artificial damage on the surface of the repair patch using (**a**) three PZT transducers and (**b**) four PZT transducers, and localization using (**c**) three PZT transducers

and (**d**) four PZT transducers at 250 kHz for the position 1 (where the "○" is the position for real damage and the "×" is the position for predicted damage).

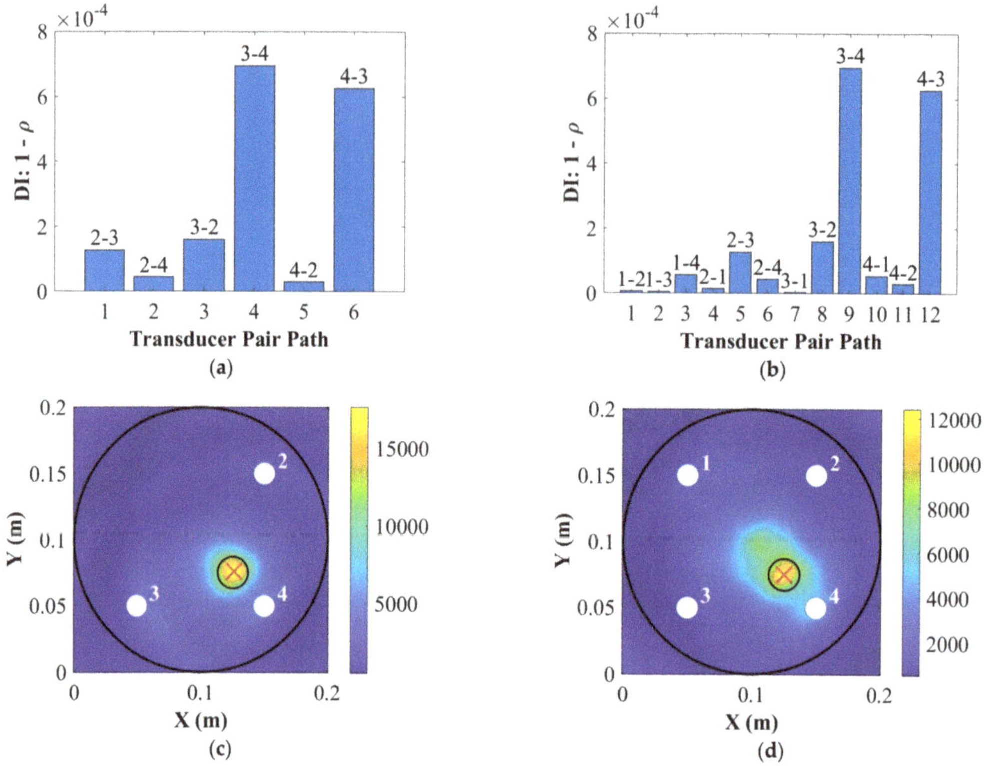

Figure 8. Detection of surface-mounted artificial damage on the surface of the repair patch using (**a**) three PZT transducers and (**b**) four PZT transducers, and localization using (**c**) three PZT transducers and (**d**) four PZT transducers at 250 kHz for the position 2 (where the "○" is the position for real damage and the "×" is the position for predicted damage).

3.3. Surface-Mounted Artificial Damage—Bottom of the Host Structure

To simulate the host structure suffering damage, a weighted blue tack was attached on the bottom of the host structure, which was the surface of the opposite side of the repair patch. Figures 9 and 10 show the damage detection and localization for the surface-mounted artificial damage attached on the bottom of the host structure. As can be seen in Figures 9 and 10, using both three and four PZT transducers can detect the damage. However, the DAS algorithm can locate the damage more accurately using three PZT transducers than using four PZT transducers. The reason why five PZT transducers were not used for damage localization is that the repair patch is relatively small and having more sensors installed generate more noise in the signals. This can be seen with the results from three and four PZT transducers that the noise in the signal was high enough to mask the scattering from the damage.

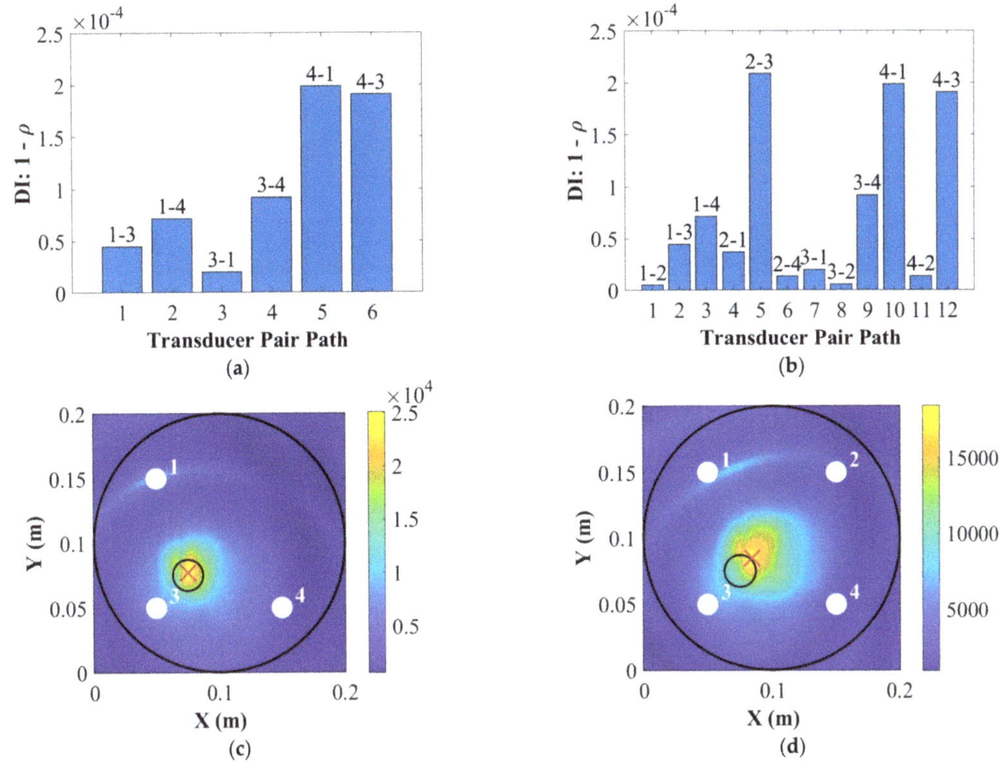

Figure 9. Detection of surface-mounted artificial damage on the bottom of the host structure using (**a**) three PZT transducers and (**b**) four PZT transducers, and localization using (**c**) three PZT transducers and (**d**) four PZT transducers at 250 kHz for the position 1 (where the "○" is the position for real damage and the "×" is the position for predicted damage).

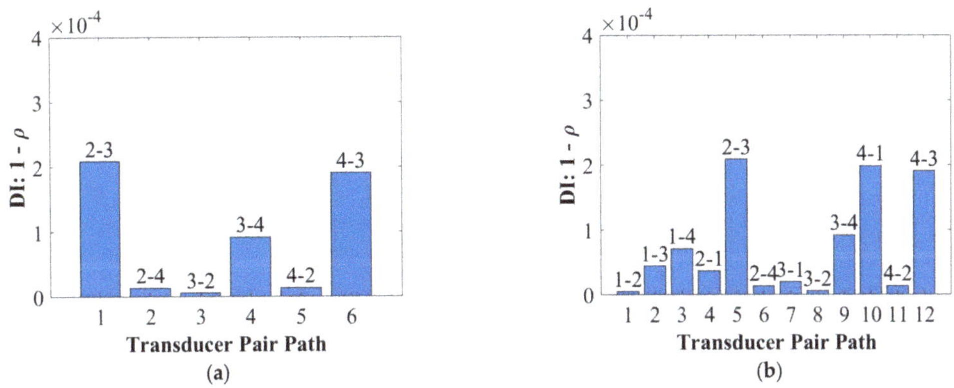

Figure 10. *Cont.*

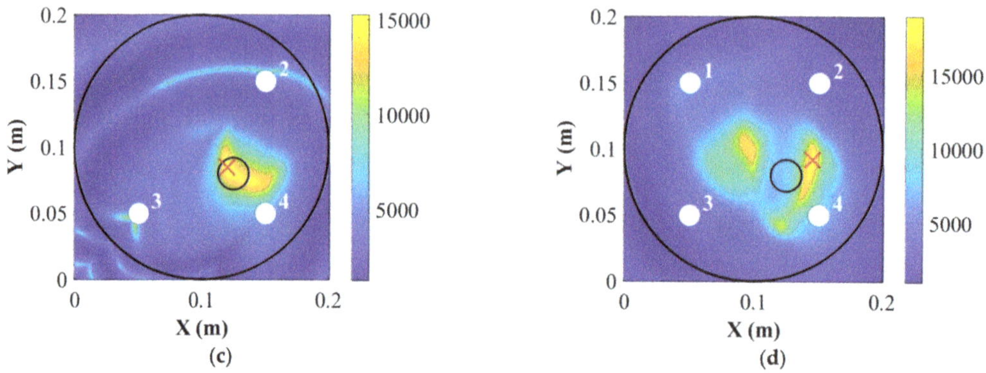

Figure 10. Detection of surface-mounted artificial damage on the bottom of the host structure using (**a**) three PZT transducers and (**b**) four PZT transducers, and localization using (**c**) three PZT transducers and (**d**) four PZT transducers at 250 kHz for the position 2 (where the "○" is the position for real damage and the "×" is the position for predicted damage).

4. Conclusions

This paper developed a smart repair patch using the diagnostic film embedded into the patch using an edge cut-out method. This film has been verified to perform well under extreme environmental and operational conditions. Electro-mechanical impedance results showed that embedded lead zirconate titanate transducers' bonding qualities were good after manufacturing. For damage detection and localization, the damage index correlation coefficient and delay-and-sum algorithm based on the active sensing technique were used. In addition, blue contact gel was used to bond the repair patch and host structure together for easier removing the patch and inserting KAPTON® films to create artificial delamination. Furthermore, the A_0 mode at 50 kHz was not applicable due to the overlap between the crosstalk and the first wave packet of measured signals.

According to the damage index results, the use of a smart repair patch can detect the artificial delamination and surface-mounted artificial damage at 250 kHz using both three and four lead zirconate titanate transducers. Furthermore, the smart repair patch can locate the delamination using three lead zirconate titanate transducers and surface-mounted artificial damage using four lead zirconate titanate transducers accurately. Since the resin film was replaced by blue contact gel for simulating the bonding and caused the inconsistency and uncertainty between the baseline and current signals due to removing the patch and inserting delamination, so the delay-and-sum results were not accurate when using four lead zirconate titanate transducers for the S_0 mode at 250 kHz. However, the use of resin film will keep the composite repair structure consistent and it can be confirmed that using four transducers will locate the delamination accurately using the delay-and-sum algorithm for actual composite repair. Therefore, the smart repair patch embedded with the developed diagnostic film and lead zirconate titanate transducers can be used to detect and locate bondline defects of repaired composite structures based on Structural Health Monitoring techniques.

Author Contributions: Experimental work, analysis and preparation of the manuscript, T.F.; supervision of the research and preparation of the manuscript, M.H.F.A. All authors have read and agreed to the published version of the manuscript.

Funding: This research received no external funding. T. Feng's scholarship is supported by the Aviation Industry Corporation of China, Ltd. (AVIC).

Institutional Review Board Statement: Not applicable.

Informed Consent Statement: Not applicable.

Data Availability Statement: Not applicable.

Acknowledgments: The first author wishes to acknowledge the funding from the Aviation Industry Corporation of China, Ltd. (AVIC), AVIC General Huanan Aircraft Industry Co., Ltd. and the China Scholarship Council (No. [2017] 5082).

Conflicts of Interest: The authors declare no conflict of interest.

References

1. Flor, F.R.; De Medeiros, R.; Tita, V. Numerical and Experimental Damage Identification in Metal-Composite Bonded Joint. *J. Adhes.* **2014**, *91*, 863–882. [CrossRef]
2. Sikdar, S.; Fiborek, P.; Malinowski, P.; Ostachowicz, W. Ultrasonic guided wave propagation in a repaired stiffened composite panel. In Proceedings of the Health Monitoring of Structural and Biological Systems XIII, Denver, CO, USA, 3–7 March 2019.
3. Lambinet, F.; Khodaei, Z.S. Damage detection & localization on composite patch repair under different environmental effects. *Eng. Res. Express* **2020**, *2*, 045032. [CrossRef]
4. Katnam, K.B.; Comer, A.J.; Roy, D.; da Silva, L.; Young, T. Composite Repair in Wind Turbine Blades: An Overview. *J. Adhes.* **2015**, *91*, 113–139. [CrossRef]
5. Spearing, S.M.; Lagace, P.A.; McManus, H.L.N. On the Role of Lengthscale in the Prediction of Failure of Composite Structures: Assessment and Needs. *Appl. Compos. Mater.* **1998**, *5*, 139–149. [CrossRef]
6. Gryzagoridis, J.; Findeis, D. Benchmarking shearographic NDT for composites. *Insight-Non-Destr. Test. Cond. Monit.* **2008**, *50*, 249–252. [CrossRef]
7. Katnam, K.B.; Dhôte, J.X.; Young, T.M. Experimental analysis of the bondline stress concentrations to characterize the influence of adhesive ductility on the composite single lap joint strength. *J. Adhes.* **2013**, *89*, 486–506. [CrossRef]
8. Gharib, H. Structural Health Monitoring of Adhesive Bond in Aircraft Repair Patches. Available online: https://www.researchgate.net/profile/Hossam-Gharib/publication/282121093_A_Review_on_Structural_Health_Monitoring_of_Adhesive_Bond_in_Aircraft_Repair_Patches/links/56036a9508ae596d2591e664/A-Review-on-Structural-Health-Monitoring-of-Adhesive-Bond-in-Aircraft-Repair-Patches.pdf (accessed on 20 March 2022).
9. Sherwin, G.R. Non-autoclave processing of advanced composite repairs. *Int. J. Adhes. Adhes.* **1999**, *19*, 155–159. [CrossRef]
10. Brostow, W.; Glass, N.M. Cure progress in epoxy systems: Dependence on temperature and time. *Mater. Res. Innov.* **2003**, *7*, 125–132. [CrossRef]
11. Katnam, K.; Comer, A.; Stanley, W.; Buggy, M.; Ellingboe, A.; Young, T. Characterising pre-preg and non-crimp-fabric composite single lap bonded joints. *Int. J. Adhes. Adhes.* **2011**, *31*, 679–686. [CrossRef]
12. Lambinet, F.; Khodaei, Z.S. Development of smart bonded composite patch repair solution. *AIP Conf. Proc.* **2020**, *2309*, 020009.
13. Lambinet, F.; Khodaei, Z.S.; Aliabadi, F.M. Effectiveness of RAPID and SSM Algorithms on Composite Scarf Repair. *Key Eng. Mater.* **2018**, *774*, 535–540. [CrossRef]
14. Zhu, J.; Wang, Y.; Qing, X. A real-time electromechanical impedance-based active monitoring for composite patch bonded repair structure. *Compos. Struct.* **2019**, *212*, 513–523. [CrossRef]
15. Rocha, H.; Semprimoschnig, C.; Nunes, J.P. Sensors for process and structural health monitoring of aerospace composites: A review. *Eng. Struct.* **2021**, *237*, 112231. [CrossRef]
16. Su, Z.; Ye, L. *Identification of Damage Using Lamb Waves: From Fundamentals to Applications*; Springer Science & Business Media: Berlin, Germany, 2009.
17. Giurgiutiu, V.; Zagrai, A.; Bao, J.J. Piezoelectric Wafer Embedded Active Sensors for Aging Aircraft Structural Health Monitoring. *Struct. Health Monit.* **2002**, *1*, 41–61. [CrossRef]
18. Kim, H.S.; Ghoshal, A.; Chattopadhyay, A.; Prosser, W.H. Development of Embedded Sensor Models in Composite Laminates for Structural Health Monitoring. *J. Reinf. Plast. Compos.* **2004**, *23*, 1207–1240. [CrossRef]
19. Dib, G.; Koricho, E.G.; Karpenko, O.; Haq, M.; Udpa, L.; Udpa, S.S. Feasibility of PZT ceramics for impact damage detection in composite structures. *AIP Conf. Proc.* **2015**, *1650*, 1072–1080. [CrossRef]
20. Qiu, Y.; Wang, Q.-B.; Zhao, H.-T.; Chen, J.-A.; Wang, Y.-Y. Review on composite structural health monitoring based on fiber Bragg grating sensing principle. *J. Shanghai Jiaotong Univ. (Science)* **2013**, *18*, 129–139. [CrossRef]
21. Giurgiutiu, V. *Structural Health Monitoring of Aerospace Composites*; Academic Press: Cambridge, MA, USA, 2015.
22. Güemes, A.; Fernández-López, A.; Díaz-Maroto, P.F.; Lozano, A.; Sierra-Perez, J. Structural health monitoring in composite structures by fiber-optic sensors. *Sensors* **2018**, *18*, 1094. [CrossRef]
23. Jinachandran, S.; Rajan, G. Fibre Bragg Grating Based Acoustic Emission Measurement System for Structural Health Monitoring Applications. *Materials* **2021**, *14*, 897. [CrossRef]
24. Zhao, X.; Gao, H.; Zhang, G.; Ayhan, B.; Yan, F.; Kwan, C.; Rose, J.L. Active health monitoring of an aircraft wing with embedded piezoelectric sensor/actuator network: I. Defect detection, localization and growth monitoring. *Smart Mater. Struct.* **2007**, *16*, 1208–1217. [CrossRef]
25. Aliabadi, M.H.F.; Khodaei, Z.S. *Structural Health Monitoring for Advanced Composite Structures*; World Scientific Publishing Europe Ltd.: London, UK, 2017.

26. Güemes, A.; Fernandez-Lopez, A.; Pozo, A.R.; Sierra-Pérez, J. Structural Health Monitoring for Advanced Composite Structures. *J. Compos. Sci.* **2020**, *4*, 13. [CrossRef]
27. Michaels, J.E. Detection, localization and characterization of damage in plates with an in situ array of spatially distributed ultrasonic sensors. *Smart Mater. Struct.* **2008**, *17*, 035035. [CrossRef]
28. Khodaei, Z.S.; Aliabadi, M.H. Assessment of delay-and-sum algorithms for damage detection in aluminium and composite plates. *Smart Mater. Struct.* **2014**, *23*, 075007. [CrossRef]
29. Rito, R.; Crocombe, A.; Ogin, S. Health monitoring of composite patch repairs using CFBG sensors: Experimental study and numerical modelling. *Compos. Part A Appl. Sci. Manuf.* **2017**, *100*, 255–268. [CrossRef]
30. Roth, W.; Giurgiutiu, V. Structural health monitoring of an adhesive disbond through electromechanical impedance spectroscopy. *Int. J. Adhes. Adhes.* **2017**, *73*, 109–117. [CrossRef]
31. Qing, X.; Beard, S.J.; Kumar, A.; Hannum, R. A real-time active smart patch system for monitoring the integrity of bonded repair on an aircraft structure. *Smart Mater. Struct.* **2006**, *15*, N66–N73. [CrossRef]
32. Bekas, D.G.; Sharif-Khodaei, Z.; Baltzis, D.; Aliabadi, M.F.; Paipetis, A.S. Quality assessment and damage detection in nanomodified adhesively-bonded composite joints using inkjet-printed interdigital sensors. *Compos. Struct.* **2019**, *211*, 557–563. [CrossRef]
33. Sánchez-Romate, X.F.; García, C.; Rams, J.; Sánchez, M.; Ureña, A. Structural health monitoring of a CFRP structural bonded repair by using a carbon nanotube modified adhesive film. *Compos. Struct.* **2021**, *270*, 114091. [CrossRef]
34. Bekas, D.G.; Sharif-Khodaei, Z.; Aliabadi, M.F. An Innovative Diagnostic Film for Structural Health Monitoring of Metallic and Composite Structures. *Sensors* **2018**, *18*, 2084. [CrossRef]
35. Feng, T.; Bekas, D.; Aliabadi, M.H.F. Active Health Monitoring of Thick Composite Structures by Embedded and Surface-Mounted Piezo Diagnostic Layer. *Sensors* **2020**, *20*, 3410. [CrossRef]
36. Lin, M.; Chang, F.-K. Composite structures with built-in diagnostics. *Mater. Today* **1999**, *2*, 18–22. [CrossRef]
37. Feng, T.; Aliabadi, M.F. Structural Integrity Assessment of Composites Plates with Embedded PZT Transducers for Structural Health Monitoring. *Materials* **2021**, *14*, 6148. [CrossRef]
38. Salmanpour, M.S.; Khodaei, Z.S.; Aliabadi, M.H. Airborne Transducer Integrity under Operational Environment for Structural Health Monitoring. *Sensors* **2016**, *16*, 2110. [CrossRef] [PubMed]
39. Yue, N.; Khodaei, Z.S.; Aliabadi, M.H. Damage detection in large composite stiffened panels based on a novel SHM building block philosophy. *Smart Mater. Struct.* **2021**, *30*, 045004. [CrossRef]
40. Xu, C.; Khodaei, Z.S. A Novel Fabry-Pérot Optical Sensor for Guided Wave Signal Acquisition. *Sensors* **2020**, *20*, 1728. [CrossRef] [PubMed]
41. Park, G.; Farrar, C.R.; Di Scalea, F.L.; Coccia, S. Performance assessment and validation of piezoelectric active-sensors in structural health monitoring. *Smart Mater. Struct.* **2006**, *15*, 1673–1683. [CrossRef]
42. Zou, F.; Benedetti, I.; Aliabadi, M.H. A boundary element model for structural health monitoring using piezoelectric transducers. *Smart Mater. Struct.* **2014**, *23*, 015022. [CrossRef]

applied sciences

MDPI

Article

Noise Reduction in the Swept Sine Identification Procedure of Nonlinear Systems

Pietro Burrascano * and Matteo Ciuffetti

Dipartimento di Ingegneria, Università di Perugia, 06125 Perugia, Italy; matteo.ciuffetti@studenti.unipg.it
* Correspondence: pietro.burrascano@unipg.it

Abstract: The Hammerstein model identification technique based on swept sine excitation signals proved in numerous applications to be particularly effective for the definition of a model for nonlinear systems. In this paper we address the problem of the robustness of this model parameter estimation procedure in the presence of noise in the measurement step. The relationship between the different functions that enter the identification procedure is analyzed to assess how the presence of additive noise affects model parameters estimation. This analysis allows us to propose an original technique to mitigate the effects of additive noise in order to improve the accuracy of model parameters estimation. The different aspects addressed in the paper and the technique for mitigating the effects of noise on the accuracy of parameter estimation are verified on both synthetic and experimental data acquired with an ultrasonic system. The results of both simulations and experiments on laboratory data confirm the correctness of the assumptions made and the effectiveness of the proposed mitigation methodology.

Keywords: nonlinear systems; Hammerstein model; pulse compression; ultrasonic systems

Citation: Burrascano, P.; Ciuffetti, M. Noise Reduction in the Swept Sine Identification Procedure of Nonlinear Systems. *Appl. Sci.* **2021**, *11*, 7273. https://doi.org/10.3390/app11167273

Academic Editor: Dimitrios G. Aggelis

Received: 25 June 2021
Accepted: 5 August 2021
Published: 7 August 2021

1. Introduction

The behavior of physical systems is very often modeled using linear techniques. In reality, only in very particular cases physical systems can behave as linear, so if we want to obtain high quality results when mathematical models of physical systems are involved, it becomes of remarkable importance to have models that can represent the behavior of the system also in a non-linear regime.

The modeling of nonlinear dynamical systems is one of the most challenging research areas in the field of system representation. Much of the research developed in recent years has increasingly focused on predominantly data-driven approaches, including neural nets or fuzzy logic [1], methodologies that assume the availability of large amounts of data. The amount of available data is not always adequate to represent the complexity of the system and the computing power is sufficient to handle them. For this reason, they continue to be widely diffused in the white-box approaches and the gray-box approaches, which place side by side to a number of data-driven only, black box methods: Such techniques integrate in different measures the information derived from the knowledge of the physics of the system and information extracted from the data produced by the nonlinear system.

The white-box approaches require an accurate knowledge of the system. These techniques rely, for example, on wave digital filters [2,3] or differential equations [4] to obtain an accurate model of the real system.

If one has detailed knowledge of the physical system, these models can be highly accurate. If it is not possible to adequately represent the overall physical system by equations, the gray and black-box approaches can be applied, as they require only partial knowledge of the system or only the knowledge of input and output signals. A partial knowledge of the physics of the system allows to fractionate it into connected subsystems, each representable in a simpler way by adopting a gray-box approach [5,6]. Among the black-box approaches, the methods derived from the Volterra series are in a prominent position, both in terms of accuracy and historically, having been defined long before the neural network-based approaches. Modeling techniques based on this infinite series are obtained by truncating them to a finite number of terms [7]. It is however important to note that, even considering a truncation to a low order of the Volterra series, the number of coefficients required to define the relevant model quickly becomes very large. This is an important limitation, because it makes it possible in practice to define adequate models by following this path only in the case where the system shows a degree of nonlinearity of small entity. This has prompted the development of simplified expansion models of the Volterra series [8]. Among these, the Hammerstein and Wiener models are among the most popular. Having available accurate models and effective techniques for their identification has allowed the successful application of these nonlinear modeling techniques of physical systems in many engineering fields, notably including acoustics and nondestructive testing [9–12].

Among the different types of black-box nonlinear models of physical structures, the parallel Hammerstein model is particularly interesting as it can be shown that the problem of identifying its parameters reduces to a linear problem, so classical least squares methods can be used for their estimate [13]. An efficient technique was proposed in recent years for the identification of the parallel Hammerstein model [14–16]: Its relative simplicity of implementation, and the excellent results obtained both in laboratory experiments and in practical applications, make it one among the most competitive techniques for the identification of the Hammerstein model: A swept sine signal is input to the nonlinear system to be modelled and the corresponding output is acquired. Processing this output signal allows to identify the parameters of the model, as will be detailed in the following sections.

An aspect of this technique, which is of great importance in practical applications, but which has been only partially addressed in the technical literature, is the analysis of the effects of noise that inevitably adds to the useful signal in the acquisition step of the identification procedure. The problem has been addressed in some particular cases such as in [17], where an iterative procedure for the identification of the Hammerstein model is defined in the particular case in which the noise, added to the useful signal at the output of the system to be identified, can be modeled as a correlated noise, assumed to be the output of a digital filter with a zero mean random sequence as input. In [18], a protocol is proposed to determine the location of the process noise in a Wiener–Hammerstein system with respect to the static nonlinearity by using a periodic, nonstationary input test signal. Hypothesizing the location of the process noise is addressed as a preliminary step in order to improve the successive model identification. In [19], the hypothesis that damage causes the structure to exhibit a nonlinear response is tested, and thus the use of Nonlinear Model Based Features is shown to increase classification performance: although the study does not directly analyze the effects of measurement noise in the model identification phase, it evaluates the performance degradation of the classification procedure in the presence of noise added to the measured signals at the output of the system under study. In [20], the identification of the nonlinear model is considered in the presence of uncertainties: Under the assumption of stationarity of the system to be modeled, the effect of uncertainties is reduced by using repetitions of the input swept sine signal, which is repeated several times in order to excite the system with a pseudo periodic signal; a period synchronized with the duration of the swept sine signal is adopted to simplify the implementation. Rebillat and Schoukens in [21] compare two methodologies of identification of the Hammerstein model by evaluating, for both, specific performance indexes on systems defined as bench test, and in the presence of noise.

In the present paper we deal with the problem of the accuracy in the identification of the Hammerstein model using the identification technique based on exponential swept sine test signals. We consider the presence of noise that is added to the useful signal when measuring the response of the nonlinear system. In the paper we propose an original analysis of the link that exists between the environmental noise superimposed on the measurement result and the uncertainty in the estimation of the kernels that identify the Hammerstein model in the case that the identification technique is based on exponential swept sine signals. This original analysis, besides clarifying the connection between the presence of measurement noise and the uncertainties in the estimation of the different parameters characterizing the identification procedure, represents the basis to define a novel technique, that we propose in the paper, to mitigate the effects of the superimposed noise on the quality of the estimation of the parameters that identify the model. The technique we propose is based on the use of lowpass filters, each designed to operate on one of the kernels, characterizing the different branches of the Hammerstein model.

The different aspects of the uncertainty reduction procedure are verified both on synthetic data and in an experimental case. The latter is an ultrasonic measurement setup with transducers operating in air, chosen because the possible effects due to measurement noise are particularly relevant: The limited impedance matching and the attenuation due to air propagation make this an experimental setup where the signal-to-noise ratio can degrade significantly.

The paper is organized as follows: In Section 2, the theoretical aspects of the procedure are detailed. In Section 3, the experimental tests carried out on both synthetic and real devices are described. Section 4 discusses the results obtained, draws conclusions, and indicates possible evolutions of the work.

2. Theoretical Aspects

In this section we will briefly describe some concepts at the basis of the procedure for reducing the effects of measurement noise in the identification of the Hammerstein model based on swept sine signals: The procedure for model parameters identification is first recalled, then the effect of noise on the assessment of model kernels $h_i(t)$ is evaluated. Finally, the frequency bands characterizing the signals involved in the different steps of the identification procedure are analyzed.

2.1. Model Identification by Swept Sine Excitation

If we assume that a valid representation of the nonlinear system is the Hammerstein model, a schematic description of which is shown in Figure 1, the input–output relation of the system will be:

$$y_H(t) = x(t) \otimes h_1(t) + x(t)^2 \otimes h_2(t) + \ldots + x(t)^{N_H} \otimes h_{N_H}(t) \tag{1}$$

where we indicated with the symbol $\otimes$ the convolution operation. The output $y_H(t)$ is the sum of the powers up to order N_H of the input signal $x(t)$, convolved with appropriate impulsive functions $h_i(t)$; such impulsive functions $h_i(t)$ are the kernels that identify the Hammerstein model.

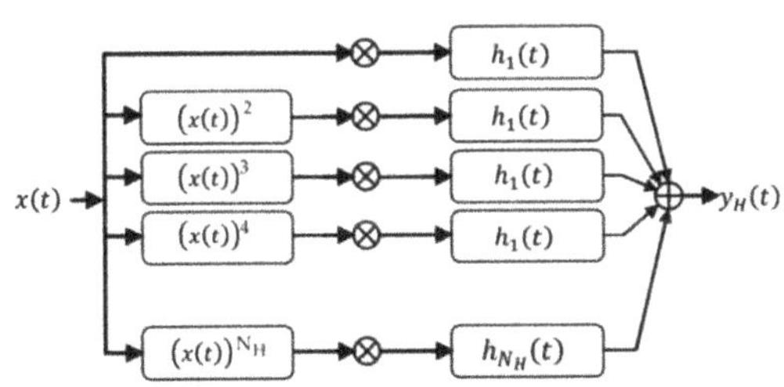

Figure 1. The Hammerstein model. The input $x(t)$ and each of its powers up to order N_H pass through a different linear filter: $(x(t))^k$ is convolved with $h_k(t)$. The order N_H and the kernels $h_k(t)$ completely characterize the model.

Assume the input is a swept sine, i.e., an harmonic signal $x(t) = R\,\mathrm{Cos}(\phi(t))$ of amplitude R, with instantaneous angular frequency $\omega(t) = d\phi(t)/dt$, which varies over time. If we represent vectors and matrices by square brackets, we can write for the output of the Hammerstein model of order N_H:

$$y_H(t) = \left[[R\ Cos(\phi(t))]^k\right]^T \otimes [h(t)] \tag{2}$$

where $[h(t)]$ is the vector of the different kernels $h_k(t)$, $k = 1, \ldots, N_H$, and $\left[[R\ Cos(\phi(t))]^k\right]^T$ is the transpose of the vector of powers of the input signal.

By means of the Chebyshev polynomials of the first kind, we can express the harmonics $Cos(k\ \phi(t))$ of the swept sine function $Cos(\phi(t))$, as functions of the powers $[R\ Cos(\phi(t))]^k$ of the input signal [16]:

$$[Cos(k\ \phi(t))] = [A_c][Cos(\phi(t))]^k = [A_c][R_c]^{-1}[R\ Cos(\phi(t))]^k \tag{3}$$

The entries of matrix $[A_c]$ are the coefficients of Chebyshev polynomials of the first type, and $[R_c]$ is the diagonal matrix in which the index term $\{i, i\}$ is the $i - th$ power of the amplitude R. From expression (3), we can derive the expression of $[R\ Cos(\phi(t))]^k$ as a function of the harmonics $Cos(k\ \phi(t))$ which, substituted in (2), gives:

$$\begin{aligned} y_H(t) &= \left[[R_c][A_c]^{-1}Cos(k\ \phi(t))\right]^T \otimes [h(t)] = [Cos(k\ \phi(t))]^T \left[[R_c][A_c]^{-1}\right]^T \otimes [h(t)] = \\ &= [Cos(k\ \phi(t))]^T \otimes \left(\left[[A_c]^{-1}\right]^T [R_c]\right)[h(t)] = [Cos(k\ \phi(t))]^T \otimes [g(t)] \end{aligned} \tag{4}$$

where we considered that $[R_c]$ is a diagonal matrix, and we defined the vector $[g(t)]$ containing the functions $g_i(t)$, $i = 1, \ldots N_H$, by means of $[g(t)] = \left[[A_c]^{-1}\right]^T [R_c]\ [h(t)]$.

A comparison between Equations (2) and (4) tells us that the output $y_H(t)$ of the nonlinear system, obtained when the swept sine signal $x(t) = R\ Cos(\phi(t))$ is input, can be expressed either as the sum of the convolutions between the powers of the harmonic signal itself and the functions in $[h(t)]$, or, alternatively, as the sum of the convolutions between the harmonics $Cos(k\ \phi(t))$ of the signal $Cos(\phi(t))$ and the impulsive functions $[g(t)]$. The functions $[g(t)]$ are related to the functions $[h(t)]$ through a linear transformation; the transformation matrix $\left[[A_c]^{-1}\right]^T [R_c]$ depends on the amplitude R of the input signal and on the matrix $[A_c]$, whose entries are the coefficients of the Chebyshev polynomials of the first kind.

The identification procedure of the Hammerstein model, and therefore of the functions $[h(t)]$, is based on this correspondence between the $[h(t)]$ and the $[g(t)]$: assume that the harmonic function: $x(t) = R\ Cos(\phi(t))$ at the input is such to modify its instantaneous frequency by following an exponential law, i.e., assume that, if T_0 is the duration of the swept sine signal ranging between the frequencies f_{MIN} and f_{MAX}, the angular frequency is defined by $\omega(t) = d\phi(t)/dt = 2\ \pi\ f_{MIN}\ Exp(t/L)$, where the constant L is defined as $L = T_0/ln(f_{MAX}/f_{MIN})$ and describes how quickly the frequency changes over time. It is easy to verify that the k − th harmonic of the input signal corresponds to a simple shift of the signal $x(t)$ by the quantity $\Delta t_k = L\ ln(k)$. In fact, we can write for the instantaneous frequency $f(t)$:

$$f(t + \Delta t_k) = f_{MIN}\ Exp[(t + \Delta t_k)/L] = f_{MIN}\ Exp[(t + L\ ln(k))/L] = k\ f_{MIN}\ Exp[t/L] = k\ f(t) \tag{5}$$

If, moreover, the input signal complies with specific constraints on its instantaneous phase, as detailed in [15], we can process the output $y_H(t)$ with the matched filter, i.e., a filter whose impulse response $\psi(t)$ is such that its convolution with the exponential swept sine signal $x(t) = R\ Cos(\phi(t))$ produces $\hat{\delta}(t)$, a band-limited approximation of Dirac's delta function $\delta(t)$. The output of the matched filter $\psi(t)$ will then be given by:

$$u(t) = y_H(t) \otimes \psi(t) = [cos[k\phi(t)]_c]^T \otimes [g(t)] \otimes \psi(t) = \left\{\left[\hat{\delta}(t + \Delta t_k)\right]^T\right\} \otimes [g(t)] \tag{6}$$

where the $k-th$ element of the vector $[\hat{\delta}(t+\Delta t_k)]$ is the band-limited approximation of the Dirac delta function shifted in time by the quantity $\Delta t_k = L\,ln(k)$ defined above.

The output $u(t)$ of the matched filter, when the signal $y_H(t)$ is input, consists therefore of a sequence of impulsive functions $g_i(t)$ starting at the times identified by the shifts $\Delta t_k = L\,ln(k)$ related only once the input swept sine signal is defined, to the order k of the harmonic. If the parameter L is large enough to keep the different $g_i(t)$ apart, each of the $g_i(t)$ functions can be obtained by simply extracting an appropriate section in time of the output signal $u(t)$, starting at a time instant identified by Δt_i. The duration of the time window beginning at Δt_k depends on the length of the impulse responses $h_k(t)$ and can only be established by trial and error.

Once the $[g(t)]$ functions have been obtained by sectioning the output $u(t)$ of the matched filter, from the $[g(t)]$ it is possible to derive the $[h(t)]$ functions that identify the Hammerstein model through a simple linear transformation:

$$[h(t)] = [R_c]^{-1}[A_c]^T[g(t)] \tag{7}$$

Figure 2 shows graphically the processing procedure to obtain the functions $[g(t)]$.

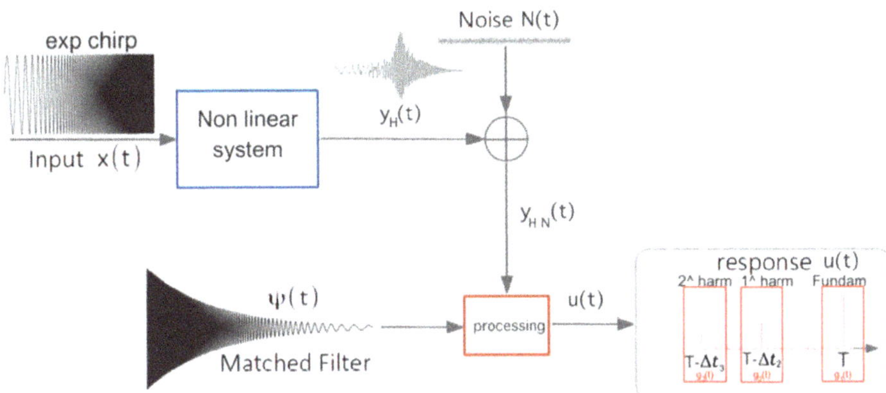

Figure 2. Description of the processing procedure for the identification of the Hammerstein model.

2.2. Noise Power on the Output $y_H(t)$ and on Functions $g(t)$ and $h(t)$

We have seen that the swept sine identification procedure of the Hammerstein model starts by considering the functions $g_i(t)$, obtained by time-windowing at appropriate positions the response $u(t)$ of the matched filter when its input is $y_H(t)$. The response $y_H(t)$ is the one that we have at the output of the nonlinear system under consideration when it has an exponential swept sine signal as input. From the functions $g_i(t)$ obtained with this procedure, through an appropriate linear transformation we can derive the functions $h_i(t)$ that characterize the Hammerstein model; the coefficients of this linear transformation are the coefficients of Chebyshev polynomials of the first type. To obtain the functions $h_i(t)$, a combination of the functions $g_i(t)$ is performed according to (7). The combination is performed through coefficients given by matrix $[R_c]^{-1}[A_c]^T$. For example, for a model of order 4, we will have:

$$\begin{bmatrix} h_1(t) \\ h_2(t) \\ h_3(t) \\ h_4(t) \end{bmatrix} = \begin{bmatrix} \frac{1}{R} & 0 & 0 & 0 \\ 0 & \frac{1}{R^2} & 0 & 0 \\ 0 & 0 & \frac{1}{R^3} & 0 \\ 0 & 0 & 0 & \frac{1}{R^4} \end{bmatrix} \begin{bmatrix} 1 & 0 & -3 & 0 \\ 0 & 2 & 0 & -8 \\ 0 & 0 & 4 & 0 \\ 0 & 0 & 0 & 8 \end{bmatrix} = \begin{bmatrix} \frac{1}{R} & 0 & \frac{-3}{R} & 0 \\ 0 & \frac{2}{R^2} & 0 & \frac{-8}{R^2} \\ 0 & 0 & \frac{4}{R^3} & 0 \\ 0 & 0 & 0 & \frac{8}{R^4} \end{bmatrix} \cdot \begin{bmatrix} g_1(t) \\ g_2(t) \\ g_3(t) \\ g_4(t) \end{bmatrix} \tag{8}$$

Assume that additive noise, superimposed on the useful signal, affects the acquisition of the output to the nonlinear system. The ideal output $y_H(t)$ of the nonlinear system will be superimposed by the noise sequence $N(t)$, supposed to be white Gaussian with power spectral density $P_N(f) = N_0/2$. The noise-affected output will be $y_{H\,N}(t) = y_H(t) + N(t)$.

According to the identification procedure, the output signal $y_{H\,N}(t)$ is filtered through the matched filter $\psi(t)$: the noise sequence $N(t)$ will thus be filtered giving rise, at the output of the matched filter, to the noise sequence $n_u(t)$ superimposed on the signal $u(t)$. If we say $\Psi(f)$ is the Fourier transform of the impulse response of the matched filter: $\Psi(f) = \mathcal{F}\{\psi(t)\}$, the power of noise signal $n_u(t)$ at the output of the matched filter can be evaluated by the:

$$P_{n_u} = \sigma^2_{n_u} = \int_{-\infty}^{\infty} P_N(f)\,|\Psi(f)|^2 df = \frac{N_0}{2}\int_{-\infty}^{\infty} |\Psi(f)|^2 df \tag{9}$$

We denote with σ^2_{nu} the variance of noise process $n_u(t)$ superimposed on $u(t)$: all functions $g_i(t)$, are extracted from $u(t)$ at different positions, and will be thus affected by superimposed noise signals $n_{g_i}(t)$ whose variance will be σ^2_{nu} for all of them: $\sigma^2_{n_{g_i}} = \sigma^2_{n_u}$. This common variance will be denoted in the following as $\sigma^2{}_g$; furthermore, the overall noise sequence $n_u(t)$ has an impulsive-like correlation function; consequently, the noise sequences $n_{g_i}(t)$, extracted at different positions from $n_u(t)$, will be uncorrelated with each other, as they refer to different portions of the overall noise sequence $n_u(t)$.

By following this reasoning, it is straightforward to evaluate the variance of noise sequences $n_{h_i}(t)$ superimposed on each one of the $h_i(t)$ kernels from the variance of the noise sequences $n_{g_i}(t)$ added to the $g_i(t)$: the noise sequences $n_{g_i}(t)$ (uncorrelated, all with zero mean and variance $\sigma^2{}_g$) are combined with each other by using the same coefficients as the $g_i(t)$; in the linear combination, the noise amplitudes $n_{g_i}(t)$ will be altered according to the coefficients $[R_c]^{-1}[A_c]^T$ seen above. The noise sequences $n_{g_i}(t)$ are not correlated with each other, so the variances of noise sequences $n_{h_i}(t)$, denoted as $\sigma^2{}_{h_i}$, will then be obtained by combining the variances $\sigma^2{}_g$ with coefficients equal to the square of those in the matrix $[R_c]^{-1}[A_c]^T$. We define the matrix $[V_c] = \left\{[R_c]^{-1}[A_c]^T\right\}^2$, where by $\{[.]\}^2$ we denote the matrix whose terms are the squares of the individual elements of the matrix $[.]$. If $\left[\sigma^2{}_h\right]$ and $\left[\sigma^2{}_g\right]$ are the vectors of $\sigma^2{}_{h_i}$ and $\sigma^2{}_g$, we have:

$$\left[\sigma^2{}_h\right] = \left\{[R_c]^{-1}[A_c]^T\right\}^2 \cdot \left[\sigma^2{}_g\right] \tag{10}$$

In the example of model order $N_H = 4$, remembering that the variances $\sigma^2{}_g$ are all equal to each other, the variances $\sigma^2{}_{h_i}$ will be obtained via the linear combination:

$$\begin{bmatrix} \sigma^2{}_{h_1} \\ \sigma^2{}_{h_2} \\ \sigma^2{}_{h_3} \\ \sigma^2{}_{h_4} \end{bmatrix} = \begin{bmatrix} \left(\frac{1}{R}\right)^2 & 0 & \left(\frac{-3}{R}\right)^2 & 0 \\ 0 & \left(\frac{2}{R^2}\right)^2 & 0 & \left(\frac{-8}{R^2}\right)^2 \\ 0 & 0 & \left(\frac{4}{R^3}\right)^2 & 0 \\ 0 & 0 & 0 & \left(\frac{8}{R^4}\right)^2 \end{bmatrix} \cdot \begin{bmatrix} \sigma^2{}_g \\ \sigma^2{}_g \\ \sigma^2{}_g \\ \sigma^2{}_g \end{bmatrix} \tag{11}$$

2.3. Frequency Band of Functions g(t) and h(t): Noise Reduction on h(t) by Lowpass Filtering

Equation (10) expresses the connection between the variance $\sigma^2{}_g$ of noise superimposed to each function $g_i(t)$ and the variance $\sigma^2{}_{h_k}$ of noise overlapping the functions $h_k(t)$ in the case where the identification procedure is the one described in the previous sections.

We already mentioned that, under the assumption of stationarity of the system under measurement [20], proposes to adopt repetitions of the input swept sine signal to reduce the amount of uncertainties arising from the presence of superimposed noise. In the present paper we propose a different way to achieve a mitigation of the effects of additive noise.

The procedure we propose deserves some further consideration about the bandwidths of the different signals involved. Equation (10) connects the variances of noise processes superimposed on functions of different types, considered all full-band, i.e., with harmonic components that can reach the Nyquist frequency. In reality, the different signals whose variances are combined in (10) relate to different frequency bands: If the excitation signal $x(t)$ has harmonic components between f_{MIN} and f_{MAX}, the kernel $h_k(t)$ present on the $k-th$ branch of the Hammerstein model will be excited in the frequency band associated with the $k-th$ power of that signal, that is, in the frequency range between f_{MIN} and $k*f_{MAX}$, having excluded the DC component, if any; it is, therefore, sufficient to characterize the kernel $h_k(t)$ up to frequency $k*f_{MAX}$, since it will not be stressed for higher frequencies. According to the swept sine identification procedure, the kernel $h_k(t)$ is obtained through Equation (7), i.e., through a weighted combination of $g_i(t)$ functions, where the $i-th$ function $g_i(t)$ is associated with the $i-th$ harmonic of the input signal, and thus with a frequency range from $i*f_{MIN}$ to $i*f_{MAX}$. The combination matrix present in Equation (7) is an upper triangular matrix, and thus the functions $g_i(t)$ that contribute to build the $k-th$ kernel $h_k(t)$ are those with indices ranging from k itself to the value N_H i.e., to the maximum order of the Hammerstein model. Thus, according to Equation (7), to obtain the functions $h_k(t)$ (that we know are defined in the frequency band limited to $k*f_{MAX}$), we linearly combine the functions $g_i(t)$ whose frequency band is wider than that of the $h_k(t)$ we are seeking. The frequency range of the combination of the $g_i(t)$ needed to obtain $h_k(t)$ has an overall harmonic content ranging from $k*f_{MIN}$ to N_H*f_{MAX}, so it extends to frequencies much higher than those we know characterize the $h_k(t)$ function we want to identify. This implies that if we lowpass filter the linear combination of the $g_i(t)$ functions contributing to $h_k(t)$, and limit the contributions to the useful band of each $h_k(t)$, i.e., to $k*f_{MAX}$, we can sensitively reduce the noise effects without altering the quality of the identification result.

This filtering operation has to be carried out in a frequency band whose range depends on the order of the function $h_k(t)$ to be identified, so for each kernel $h_k(t)$ an ad hoc lowpass filter will have to be defined.

3. Experimental Results

The experiments reported in this section are aimed at verifying the different aspects highlighted in the previous section of the paper. A first aspect is the relationship between the noise superimposed on the $g_i(t)$ functions and the noise superimposed on the kernel $h_k(t)$ that identify the Hammerstein model: This aspect is tested in the following Section 3.1. The subsequent sections are instead aimed at verifying the effectiveness of the technique of lowpass filtering to mitigate uncertainty in the identification of kernels $h_k(t)$ due to the effects of noise superimposed in the measurement phase: Section 3.2 reports the results of noise effects mitigation obtained in a simulated experiment, while Section 3.3 reports the results obtained when measuring real data collected in a measurement bench with ultrasonic probes operating in air.

3.1. Noise on the $g_i(t)$ Versus Noise on the $h_k(t)$

In this section, we verify expression (10), which we have shown is the functional link between the power of noise superimposed on functions $g_i(t)$ and the corresponding power of noise superimposed on the $h_k(t)$ estimated by combining the $g_i(t)$. We start with the simulation of the noise sequence superimposed on functions $g_i(t)$ and estimate their variances; we then operate the linear transformation (7) and obtain the corresponding noise sequences affecting the $h_k(t)$ kernels; in this case too we estimate the variance vector $\left[\sigma^2{}_h\right]$ characterizing these noise sequences. We thus compare the latter estimates $\left[\sigma^2{}_h\right]$ with those obtained by means of Equation (10).

Figure 3 plots the noise sequences: The first line of Figure 3 reports an example of the noise sequences obtained by truncating at different Δt_k the function $u(t)$, in the absence of input excitations: The value of the noise powers in the four segments, estimated by averaging over 50 trials, is $\left[\sigma^2{}_g\right] = \left[\begin{array}{cccc} 0.0098 & 0.0090 & 0.0101 & 0.0099 \end{array}\right]^T$. The second row of Figure 3 shows the noise sequences obtained by the transformation (7). Noise powers, estimated from the sequences, are $\left[\hat{\sigma}^2{}_h\right] = \left[\begin{array}{cccc} 0.0157 & 0.0048 & 0.0006 & 0.0002 \end{array}\right]^T$.

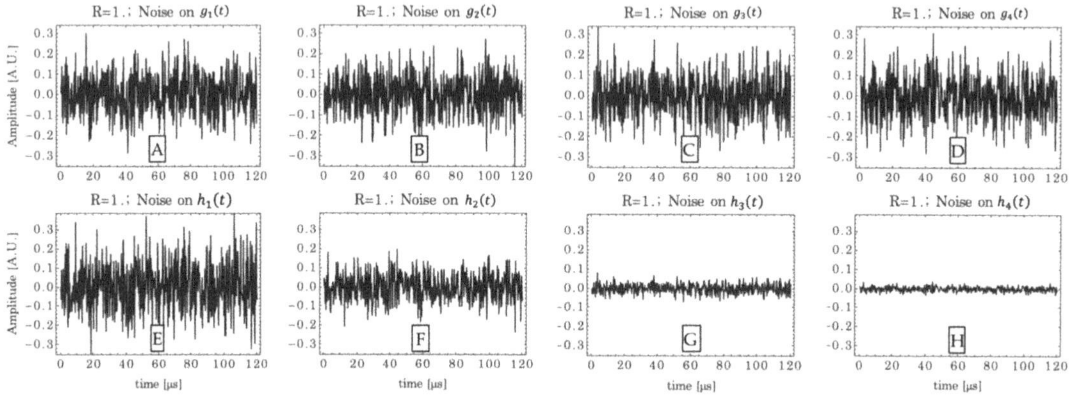

Figure 3. Correspondence between noise sequences superimposed on $g(t)$ and $h(t)$. Plots (**A–D**): noise sequences on $g_1(t)$, $g_2(t)$, $g_3(t)$, $g_4(t)$, respectively. Plots (**E–H**): noise sequences on $h_1(t)$, $h_2(t)$, $h_3(t)$, $h_4(t)$, respectively.

We then compare these estimates with the corresponding variance values predicted by using equation (10), which gives the estimates of the same $\left[\sigma^2{}_{h(t)}\right]$ starting from the $\left[\sigma^2{}_g\right]$: we obtain from (10): $\left[\hat{\hat{\sigma}}^2{}_h\right] = \left[\begin{array}{cccc} 0.0154 & 0.0047 & 0.0006 & 0.0002 \end{array}\right]^T$.

The percentage estimate error made adopting (10), evaluated by means of $\sigma^2{}_{i\_ERR\%} = 100\left(\hat{\hat{\sigma}}^2{}_{h_k} - \hat{\sigma}^2{}_{h_k}\right)/\hat{\sigma}^2{}_{h_k}$, in the case of the above example gives:

$$\left[\sigma^2{}_{ERR\%}\right] = \left[\begin{array}{cccc} -1.54\% & -1.87\% & 0.0\% & 0.0\% \end{array}\right]^T$$

The low error values confirm that the variance $\left[\sigma^2{}_h\right]$ of the noise superimposed on the kernels $[h(t)]$ can be accurately obtained from an estimate of the power of the noise $\left[\sigma^2{}_g\right]$ superimposed on the functions $[g(t)]$ by adopting Equation (10).

3.2. Mitigation of Noise Effects: Simulation Tests

We begin testing the effectiveness of the procedure proposed for mitigating the effects of noise in a simulation experiment: The importance of performing simulation tests is to have a complete feedback of the improvement we achieve in the estimation of the kernels $h_i(t)$: Their reference trends are known, since they have been analytically defined to be included in the synthetic system. Consequently, a direct comparison with the result of the identification procedure in the presence of noise is possible, and we can quantify the benefits arising from the proposed mitigation procedure.

We have generated a synthetic nonlinear system of the fourth order, defined according to the Hammerstein model. In the different branches of the model we adopted as kernels $h_i(t)$, the four functions are represented in Figure 4. They all have typical behavior of the impulsive response of a filter of the second order and differ for the central frequency and for the attack delay time, which both increase as the order increases. The values adopted in our simulations to characterize the four kernels of the synthetic model are reported in Table 1.

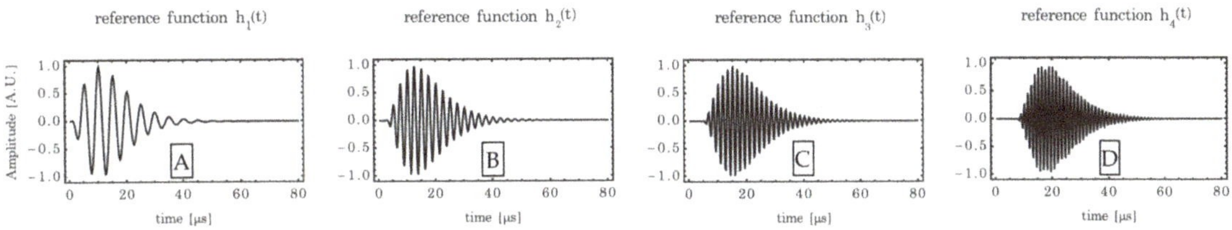

Figure 4. Kernels $h_i(t)$ adopted in the synthetic Hammerstein model of the fourth order. Plots (**A–D**): reference functions $h_1(t)$, $h_2(t)$, $h_3(t)$, $h_4(t)$, respectively.

Table 1. Values that characterize the four kernels of the synthetic model.

	$h_1(t)$	$h_2(t)$	$h_3(t)$	$h_4(t)$
f_0 [kHz]	200	400	600	800
Delay [μs]	0.0	2.5	5.0	7.5

For the purpose of verifying the accuracy in model identification, the system is elicited with an exponential swept sine signal of amplitude R = 1 whose frequency range $f_{MIN} = 40$ [kHz] to $f_{MAX} = 1600$ [kHz] contains the frequency bands that characterize all the $h_i(t)$ kernels.

A Gaussian white noise is added to the response $y_H(t)$; the value of its standard deviation is such that the signal-to-noise ratio (SNR) is brought to the desired level: The values $SNR_1 = 10$ [dB] and $SNR_2 = 5$ [dB] were considered. Let us say $y_{H\ N}(t)$ is the noisy response. It is convolved with the filter matched to the input signal to obtain the output signal $u(t)$. The functions $g_i(t)$ are obtained by taking portions of $u(t)$ at the time instants Δt_i. From $g_i(t)$, the estimates of the kernels $h_k(t)$ are obtained. To verify the effectiveness of the noise reduction technique, we report both the results of the estimates in the absence of lowpass filtering and the results obtained after the additional lowpass filtering step of each of the estimated $h_k(t)$ kernels. The impulse response of the FIR (Finite Impulse Response) filter is defined using a frequency sampling method, starting with the amplitude values that define the required low-pass amplitude response in the frequency domain.

Figures 5 and 6 show the superposition between the estimated $h_k(t)$ kernels and the corresponding reference trends, respectively, for the case $SNR_1 = 10$ [dB], and for the case $SNR_2 = 5$ [dB]: The upper row of each figure refers to the estimate without filtering and the lower row shows the trend of the estimated $h_k(t)$ after the additional lowpass filtering step. The figures clearly show that the lowpass filters defined ad hoc for each order of the kernel greatly reduce the effects of superimposed noise without deteriorating the time course of the function we want to estimate.

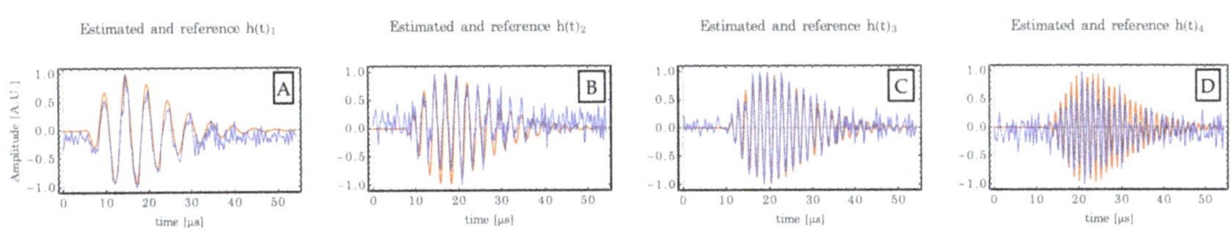

Figure 5. *Cont.*

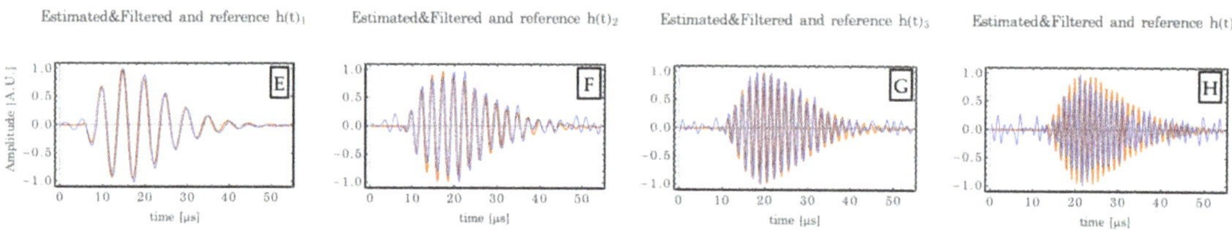

Figure 5. Estimated $h_k(t)$ kernels in the case $SNR_1 = 10$ [dB] (blue lines) and reference kernels (red lines). Plots (**A–D**): $h_1(t)$, $h_2(t)$, $h_3(t)$, $h_4(t)$ estimates before lowpass filtering and reference. Plots (**E–H**): $h_1(t)$, $h_2(t)$, $h_3(t)$, $h_4(t)$ estimates after lowpass filtering and reference.

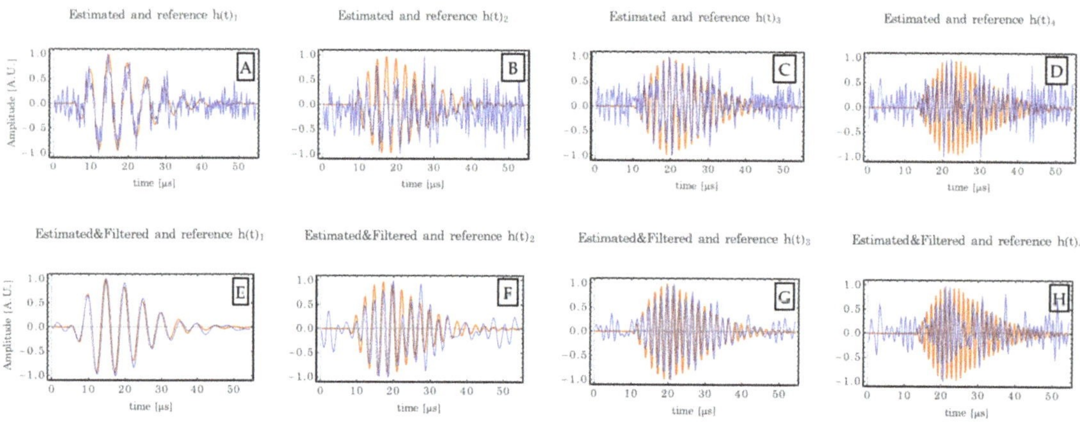

Figure 6. Estimated $h_k(t)$ kernels in the case $SNR_1 = 5$ [dB] (blue lines) and reference kernels (red lines). Plots (**A–D**): $h_1(t)$, $h_2(t)$, $h_3(t)$, $h_4(t)$ estimates before lowpass filtering and reference. Plots (**E–H**): $h_1(t)$, $h_2(t)$, $h_3(t)$, $h_4(t)$ estimates after lowpass filtering and reference.

In order to give a quantitative evaluation of the improvement effect allowed by the filtering technique proposed in this paper, we adopted the performance index proposed in [21]:

$$PI_{h_k} = \frac{\sum_{j=0}^{N_s}\left[\hat{h}_k[j]-h_k[j]\right]^2}{\sum_{j=0}^{N_s}[h_k[j]]^2} \qquad k = 1, \ldots, N_H$$
$$PI_3 = \frac{1}{N_H}\sum_{k=1}^{N_H}\left(PI_{h_k}\right) = \frac{1}{N_H}\sum_{k=1}^{N_H}\left(\frac{\sum_{j=0}^{N_s}\left[\hat{h}_k[j]-h_k[j]\right]^2}{\sum_{j=0}^{N_s}[h_k[j]]^2}\right) \tag{12}$$

where $h_k[j]$ is the $k-th$ reference kernel and $\hat{h}_k[j]$ is its estimate, N_s is the number of samples of both the actual and estimated kernels, and N_H is the order of the Hammerstein model.

The values of the indices calculated both before and after filtering for the two cases of signal to noise ratio equal to 10 [dB] and 5 [dB] are shown in Tables 2 and 3.

Table 2. Values [dB] of the performance indexes defined by (12): case of SNR = 10 [dB].

***SNR* = 10 dB**	PI_{h_1}	PI_{h_2}	PI_{h_3}	PI_{h_4}	PI_3
pre filter	−7.37	−0.61	−0.58	2.92	−0.12
filtered	−14.74	−8.91	−10.76	−4.13	−8.00

Table 3. Values [dB] of the performance indexes defined by (12): case of SNR = 5 [dB].

SNR = 5 **dB**	PI_{h_1}	PI_{h_2}	PI_{h_3}	PI_{h_4}	PI_3
pre filter	−0.99	2.02	2.80	4.07	2.33
filtered	−8.88	−2.78	−4.58	0.10	−2.95

It is evident from the values shown in Tables 2 and 3 that both in the less noisy case and in the noisier one, the action of filtering is such as to significantly reduce the index value, and, therefore, improve the signal to noise ratio obtained.

3.3. Mitigation of Noise Effects: Ultrasonic System with in-Air Propagation

A verification of the proposed procedures was also carried out in the case of laboratory-acquired data. The measurements were performed on an ultrasonic acquisition system designed for in-air propagation, as shown in Figure 7. The measurement setup consists, in sequence, of a personal computer (PC) for management and supervision; an HS5 TiePie handyscope used as a generator of arbitrary waveforms; a Falco-System power amplifier; a pair of identical non-contact point focused ultrasonic transducers by Ultran, used in transmission and reception; again the TiePie, used this time as a data logger; and the PC for data storage. The ultrasonic probes operate at a center frequency of 200 KHz, and the emitted beam is focused at a single point (F = 10 cm). The probes allow a maximum input voltage of 150 V. The Falco-System power amplifier was used with an amplification factor of 50×. The probes, a pair of ULTRAN NCG 200-D25 P100 focused probes, were mounted in a through-transmission configuration (through air alone) on precision mounts and placed at a distance of approximately 21 cm, the distance at which the received signal was maximized.

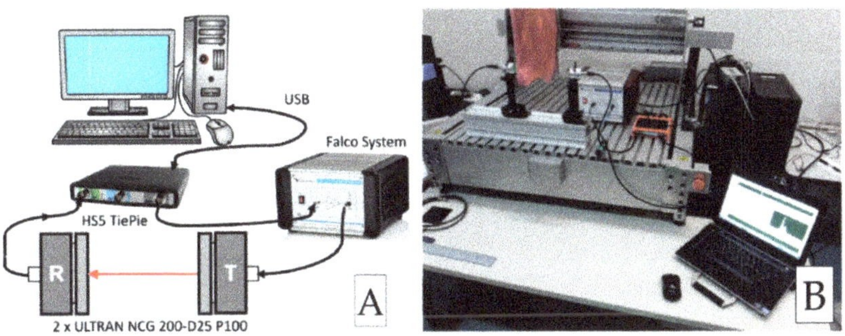

Figure 7. Schematic representation (**A**) and image (**B**) of the Ultrasonic non-contact bench.

The identification step was performed using an exponential swept sine signal of amplitude R = 1. It was assumed that the system could be usefully characterized by a swept sine signal of duration 9.3 [ms] operating in the range between 50 kHz and 400 kHz, sampled at 5 Ms/s. The order adopted for the Hammerstein model was $N_H = 4$. Several measurements were carried out by slightly modifying the distance of the probes in order to change the level of useful signal received, and therefore the SNR ratio through slight defocusing. We report in the following the results of two experiments, which differ only in the value of the SNR ratio that characterizes them.

Figure 8 shows, in an expanded scale, the trend of the $g_i(t)$ functions for the four orders considered in the case of the lower SNR value (estimated Peak Signal to Noise ratio on $y_{H\,N}(t)$ = 20 [dB]).

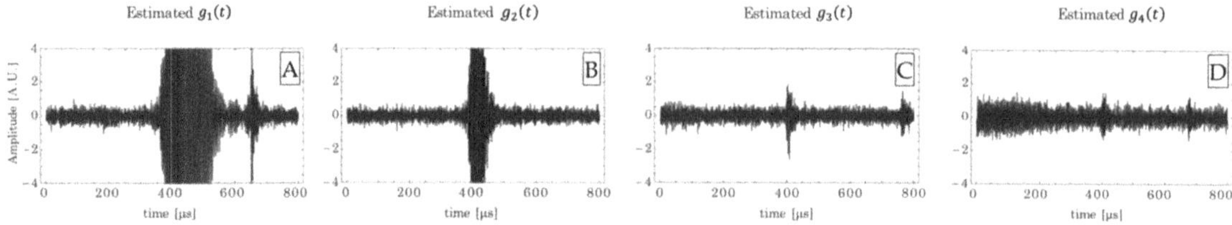

Figure 8. Non-contact point focused ultrasonic probes experiment: $g_i(t)$ functions for the four orders considered (expanded representation scale) in the case of the lower SNR value (20 [dB]). Plots (**A–D**): functions $g_1(t)$, $g_2(t)$, $g_3(t)$, $g_4(t)$, respectively.

They were obtained by taking the appropriate sections of the signal at the output to the matched filter. A first observation concerns the fact that even in the real data case, the hypothesis seems to be verified—the noise power on the function $g_i(t)$ is independent of the order of the function. The variance of the noise superimposed on the $g_i(t)$ functions, estimated in segments of the output signal $u(t)$ preceding the instants Δt_k and averaged over 50 trials, gives the following values:

$$\left[\sigma^2{}_g\right] = \begin{bmatrix} 0.47 & 0.49 & 0.43 & 0.44 \end{bmatrix}^T$$

The corresponding functions $h_k(t)$ are shown in Figure 9: In analogy with Figures 6 and 7, the first line of Figure 9 shows the trend of the functions $h_k(t)$ before the lowpass filtering operation. The second line shows the trends of the same functions after they have been lowpass filtered: Even at visual inspection, the improvement of the signal on the first three functions clearly appears, while it is also evident from $h_4(t)$, that if the noise level exceeds the useful signal, the filtering procedure can remove part of the noise, but cannot fully highlight the trend of the desired function, where present. We observe also in this real data case, that the filtering operation has not altered the temporal trend of the $h_k(t)$ functions, as far as it is possible to appreciate in cases where the noise level is not excessively high. The variance of the noise superimposed on the $h_k(t)$ functions in the case of the lower SNR value, estimated in the initial portion of the sequences and averaged over 50 trials, gives the values reported in Table 4: The positive effect of filtering is evident in all four orders. Improvements of different magnitudes are apparent for different orders: this will be discussed in the following section.

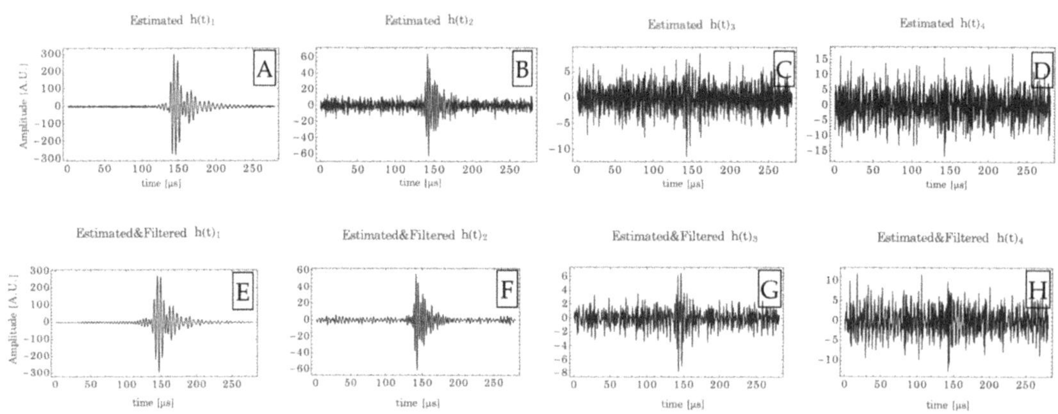

Figure 9. Estimated $h_k(t)$ kernels in the case of the ultrasonic acquisition system designed for in-air propagation in the case of the lower SNR value (20 [dB]). Plots (**A–D**): $h_1(t)$, $h_2(t)$, $h_3(t)$, $h_4(t)$ estimates before lowpass filtering. Plots (**E–H**): $h_1(t)$, $h_2(t)$, $h_3(t)$, $h_4(t)$ estimates after lowpass filtering.

Table 4. Noise variance estimates on the $h_k(t)$ in the absence and presence of low-pass filtering in the case of the lower SNR value.

	$h_1(t)$	$h_2(t)$	$h_3(t)$	$h_4(t)$
No filter	4.69	29.15	5.25	25.95
Low pass filter	3.70	5.31	1.22	11.91

Figure 10 shows, in an expanded scale, the trend of the $g_i(t)$ functions for the four orders considered in the case of the higher SNR value (estimated Peak Signal To Noise ratio on $y_{H\ N}(t)$ = 31 [dB]).

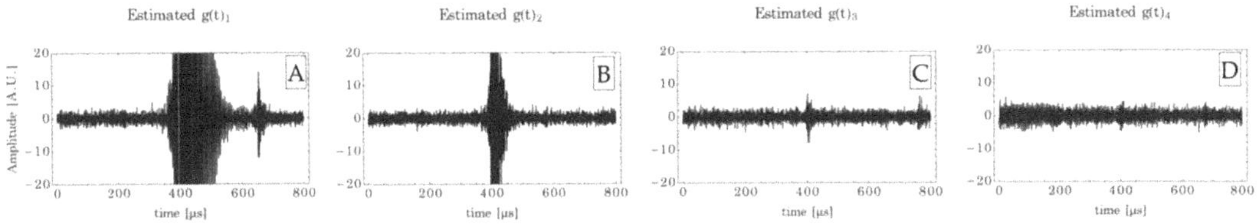

Figure 10. Non-contact point focused ultrasonic probes experiment: $g_i(t)$ functions for the four orders considered (expanded representation scale) in the case of the higher SNR value (31 [dB]). Plots (**A–D**): functions $g_1(t)$, $g_2(t)$, $g_3(t)$, $g_4(t)$, respectively.

The corresponding functions $h_k(t)$ are shown in Figure 11, which is organized as in the case of Figure 9.

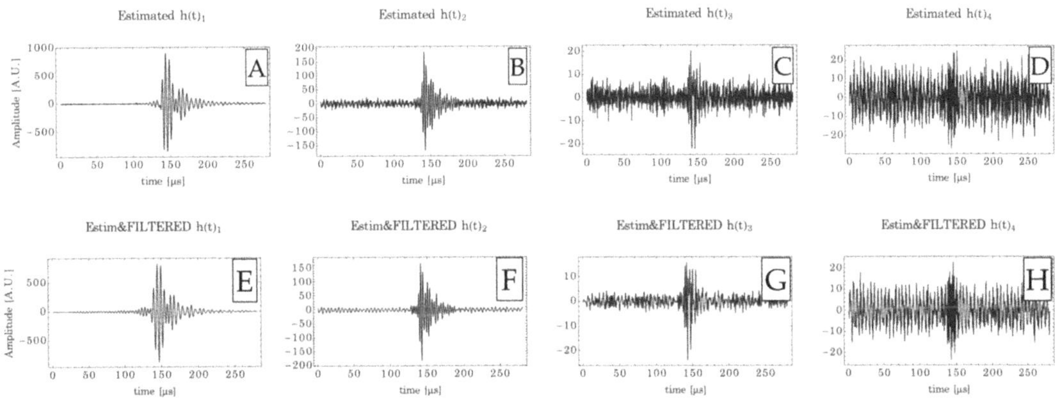

Figure 11. Estimated $h_k(t)$ kernels in the case of the ultrasonic acquisition system designed for in-air propagation in the case of the higher SNR value (31 [dB]). Plots (**A–D**): $h_1(t)$, $h_2(t)$, $h_3(t)$, $h_4(t)$ estimates before lowpass filtering. Plots (**E–H**): $h_1(t)$, $h_2(t)$, $h_3(t)$, $h_4(t)$ estimates after lowpass filtering.

The variance of the noise superimposed on the $g_i(t)$ functions in the case of the higher SNR value, estimated in segments of the output signal $u(t)$ preceding the instants Δt_k and averaged over 50 trials, gives the following values:

$$\left[\sigma^2{}_g\right] = \left[\begin{array}{cccc} 1.28 & 1.14 & 1.65 & 3.81 \end{array}\right]^T$$

The variance of the noise superimposed on the $h_k(t)$ functions in the case of the higher SNR value, estimated in the initial portion of the sequences and averaged over 50 trials, gives the values reported in Table 5: The positive effect of filtering is evident in all four orders. Improvements of different magnitude are apparent for different orders: This will be discussed in the following section.

Table 5. Noise variance estimates on the $h_k(t)$ in the absence and presence of low-pass filtering in the case of the higher SNR value (31 [dB]).

	$h_1(t)$	$h_2(t)$	$h_3(t)$	$h_4(t)$
No filter	8.05	68.60	12.21	63.62
Low pass filter	5.84	35.59	7.20	45.04

It can be interesting to compare the results obtained using the method we propose in the present paper with those obtained using the method proposed in [20]. Table 6 shows the results obtained by applying, on the same experimental setup, the noise reduction technique proposed in [20] in the case of 20 repetitions. We first observe that the comparison assumes the applicability of the method proposed in [20], and thus, that the nonlinear system is stationary, at least for the duration of the repetitions.

Table 6. Noise variance estimates on the $h_k(t)$ in the absence of processing and averaging the estimates over 20 repetitions; case of the higher SNR value (31 [dB]).

	$h_1(t)$	$h_2(t)$	$h_3(t)$	$h_4(t)$
No processing	8.05	68.60	12.21	63.62
20 repetitions	6.05	54.28	8.72	51.36

Table 6 shows that in the cases of the estimation of $h_1(t)$ and $h_3(t)$, the results obtained are comparable with those of the method proposed in this paper, but considerably lower than what should be theoretically expected in the presence of Gaussian white noise.

In the cases of $h_2(t)$ and $h_4(t)$ estimation, the improvement obtained with the averaging technique is in line with theoretical expectations. This can be explained by analyzing the trends over time, for example in Figure 11: The residual variability is related to an oscillation and not to white noise; as shown in [22], this variability of the signal results directly from the limited bandwidth of $\hat{\delta}(t)$, the approximated delta function, and thus from limitations of the identification technique, and is not associated with random noise; therefore, it cannot be removed either with filtering techniques such as those proposed in this paper or by averaging over repetitions of the experiment. If we want to draw conclusions about the comparison between the two methods, we can say that, where the noise level is very low, the two methods are equivalent; the method proposed in this work, in addition to not assuming characteristics of stationarity of the system, allows to obtain better results in the presence of high noise variance compared to the signal level.

4. Discussion and Conclusions

The various experiments we made were aimed at verifying the different aspects that had been highlighted theoretically in the paper. The correspondence between the noise superimposed on the $g_i(t)$ functions and the noise superimposed on the $h_k(t)$ functions, estimated from the $g_i(t)$ functions, was verified through the tests performed in Section 3.1: The estimates made on the noise sequences superimposed on the $g_i(t)$ functions and the $h_k(t)$ functions showed to be in line with what could be theoretically predicted using relation (10), confirming the validity of the proposed technique.

The subsequent Sections 3.2 and 3.3 had instead the purpose of verifying the theoretical hypotheses made on a possible improvement in the estimation of the $h_k(t)$ parameters, in the presence of noise, through a low-pass filtering operation. As a preliminary remark, since we wanted to consider the time behavior of the $h_k(t)$ functions, we chose to perform low-pass filtering through linear phase FIR filters, each with a specifically defined cutoff frequency related to the kernel order k, in order to preserve, as far as possible, the time course of the $h_k(t)$ functions linked to the harmonic contents present in the pass band of each low-pass filter. Both in the case of synthetic data and in the case of real data it was possible to find that the filtering operations significantly improve the quality of the estimate and do not significantly alter the $h_k(t)$ functions.

The results in Tables 4 and 5 deserve specific comment. The variance of the noise superimposed on the functions $h_k(t)$, estimated in the initial portion of the sequences and averaged over 50 trials, gives the values reported in the table: The positive effect of filtering is evident in all four cases and for both SNR levels, but especially in the cases of order 2 and 4. This effect, so different among the different orders, can be easily explained: In the calculation of $h_2(t)$ and $h_4(t)$, the transformation (7) implies a strong emphasis (with coefficient $|8|$) of the function $g_4(t)$ (the one of highest order $N_H = 4$) and therefore of the noise associated with it: The high variance value of the noise on $h_2(t)$ and $h_4(t)$ before filtering can be ascribed to this aspect. The same function $g_4(t)$ is the one associated to the fourth harmonic, so the corresponding noise components have an harmonic content that includes high frequencies: They can be easily removed by the low pass filtering operation. This also explains the strong reduction, as a consequence of filtering, of the estimated variance.

The very good results we presented encourage us to work on further extensions of the research activity; we plan to use the techniques proposed in this paper in new experimental setups for further verifications. In addition, we want to test how these techniques combine with the filtering techniques proposed in [22] for the elimination of oscillations due to the band-limited swept sine excitation signal. Extending those techniques to the case of nonlinear device identification in the presence of noise could bring further improvements in the Hammerstein model identification step of a nonlinear system.

Author Contributions: P.B. and M.C. equally contributed to all aspects of the paper. All authors have read and agreed to the published version of the manuscript.

Funding: The authors acknowledge the support of the *Fondazione Cassa di Risparmio di Terni* for the partial financing of the research through the project *"HydroTOUR-Hydrogen Terni Orizzonte Università e Ricerca"*.

Institutional Review Board Statement: Not applicable.

Informed Consent Statement: Not applicable.

Data Availability Statement: The corresponding author can be contacted.

Acknowledgments: The authors acknowledge the support of L. Senni and S. Laureti for the acquisition of laboratory data. The authors thank the anonymous reviewers for their comments, which helped to make the paper more complete and clear.

Conflicts of Interest: The authors declare no conflict of interest.

References

1. Murray-Smith, R.; Johansen, T. (Eds.) *Multiple Model Approaches to Nonlinear Modelling and Control*; CRC Press: Boca Raton, FL, USA, 2020.
2. Karjalainen, M.; Pakarinen, J. Wave digital simulation of a vacuum-tube amplifier. In Proceedings of the 2006 IEEE International Conference on Acoustics Speech and Signal Processing Proceedings, Toulouse, France, 14–19 May 2006; Volume 5, pp. 153–156.
3. Dunkel, W.R.; Rest, M.; Werner, K.J.; Olsen, M.J.; Smith, J.O. The fender Bassman 5F6-A family of preamplifier circuits—A wave digital filter case study. In Proceedings of the 19th International Conference on Digital Audio Effects (DAFx-16), Brno, Czech Republic, 5–9 September 2016; pp. 263–270.
4. Karimov, T.I.; Butusov, D.N.; Karimov, A.I. Computer simulation of audio circuits with vacuum tubes. In Proceedings of the 2016 XIX IEEE International Conference on Soft Computing and Measurements (SCM), Saint Petersburg, Russia, 25–27 May 2016; pp. 114–116.
5. Romoli, L.; Gasparini, M.; Cecchi, S.; Primavera, A.; Piazza, F. Adaptive identification of nonlinear models using orthogonal nonlinear functions. In Proceedings of the 48th International Conference: Automotive Audio, Munich, Germany, 21–23 September 2012; pp. 1–7.
6. Gasparini, M.; Primavera, A.; Romoli, L.; Cecchi, S.; Piazza, F. System identification based on hammerstein models using cubic splines. In Proceedings of the Audio Engineering Society Convention 134, Rome, Italy, 4–7 May 2013; pp. 1–8.

7. Orcioni, S.; Terenzi, A.; Cecchi, S.; Piazza, F.; Carini, A. Identification of volterra models of tube audio devices using multiple-variance method. *J. Audio Eng. Soc.* **2018**, *66*, 823–838. [CrossRef]
8. Boyd, S.; Chua, L. Fading memory and the problem of approximating nonlinear operators with volterra series. *IEEE Trans. Circuits Syst.* **1985**, *32*, 1150–1161. [CrossRef]
9. Burrascano, P.; Terenzi, A.; Cecchi, S.; Ciuffetti, M.; Spinsante, S. A swept-sine type single measurement to estimate intermodulation distortion in a dynamic range of audio signal amplitudes. *IEEE Trans. Instrum. Meas.* **2021**, *70*, 1–11. [CrossRef]
10. Burrascano, P.; Laureti, S.; Ricci, M.; Terenzi, A.; Cecchi, S.; Spinsante, S.; Piazza, F. A swept-sine pulse compression procedure for an effective measurement of intermodulation distortion. *IEEE Trans. Instrum. Meas.* **2019**, *69*, 1708–1719. [CrossRef]
11. Burrascano, P.; Ciuffetti, M. Early detection of defects through the identification of distortion characteristics in ultrasonic responses. *Mathematics* **2021**, *9*, 850. [CrossRef]
12. Burrascano, P.; Di Capua, G.; Femia, N.; Laureti, S.; Ricci, M. A pulse compression procedure for power inductors modeling up to moderate non-linearity. *Integration* **2019**, *66*, 16–23. [CrossRef]
13. Schoukens, M.; Tiels, K. Identification of block-oriented nonlinear systems starting from linear approximations: A survey. *Automatica* **2017**, *85*, 272–292. [CrossRef]
14. Farina, A. Simultaneous measurement of impulse response and distortion with a swept-sine technique. In Proceedings of the Audio Engineering Society Convention 108, Paris, France, 19–22 February 2000.
15. Novak, A.; Simon, L.; Kadlec, F.; Lotton, P. Nonlinear system identification using exponential swept-sine signal. *IEEE Trans. Instrum. Meas.* **2009**, *59*, 2220–2229. [CrossRef]
16. Rébillat, M.; Hennequin, R.; Corteel, E.; Katz, B.F. Identification of cascade of Hammerstein models for the description of nonlinearities in vibrating devices. *J. Sound Vib.* **2011**, *330*, 1018–1038. [CrossRef]
17. Haist, N.; Chang, F.; Luus, R. Nonlinear identification in the presence of correlated noise using a Hammerstein model. *IEEE Trans. Autom. Control.* **1973**, *18*, 552–555. [CrossRef]
18. Zhang, E.; Schoukens, M.; Schoukens, J. Structure detection of Wiener–Hammerstein systems with process noise. *IEEE Trans. Instrum. Meas.* **2017**, *66*, 569–576. [CrossRef]
19. Ghrib, M.; Rébillat, M.; Des Roches, G.V.; Mechbal, N. Automatic damage type classification and severity quantification using signal based and nonlinear model based damage sensitive features. *J. Process. Control.* **2019**, *83*, 136–146. [CrossRef]
20. Rebillat, M.; Ege, K.; Gallo, M.; Antoni, J. Repeated exponential swept sines for the autonomous estimation of nonlinearities and bootstrap assessment of uncertainties. *Proc. Inst. Mech. Eng. Part C J. Mech. Eng. Sci.* **2016**, *230*, 1007–1018. [CrossRef]
21. Rebillat, M.; Schoukens, M. Comparison of least squares and exponential swept sine methods for parallel Hammerstein models estimation. *Mech. Syst. Signal Process.* **2018**, *104*, 851–865. [CrossRef]
22. Burrascano, P.; Laureti, S.; Senni, L.; Ricci, M. Pulse compression in nondestructive testing applications: Reduction of near sidelobes exploiting reactance transformation. *IEEE Trans. Circuits Syst. I Regul. Pap.* **2018**, *66*, 1886–1896. [CrossRef]

applied sciences

MDPI

Article

Marchenko Green's Function Retrieval in Layered Elastic Media from Two-Sided Reflection and Transmission Data

Joost Van der Neut [1,*], Joeri Brackenhoff [2], Giovanni Angelo Meles [3], Evert Slob [1] and Kees Wapenaar [1]

[1] Department of Geoscience and Engineering, Delft University of Technology, Stevinweg 1, 2628 CN Delft, The Netherlands
[2] Institut für Geophysik, ETH Zürich, Sonneggstrasse 5, 8092 Zürich, Switzerland
[3] Institute of Earth Sciences, University of Lausanne, 1015 Lausanne, Switzerland
* Correspondence: j.r.vanderneut@tudelft.nl

Abstract: By solving a Marchenko equation, Green's functions at an arbitrary (inner) depth level inside an unknown elastic layered medium can be retrieved from single-sided reflection data, which are collected at the top of the medium. To date, it has only been possible to obtain an exact solution if the medium obeyed stringent monotonicity conditions and if all forward-scattered (non-converted and converted) transmissions between the acquisition level and the inner depth level were known a priori. We introduce an alternative Marchenko equation by revising the window operators that are applied in its derivation. We also introduce an auxiliary equation for transmission data, which are collected at the bottom of the medium, and a coupled equation, which is based on both reflection and transmission data. We show that the joint system of the Marchenko equation, the auxiliary equation and the coupled equation can be succesfully inverted when broadband reflection and transmission data are available. This results in a novel methodology for elastodynamic Green's function retrieval from two-sided data. Apart from these data, our approach requires P- and S-wave transmission times between the inner depth level and the top of the medium, as well as two angle-dependent amplitude scaling factors, which can be estimated from the data by enforcing energy conservation.

Keywords: Marchenko equation; Green's function retrieval; elastodynamic wave propagation

Citation: Van der Neut, J.; Brackenhoff, J.; Meles, G. A.; Slob, E.; Wapenaar, K. Marchenko Green's Function Retrieval in Layered Elastic Media from Two-Sided Reflection and Transmission Data. *Appl. Sci.* **2022**, *12*, 7824. https://doi.org/10.3390/app12157824

Academic Editor: Michel Darmon

Received: 8 July 2022
Accepted: 1 August 2022
Published: 4 August 2022

1. Introduction

Inversion of the Marchenko equation has proven to be an effective tool for the retrieval of Green's functions in an unknown acoustic medium from single-sided reflection data [1,2]. For an introduction to this subject, the numerical implementation of the Marchenko equation, field data applications and recent developments, see [3–11], respectively. An equivalent (Marchenko) equation has also been derived for wave propagation in elastic media [12–14]. Inversion of this equation requires a priori knowledge of all forward-scattered (non-converted and converted) waveforms [15]. Moreover, a unique solution can only be obtained if the medium obeys stringent monotonicity conditions [16], which are often not met in realistic scenarios. Once a solution to the Marchenko equation is found, it can be used for various purposes, such as wavefield retrieval inside an unknown medium [17], the imaging of elastic medium properties [18] or the suppression of multiple undesired reflections in reflection data [19].

In this paper, we show that the conditions for elastodynamic Green's function retrieval are significantly better when an elastic volume can be accessed from two sides, as is the case in particular laboratory experiments [20], non-destructive testing [21,22], brain imaging [23,24], transcranial ultrasound focusing [25,26], transcranial photoacoustics [27,28] and when using auxiliary downhole receivers in seismic data acquisition [29,30]. Although the underlying representations of our work could be extended to account for lateral variations [19], the presence of a free surface [31,32] and intrinsic attenuation [33,34], we restrict ourselves to a layered lossless medium for simplicity.

In Section 2, we derive a system of forward equations that relate the (multi-component) focusing function at a specified focal depth z_I to observed reflection and transmission data, which are to be acquired at depths of $z_U < z_I$ and $z_L > z_I$. In Section 3, we show how the system can be inverted for the focusing function and two unknown amplitude scaling factors, α and β, which are related to transmission losses of (non-converted) P- and S-waves, respectively, between depth levels z_I and z_U. Apart from the recorded (reflection and transmission) data, our scheme requires two direct arrivals, which are represented by pulses of unit amplitude, delayed with the (non-converted) P- and S-wave travel times from z_I to z_U. In this way, we can apply exact (data-driven) Marchenko redatuming of two-sided data in a layered elastic medium, which is the main contribution of this paper. The retrieved focusing functions can be transformed into Green's functions as if there were virtual P- or S-wave sources at z_I, which could eventually be used for imaging and inversion of the elastic medium's properties. We close the paper with a discussion in Section 4.

2. Forward Equations

After providing some preliminaries in Section 2.1, we discuss the causality cones of multi-component Green's functions and focusing functions in Section 2.2. We propose novel window operators for Green's functions (based on non-converted P-wave travel times) and focusing functions (based on non-converted S-wave travel times). With the help of these operators, we derive (reflection-based) Marchenko equations in Section 2.3, (transmission-based) auxiliary equations in Section 2.4 and (transmission- and reflection-based) coupled equations in Section 2.5. In Section 2.6, we take these equations together, leading to a joint system. Finally, we present a relation to convert focusing functions into Green's functions in Section 2.7.

2.1. Preliminaries

Let (x, y, z) be an Euclidean coordinate system with the z-axis pointing downwards, whereas t denotes time. We consider a layered lossless isotropic elastic medium, which is characterized by P-wave velocity $c^P(z)$, S-wave velocity $c^S(z)$ and mass density $\rho(z)$. Let z_U and z_L be two depth levels, which are located above and below all heterogeneties in the medium, respectively (hence, constant medium properties are assumed above z_U and below z_L). Elastodynamic wave propagation is considered in the (x, z)-plane, where wavefields are assumed to be constant in the y-direction. All wavefields are decomposed into flux-normalized up- and downgoing P-, Sv- and Sh-components in the (p, τ)-domain [35–37], where p is the rayparameter and τ is the intercept time. The Sh-components are decoupled from the P- and Sv-components and will not be considered in this paper (for notational convenience, component Sv will be referred to as S).

In Figure 1, we show a layered elastic medium, which will be used throughout our paper as a running example. For convenience, we have chosen the vertical dimension of the model to be 1 m. However, all quantities can be rescaled to fit a particular application in, e.g., ultrasound or seismology applications.More information on the design of the medium and the parameters that are used for modeling are provided in Appendix A. Our objective is to retrieve the Green's responses at z_U and z_L to a Green's source at a specified depth level z_I (see the dashed magenta line in Figure 1) from recorded (reflection and transmission) data. These Green's functions are represented by the following matrix:

$$\mathbf{G} = \begin{pmatrix} -\mathbf{G}_U^{-+} \\ -\mathbf{G}_U^{--} \\ \mathbf{G}_L^{+-} \\ \mathbf{G}_L^{++} \end{pmatrix}. \tag{1}$$

Here, $\mathbf{G}_U^{-+}$ and $\mathbf{G}_U^{--}$ are the upgoing (indicated by the first superscript $-$) Green's functions at z_U from a down- and upwards radiating (indicated by the second superscript

+ or −) virtual source at z_I, respectively. These quantities contain distinguished PP-, PS-, SP- and SS-components and are organized as

$$\mathbf{G}_U^{-\pm} = \begin{pmatrix} G_{U,PP}^{-\pm}(p,z,\tau) & G_{U,PS}^{-\pm}(p,z,\tau) \\ G_{U,SP}^{-\pm}(p,z,\tau) & G_{U,SS}^{-\pm}(p,z,\tau) \end{pmatrix}. \tag{2}$$

Here, subscripts PP and SP indicate the P- and S-wave responses to a P-wave source, whereas subscripts PS and SS indicate the P- and S-wave responses to an S-wave source. Matrices $\mathbf{G}_L^{+-}$ and $\mathbf{G}_L^{++}$ in Equation (1) represent the downgoing Green's functions at z_L from an up- and downwards-radiatingvirtual source at z_I, and are organized akin to Equation (2).

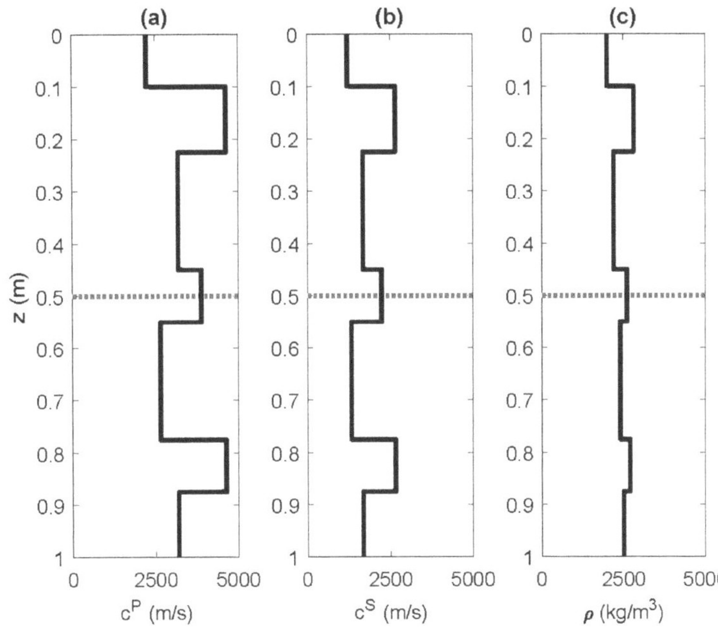

Figure 1. Example of a layered elastic medium with (**a**) P-wave velocity c^P (in m·s^{-1}), (**b**) S-wave velocity c^S (in m·s^{-1}) and (**c**) density ρ (in kg·m^{-3}) as a function of depth z (in m). Above z_U = 0 m and below z_L = 1 m, the medium is homogeneous. The dashed magenta line indicates the focusing depth z_I = 0.5 m, where a virtual source is to be constructed.

For the representations of Green's functions, we make use of so-called focusing functions [13], which are represented by the matrix

$$\mathbf{F} = \begin{pmatrix} \mathbf{F}_U^- \\ \mathcal{Z}\mathbf{F}_U^+ \\ \mathbf{F}_L^+ \\ \mathcal{Z}\mathbf{F}_L^- \end{pmatrix}. \tag{3}$$

Here, $\mathbf{F}_U^-$ and $\mathbf{F}_U^+$ are the up- and downgoing focusing functions (organized akin to Equation (2) at z_U. These functions are defined in a fictitious medium where the halfspace below z_I is homogeneous. They focus 'from above' at z_I and continue as a downgoing wavefield below this depth level (see [15] for details). Similarly, $\mathbf{F}_L^+$ and $\mathbf{F}_L^-$ are the down- and upgoing focusing functions at z_L. These functions are defined in a medium where the halfspace above z_I is homogeneous. These functions focus 'from below' at z_I and continue as an upgoing wavefield above this depth level. Finally, $\mathcal{Z}$ is an operator that reverses the signs of p and τ. For example, applying this operator to $\mathbf{F}_U^+$ yields

$$\mathcal{Z}\mathbf{F}_U^+ = \mathcal{Z}\begin{pmatrix} F_{U,PP}^+(p,z,\tau) & F_{U,PS}^+(p,z,\tau) \\ F_{U,SP}^+(p,z,\tau) & F_{U,SS}^+(p,z,\tau) \end{pmatrix} = \begin{pmatrix} F_{U,PP}^+(-p,z,-\tau) & F_{U,PS}^+(-p,z,-\tau) \\ F_{U,SP}^+(-p,z,-\tau) & F_{U,SS}^+(-p,z,-\tau) \end{pmatrix}. \quad (4)$$

Our objective is to retrieve the focusing functions and Green's functions from reflection and transmission data, to be acquired at z_U and z_L. Let $R_{U,PP}$, $R_{U,PS}$, $R_{U,SP}$ and $R_{U,SS}$ be the PP-, PS-, SP- and SS-reflection responses at z_U (for their definitions, see Appendix B.1). Based on these recordings, we can construct an operator $\mathcal{R}_U$ that convolves a wavefield with the reflection response. When applied to $\mathbf{F}_U^+$, this multidimensional convolution is defined as

$$\mathcal{R}_U\mathbf{F}_U^+ = \int_0^\tau \begin{pmatrix} R_{U,PP}(p,\tau-\tau') & R_{U,PS}(p,\tau-\tau') \\ R_{U,SP}(p,\tau-\tau') & R_{U,SS}(p,\tau-\tau') \end{pmatrix} \begin{pmatrix} F_{U,PP}^+(p,z_U,\tau') & F_{U,PS}^+(p,z_U,\tau') \\ F_{U,SP}^+(p,z_U,\tau') & F_{U,SS}^+(p,z_U,\tau') \end{pmatrix} \mathrm{d}\tau'. \quad (5)$$

Similar operators $\mathcal{R}_L$, $\mathcal{T}_{LU}$ and $\mathcal{T}_{UL}$ can be constructed for convolution with the reflection response at z_L, the transmission response from z_U to z_L and the transmission response from z_L to z_U, respectively. Apart from $\mathcal{R}_U$, $\mathcal{R}_L$, $\mathcal{T}_{LU}$, $\mathcal{T}_{UL}$ and $\mathcal{Z}$, we make use of two window operators that will be defined in the following section.

2.2. Causality Cones of Green's Functions and Focusing Functions

In moderately inhomogeneous acoustic media, the Green's function and the focusing function are separated in time, except for a single overlapping event, which is commonly referred to as the *direct wave*. In the derivation of the acoustic Marchenko equation, this observation is exploited by truncating wavefields either before [2,11] or after [38,39] the direct wave. In elastic media, there can be a multitude of overlapping events, making the situation significantly more cumbersome [16]. To illustrate this problem, we show the (symmetrized) causality cones of multi-component Green's functions and focusing functions in Figure 2. In particular, we refer the reader to the orange areas, where the Green's functions and focusing functions may overlap. Because of this potential overlap, we have designed two distinct time window operators: one for Green's functions, which is based on the (non-converted) direct P-wave travel time τ_d^P and one for focusing functions, which is based on the (non-converted) direct S-wave travel time τ_d^S.

First, we discuss the window operator for Green's functions, which is based on the travel time τ_d^P of the (non-converted) P-wave, propagating from z_I outwards. In Figure 2a, we can see that the Green's function and its time-reversed counterpart vanish in the interval $(-\tau_d^P, \tau_d^P)$. Let $\mathbf{G}_{Ud}^{--} = \begin{pmatrix} G_{Ud,PP}^{--} & 0 \\ 0 & 0 \end{pmatrix}$ and $\mathbf{G}_{Ld}^{++} = \begin{pmatrix} G_{Ld,PP}^{++} & 0 \\ 0 & 0 \end{pmatrix}$ be the components of $\mathbf{G}_U^{--}$ and $\mathbf{G}_L^{++}$ that reside at the boundary of the interval $[-\tau_d^P, \tau_d^P]$ (corresponding to the direct non-converted P-wave transmissions). Now, we can partition the Green's function that we defined earlier in Equation (1) as

$$\underbrace{\begin{pmatrix} -\mathbf{G}_U^{-+} \\ -\mathbf{G}_U^{--} \\ \mathbf{G}_L^{+-} \\ \mathbf{G}_L^{++} \end{pmatrix}}_{\mathbf{G}} = \underbrace{\begin{pmatrix} \mathbf{O} \\ -\mathbf{G}_{Ud}^{--} \\ \mathbf{O} \\ \mathbf{G}_{Ld}^{++} \end{pmatrix}}_{\mathbf{G}_d} + \underbrace{\begin{pmatrix} -\mathbf{G}_U^{-+} \\ -\mathbf{G}_{Um}^{--} \\ \mathbf{G}_L^{+-} \\ \mathbf{G}_{Lm}^{++} \end{pmatrix}}_{\mathbf{G}_m}, \quad (6)$$

where $\mathbf{G}_{Um}^{--} = \mathbf{G}_U^{--} - \mathbf{G}_{Ud}^{--}$ and $\mathbf{G}_{Lm}^{++} = \mathbf{G}_L^{++} - \mathbf{G}_{Ld}^{++}$ are referred to as the Green's function codasand $\mathbf{O}$ is a zero matrix. We design a window matrix $\mathbf{\Theta}^{[P]}$ that removes all data outside the interval $[-\tau_d^P, \tau_d^P]$. When we apply this matrix to the Green's function in Equation (6), it follows that

$$\underbrace{\begin{pmatrix} \Theta_U^{[P]} & \mathbf{O} & \mathbf{O} & \mathbf{O} \\ \mathbf{O} & \Theta_U^{[P]} & \mathbf{O} & \mathbf{O} \\ \mathbf{O} & \mathbf{O} & \Theta_L^{[P]} & \mathbf{O} \\ \mathbf{O} & \mathbf{O} & \mathbf{O} & \Theta_L^{[P]} \end{pmatrix}}_{\mathbf{\Theta}^{[P]}} \underbrace{\begin{pmatrix} -\mathbf{G}_U^{-+} \\ -\mathbf{G}_U^{--} \\ \mathbf{G}_L^{+-} \\ \mathbf{G}_L^{++} \end{pmatrix}}_{\mathbf{G}} = \underbrace{\begin{pmatrix} \mathbf{O} \\ -\mathbf{G}_{Ud}^{--} \\ \mathbf{O} \\ \mathbf{G}_{Ld}^{++} \end{pmatrix}}_{\mathbf{G_d}}. \quad (7)$$

In this formulation, $\Theta_U^{[P]}$ and $\Theta_L^{[P]}$ are operators that remove all data outside the intervals $[-\tau_{Ud}^P, \tau_{Ud}^P]$ and $[-\tau_{Ld}^P, \tau_{Ld}^P]$, respectively. Here, τ_{Ud}^P and τ_{Ld}^P are the direct P-wave travel times for propagation from z_I to z_U and z_L, respectively.

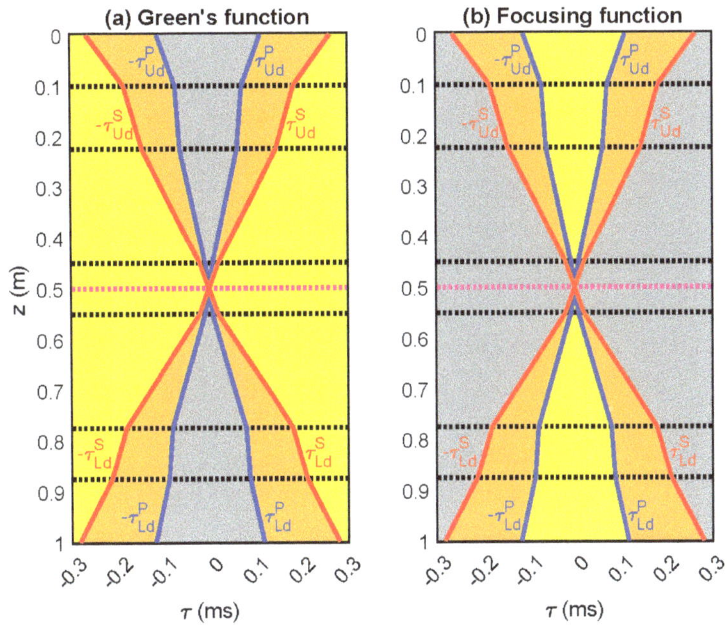

Figure 2. Symmetrized causality cone of (**a**) a Green's function (**G** or $\mathcal{Z}$**G**) and (**b**) a focusing function for the medium in Figure 1 at p = 0.2 ms·m^{-1}, with the source/focal depth at z_I = 0.5 m (indicated by the magenta dashed line; the black dashed lines indicate layer boundaries). The blue lines denote the travel times $\pm\tau_{Ud}^P$ and $\pm\tau_{Ld}^P$ of the direct (non-converted) P-wave transmissions. The red lines denote the travel times $\pm\tau_{Ud}^S$ and $\pm\tau_{Ld}^S$ of the direct (non-converted) S-wave transmissions. All wavefields are stricly zero in the gray areas (whereas they may be non-zero in the yellow and orange areas). The areas where the Green's functions and focusing functions can overlap are indicated in orange.

We proceed with the window operator for focusing functions, which is based on the travel time τ_d^S of the (non-converted) direct S-wave, propagating from z_I outwards. As illustrated in Figure 2b, the focusing functions and their time-reversed counterparts vanish outside $[-\tau_d^S, \tau_d^S]$. Let $\mathbf{F}_{Ud}^+ = \begin{pmatrix} 0 & 0 \\ 0 & F_{Ud}^{+SS} \end{pmatrix}$ and $\mathbf{F}_{Ld}^- = \begin{pmatrix} 0 & 0 \\ 0 & F_{Ld}^{-SS} \end{pmatrix}$ be the components of $\mathbf{F}_U^+$ and $\mathbf{F}_L^-$ that reside at the boundary of the interval $(-\tau_d^S, \tau_d^S)$ (corresponding to the direct non-converted S-wave transmissions). Now, we may partition the focusing function that we defined earlier in Equation (3) as

$$\underbrace{\begin{pmatrix} \mathbf{F}_U^- \\ \mathcal{Z}\mathbf{F}_U^+ \\ \mathbf{F}_L^+ \\ \mathcal{Z}\mathbf{F}_L^- \end{pmatrix}}_{\mathbf{F}} = \underbrace{\begin{pmatrix} \mathbf{O} \\ \mathcal{Z}\mathbf{F}_{Ud}^+ \\ \mathbf{O} \\ \mathcal{Z}\mathbf{F}_{Ld}^- \end{pmatrix}}_{\mathbf{F}_d} + \underbrace{\begin{pmatrix} \mathbf{F}_U^- \\ \mathcal{Z}\mathbf{F}_{Um}^+ \\ \mathbf{F}_L^+ \\ \mathcal{Z}\mathbf{F}_{Lm}^- \end{pmatrix}}_{\mathbf{F}_m}, \tag{8}$$

where $\mathbf{F}_{Um}^+ = \mathbf{F}_U^+ - \mathbf{F}_{Ud}^+$ and $\mathbf{F}_{Lm}^- = \mathbf{F}_L^- - \mathbf{F}_{Ld}^-$ are referred to as the focusing function codas. We design a window matrix $\mathbf{\Theta}^{(S)}$ that removes all data outside the interval $(-\tau_d^S, \tau_d^S)$. During the inversion that will be applied later in this paper, we wish to restrict $\mathbf{F}_m$ to the interval $(-\tau_d^S, \tau_d^S)$. To enforce this in practice, we replace $\mathbf{F}_m$ in Equation (8) with $\mathbf{\Theta}^{(S)}\mathbf{F}_m$, leading to

$$\underbrace{\begin{pmatrix} \mathbf{F}_U^- \\ \mathcal{Z}\mathbf{F}_U^+ \\ \mathbf{F}_L^+ \\ \mathcal{Z}\mathbf{F}_L^- \end{pmatrix}}_{\mathbf{F}} = \underbrace{\begin{pmatrix} \mathbf{O} \\ \mathcal{Z}\mathbf{F}_{Ud}^+ \\ \mathbf{O} \\ \mathcal{Z}\mathbf{F}_{Ld}^- \end{pmatrix}}_{\mathbf{F}_d} + \underbrace{\begin{pmatrix} \Theta_U^{(S)} & \mathbf{O} & \mathbf{O} & \mathbf{O} \\ \mathbf{O} & \Theta_U^{(S)} & \mathbf{O} & \mathbf{O} \\ \mathbf{O} & \mathbf{O} & \Theta_L^{(S)} & \mathbf{O} \\ \mathbf{O} & \mathbf{O} & \mathbf{O} & \Theta_L^{(S)} \end{pmatrix}}_{\mathbf{\Theta}^{(S)}} \underbrace{\begin{pmatrix} \mathbf{F}_U^- \\ \mathcal{Z}\mathbf{F}_{Um}^+ \\ \mathbf{F}_L^+ \\ \mathcal{Z}\mathbf{F}_{Lm}^- \end{pmatrix}}_{\mathbf{F}_m}. \tag{9}$$

In this formulation, $\Theta_U^{(S)}$ and $\Theta_L^{(S)}$ are operators that remove all data outside the intervals $(-\tau_{Ud}^S, \tau_{Ud}^S)$ and $(-\tau_{Ld}^S, \tau_{Ld}^S)$, respectively. Here, τ_{Ud}^S and τ_{Ld}^S are the direct S-wave travel times for propagation from z_I to z_U and z_L, respectively.

2.3. Marchenko Equations

In Appendix B.1, we derive the following system of Green's function representations that are based on reflection data:

$$\underbrace{\begin{pmatrix} -\mathbf{G}_U^{-+} \\ -\mathbf{G}_U^{--} \\ \mathbf{G}_L^{+-} \\ \mathbf{G}_L^{++} \end{pmatrix}}_{\mathbf{G}} = \begin{pmatrix} \mathbf{I} & -\mathcal{R}_U\mathcal{Z} & \mathbf{O} & \mathbf{O} \\ -\mathcal{R}_U\mathcal{Z} & \mathbf{I} & \mathbf{O} & \mathbf{O} \\ \mathbf{O} & \mathbf{O} & \mathbf{I} & -\mathcal{R}_L\mathcal{Z} \\ \mathbf{O} & \mathbf{O} & -\mathcal{R}_L\mathcal{Z} & \mathbf{I} \end{pmatrix} \underbrace{\begin{pmatrix} \mathbf{F}_U^- \\ \mathcal{Z}\mathbf{F}_U^+ \\ \mathbf{F}_L^+ \\ \mathcal{Z}\mathbf{F}_L^- \end{pmatrix}}_{\mathbf{F}}, \tag{10}$$

where $\mathbf{I}$ is a 2×2 identity matrix. When we apply the operator $\mathbf{\Theta}^{[P]}$ to both sides of this equation, it follows, with the help of Equation (7), that

$$\underbrace{\begin{pmatrix} \mathbf{O} \\ -\mathbf{G}_{Ud}^{--} \\ \mathbf{O} \\ \mathbf{G}_{Ld}^{++} \end{pmatrix}}_{\mathbf{G}_d} = \mathbf{\Theta}^{[P]} \begin{pmatrix} \mathbf{I} & -\mathcal{R}_U\mathcal{Z} & \mathbf{O} & \mathbf{O} \\ -\mathcal{R}_U\mathcal{Z} & \mathbf{I} & \mathbf{O} & \mathbf{O} \\ \mathbf{O} & \mathbf{O} & \mathbf{I} & -\mathcal{R}_L\mathcal{Z} \\ \mathbf{O} & \mathbf{O} & -\mathcal{R}_L\mathcal{Z} & \mathbf{I} \end{pmatrix} \underbrace{\begin{pmatrix} \mathbf{F}_U^- \\ \mathcal{Z}\mathbf{F}_U^+ \\ \mathbf{F}_L^+ \\ \mathcal{Z}\mathbf{F}_L^- \end{pmatrix}}_{\mathbf{F}}. \tag{11}$$

Next, we may substitute Equation (9) and rewrite the result as

$$\underbrace{\begin{pmatrix} \mathbf{\Theta}_U^{[P]}\mathcal{R}_U\mathbf{F}_{Ud}^+ \\ -\mathbf{G}_{Ud}^{--} \\ \mathbf{\Theta}_L^{[P]}\mathcal{R}_L\mathbf{F}_{Ld}^- \\ \mathbf{G}_{Ld}^{++} \end{pmatrix}}_{\mathbf{B}_{Mar}} = \underbrace{\mathbf{\Theta}^{[P]} \begin{pmatrix} \mathbf{I} & -\mathcal{R}_U\mathcal{Z} & \mathbf{O} & \mathbf{O} \\ -\mathcal{R}_U\mathcal{Z} & \mathbf{I} & \mathbf{O} & \mathbf{O} \\ \mathbf{O} & \mathbf{O} & \mathbf{I} & -\mathcal{R}_L\mathcal{Z} \\ \mathbf{O} & \mathbf{O} & -\mathcal{R}_L\mathcal{Z} & \mathbf{I} \end{pmatrix} \mathbf{\Theta}^{(S)}}_{\mathbf{A}_{Mar}} \underbrace{\begin{pmatrix} \mathbf{F}_U^- \\ \mathcal{Z}\mathbf{F}_{Um}^+ \\ \mathbf{F}_L^+ \\ \mathcal{Z}\mathbf{F}_{Lm}^- \end{pmatrix}}_{\mathbf{F}_m}. \tag{12}$$

Here, we have used the fact that $\tau_d^S \notin [-\tau_d^P, \tau_d^P]$, such that $\Theta_U^{[P]}\mathcal{Z}\mathbf{F}_{Ud}^+ = \mathbf{O}$ and $\Theta_L^{[P]}\mathcal{Z}\mathbf{F}_{Ld}^- = \mathbf{O}$. We refer to Equation (12) as a system of reflection-based Marchenko equations, which could be inverted for the unknown components of the focusing function $\mathbf{F}_m$. The block-diagonal structure of this system reveals that the Marchenko equations at z_U and z_L are decoupled. Operator $\mathbf{\Theta}^{(S)}$ restricts the unknown focusing function $\mathbf{F}_m$ to the interval $(-\tau_d^S, \tau_d^S)$. Matrix $\mathbf{A}_{Mar}$ projects $\mathbf{F}_m$ to the interval $[-\tau_d^P, \tau_d^P]$. As $\tau_d^P < \tau_d^S$, this leads to an underdetermined system of equations, which cannot be unconditionally inverted. We illustrate this by plotting the singular values of $\mathbf{A}_{Mar}$ in Figure 3a (red curve) for data from the model that we presented above in Figure 1. In this case, $\mathbf{A}_{Mar}$ contains 2192 columns but only 928 independent rows. To increase the rank of this matrix, we propose to use auxiliary transmission data, for which we derive a similar system of equations in the following section.

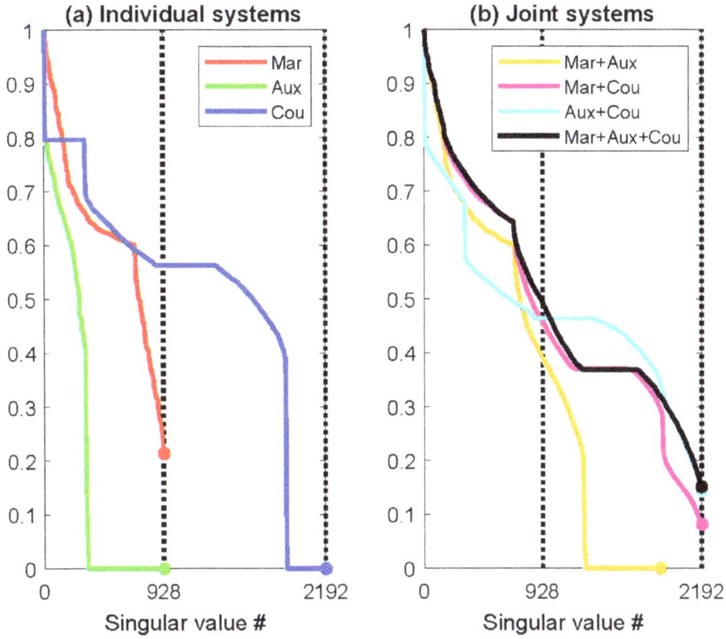

Figure 3. (**a**) Singular values of the matrices $\mathbf{A}_{Mar}$ (size: 928 × 2192), $\mathbf{A}_{Aux}$ (size: 928 × 2192) and $\mathbf{A}_{Cou}$ (size: 16,384 ×2192) for the model in Figure 1 at p = 0.2 ms·m^{-1}. (**b**) Singular values after concatenating various combinations of the matrices in (**a**). All curves have been normalized with respect to the highest singular value. The dots indicate the lowest singular values in the matrices.

2.4. Auxiliary Equations

In Appendix B.2, we derive the following system of Green's function representations that are based on transmission data:

$$\underbrace{\begin{pmatrix} -\mathbf{G}_U^{-+} \\ -\mathbf{G}_U^{--} \\ \mathbf{G}_L^{+-} \\ \mathbf{G}_L^{++} \end{pmatrix}}_{\mathbf{G}} = \begin{pmatrix} \mathbf{O} & \mathbf{O} & \mathcal{T}_{UL}\mathcal{Z} & \mathbf{O} \\ \mathbf{O} & \mathbf{O} & \mathbf{O} & \mathcal{T}_{UL}\mathcal{Z} \\ \mathcal{T}_{LU}\mathcal{Z} & \mathbf{O} & \mathbf{O} & \mathbf{O} \\ \mathbf{O} & \mathcal{T}_{LU}\mathcal{Z} & \mathbf{O} & \mathbf{O} \end{pmatrix} \underbrace{\begin{pmatrix} \mathbf{F}_U^- \\ \mathcal{Z}\mathbf{F}_U^+ \\ \mathbf{F}_L^+ \\ \mathcal{Z}\mathbf{F}_L^- \end{pmatrix}}_{\mathbf{F}}. \tag{13}$$

When we apply operator $\mathbf{\Theta}^{[P]}$ to both sides of this equation, it follows, with the help of Equation (7), that

$$\underbrace{\begin{pmatrix} \mathbf{O} \\ -\mathbf{G}_{Ud}^{--} \\ \mathbf{O} \\ \mathbf{G}_{Ld}^{++} \end{pmatrix}}_{\mathbf{G}_d} = \mathbf{\Theta}^{[P]} \begin{pmatrix} \mathbf{O} & \mathbf{O} & \mathcal{T}_{UL}\mathcal{Z} & \mathbf{O} \\ \mathbf{O} & \mathbf{O} & \mathbf{O} & \mathcal{T}_{UL}\mathcal{Z} \\ \mathcal{T}_{LU}\mathcal{Z} & \mathbf{O} & \mathbf{O} & \mathbf{O} \\ \mathbf{O} & \mathcal{T}_{LU}\mathcal{Z} & \mathbf{O} & \mathbf{O} \end{pmatrix} \underbrace{\begin{pmatrix} \mathbf{F}_U^- \\ \mathcal{Z}\mathbf{F}_U^+ \\ \mathbf{F}_L^+ \\ \mathcal{Z}\mathbf{F}_L^- \end{pmatrix}}_{\mathbf{F}}. \tag{14}$$

When we substitute Equation (9), we find eventually that

$$\underbrace{\begin{pmatrix} \mathbf{O} \\ -\mathbf{G}_{Ud}^{--} - \Theta_U^{[P]}\mathcal{T}_{UL}\mathbf{F}_{Ld}^- \\ \mathbf{O} \\ \mathbf{G}_{Ld}^{++} - \Theta_L^{[P]}\mathcal{T}_{LU}\mathbf{F}_{Ud}^+ \end{pmatrix}}_{\mathbf{B}_{Aux}} = \underbrace{\mathbf{\Theta}^{[P]} \begin{pmatrix} \mathbf{O} & \mathbf{O} & \mathcal{T}_{UL}\mathcal{Z} & \mathbf{O} \\ \mathbf{O} & \mathbf{O} & \mathbf{O} & \mathcal{T}_{UL}\mathcal{Z} \\ \mathcal{T}_{LU}\mathcal{Z} & \mathbf{O} & \mathbf{O} & \mathbf{O} \\ \mathbf{O} & \mathcal{T}_{LU}\mathcal{Z} & \mathbf{O} & \mathbf{O} \end{pmatrix} \mathbf{\Theta}^{(S)}}_{\mathbf{A}_{Aux}} \underbrace{\begin{pmatrix} \mathbf{F}_U^- \\ \mathcal{Z}\mathbf{F}_{Um}^+ \\ \mathbf{F}_L^+ \\ \mathcal{Z}\mathbf{F}_{Lm}^- \end{pmatrix}}_{\mathbf{F}_m}. \tag{15}$$

We refer to 15 as a system of auxiliary equations, which can be interpreted as a transmission-based inverse problem for $\mathbf{F}_m$. In Figure 3a (green curve), we show that the governing matrix $\mathbf{A}_{Aux}$ of this problem is rank-deficient (at least for the medium in Figure 1). Nevertheless, this matrix can provide complementary information to $\mathbf{A}_{Mar}$, as illustrated in Figure 3b (orange curve). In this case, we have concatenated the rows of $\mathbf{A}_{Mar}$ and $\mathbf{A}_{Aux}$, leading to a matrix of rank > 928. However, the rank is still far below 2192, which is the number of unknowns for this problem. In the next section, we show how we can improve on this by coupling reflection and transmission data, leading to yet another system of equations.

2.5. Coupled Equations

It is observed that the Green's function matrix $\mathbf{G}$ can be eliminated from the reflection- and transmission-based representations by subtracting Equation (13) from Equation (10). This leads to

$$\underbrace{\begin{pmatrix}\mathbf{O}\\\mathbf{O}\\\mathbf{O}\\\mathbf{O}\end{pmatrix}}_{\mathbf{O}} = \frac{1}{2}\begin{pmatrix}\mathbf{I} & -\mathcal{R}_U\mathcal{Z} & -\mathcal{T}_{UL}\mathcal{Z} & \mathbf{O}\\ -\mathcal{R}_U\mathcal{Z} & \mathbf{I} & \mathbf{O} & -\mathcal{T}_{UL}\mathcal{Z}\\ -\mathcal{T}_{LU}\mathcal{Z} & \mathbf{O} & \mathbf{I} & -\mathcal{R}_L\mathcal{Z}\\ \mathbf{O} & -\mathcal{T}_{LU}\mathcal{Z} & -\mathcal{R}_L\mathcal{Z} & \mathbf{I}\end{pmatrix}\underbrace{\begin{pmatrix}\mathbf{F}_U^-\\ \mathcal{Z}\mathbf{F}_U^+\\ \mathbf{F}_L^+\\ \mathcal{Z}\mathbf{F}_L^-\end{pmatrix}}_{\mathbf{F}}. \tag{16}$$

Here, we divided by a factor 2 to achieve a better amplitude balance with the matrices that were derived in the previous sections. After the substitution of Equation (9), we obtain

$$\underbrace{\frac{1}{2}\begin{pmatrix}\mathcal{R}_U\mathbf{F}_{Ud}^+\\ -\mathcal{Z}\mathbf{F}_{Ud}^+ + \mathcal{T}_{UL}\mathbf{F}_{Ld}^-\\ \mathcal{R}_L\mathbf{F}_{Ld}^-\\ -\mathcal{Z}\mathbf{F}_{Ld}^- + \mathcal{T}_{LU}\mathbf{F}_{Ud}^+\end{pmatrix}}_{\mathbf{B}_{Cou}} = \underbrace{\frac{1}{2}\begin{pmatrix}\mathbf{I} & -\mathcal{R}_U\mathcal{Z} & -\mathcal{T}_{UL}\mathcal{Z} & \mathbf{O}\\ -\mathcal{R}_U\mathcal{Z} & \mathbf{I} & \mathbf{O} & -\mathcal{T}_{UL}\mathcal{Z}\\ -\mathcal{T}_{LU}\mathcal{Z} & \mathbf{O} & \mathbf{I} & -\mathcal{R}_L\mathcal{Z}\\ \mathbf{O} & -\mathcal{T}_{LU}\mathcal{Z} & -\mathcal{R}_L\mathcal{Z} & \mathbf{I}\end{pmatrix}\mathbf{\Theta}^{(S)}}_{\mathbf{A}_{Cou}}\underbrace{\begin{pmatrix}\mathbf{F}_U^-\\ \mathcal{Z}\mathbf{F}_{Um}^+\\ \mathbf{F}_L^+\\ \mathcal{Z}\mathbf{F}_{Lm}^-\end{pmatrix}}_{\mathbf{F}_m}. \tag{17}$$

We refer to 17 as a system of coupled equations, which can be interpreted as another inverse problem for $\mathbf{F}_m$. Although we have increased the number of rows signficantly (up to 16,384 in our running example) by not applying the window operator $\Theta^{[P]}$, the matrix $\mathbf{A}_{Cou}$ is still rank-deficient, as shown in Figure 3a (blue curve). However, adding the rows of either $\mathbf{A}_{Mar}$ or $\mathbf{A}_{Aux}$ to the rows of $\mathbf{A}_{Cou}$ results in a full-rank matrix, as illustrated by the magenta and cyan curves in Figure 3b. An intuitive understanding of this observation is that the subtraction of Equation (13) from Equation (10) (which was required for the construction of $\mathbf{A}_{Cou}$) has reduced the row space of our system matrix, which can be compensated for by adding complementary rows from either the Marchenko or auxiliary system.

2.6. Joint System of Equations

Although the concatenation of matrix $\mathbf{A}_{Cou}$ and either $\mathbf{A}_{Aux}$ or $\mathbf{A}_{Mar}$ seems sufficient by itself to construct a full-rank matrix, we choose to merge all three matrices, leading to the overall system

$$\underbrace{\begin{pmatrix}\mathbf{B}_{Mar}\\ \mathbf{B}_{Aux}\\ \mathbf{B}_{Cou}\end{pmatrix}}_{\mathbf{B}} = \underbrace{\begin{pmatrix}\mathbf{A}_{Mar}\\ \mathbf{A}_{Aux}\\ \mathbf{A}_{Cou}\end{pmatrix}}_{\mathbf{A}}\mathbf{F}_m. \tag{18}$$

As indicated by the black curve in Figure 3b, matrix $\mathbf{A}$ has full rank, and hence can be inverted. When we apply singular-value decomposition $\mathbf{A} = \mathbf{U}\mathbf{\Sigma}\mathbf{V}^t$ and define the pseudo-inverse as $\mathbf{A}^\ddagger = \mathbf{V}\mathbf{\Sigma}^\ddagger\mathbf{U}^t$ (where $\mathbf{\Sigma}^\ddagger$ contains the reciprocals of all non-zero singular values), we may now write $\mathbf{F}_m = \mathbf{A}^\ddagger\mathbf{B}$. Akin to the acoustic Marchenko problem, a range of alternative solvers can be used to compute the pseudo-inverse [40,41].

2.7. Construction of Green's Functions from Focusing Functions

Either (the reflection-based) Equation (10) or (the transmission-based) Equation (13) can be used to convert focusing functions into Green's functions. Alternatively, we may take the average of both approaches, leading to

$$\underbrace{\begin{pmatrix} -\mathbf{G}_U^{-+} \\ -\mathbf{G}_U^{--} \\ \mathbf{G}_L^{+-} \\ \mathbf{G}_L^{++} \end{pmatrix}}_{\mathbf{G}} = \frac{1}{2}\begin{pmatrix} \mathbf{I} & -\mathcal{R}_U\mathcal{Z} & \mathcal{T}_{UL}\mathcal{Z} & \mathbf{O} \\ -\mathcal{R}_U\mathcal{Z} & \mathbf{I} & \mathbf{O} & \mathcal{T}_{UL}\mathcal{Z} \\ \mathcal{T}_{LU}\mathcal{Z} & \mathbf{O} & \mathbf{I} & -\mathcal{R}_L\mathcal{Z} \\ \mathbf{O} & \mathcal{T}_{LU}\mathcal{Z} & -\mathcal{R}_L\mathcal{Z} & \mathbf{I} \end{pmatrix}\underbrace{\begin{pmatrix} \mathbf{F}_U^{-} \\ \mathcal{Z}\mathbf{F}_U^{+} \\ \mathbf{F}_L^{+} \\ \mathcal{Z}\mathbf{F}_L^{-} \end{pmatrix}}_{\mathbf{F}}. \tag{19}$$

We use this result later in this paper to construct Green's functions from (retrieved) focusing functions.

3. Inversion

In order to construct matrix $\mathbf{B}$ in Equation (18), we require a priori knowledge of four direct arrivals: G_{Ud}^{--PP}, G_{Ld}^{++PP}, F_{Ud}^{+SS} and F_{Ld}^{-SS}. In Section 3.1, we show that these arrivals can be expressed in terms of two travel times, τ_{Ud}^{P} (for P-wave transmission from z_l to z_U) and τ_{Ud}^{S} (for S-wave transmission from z_l to z_U), as well as two amplitude scaling factors, α and β. In Section 3.2, we present a procedure to estimate these scaling factors. In Section 3.3, we apply this procedure to retrieve focusing functions and Green's functions from numerical data.

3.1. Initialization

The direct arrival G_{Ud}^{--PP} that is required for the construction of $\mathbf{G}_{Ud}^{--}$ can be expressed in terms of a delayed unit pulse $\delta(\tau-\tau_{Ud}^{P})$ and an amplitude scaling factor, which we parameterize strategically as $-\alpha^{\frac{1}{2}}$ for some (yet-unknown) α. This leads us to obtain

$$\mathbf{G}_{Ud}^{--} = \begin{pmatrix} G_{Ud}^{--PP} & 0 \\ 0 & 0 \end{pmatrix} = -\underbrace{\begin{pmatrix} \delta(\tau-\tau_{Ud}^{P}) & 0 \\ 0 & 0 \end{pmatrix}}_{\mathbf{E}_{Ud}^{P}}\alpha^{\frac{1}{2}}. \tag{20}$$

The direct arrival, G_{Ld}^{++PP}, that is required for the construction of $\mathbf{G}_{Ld}^{++}$ can be related to G_{Ud}^{--PP} by the $1D$ convolutional model $T_{LUd}^{+PP}(\tau) = -\int_{-\infty}^{+\infty} G_{Ld}^{++PP}(\tau-\tau')G_{Ud}^{--PP}(\tau')\mathrm{d}\tau'$. Here, T_{LUd}^{+PP} is the first event of the PP-component of the recorded transmission response $\mathbf{T}_{LU}^{+}$ from z_U to z_L. We assume that this event can be isolated from the transmission data by means of a time gate. Next, we define operator $\mathcal{T}_d^{P}$ for the $1D$ convolution of any signal with T_{LUd}^{+PP}. The direct Green's function at the lower level G_{Ld}^{++PP} may now be obtained by applying $\mathcal{T}_d^{P}$ to the inverse of $-G_{Ud}^{--PP}$, which can be expressed in our notation as $\mathcal{Z}\mathbf{E}_{Ud}^{P}\alpha^{-\frac{1}{2}}$ (with $\mathbf{E}_{Ud}^{P}$ as defined in Equation (20). This leads to

$$\mathbf{G}_{Ld}^{++} = \begin{pmatrix} G_{Ld}^{++PP} & 0 \\ 0 & 0 \end{pmatrix} = \mathcal{T}_d^{P}\mathcal{Z}\mathbf{E}_{Ud}^{P}\alpha^{-\frac{1}{2}}. \tag{21}$$

Similarly, the direct wave F_{Ud}^{+SS} that is required for the construction of $\mathbf{F}_{Ud}^{+}$ can be expressed in terms of a time-advanced unit pulse $\delta(\tau+\tau_{Ud}^{S})$ and an amplitude scaling factor, which we parameterize strategically as $\beta^{\frac{1}{2}}$ for some (yet-unknown) β. This leads to

$$\mathbf{F}_{Ud}^{+} = \begin{pmatrix} 0 & 0 \\ 0 & F_{Ud}^{+SS} \end{pmatrix} = \underbrace{\begin{pmatrix} 0 & 0 \\ 0 & \delta(\tau+\tau_{Ud}^{S}) \end{pmatrix}}_{\mathcal{Z}\mathbf{E}_{Ud}^{S}}\beta^{\frac{1}{2}}. \tag{22}$$

The direct arrival F_{Ld}^{-SS} that is required for the construction of $\mathbf{F}_{Ld}^{-}$ can be related to F_{Ud}^{+SS} by the $1D$ convolutional model $H_{LUd}^{+SS}(\tau) = \int_{-\infty}^{+\infty} F_{Ld}^{-SS}(\tau-\tau')F_{Ud}^{+SS}(\tau')\mathrm{d}\tau'$. Here, H_{LUd}^{+SS} is the first event of the SS-component of the inverse transmission response $\mathbf{H}_{LU}^{+}$ from z_U to z_L, which can be obtained via the inversion of $\mathbf{I}S(\tau) = \int_{\infty}^{+\infty} \mathbf{T}_{LU}^{+}(\tau - \tau')\mathbf{H}_{LU}^{+}(\tau')\mathrm{d}\tau'$, where $S(\tau) = \int_{-\infty}^{\infty} s(\tau + \tau')s(\tau')d\tau'$ is the autocorrelation of the source signal $s(\tau)$, as defined in Appendix A. We assume that H_{LUd}^{+SS} can be isolated from H_{LU}^{+SS} by means of a time gate. Next, we define an operator $\mathcal{H}_d^S$ for $1D$ convolution with H_{LUd}^{+SS}. The direct part of the focusing function at the lower level F_{Ld}^{-SS} may now be obtained by applying this operator to the inverse of F_{Ud}^{+SS}, which can be expressed in our notation as $\mathbf{E}_{Ud}^{S}\beta^{-\frac{1}{2}}$ (with $\mathbf{E}_{Ud}^{S} = \mathcal{Z}\mathcal{Z}\mathbf{E}_{Ud}^{S}$ as defined in Equation (22). This leads to

$$\mathbf{F}_{Ld}^{-} = \begin{pmatrix} 0 & 0 \\ 0 & F_{Ld}^{-SS} \end{pmatrix} = \mathcal{H}_d^S \mathbf{E}_{Ud}^{S}\beta^{-\frac{1}{2}}. \tag{23}$$

We assume that the travel times τ_d^P and τ_d^S are known. The amplitude scaling factors α and β can be estimated from the data, as we discuss in the following section.

3.2. Estimation of Amplitude Scaling Factors α and β

In Appendix C, we show that the focusing function matrix $\mathbf{F}$ can be written as an explicit function of τ, α and β, according to

$$\underbrace{\begin{pmatrix} \mathbf{F}_U^{-}(\tau,\alpha,\beta) \\ \mathcal{Z}\mathbf{F}_U^{+}(\tau,\alpha,\beta) \\ \mathbf{F}_L^{+}(\tau,\alpha,\beta) \\ \mathcal{Z}\mathbf{F}_L^{-}(\tau,\alpha,\beta) \end{pmatrix}}_{\mathbf{F}(\tau,\alpha,\beta)} = \underbrace{\begin{pmatrix} \mathbf{k}_{U1}^{P}(\tau) & \mathbf{k}_{U1}^{S}(\tau) \\ \mathbf{k}_{U2}^{P}(\tau) & \mathbf{k}_{U2}^{S}(\tau) \\ \mathbf{k}_{L1}^{P}(\tau) & \mathbf{k}_{L1}^{S}(\tau) \\ \mathbf{k}_{L2}^{P}(\tau) & \mathbf{k}_{L2}^{S}(\tau) \end{pmatrix}}_{\mathbf{K}(\tau)} \begin{pmatrix} \alpha & 0 \\ 0 & \beta \end{pmatrix}^{\frac{1}{2}} + \underbrace{\begin{pmatrix} \mathbf{l}_{U1}^{P}(\tau) & \mathbf{l}_{U1}^{S}(\tau) \\ \mathbf{l}_{U2}^{P}(\tau) & \mathbf{l}_{U2}^{S}(\tau) \\ \mathbf{l}_{L1}^{P}(\tau) & \mathbf{l}_{L1}^{S}(\tau) \\ \mathbf{l}_{L2}^{P}(\tau) & \mathbf{l}_{L2}^{S}(\tau) \end{pmatrix}}_{\mathbf{L}(\tau)} \begin{pmatrix} \alpha & 0 \\ 0 & \beta \end{pmatrix}^{-\frac{1}{2}}. \tag{24}$$

In this formulation, $\mathbf{k}(\tau)$ and $\mathbf{l}(\tau)$ represent 2×1 vectors, which can be explicitly computed for each value of τ from the recorded data (see Appendix C for details). To estimate the amplitude scaling factors α and β, we make use of a relation for energy conservation [42,43], which has been used earlier for the estimation of amplitude scaling factors in acoustic media [44]. This criterion can be written as

$$\int_{-\infty}^{\infty} [\{\mathbf{F}_U^{+}(\tau + \tau')\}^t \mathbf{F}_U^{+}(\tau') - \{\mathbf{F}_U^{-}(\tau + \tau')\}^t \mathbf{F}_U^{-}(\tau')]\mathrm{d}\tau' = \mathbf{I}S(\tau). \tag{25}$$

When we substitute the expressions for $\mathbf{F}_U^{\pm}(\tau,\alpha,\beta)$ from (24) into (25), subtract $\mathbf{I}S(\tau)$ on both sides and evaluate the result at $\tau = 0$, we find four expressions for α and β (i.e., the four entries of the 2×2 matrix equation). The first of these expressions (corresponding to the first diagonal entry of the matrix) is independent of β. We multiply this expression by α and define the left-hand side of the result as $h_U^P(\alpha)$. We find that

$$\begin{aligned} h_U^P(\alpha) = & \left[\int_{-\infty}^{\infty} [\{\mathbf{k}_{U2}^{P}(\tau')\}^t \mathbf{k}_{U2}^{P}(\tau') - \{\mathbf{k}_{U1}^{P}(\tau')\}^t \mathbf{k}_{U1}^{P}(\tau')]\mathrm{d}\tau'\right]\alpha^2 \\ & + \left[2\int_{-\infty}^{\infty} [\{\mathbf{k}_{U2}^{P}(\tau')\}^t \mathbf{l}_{U2}^{P}(\tau') - \{\mathbf{k}_{U1}^{P}(\tau')\}^t \mathbf{l}_{U1}^{P}(\tau')]\mathrm{d}\tau' - S(0)\right]\alpha \\ & + \left[\int_{-\infty}^{\infty} [\{\mathbf{l}_{U2}^{P}\}^t(\tau')\mathbf{l}_{U2}^{P}(\tau') - \{\mathbf{l}_{U1}^{P}(\tau')\}^t \mathbf{l}_{U1}^{P}(\tau')]\mathrm{d}\tau'\right] = 0. \end{aligned} \tag{26}$$

In a similar way, the last of our four expressions (corresponding to the last diagonal entry of the matrix) is independent of α. We multiply this expression by β and define the left-hand side of the result as $h_U^S(\beta)$. This leads to

$$h_U^S(\beta) = \left[\int_{-\infty}^{\infty} [\{\mathbf{k}_{U2}^S\}^t(\tau')\mathbf{k}_{U2}^S(\tau') - \{\mathbf{k}_{U1}^S(\tau')\}^t\mathbf{k}_{U1}^S(\tau')]\mathrm{d}\tau'\right]\beta^2$$
$$+ \left[2\int_{-\infty}^{\infty} [\{\mathbf{k}_{U2}^S(\tau')\}^t\mathbf{l}_{U2}^S(\tau') - \{\mathbf{k}_{U1}^S(\tau')\}^t\mathbf{l}_{U1}^S(\tau')]\mathrm{d}\tau' - S(0)\right]\beta \tag{27}$$
$$+ \left[\int_{-\infty}^{\infty} [\{\mathbf{l}_{U2}^S(\tau')\}^t\mathbf{l}_{U2}^S(\tau') - \{\mathbf{l}_{U1}^S(\tau')\}^t\mathbf{l}_{U1}^S(\tau')]\mathrm{d}\tau'\right] = 0.$$

Two more expressions can be obtained by enforcing the energy conservation of the focusing function $\mathbf{F}_L$. We find, akin to Equation (25), that

$$\int_{-\infty}^{\infty} [\{\mathbf{F}_L^-(\tau+\tau')\}^t\mathbf{F}_L^-(\tau') - \{\mathbf{F}_L^+(\tau+\tau')\}^t\mathbf{F}_L^+(\tau')]\mathrm{d}\tau' = \mathbf{I}S(\tau). \tag{28}$$

When we substitute the expressions for $\mathbf{F}_L^{\pm}(\tau,\alpha,\beta)$ from Equation (24) into this result and repeat the abovementioned steps, we arrive at expressions for $h_L^P(\alpha)$ and $h_L^S(\beta)$ (which are equivalent to Equations (26) and (27) with the subscript U replaced by L). The scaling factors α and β could be found by evaluating the roots of $h_U^P(\alpha)$, $h_L^P(\alpha)$, $h_U^S(\beta)$ and $h_L^S(\beta)$. However, we have chosen an alternative approach based on minimizing the cost functions

$$J^P(\alpha) = \{h_U^P(\alpha)\}^2 + \{h_L^P(\alpha)\}^2, \tag{29}$$

and

$$J^S(\beta) = \{h_U^S(\beta)\}^2 + \{h_L^S(\beta)\}^2. \tag{30}$$

In Figure 4, we show both cost functions as computed from the numerical data of our running example. We can use Matlab's *fminbnd* routine to minimize these functions, yielding the estimates $\alpha \approx 0.4716$ and $\beta \approx 0.7946$ (their true values being 0.4694 and 0.7934, respectively).

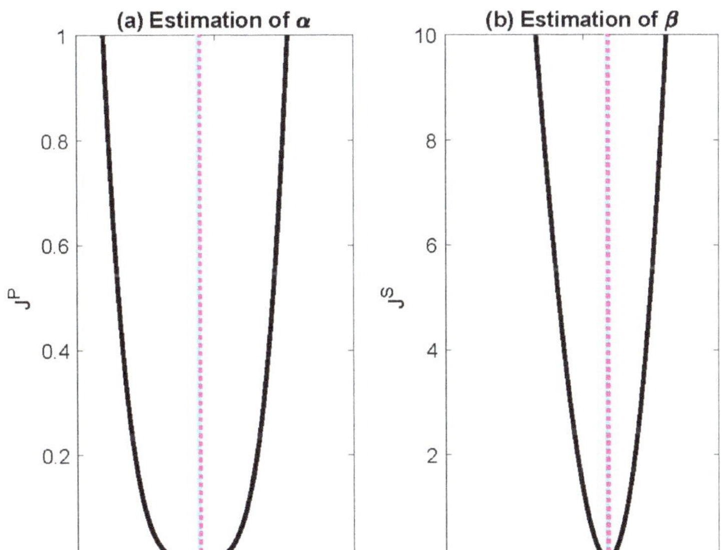

Figure 4. Cost functions (**a**) $J^P(\alpha)$ and (**b**) $J^S(\beta)$ for our numerical data. The dashed magenta lines denote the minima, 0.4716 (for α) and 0.7946 (for β), that were found using Matlab's *fminbnd* routine. The solid cyan lines denote the exact values 0.4694 (for α) and 0.7934 (for β), as extracted from the reference data.

3.3. Results

Now that α and β are resolved, the focusing function can be computed for our running example with the help of Equation (24). In Figures 5–8 we compare the *PP*-, *SP*-, *PS*- and *SS*-components of the retrieved focusing functions with the results of direct modeling. We observe that all events have been recovered well, where the most significant differences (which can hardly be observed in the figure) can be attributed to the (small) errors in our estimates of α and β. Next, we compute the Green's functions from the retrieved focusing functions with the help of Equation (19). In Figures 9–12, we compare the *PP*-, *SP*-, *PS*- and *SS*-components of the retrieved Green's functions with the results of direct modeling. Once more, we report an acceptable match, where the main differences (which can hardly be observed in the figures) can be attributed to errors in our estimates of α and β.

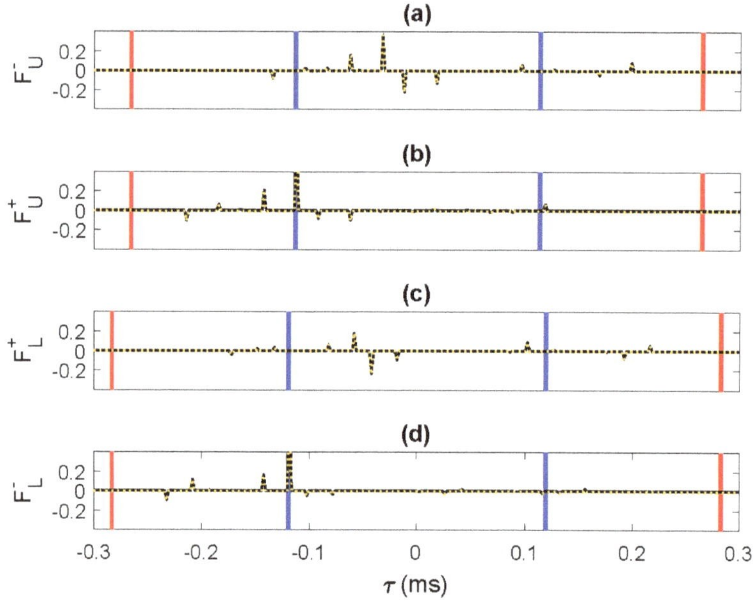

Figure 5. *PP*-component of the retrieved focusing functions: (**a**) F_U^-, (**b**) F_U^+, (**c**) F_L^+ and (**d**) F_L^-. The solid black traces were computed via direct modeling. The dashed orange traces were retrieved by means of our methodology. The blue lines have been drawn at $\pm\left(\tau_d^P + \frac{d\tau}{2}\right)$ (where $d\tau = 2$ µs denotes the intercept time sampling) to visualize the interval $[-\tau_d^P, \tau_d^P]$. The red lines have been drawn at $\pm\left(\tau_d^S - \frac{d\tau}{2}\right)$ to visualize the interval (τ_d^S, τ_d^S).

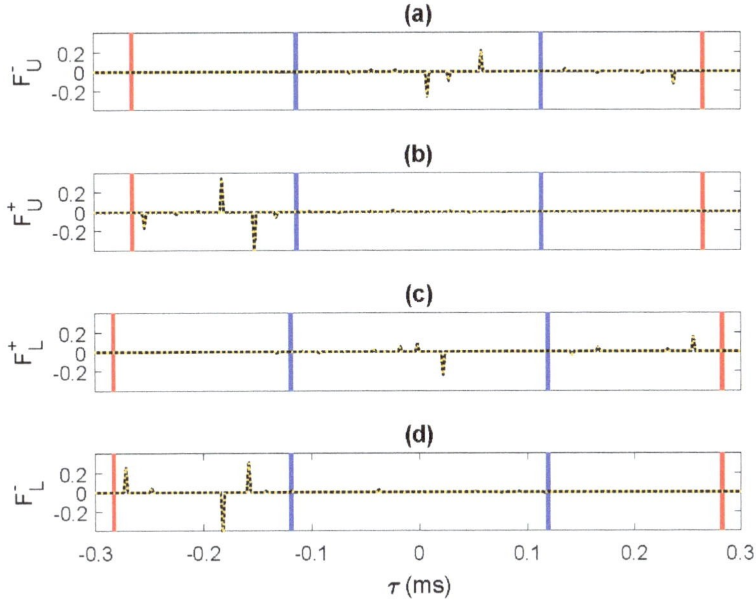

Figure 6. SP-component of the retrieved focusing functions (organized as in Figure 5).

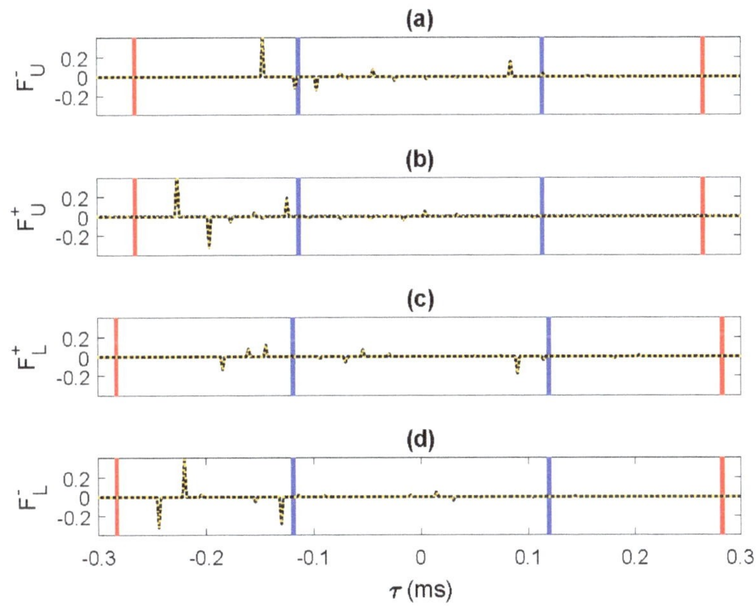

Figure 7. PS-component of the retrieved focusing functions (organized as in Figure 5).

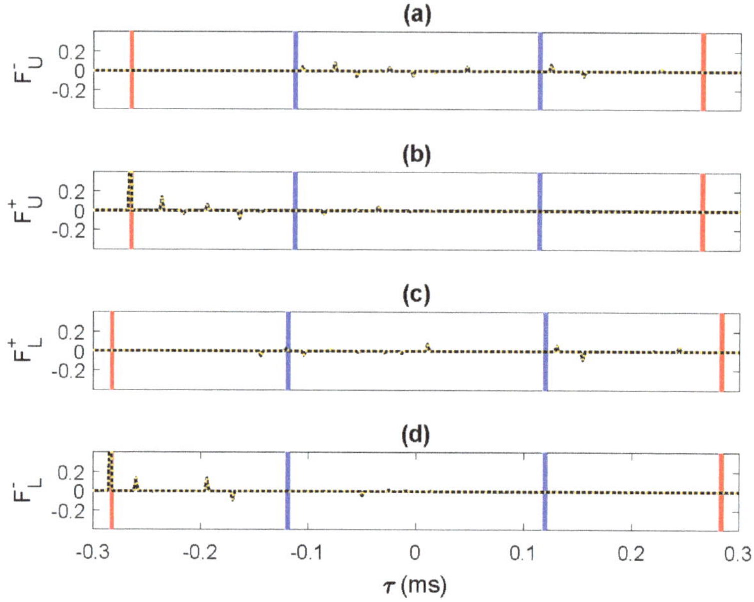

Figure 8. *SS*-component of the retrieved focusing functions (organized as in Figure 5).

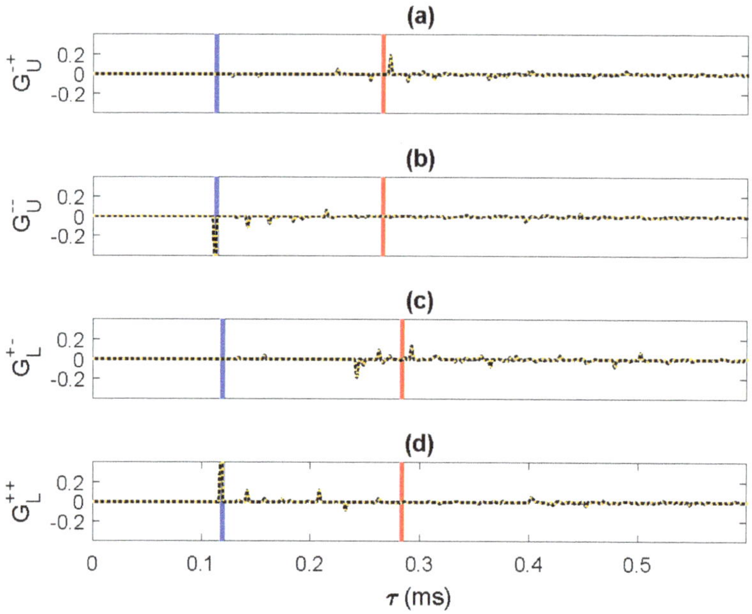

Figure 9. *PP*-component of the retrieved Green's functions: (**a**) G_U^{-+}, (**b**) G_U^{--}, (**c**) G_L^{+-} and (**d**) G_L^{++}. The solid black traces were computed via direct modeling. The dashed orange traces were retrieved using our methodology. The blue and red lines have been drawn at $\tau_d^p + \frac{d\tau}{2}$ and $\tau_d^s - \frac{d\tau}{2}$, respectively.

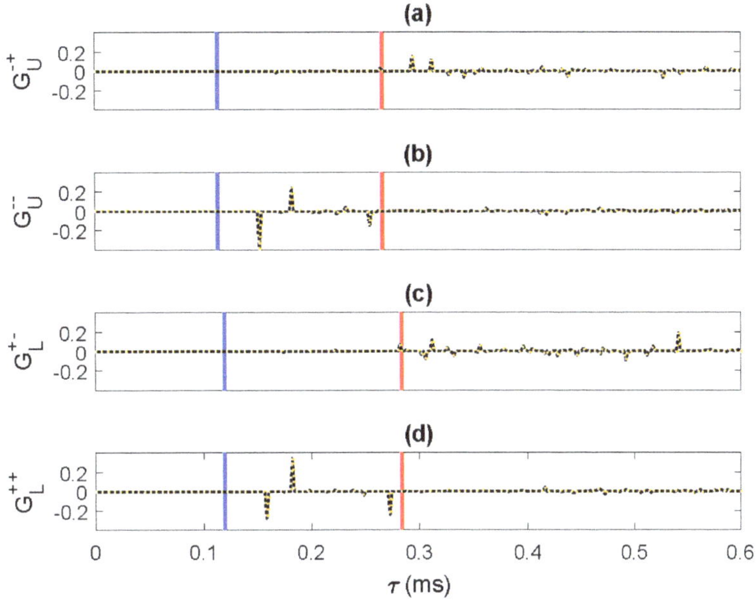

Figure 10. *SP*-component of the retrieved Green's functions (organized as in Figure 9).

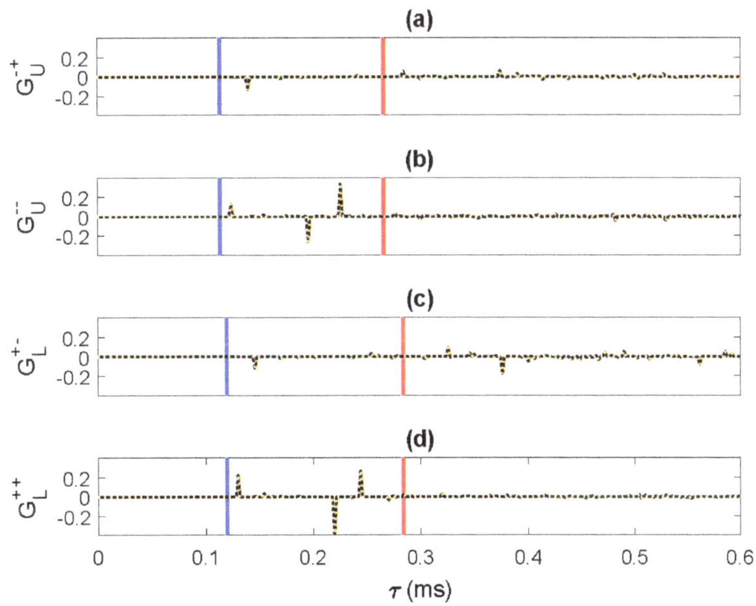

Figure 11. *PS*-component of the retrieved Green's functions (organized as in Figure 9).

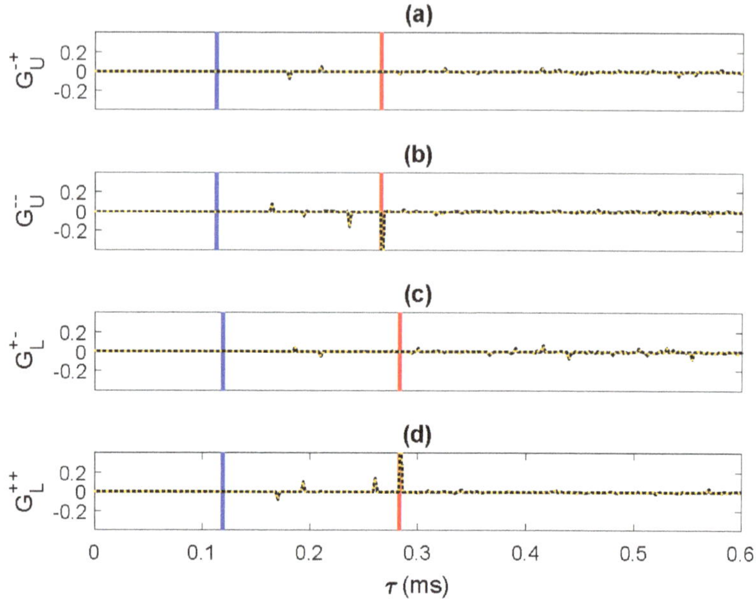

Figure 12. SS-component of the retrieved Green's functions (organized as in Figure 9).

4. Discussion

Our methodology requires knowledge of the intercept times τ_{Ud}^{P} and τ_{Ud}^{S}, whereas the focal depth z_I may be unknown. Hence, we can effectively retrieve Green's functions at a desired (intercept) time, even in the absence of velocity information [2]. A velocity model is only required to convert intercept times into depths, akin to acoustic Marchenko imaging [45]. An important observation in this context is that the construction of a virtual P-wave source is intrinsically decoupled from the construction of a virtual S-wave source in our formalism and both tasks may even be processed independently. Consequently, we may choose τ_{Ud}^{P} and τ_{Ud}^{S} at mutually different focal depths without affecting the accuracy of our results. Hence, we may conclude that neither $c^P(z)$ and $c^S(z)$, nor the ratio $\frac{c^P}{c^S}(z)$, is intrinsically required for the application of our methodology.

For our numerical simulations, we have designed a medium such that the arrival times of all waveforms coincide with exact time samples (see Appendix A). In this way, we could avoid problems related to discretization. In practical applications, data are recorded within a finite frequency band only, posing limitations to our resolution, especially in the presence of thin layers [2]. It has been shown previously that some of these limitations can be overcome by enforcing energy conservation and minimum-phase conditions in the single-sided Marchenko equation [16,42,43]. Similar strategies may be applied to the system of equations that we presented in this paper.

Although our methodology has been derived for a layered lossless medium with homogeneous halfspaces above z_U and below z_L, it could potentially be applied to a broader range of problems. Mild lateral variations of the medium's properties may be tolerable, akin to elastodynamic Marchenko imaging of single-sided data [12,18]. The effects of dissipation might be incorporated by computing all correlation-based reflection and transmission operators in an effectual medium, akin to the equivalent two-sided acoustic problem [33,34]. Heterogeneities above z_U might be accounted for by convolving our representations with areal sources that take interactions with this part of the medium into account, akin to Rayleigh–Marchenko redatuming [46,47]. A similar strategy may allow us to account for heterogeneities below z_L.

The invertibility of matrix **A** in our formalism is likely to depend on the medium's properties, the available bandwidth and the wavefield components that can be emitted and recorded in practice. For transcranial applications, we may modify the theory [48] or apply redatuming [49] to account for spherical arrays. Moreover, it might be necessary to place our transducers in a water layer, which is impenetrable for S-waves. In terms of matrix algebra, such a configuration induces a projection of our multi-component wavefields to a reduced domain of P-waves only [50]. It seems plausible that such a projection would affect the invertibility of **A**, but this remains to be investigated.

5. Conclusions

We have revised the window operators in the elastodynamic Marchenko equation. This leads to a system of equations that is intrinsically rank-deficient and hence cannot be solved without additional constraints. To overcome this issue, we have introduced an auxiliary equation (based on transmission data) and a coupled equation (based on reflection and transmission data). By concatenating these equations, we can construct a joint system for two-sided data that is invertible. Apart from the reflection and transmission data, this approach requires the direct (non-converted) P- and S-wave transmission times from the focal level to the upper acquisition array and two amplitude scaling factors, which can be retrieved by enforcing energy conservation. This leads to a methodology for velocity-independent true-amplitude Green's function retrieval in a lossless layered isotropic elastic medium from two-sided data. The methodology could potentially be extended to account for mild lateral variations, dissipation, anisotropy and heterogeneities above the upper acquisition array, as well as below the lower acquisition array.

Author Contributions: Conceptualization, J.V.d.N., J.B., G.A.M., E.S. and K.W.; Investigation, J.V.d.N.; Methodology, J.V.d.N., J.B., G.A.M., E.S. and K.W.; Supervision, E.S. and K.W.; Validation, J.V.d.N.; Visualization, J.V.d.N.; Writing—original draft, J.V.d.N.; Writing—review & editing, J.V.d.N. All authors have read and agreed to the published version of the manuscript.

Funding: This research has received funding from the European Research Council (grant no. 742703).

Institutional Review Board Statement: Not applicable.

Informed Consent Statement: Not applicable.

Data Availability Statement: Not applicable.

Conflicts of Interest: The authors declare no conflict of interest.

Appendix A. Numerical Settings

In this Appendix, we provide more information on the layered elastic medium depicted in Figure 1. We also provide the parameters that were used to generate numerical data for the running example that is discussed in the main text. Our model consisted of 41 depth samples with spacing dz = 25 mm. Data records were generated by modeling at a single ray parameter p = 0.2 ms·m^{-1} [36]. Our traces consisted of 2048 (intercept) time samples, which were sampled with $d\tau$ = 2 μs. For the source signal $s(\tau)$, we used a discretized delta function, where the sample at $\tau = 0$ equaled one and all remaining samples were zero. Inspired by [16], we chose model parameters that generated on-sample data at p = 0.2 ms·m^{-1}. This was achieved by choosing velocity values as elements of the set

$$\mathbb{V} = \left\{ c(n) = \sqrt{\frac{(dz)^2}{(dz)^2 p^2 + (d\tau)^2 n^2}} : n \in \mathbb{N} \right\}. \tag{A1}$$

In Table A1, we show the specific integers n^P and n^S that were used for each layer k of our model, to compute the velocities $c^P = c(n^P)$ and $c^S = c(n^S)$ as elements of $\mathbb{V}$. This procedure ensured that all (primary and multiple) reflections and transmissions arrived at exact sample values in our records. We also indicate the layer thicknesses Δz, densities

ρ, as well as the travel times $\Delta\tau^P = \sqrt{(c^P)^{-2} - p^2} \cdot \Delta z$ and $\Delta\tau^S = \sqrt{(c^S)^{-2} - p^2} \cdot \Delta z$ that were required for P- and S-waves to traverse each layer when p = 0.2 ms·m^{-1}. We note that the medium broke the separability conditions of [16] at both $z_U = 0$ and z_L = 1 m, given the focal depth z_I = 0.5 m (which is in layer $k = 3$ in the model). This was due to the fact that $\sum_{k=1}^{2}(\Delta\tau^S(k) - \Delta\tau^P(k)) = 102$ µs exceeded both $2\Delta\tau^P(3) = 32$ µs and $2\min\{\Delta\tau^P(k) : k \in \{1,2,3\}\} = 20$ µs, whereas $\sum_{k=4}^{5}(\Delta\tau^S(k) - \Delta\tau^P(k)) = 114$ µs exceeded both $2\Delta\tau^P(3) = 32$ µs and $2\min\{\Delta\tau^P(k) : k \in \{3,4,5\}\} = 16$ µs.

Table A1. Parameters used for the model shown in Figure 1.

k	n^P	n^S	Δz (mm)	ρ (kg· m^{-3})	$\Delta\tau^P$ (µs)	$\Delta\tau^S$ (µs)
0	5	10	100	2000	40	80
1	1	4	125	2800	10	40
2	3	7	225	2200	54	126
3	2	5	100	2600	16	40
4	4	9	225	2400	72	162
5	1	4	100	2700	8	32
6	3	7	125	2500	30	70

Appendix B. Derivations

In this Appendix, we derive several representations that were used in the main text. We conduct these derivations in the (p, z, ω)-domain (which is indicated by a hat), where ω denotes the angular frequency. We define the Fourier transform of an arbitrary (down- or upgoing) wavefield $\mathbf{P}^{\pm}(p, z, \tau)$ as

$$\hat{\mathbf{P}}^{\pm}(p, z, \omega) = \int_{-\infty}^{\infty} \mathbf{P}^{\pm}(p, z, \tau) \exp(-j\omega\tau) \mathrm{d}\tau, \quad \text{(A2)}$$

whereas the associated inverse Fourier transform is given by

$$\mathbf{P}^{\pm}(p, z, \tau) = \frac{1}{\pi} \Re\left[\int_{0}^{\infty} \hat{\mathbf{P}}^{\pm}(p, z, \omega) \exp(j\omega\tau) \mathrm{d}\omega\right]. \quad \text{(A3)}$$

Here, it is assumed that the signals are real-valued in the (p, z, τ)-domain and $\Re$ denotes the real part. Our derivations are based on two reciprocity theorems for flux-normalized wave fields, which we present for a source-free volume that is enclosed by depth levels z_m (at the top) and z_n (at the bottom). First, we have the reciprocity theorem of the convolution type [15]

$$\begin{aligned} &\{\hat{\mathbf{P}}_A^{+}(-p, z_m, \omega)\}^t \hat{\mathbf{P}}_B^{-}(p, z_m, \omega) - \{\hat{\mathbf{P}}_A^{-}(-p, z_m, \omega)\}^t \hat{\mathbf{P}}_B^{+}(p, z_m, \omega) \\ &= \{\hat{\mathbf{P}}_A^{+}(-p, z_n, \omega)\}^t \hat{\mathbf{P}}_B^{-}(p, z_n, \omega) - \{\hat{\mathbf{P}}_A^{-}(-p, z_n)\}^t \hat{\mathbf{P}}_B^{+}(p, z_n, \omega). \end{aligned} \quad \text{(A4)}$$

Here, $\hat{\mathbf{P}}_A^{+}$ and $\hat{\mathbf{P}}_A^{-}$ are down- and upgoing wavefields in state A, whereas $\hat{\mathbf{P}}_B^{+}$ and $\hat{\mathbf{P}}_B^{-}$ are equivalent wavefields in state B. Furthermore, ^{t} denotes matrix transposition. We have an equivalent reciprocity theorem of the correlation type [15]

$$\begin{aligned} &\{\hat{\mathbf{P}}_A^{+}(p, z_m, \omega)\}^{\dagger} \hat{\mathbf{P}}_B^{+}(p, z_m, \omega) - \{\hat{\mathbf{P}}_A^{-}(p, z_m, \omega)\}^{\dagger} \hat{\mathbf{P}}_B^{-}(p, z_m, \omega) \\ &= \{\hat{\mathbf{P}}_A^{+}(p, z_n, \omega)\}^{\dagger} \hat{\mathbf{P}}_B^{+}(p, z_n, \omega) - \{\hat{\mathbf{P}}_A^{-}(p, z_n, \omega)\}^{\dagger} \hat{\mathbf{P}}_B^{-}(p, z_n, \omega), \end{aligned} \quad \text{(A5)}$$

where † denotes the adjoint. In the following, we use Equations (A4) and (A5) to derive representations that are based on reflection and transmission data.

Appendix B.1. Reflection-Based Representations

First, we derive a convolution-based representation for reflection data at z_U. For this purpose, we set $m = U$ and $n = I$ in Equation (A4). In state A, we use the properties of the actual medium and we place a unit source just above z_U, such that $\hat{\mathbf{P}}_A^+(p, z_U, \omega) = \mathbf{I}$ (a 2×2 identity matrix), $\hat{\mathbf{P}}_A^-(p, z_U, \omega) = \hat{\mathbf{R}}_U^\cup(p)$ (the reflection response 'from above' at z_U), $\hat{\mathbf{P}}_A^+(p, z_I, \omega) = \hat{\mathbf{G}}_{IU}^{++}(p)$ (the downwgoing Green's function at z_I due to a downgoing source at z_U) and $\hat{\mathbf{P}}_A^-(p, z_I, \omega) = \hat{\mathbf{G}}_{IU}^{-+}(p)$ (the upgoing Green's function at z_I due to a downgoing source at z_U). In state B, we truncate the medium at z_I and choose a homogeneous halfspace below this level. For the wavefield in this state, we choose a focusing function with a focal point at z_I, which is purely downgoing below this level [15]. This leads to $\hat{\mathbf{P}}_B^+(p, z_U, \omega) = \hat{\mathbf{F}}_{UI}^+(p)$ (the downgoing focusing function at z_U), $\hat{\mathbf{P}}_B^-(p, z_U, \omega) = \hat{\mathbf{F}}_{UI}^-(p)$ (the upgoing focusing function at z_U), $\hat{\mathbf{P}}_B^+(p, z_I, \omega) = \mathbf{I}$ (the focused field at z_I) and $\hat{\mathbf{P}}_B^-(p, z_I, \omega) = \mathbf{O}$. Substitution of these quantities into Equation (A4) yields

$$-\hat{\mathbf{G}}_{UI}^{-+}(p) = \hat{\mathbf{F}}_{UI}^-(p) - \hat{\mathbf{R}}_U^\cup(p)\hat{\mathbf{F}}_{UI}^+(p), \tag{A6}$$

where we have used $\{\hat{\mathbf{R}}_U^\cup(-p)\}^t = \hat{\mathbf{R}}_U^\cup(p)$ and $\{\hat{\mathbf{G}}_{IU}^{-+}(-p)\}^t = \hat{\mathbf{G}}_{UI}^{-+}(p)$ [51]. We can derive an equivalent correlation-based representation by substituting the same quantities into Equation (A5), leading to

$$-\{\hat{\mathbf{G}}_{UI}^{--}(-p)\}^\star = \hat{\mathbf{F}}_{UI}^+(p) - \{\hat{\mathbf{R}}_U^\cup(-p)\}^\star\hat{\mathbf{F}}_{UI}^-(p), \tag{A7}$$

where we have used $\{\hat{\mathbf{R}}_U^\cup(p)\}^\dagger = \{\hat{\mathbf{R}}_U^\cup(-p)\}^\star$ and $\{\hat{\mathbf{G}}_{IU}^{++}(p)\}^\dagger = -\{\hat{\mathbf{G}}_{UI}^{--}(-p)\}^\star$ [51]. Two more representation can be derived for reflection data at z_L by choosing $m = I$ and $n = L$ in Equations (A4) and (A5) and placing a source just below z_L. Once again, we choose the actual medium properties in state A, such that $\hat{\mathbf{P}}_A^+(p, z_I, \omega) = \hat{\mathbf{G}}_{IL}^{+-}(p)$ (the downgoing Green's function at z_I due to an upgoing source at z_L), $\hat{\mathbf{P}}_A^-(p, z_I, \omega) = \hat{\mathbf{G}}_{IL}^{--}(p)$ (the upgoing Green's function at z_I due to an upgoing source at z_L), $\hat{\mathbf{P}}_A^+(p, z_L, \omega) = -\hat{\mathbf{R}}_L^\cap(p)$ (the reflection response 'from below' at z_L) and $\hat{\mathbf{P}}_A^-(p, z_L, \omega) = -\mathbf{I}$ (a 2×2 identity matrix). In state B, we truncate the medium at z_I and choose a homogeneous halfspace above this level. For the wavefield in this state, we choose a focusing function with a focal point at z_I, which is purely upgoing above this level. This leads to $\hat{\mathbf{P}}_B^+(p, z_I, \omega) = \mathbf{O}$, $\hat{\mathbf{P}}_B^-(p, z_I, \omega) = \mathbf{I}$ (the focused field at z_I), $\hat{\mathbf{P}}_B^+(p, z_L, \omega) = \hat{\mathbf{F}}_{LI}^+(p)$ (the downgoing focusing function at z_L) and $\hat{\mathbf{P}}_B^-(p, z_L, \omega) = \hat{\mathbf{F}}_{LI}^-(p)$ (the upgoing focusing function at z_L). Substitution of these quantities into Equation (A4) yields

$$\hat{\mathbf{G}}_{LI}^{+-}(p) = \hat{\mathbf{F}}_{LI}^+(p) - \hat{\mathbf{R}}_L^\cap(p)\hat{\mathbf{F}}_{LI}^-(p), \tag{A8}$$

where we have used $\{\hat{\mathbf{R}}_L^\cap(-p)\}^t = \hat{\mathbf{R}}_L^\cap(p)$ and $\{\hat{\mathbf{G}}_{IL}^{+-}(-p)\}^t = \hat{\mathbf{G}}_{LI}^{+-}(p)$ [51]. Alternatively, we may substitute the quantities into Equation (A5), leading to

$$\{\hat{\mathbf{G}}_{LI}^{++}(-p)\}^\star = \hat{\mathbf{F}}_{LI}^-(p) - \{\hat{\mathbf{R}}_L^\cap(-p)\}^\star\hat{\mathbf{F}}_{LI}^+(p), \tag{A9}$$

where we have used $\{\hat{\mathbf{R}}_L^\cap(p)\}^\dagger = \{\hat{\mathbf{R}}_L^\cap(-p)\}^\star$ and $\{\hat{\mathbf{G}}_{IL}^{--}(p)\}^\dagger = -\{\hat{\mathbf{G}}_{LI}^{++}(-p)\}^\star$ [51]. The system of Equations (A6)–(A9) can be rewritten as

$$\begin{pmatrix} -\hat{\mathbf{G}}_U^{-+}(p) \\ -\{\hat{\mathbf{G}}_U^{--}(-p)\}^\star \\ \hat{\mathbf{G}}_L^{+-}(p) \\ \{\hat{\mathbf{G}}_L^{++}(-p)\}^\star \end{pmatrix} = \begin{pmatrix} \mathbf{I} & -\hat{\mathbf{R}}_U^\cup(p) & \mathbf{O} & \mathbf{O} \\ -\{\hat{\mathbf{R}}_U^\cup(-p)\}^\star & \mathbf{I} & \mathbf{O} & \mathbf{O} \\ \mathbf{O} & \mathbf{O} & \mathbf{I} & -\hat{\mathbf{R}}_L^\cap(p) \\ \mathbf{O} & \mathbf{O} & -\{\hat{\mathbf{R}}_L^\cap(-p)\}^\star & \mathbf{I} \end{pmatrix} \begin{pmatrix} \hat{\mathbf{F}}_U^-(p) \\ \hat{\mathbf{F}}_U^+(p) \\ \hat{\mathbf{F}}_L^+(p) \\ \hat{\mathbf{F}}_L^-(p) \end{pmatrix}, \tag{A10}$$

where we have dropped subscript I (denoting the focal depth) for notational convenience. After taking the inverse Fourier transform (as defined in Equation (A3) and applying operator $\mathcal{Z}$ to the second and fourth rows, we obtain Equation (10), presented in the main text.

Appendix B.2. Transmission-Based Representations

We can derive a system of equivalent equations that are based on transmission data. First, we set $m = U$ and $n = I$. We choose the actual medium in state A and place a source just below z_L, leading to $\hat{\mathbf{P}}_A^+(p,z_U,\omega) = \mathbf{O}$, $\hat{\mathbf{P}}_A^-(p,z_U,\omega) = -\hat{\mathbf{T}}_{UL}^-(p)$ (the transmission response from z_L to z_U), $\hat{\mathbf{P}}_A^+(p,z_I,\omega) = \hat{\mathbf{G}}_{IL}^{+-}(p)$ and $\hat{\mathbf{P}}_A^-(p,z_I,\omega) = \hat{\mathbf{G}}_{IL}^{--}(p)$. In state B, we truncate the medium at z_I and choose a homogeneous halfspace below this level. For the wavefield in this state, we choose a focusing function that focuses at z_I and is purely downgoing below this level. This leads to $\hat{\mathbf{P}}_B^+(p,z_U,\omega) = \hat{\mathbf{F}}_{UI}^+(p)$, $\hat{\mathbf{P}}_B^-(p,z_U,\omega) = \hat{\mathbf{F}}_{UI}^-(p)$, $\hat{\mathbf{P}}_B^+(p,z_I,\omega) = \mathbf{I}$ and $\hat{\mathbf{P}}_B^-(p,z_I,\omega) = \mathbf{O}$. Substituting these quantities into Equation (A4) yields

$$\hat{\mathbf{G}}_{LI}^{++}(p) = \hat{\mathbf{T}}_{LU}^+(p)\hat{\mathbf{F}}_{UI}^+(p), \tag{A11}$$

where we have used $\{\hat{\mathbf{T}}_{UL}^-(-p)\}^t = \hat{\mathbf{T}}_{LU}^+(p)$ and $\{\hat{\mathbf{G}}_{IL}^{--}(-p)\}^t = -\hat{\mathbf{G}}_{LI}^{++}(p)$ [51]. We may substitute the same quantities into Equation (A5), leading to

$$\{\hat{\mathbf{G}}_{LI}^{+-}(-p)\}^\star = \{\hat{\mathbf{T}}_{LU}^+(-p)\}^\star\hat{\mathbf{F}}_{UI}^-(p), \tag{A12}$$

where we have used $\{\hat{\mathbf{T}}_{UL}^-(p)\}^\dagger = \{\hat{\mathbf{T}}_{LU}^+(-p)\}^\star$ and $\{\hat{\mathbf{G}}_{IL}^{+-}(p)\}^\dagger = \{\hat{\mathbf{G}}_{LI}^{+-}(-p)\}^\star$ [51]. We can derive two more representations by setting $m = I$ and $n = L$, and placing a source just above z_U. In state A, we use the actual medium properties, leading to: $\hat{\mathbf{P}}_A^+(p,z_I,\omega) = \hat{\mathbf{G}}_{IU}^{++}(p)$, $\hat{\mathbf{P}}_A^-(p,z_I,\omega) = \hat{\mathbf{G}}_{IU}^{-+}(p)$, $\hat{\mathbf{P}}_A^+(p,z_L,\omega) = \hat{\mathbf{T}}_{LU}^+(p)$ (the transmission response from z_U to z_L) and $\hat{\mathbf{P}}_A^-(p,z_L,\omega) = \mathbf{O}$. In state B, we truncate the medium at z_I and choose a homogeneous halfspace above this level. For the wavefield in this state, we choose a focusing function that focuses at z_I and is purely upgoing above this level. This leads to $\hat{\mathbf{P}}_B^+(p,z_I,\omega) = \mathbf{O}$, $\hat{\mathbf{P}}_B^-(p,z_I,\omega) = \mathbf{I}$, $\hat{\mathbf{P}}_B^+(p,z_L,\omega) = \hat{\mathbf{F}}_{LI}^+(p)$ and $\hat{\mathbf{P}}_B^-(p,z_L,\omega) = \hat{\mathbf{F}}_{LI}^-(p)$ Substituting these quantities into Equation (A4) yields

$$-\hat{\mathbf{G}}_{UI}^{--}(p) = \hat{\mathbf{T}}_{UL}^-(p)\hat{\mathbf{F}}_{LI}^-(p), \tag{A13}$$

where we have used $\{\hat{\mathbf{T}}_{LU}^+(-p)\}^t = \hat{\mathbf{T}}_{UL}^-(p)$ and $\{\hat{\mathbf{G}}_{IU}^{++}(-p)\}^t = -\hat{\mathbf{G}}_{UI}^{--}(p)$ [15]. Alternatively, the quantities can be substituted into Equation (A5), leading to

$$-\{\hat{\mathbf{G}}_{UI}^{-+}(-p)\}^\star = \{\hat{\mathbf{T}}_{UL}^-(-p)\}^\star\hat{\mathbf{F}}_{LI}^+(p), \tag{A14}$$

where we have used $\{\hat{\mathbf{T}}_{LU}^+(p)\}^\dagger = \{\hat{\mathbf{T}}_{UL}^-(-p)\}^\star$ and $\{\mathbf{G}_{IU}^{-+}(p)\}^\dagger = \{\hat{\mathbf{G}}_{UI}^{-+}(-p)\}^\star$ [51]. The system of Equations (A11)–(A14) can be rewritten as

$$\begin{pmatrix} -\{\hat{\mathbf{G}}_U^{-+}(-p)\}^\star \\ -\hat{\mathbf{G}}_U^{--}(p) \\ \{\mathbf{G}_L^{+-}(-p)\}^\star \\ \hat{\mathbf{G}}_L^{++}(p) \end{pmatrix} = \begin{pmatrix} \mathbf{O} & \mathbf{O} & \{\hat{\mathbf{T}}_{UL}^-(-p)\}^\star & \mathbf{O} \\ \mathbf{O} & \mathbf{O} & \mathbf{O} & \hat{\mathbf{T}}_{UL}^-(p) \\ \{\hat{\mathbf{T}}_{LU}^+(-p)\}^\star & \mathbf{O} & \mathbf{O} & \mathbf{O} \\ \mathbf{O} & \hat{\mathbf{T}}_{LU}^+(p) & \mathbf{O} & \mathbf{O} \end{pmatrix} \begin{pmatrix} \hat{\mathbf{F}}_U^-(p) \\ \hat{\mathbf{F}}_U^+(p) \\ \hat{\mathbf{F}}_L^+(p) \\ \hat{\mathbf{F}}_L^-(p) \end{pmatrix}, \tag{A15}$$

where we have dropped the subscript I once again for notational convenience. After taking the inverse Fourier transform (as defined in Equation (A3) and applying operator $\mathcal{Z}$ to the first and third row, we obtain Equation (13), presented in the main text.

Appendix C. Expression for F as a Function of τ, α and β

In this Appendix, we write $\mathbf{F}$ explicitly as a function of τ, α and β. We start with the substitution of Equations (20)–(23) into the definition of $\mathbf{B}_{Mar}$, which is given in the left-hand side of Equation (12). We write the result as

$$\mathbf{B}_{Mar} = \underbrace{\begin{pmatrix} \mathbf{\Theta}_U^{[P]}\mathcal{R}_U\mathcal{Z}\mathbf{E}_{Ud}^S \\ \mathbf{E}_{Ud}^P \\ \mathbf{O} \\ \mathbf{O} \end{pmatrix}}_{\mathbf{M}_{Mar}^I} \begin{pmatrix} \alpha & 0 \\ 0 & \beta \end{pmatrix}^{\frac{1}{2}} + \underbrace{\begin{pmatrix} \mathbf{O} \\ \mathbf{O} \\ \mathbf{\Theta}_L^{[P]}\mathcal{R}_L\mathcal{H}_d^S\mathbf{E}_{Ud}^S \\ \mathcal{T}_d^P\mathcal{Z}\mathbf{E}_{Ud}^P \end{pmatrix}}_{\mathbf{M}_{Mar}^{II}} \begin{pmatrix} \alpha & 0 \\ 0 & \beta \end{pmatrix}^{-\frac{1}{2}}, \tag{A16}$$

where $\mathbf{M}^{I}_{Mar}$ and $\mathbf{M}^{II}_{Mar}$ have been used to identify the constructed matrices. In a similar way, we can substitute Equations (20)–(23) into the definition of $\mathbf{B}_{Aux}$, which is given in the left-hand side of Equation (15). This leads to

$$\mathbf{B}_{Aux} = \underbrace{\begin{pmatrix} \mathbf{O} \\ \mathbf{E}^{P}_{Ud} \\ \mathbf{O} \\ -\boldsymbol{\Theta}^{[P]}_{L}\mathcal{T}_{LU}\mathcal{Z}\mathbf{E}^{S}_{Ud} \end{pmatrix}}_{\mathbf{M}^{I}_{Aux}} \begin{pmatrix} \alpha & 0 \\ 0 & \beta \end{pmatrix}^{\frac{1}{2}} + \underbrace{\begin{pmatrix} \mathbf{O} \\ -\boldsymbol{\Theta}^{[P]}_{U}\mathcal{T}_{UL}\mathcal{H}^{S}_{d}\mathbf{E}^{S}_{Ud} \\ \mathbf{O} \\ \mathcal{T}^{P}_{d}\mathcal{Z}\mathbf{E}^{P}_{Ud} \end{pmatrix}}_{\mathbf{M}^{II}_{Aux}} \begin{pmatrix} \alpha & 0 \\ 0 & \beta \end{pmatrix}^{-\frac{1}{2}}, \tag{A17}$$

where $\mathbf{M}^{I}_{Aux}$ and $\mathbf{M}^{II}_{Aux}$ have been used to identify the constructed matrices. Finally, we may substitute Equations (22) and (23) into the definition of $\mathbf{B}_{Cou}$, which is given in the left-hand side of Equation (17). This yields

$$\mathbf{B}_{Cou} = \underbrace{\frac{1}{2}\begin{pmatrix} \mathcal{R}_{U}\mathcal{Z}\mathbf{E}^{S}_{Ud} \\ -\mathbf{E}^{S}_{Ud} \\ \mathbf{O} \\ \mathcal{T}_{LU}\mathcal{Z}\mathbf{E}^{S}_{Ud} \end{pmatrix}}_{\mathbf{M}^{I}_{Cou}} \begin{pmatrix} \alpha & 0 \\ 0 & \beta \end{pmatrix}^{\frac{1}{2}} + \underbrace{\frac{1}{2}\begin{pmatrix} \mathbf{O} \\ \mathcal{T}_{UL}\mathcal{H}^{S}_{d}\mathbf{E}^{S}_{Ud} \\ \mathcal{R}_{L}\mathcal{H}^{S}_{d}\mathbf{E}^{S}_{Ud} \\ -\mathcal{Z}\mathcal{H}^{S}_{d}\mathbf{E}^{S}_{Ud} \end{pmatrix}}_{\mathbf{M}^{II}_{Cou}} \begin{pmatrix} \alpha & 0 \\ 0 & \beta \end{pmatrix}^{-\frac{1}{2}}, \tag{A18}$$

where $\mathbf{M}^{I}_{Mar}$ and $\mathbf{M}^{II}_{Mar}$ have been used to identify the constructed matrices. Next, we substitute Equations (A16)–(A18) into (18) and apply the pseudo-inverse of $\mathbf{A}$ to both sides of the result. This eventually leads to

$$\mathbf{F}_{m} = \mathbf{A}^{\ddagger}\underbrace{\begin{pmatrix} \mathbf{M}^{I}_{Mar} \\ \mathbf{M}^{I}_{Aux} \\ \mathbf{M}^{I}_{Cou} \end{pmatrix}}_{\mathbf{M}^{I}} \begin{pmatrix} \alpha & 0 \\ 0 & \beta \end{pmatrix}^{\frac{1}{2}} + \mathbf{A}^{\ddagger}\underbrace{\begin{pmatrix} \mathbf{M}^{II}_{Mar} \\ \mathbf{M}^{II}_{Aux} \\ \mathbf{M}^{II}_{Cou} \end{pmatrix}}_{\mathbf{M}^{II}} \begin{pmatrix} \alpha & 0 \\ 0 & \beta \end{pmatrix}^{-\frac{1}{2}}. \tag{A19}$$

The direct focusing function matrix $\mathbf{F}_{d}$ can be written in a similar form. This is achieved by substituting Equations (22) and (23) into the definition of $\mathbf{F}_{d}$, which is given in the right-hand side of Equation (8). The result can strategically be written as

$$\underbrace{\begin{pmatrix} \mathbf{O} \\ \mathcal{Z}\mathbf{F}^{+}_{Ud} \\ \mathbf{O} \\ \mathcal{Z}\mathbf{F}^{-}_{Ld} \end{pmatrix}}_{\mathbf{F}_{d}} = \underbrace{\begin{pmatrix} \mathbf{O} \\ \mathbf{E}^{S}_{Ud} \\ \mathbf{O} \\ \mathbf{O} \end{pmatrix}}_{\mathbf{D}^{I}} \begin{pmatrix} \alpha & 0 \\ 0 & \beta \end{pmatrix}^{\frac{1}{2}} + \underbrace{\begin{pmatrix} \mathbf{O} \\ \mathbf{O} \\ \mathbf{O} \\ \mathcal{Z}\mathcal{H}^{S}_{d}\mathbf{E}^{S}_{Ud} \end{pmatrix}}_{\mathbf{D}^{II}} \begin{pmatrix} \alpha & 0 \\ 0 & \beta \end{pmatrix}^{-\frac{1}{2}}. \tag{A20}$$

Adding Equations (A19) and (A20) yields

$$\mathbf{F}(\tau,\alpha,\beta) = \underbrace{\left(\mathbf{D}^{I}(\tau) + \mathbf{A}^{\ddagger}(\tau)\mathbf{M}^{I}(\tau)\right)}_{\mathbf{K}(\tau)} \begin{pmatrix} \alpha & 0 \\ 0 & \beta \end{pmatrix}^{\frac{1}{2}} + \underbrace{\left(\mathbf{D}^{II}(\tau) + \mathbf{A}^{\ddagger}(\tau)\mathbf{M}^{II}(\tau)\right)}_{\mathbf{L}(\tau)} \begin{pmatrix} \alpha & 0 \\ 0 & \beta \end{pmatrix}^{-\frac{1}{2}}. \tag{A21}$$

Here, we have indicated the arguments of all matrices for convenience to emphasize that we have expressed $\mathbf{F}$ explicitly as a function of τ, α and β. Note that $\mathbf{K}(\tau)$ and $\mathbf{L}(\tau)$ are independent ofthe scaling factors and hence can be computed from the recorded data, τ^{P}_{d} and τ^{S}_{d}. Finally, we can express our result as Equation (24), presented in the main text, by renaming the quantities that constitute matrices $\mathbf{K}(\tau)$ and $\mathbf{L}(\tau)$.

References

1. Broggini, F.; Snieder, R. Connection of scattering principles: A visual and mathematical tour. *Eur. J. Phys.* **2012**, *33*, 593-613. [CrossRef]
2. Slob, E.; Wapenaar, K.; Broggini, F.; Snieder, R. Seismic reflector imaging using internal multiples with Marchenko-type equations. *Geophysics* **2014**, *79*, S63–S76. [CrossRef]
3. Cui, T.; Vasconcelos, I.; van Manen, D.-J.; Wapenaar, K. A tour of Marchenko redatuming: Focusing the subsurface wavefield. *Lead. Edge* **2018**, *37*, 67a1–67a6. [CrossRef]
4. Lomas, A.; Curtis, A. An introduction to Marchenko methods for imaging. *Geophysics* **2019**, *84*, F35–F45. [CrossRef]
5. Thorbecke, J.; Slob, E.; Brackenhoff, J.; van der Neut, J.; Wapenaar, K. Implementation of the Marchenko method. *Geophysics* **2017**, *82*, WB29–WB45. [CrossRef]
6. Ravasi, M.; Vasconcelos, I. An open-source framework for the implementation of large-scale integral operators with flexible, modern HPC solutions-enabling 3D Marchenko imaging by least squares inversion. *Geophysics* **2021**, *86*, WC177–WC194. [CrossRef]
7. Brackenhoff, J.; Thorbecke, J.; Meles, G.; Koehne, V.; Barrera, D.; Wapenaar, K. 3D Marchenko applications: Implementation and examples. *Geophys. Prospect.* **2022**, *70*, 35–56. [CrossRef]
8. Staring, M.; Pereira, R.; Douma, H.; van der Neut, J.; Wapenaar, K. Source-receiver Marchenko redatuming on field data using an adaptive double-focusing method. *Geophysics* **2018**, *83*, S579–S590. [CrossRef]
9. Zhang, L; Slob, E. A field data example of Marchenko multiple elimination. *Geophysics* **2020**, *85*, S65–S70. [CrossRef]
10. Jia, X.; Baumstein, A.; Jing, C.; Neumann, E.; Snieder, R. Subsalt Marchenko imaging with offshore Brazil field data. *Geophysics* **2021**, *86*, WC31–WC40. [CrossRef]
11. Wapenaar, K.; Brackenhoff. J.; Dukalski, M.; Meles, G.; Reinicke, C.; Slob, E.; Staring, M.; Thorbecke, J.; van der Neut, J.; Zhang, L. Marchenko redatuming, imaging, and multiple elimination and their mutual relations. *Geophysics* **2021**, *86*, WC117–WC140. [CrossRef]
12. da Costa Filo, C.A.; Ravasi, M.; Curtis, A.; Meles, G.A. Elastodynamic Green's function retrieval through single-sided Marchenko inverse scattering. *Phys. Rev. E* **2014**, *90*, 063201. [CrossRef] [PubMed]
13. Wapenaar, K. Single-sided Marchenko focusing of compressional and shear waves. *Phys. Rev. E* **2014**, *90*, 063202. [CrossRef]
14. Wapenaar, K.; Snieder, R.; de Ridder, S.; Slob, E. Green's function representation for Marchenko imaging without up/down decomposition. *Geophys. J. Int.* **2021**, *227*, 184–203. [CrossRef]
15. Wapenaar, K; Slob, E. On the Marchenko equation for multicomponent single-sided reflection data. *Geophys. J. Int.* **2014**, *199*, 1367–1371. [CrossRef]
16. Reinicke, C.; Dukalski, M.; Wapenaar, K. Comparison of monotonicity challenges encountered by the inverse scattering series and the Marchenko demultiple method for elastic waves. *Geophysics* **2020**, *85*, Q11–Q26. [CrossRef]
17. Reinicke, C.; Wapenaar, K. Elastodynamic single-sided homogeneous Green's function representation: Theory and numerical examples. *Wave Motion* **2019**, *89*, 245–264. [CrossRef]
18. da Costa Filho, C.; Ravasi, M.; Curtis, C. Elastic P- and S-wave autofocus imaging with primaries and internal multiples. *Geophysics* **2015**, *80*, S187–S202. [CrossRef]
19. da Costa Filho, C.; Meles, G.A.; Curtis, C. Elastic internal multiple analysis and attenuation using Marchenko and interferometric medhods. *Geophysics* **2017**, *82*, Q1–Q12. [CrossRef]
20. Thomsen, H.R.; Molerón, M.; Haag, T.; van Manen, D.-J.; Robertsson, J.O.A. Elastic immersive wave experimentation: Theory and physical implementation. *J. Phys. Rev. Res.* **2019**, *1*, 033203. [CrossRef]
21. Niederleithinger, E.; Wolf, J.; Mielentz, F.; Wiggenhauser, H.; Pirskawetz, S. Embedded ultrasonic transducers for active and passive concrete monitoring. *Sensors* **2015**, *15*, 9756–9772. [CrossRef] [PubMed]
22. Bazulin, E.; Goncharsky, A.; Romanov, S.; Seryozhnikov, S. Ultrasound transmission and reflection tomography for nondestructive testing using experimental data. *Ultrasonics* **2022**, *124*, 106765. [CrossRef] [PubMed]
23. Taskin, U; Eikrem, K.S.; Nævdal, G.; Jakobsen, M.; Verschuur, D.J.; van Dongen, K.W.A. Ultrasound imaging of the brain using full-waveform inversion. In Proceedings of the 2020 IEEE International Ultrasonics Symposium (IUS), Las Vegas, NV, USA, 7–11 September 2020.
24. Marty, P.; Boehm, C.; Fichtner, A. Acoustoelastic full-waveform inversion for transcranial ultrasound computed tomography. In Proceedings of the SPIE Medical Imaging 2021, Online, 15–20 February 2021.
25. Meles, G.A.; van der Neut, J.; van Dongen, K.W.A.; Wapenaar, K. Wavefield finite time focusing with reduced spatial exposure. *J. Acoust. Soc. Am.* **2019**, *145*, 3521–3530. [CrossRef] [PubMed]
26. Brackenhoff, J.; van der Neut, J.; Meles, G.; Marty, P.; Boehm, C. Virtual ultrasound transducers in the human brain. In Proceedings of the SPIE Medical Imaging 2022, San Diego, CA, USA, 20 February–28 March 2022.
27. Na, S.; Yuan, X.; Lin, L.; Isla, J.; Garrett, D.; Wang, L.V. Transcranial photoacoustic computed tomography based on a layered back-projection method. *Photoacoustics* **2020**, *20*, 100213. [CrossRef] [PubMed]
28. Poudel, J.; Na, S.; Wang, L.V.; Anastasio, M.A. Iterative image reconstruction in transcranial photoacoustic tomography based on the elastic wave equation. *Phys. Med. Biol.* **2020**, *65*, 055009. [CrossRef]
29. Liu, Y; van der Neut, J; Arntsen, B; Wapenaar, K. Combination of surface and borehole seismic data for robust target-oriented imaging. *Geoph. J. Int.* **2016**, *205*, 758–775. [CrossRef]
30. Lomas, A.; Singh, S.; Curtis, A. Imaging vertical structures using Marchenko methods with vertical seismic-profile data. *Geophysics* **2020**, *85*, S103–S113. [CrossRef]

31. Singh, S.; Snieder, R.; van der Neut, J.; Thorbecke, J.; Slob, E.; Wapenaar, K. Accounting for free-surface multiples in Marchenko imaging. *Geophysics* **2017**, *82*, R19–R30. [CrossRef]
32. Dukalski, M.; de Vos, K. Marchenko inversion in a strong scattering regime including surface-related multiples. *Geophys. J. Int.* **2018**, *212*, 760–776. [CrossRef]
33. Slob, E. Green's function retrieval and Marchenko imaging in a dissipative acoustic medium. *Phys. Rev. Lett.* **2016**, *116*, 164301. [CrossRef]
34. Cui, T.; Becker, T.S.; van Manen, D.J.; Rickett, J.E.; Vasconcelos, I. Marchenko redatuming in a dissipative medium: Numerical and experimental implementation. *Phys. Rev. Appl.* **2018**, *19*, 044022. [CrossRef]
35. Frasier, C.W. Discrete time solution of plane P-SV waves in a plane layered medium. *Geophysics* **1970**, *35*, 197–219. [CrossRef]
36. Kennett, B.L.N. *Seismic Wave Propagation in Stratified Media*; Cambridge University Press: Cambridge, UK, 1983.
37. Schalkwijk, K.M.; Wapenaar, C.P.A.; Verschuur, D.J. Adaptive decomposition of multicomponent ocean-bottom seismic data into downgoing and upgoing P- and S-waves. *Geophysics* **2003**, *68*, 1091–1102. [CrossRef]
38. Zhang, L; Slob, E.; van der Neut, J.; Wapenaar, K. Artifact-free reverse time migration. *Geophysics* **2018**, *83*, A65–A68. [CrossRef]
39. Diekmann, L.; Vasconcelos, I. Focusing and Green's function retrieval in three-dimensional inverse scattering revisited: A single-sided Marchenko integral for the full wave field. *Phys. Rev. Res.* **2021**, *3*, 013206. [CrossRef]
40. Slob, E.; Zhang, L. Unified elimination of 1D acoustic multiple reflection. *Geophys. Prospect.* **2021**, *69*, 327–348. [CrossRef]
41. Santos, R.S.; Revelo, D.E.; Pestana, R.C.; Koehne, V.; Barrera, D.F.; Souza, M.S. An application of the Marchenko internal multiple elimination scheme formulated as a least-squares problem. *Geophysics* **2021**, *86*, WC105–WC116. [CrossRef]
42. Dukalski, M.; Mariani, E.; de Vos, K. Handling short-period scattering using augmented Marchenko autofocusing. *Geophys. J. Int.* **2019**, *216*, 2129–2133. [CrossRef]
43. Elison, P.; Dukalski, M.S.; de Vos, K.; van Manen, D.J.; Robertsson, J.O.A. Data-driven control over short-period internal multiples in media with a horizontally layered overburden. *Geophys. J. Int.* **2020**, *221*, 769–787. [CrossRef]
44. Mildner, C.; Dukalski, M.; Elison, P.; de Vos, K.; Broggini, F.; Robertsson, J.O.A. True-amplitude-versus-offset Green's function retrieval using augmented Marchenko focusing. In Proceedings of the 81st EAGE Conference & Exhibition 2019, London, UK, 3–6 June 2019.
45. Sripanich, Y.; Vasconcelos, I; Wapenaar, K. Velocity-independent Marchenko focusing in time- and depth-imaging domains for media with mild lateral heterogeneity. *Geophysics* **2019**, *84*, Q57–Q72. [CrossRef]
46. Ravasi, M. Rayleigh-Marchenko redatuming for target-oriented, true-amplitude imaging. *Geophysics* **2017**, *82*, S439–S452. [CrossRef]
47. Vargas, D.; Vasconcelos, I.; Sripanich, Y.; Ravasi, M. Scattering-based focusing for imaging in highly complex media from band-limited, multicomponent data. *Geophysics* **2021**, *86*, WC141–WC157. [CrossRef]
48. Brackenhoff, J.; van der Neut, J.; Marty, P. Creating virtual measurements inside a medium using spherical recording arrays. In Proceedings of the AGU Fall Meeting 2021, New Orleans, LA, USA, 13–17 December 2021.
49. Taskin, U; van der Neut, J.; Gemmeke, H.; van Dongen, K.W.A. Redatuming of 2-D wavefields measured on an arbitrary-shaped closed aperture. *IEEE Trans. Ultrason. Ferroelectr. Freq. Control* **2019**, *67*, 173–179. [CrossRef] [PubMed]
50. Reinicke, C.; Dukalski, M; Wapenaar, K. Internal multiple elimination: Can we trust an acoustic approximation? *Geophysics* **2021**, *86*, WC41–WC54. [CrossRef]
51. Wapenaar, C.P.A. One-way representations of seismic data. *Geophys. J. Int.* **1996**, *127*, 178–188. [CrossRef]

applied sciences

MDPI

Article

Implementing Data-Driven Approach for Modelling Ultrasonic Wave Propagation Using Spatio-Temporal Deep Learning (SDL)

Thulsiram Gantala * and Krishnan Balasubramaniam

Centre for Non-Destructive Evaluation, Department of Mechanical Engineering, Indian Institute of Technology Madras, Chennai 600036, India; balas@iitm.ac.in
* Correspondence: me18d040@smail.iitm.ac.in

Abstract: In this paper, we proposed a data-driven spatio-temporal deep learning (SDL) model, to simulate forward and reflected ultrasonic wave propagation in the 2D geometrical domain, by implementing the convolutional long short-term memory (ConvLSTM) algorithm. The SDL model learns underlying wave physics from the spatio-temporal datasets. Two different SDL models are trained, with the following time-domain finite element (FE) simulation datasets, by applying: (1) multi-point excitation sources inside the domain and (2) single-point excitation sources on the edge of the different geometrical domains. The proposed SDL models simulate ultrasonic wave dynamics, for the forward ultrasonic wave propagation in the different geometrical domains and reflected wave propagation phenomenon, from the geometrical boundaries such as curved, T-shaped, triangular, and rectangular domains, with varying frequencies and cycles. The SDL is a reliable model, which generates simulations faster than the conventional finite element solvers.

Keywords: data-driven modeling; spatio-temporal datasets; ultrasonic wave propagation; deep learning; RNN; ConvLSTM; finite element

Citation: Gantala, T.; Balasubramaniam, K. Implementing Data-Driven Approach for Modelling Ultrasonic Wave Propagation Using Spatio-Temporal Deep Learning (SDL). *Appl. Sci.* **2022**, *12*, 5881. https://doi.org/10.3390/app12125881

Academic Editor: Michel Darmon

Received: 29 March 2022
Accepted: 19 May 2022
Published: 9 June 2022

Publisher's Note: MDPI stays neutral with regard to jurisdictional claims in published maps and institutional affiliations.

1. Introduction

Ultrasonic wave propagation is used in various applications, including biomedical imaging [1], nondestructive evaluation [2,3], seismic and geological studies [4], etc. The numerical modeling of the ultrasonic waves is critical, for improving the understanding of the underlying physics of these applications. The ultrasonic wave propagation phenomenon is widely modeled using Finite Element [5–8], Finite Difference [9], and Finite Volume [10] techniques. Ultrasonic wave propagation is modeled by the widely used FE analysis, through solving the partial differential equations on discrete nodes, using iterative time-steeping schemes. Due to the transient nature of the wave propagation and the three-dimensional volume in which the ultrasonic wave is modeled, the computational resources and the time to complete the calculations are often extensive, thus limiting the utilization of modeling. We can develop a data-driven solver for modelling wave propagation, by capturing the underlying physics from numerical simulation datasets, as an alternative approach [11].

In recent years, modern deep-learning techniques, such as long short-term memory (LSTM) [12,13], have been successful in many domains [14,15]. These techniques are used in many applications, such as propagating the latent space to the future [16,17], using [18] the LSTM network to realize gesture recognition, wave propagation [19], automating tumor segmentation in whole breast [20], and shear wave elastography [21]. Since the ultrasonic wave propagation prediction is a sequence of spatio-temporal images data, so, typically, the LSTM approach will not provide and capture the desired results [22]. The convolutional long short-term memory (ConvLSTM) networks, effectively, address the long-range simulation prediction [23–26]. The ConvLSTM approach uses the convolutional operation inside the LSTM cell, to pair the temporal state with spatial information. On the other hand, the wave propagation simulation solver can be built using a convolutional neural network (CNN) architecture [27,28]. Another group in the research community is adapting physics-informed neural networks to solve the partial differential equations [29–31].

In the current work, the authors have developed two separate spatio-temporal deep learning (SDL) models. In the prior work, [26] shows the modeling forward wave propagation simulation in a 2D domain with different physical settings. The authors, further, want to generalize the SDL network for modelling forward wave propagation in the different geometrical domains (i.e., other than the trained domains) and reflecting wave propagation simulation in the different geometrical boundaries. In the first SDL model, the network algorithm is tuned to incorporate geometrical domain information and trained using datasets containing numerous ultrasonic wave simulations, with single-point excitation to five-point excitation sources inside the solid medium, without reflection from the edges. This trained SDL model can be deployed for modelling real-time forward ultrasonic wave propagation in the different geometrical shapes. Whereas in the second SDL model, the learnable hyperparameters are tuned to generate wave propagation, without changing network architecture. The datasets used for training the second model have three distinct geometrical domains, with a single-point excitation source applied on the edges. This trained SDL model can be used to study the reflected wave propagation phenomenon from different geometrical boundaries, such as curved, T-shaped, triangular, and rectangular domains, with varying frequencies and cycles. Thus, depending on the availability of the datasets's structure, we should employ one of these models for generating wave propagation. These training datasets have spatial and temporal features; hence, algorithms based on the recurrent neural network (RNN) are treated as the most suitable for modelling ultrasonic wave propagation simulations [19].

This paper is organized as follows: Section 2 describes the procedure for generating training datasets, and Section 3 details the SDL model formulation for the wave propagation phenomenon. Section 4 discusses the SDL model implementation and results, to generalize for different domains. Section 5 brings the summary of the work, along with concluding remarks.

2. Training Datasets Generation through Finite Element Simulation

To generate wave propagation in solid media using AI algorithms requires large volumes of training datasets. The following section, comprehensively, discusses the methodology adopted for generating the FE training datasets.

2.1. Finite Element Modelling of Ultrasonic Wave Propagation in Solid Media

The datasets to train the AI model are created by modelling numerous time-domain 2D FE simulations, by solving the governing partial differential equation, by employing the commercial FE Abaqus/Explicit (v. 18.0) solver (see ABAQUS User Manual v. 6.11, Dassault system, Providence, RI, USA). The two-dimensional FE CAD models are created using carbon steel isotropic material properties, as shown in Figure 1a,b. The mechanical properties of carbon steel are a mass density of ρ = 7850 kg/m^3, Young's modulus of E = 200 GPa, and Poisson's ratio of ν = 0.29. The 2D part is discretized with a four-noded quadrilateral mesh size of 3.2×10^{-5} mm, which is about 22 elements per wavelength λ_{shear}, for 5 MHz in carbon steel for mesh convergence. A similar approach is followed from our prior work for modelling [3]. The two cycles of the Hanning-windowed tone burst signal of 5 MHz frequency are used to trigger an incident wave. To avoid the undesired reflection from the boundaries's [32] absorbing, boundary conditions are applied using the ALID (absorbing layers using increasing damping) method, as shown in Figure 1a. After completing the simulation, the spatial and temporal scale is adjusted during the post-processing, to ensure that the necessary temporal and spatial information is captured in the displacement plots.

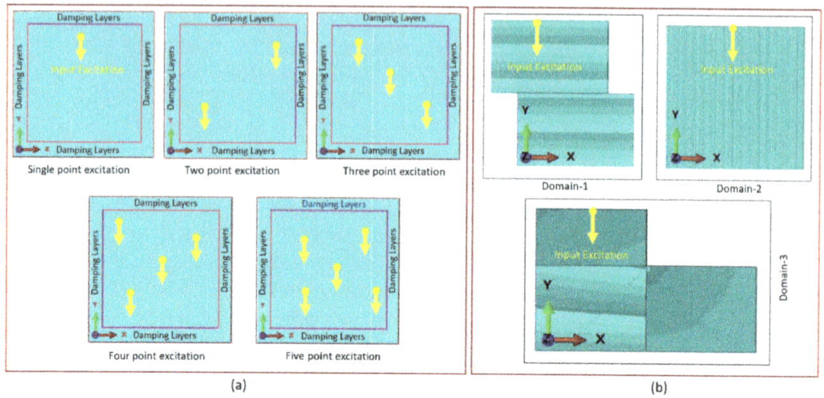

Figure 1. (**a**) Type-1 datasets: finite-element models: FE models with single to multi-point sources excitation are triggered in the Y-direction, and an absorbing boundary condition is applied on all the domain edges, to avoid undesired reflection. (**b**) Type-2 datasets: finite-element models: FE models with single-point source excitation are triggered in the Y-direction on the top edge of the model and a traction-free boundary condition is applied on all other edges.

2.2. Type-1 Datasets Creation: Forward Wave Propagation Simulation

The type 1 training datasets is created by modelling several 2D FE simulations, by changing the following parameters: (a) the incident wave is triggered in the X-direction or in the Y-direction inside the domain in each simulation, (b) the excitation point sources are distributed randomly throughout the domain using a uniform probability distribution, and (c) the excitation sources vary from single to five-point excitation sources, as shown in Figure 1a. A total of 1250 simulations are generated for the training, consisting of 250 CAD models, with 35 × 35 mm physical dimensions for each (a) single-point excitation, (b) two-point excitation, (c) three-point excitation, (d) four-point excitation, and (e) five-point excitation source, respectively. These displacement plots are saved as frames, with the time interval of 0.0074 µs, for the total simulation.

2.3. Type-2 Datasets Creation: Reflection Wave Propagation Simulation

For the type-2 training datasets generation, we have created three different geometrical 2D CAD models, as shown in Figure 1b. The overall physical dimension of each model is 40 mm in length and 40 mm in height. A total of 1500 CAD models are created, of which 500 are of the same geometrical shape. The excitation load is applied on the top-edge nodes of the CAD models, as shown in Figure 1b, and on the remaining edges, a traction-free boundary condition is applied to allow the sidewall and back wall reflections. After performing each FE simulation, the displacement plots are saved as a sequence of images, with a time interval of 0.0125 µs.

One FE simulation generation executed using Dual Intel Xeon Platinum 8168 processor with 48 cores machine (Figure 1b) took 3600 s, using time-domain FE analysis, which is a time-consuming process. We have introduced the SDL model to overcome this limitation, which generates wave propagation simulations in minimal time once trained.

3. The Formulation of Spatio-Temporal Deep Learning Model for Ultrasonic Wave Propagation

The LSTM is a unique RNN structure modeled for addressing vanishing gradients and learning long-range dependencies in the deep machine learning framework. The LSTM shown in Figure 2 consists of the following elements: a memory cell C_t, which can accumulate, and the forgetthe state being tackled, time step to time step. i_t, f_t, and o_t are the input, forget, and output gate, respectively. The i_t monitors the flow of new input into the C_t, the responsibility of f_t is to remove irrelevant information from C_t, and the work of o_t is to send information from C_t to hidden state H_t. The H_t could retain the memory of past information from the sequence data. Since the nature of the training datasets is in the form of spatio-temporal sequence images, the LSTM approach suffers from capturing the spatial features from the 2D images. The ConvLSTM network, effectively, learns the spatial and temporal features from the sequential images because it uses the convolutional operation inside the LSTM cell to pair the temporal state with spatial information. The formulation of the ConvLSTM model is followed from [23].

$$i_t = \sigma(W_{xi} * X_t + W_{hi} * H_{t-1} + W_{ci} \circ C_{t-1} + b_i) \tag{1}$$

$$f_t = \sigma(W_{xf} * X_t + W_{hf} * H_{t-1} + W_{cf} \circ C_{t-1} + b_f) \tag{2}$$

$$C_t = f_t \circ C_{t-1} + i_t \circ Tanh(W_{xc} * X_t + W_{hc} * H_{t-1} + b_c) \tag{3}$$

$$o_t = \sigma(W_{xo} * X_t + W_{ho} * H_{t-1} + W_{co} \circ C_t + b_o) \tag{4}$$

$$H_t = o_t \circ Tanh(C_t) \tag{5}$$

here, 'W' denotes the weights, and 'b' represents the network's biases. The convolutional operation is marked with '$*$' and '$\circ$', representing the Hadamard product.

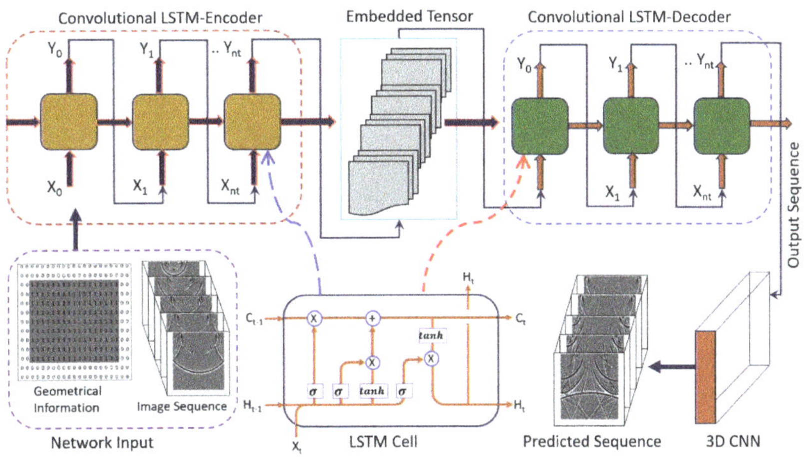

Figure 2. Typical spatio-temporal deep learning (SDL) model encoder–decoder architecture, for modelling wave propagation phenomena.

We have used the proposed SDL network for our spatial-temporal sequence generation problem, as shown in Figure 2, for two types of different training datasets. First, for type-1 datasets, we have modified the existing ConvLSTM algorithm to fuse the geometrical information with each hidden dimension output, which is similar to our previous study [26]. So, for the type-2 datasets, we have modified the suitable network parameters, such as hidden dimensions and kernel size, instead of modifying the network architecture.

These models consist of an encoder and a decoder network, containing a series of convolutional LSTM cells stacked together, since a single ConvLSTM layer will not capture the forward and reflected wave dynamics from training datasets, due to the spatial-temporal nature. Each input data sequence is fed into each encoder ConvLSTM cell, to learn the wave propagation physics from the datasets. The ConvLSTM layer takes the batch of the 3D input tensor (length × width × time step) and outputs the hidden state (H_t). Then, the following input tensor, similarly with H_t, is given input to the next layer of the ConvLSTM cell. Due to the convolutional operation, the model captures the required features from the datasets, during the processing into layers of ConvLSTM cells. The encoder, iteratively, processes input sequences through various ConvLSTM cells and outputs the embedding tensor, representing wave propagation. The output of the encoder-embedded tensor is fed to the decoder network, to produce predicted wave propagation simulations. The outcome of the decoder cell's sequence is passed to the 3D CNN layers, with a sigmoid activation function, to transform into an actual wave propagation prediction.

4. Results and Discussion

The implementation of the SDL model, using type-1 and type-2 training datasets, as well as testing results, are described in detail in this section.

4.1. SDL Model Implementation on Type-1 Datasets: Forward Wave Propagation Simulation

The proposed SDL network is implemented in PyTorch Lightning, an open-source Python library (see https://www.pytorchlightning.ai (accessed on 28 March 2022)), by integrating multi-GPU, and trained using simulation-assisted FE datasets of type-1 (Section 2.2). These training datasets from the FE analysis form a sequence of images in each simulation. Each simulation contains 675 images, and a mini-batch of 15 images is randomly selected. A sequence of 5 images is used for input to the network from the mini-batch, and the next successive 10 images are used to compare the network-predicted output. These images are in grayscale, with a size of 128 × 128 pixels. The SDL architecture contains the four layers of an encoder structure, and the four layers of the decoder structure are stacked together. Each layer contains 256 hidden states, and the convolutional operation is performed on each input image, with a kernel size of 5 × 5 with the same padding. The geometrical information from the CAD model is captured in the binary matrix, by assigning '1' for the inside and '0' for an outside domain, as shown in Figure 2. The binary matrix and the sequence of frames are used as input to the network. This binary matrix is multiplied with an output of a hidden state, before updating the next hidden state. In total, 80% of the simulations are used for training and the remaining for testing the model.

To predict the more accurate simulations from the SDL model, the suitable learnable parameters and hyperparameters are selected. The hyperparameters are selected before the training process and cannot be altered, and the weights and basis are learnable parameters that must be updated during training. First, the input datasets are fed and propagated to compute the output through the network. This predicted output is compared with the ground truth, to determine the error and, then, calculate the derivative of the error function, with respect to the network learnable parameters. Now, network weights and basis are updated to minimize the error; this process is well known as the back-propagation algorithm. The back-propagation algorithm is suitable for fixed-size input-output pairs in feed-forward neural networks [33]. However, in the current work, the training datasets are spatiotemporal, so the back-propagation through time (BPTT) algorithm is employed, instead of the back-propagation algorithm. The mean square error (MSE) loss function is minimized during the training process, using the back-propagation through time (BPTT) algorithm. The MSE loss function used in this network is as follows:

$$MSE = \frac{1}{n}\sum_{t=1}^{n}(\hat{y}_i - y_i)^2 \tag{6}$$

here, $\hat{y}$ is the predicted output sequence, y is the ground truth sequence, and n = the number of instances. While training, the ConvLSTM network, as shown in Figure 2, predicts the output for one input in each time step. So, the BPTT works by unrolling all input time steps. Furthermore, each time step has one input time step, one output time step, and a copy of the network. Then, the error is calculated for every time step and accumulated for each time step. The network is unrolled back, and learnable parameters are updated. The network learning rate is set to a constant value of 1.0×10^{-4} in the Adam optimizer and trained for 150 epochs. The network took a total of 36 h to train, using two NVIDIA GeForce RTX 3090 GPU processors. The average loss value computed by comparing network-predicted output with ground truth is used to determine the model learning efficiency. Figure 3a shows the average loss values over the number of epochs, and training loss is determined to be 6.0×10^{-4}, whereas testing loss is 2.0×10^{-4}. The performance of the train SDL model is validated, by feeding the remaining 20% of testing datasets of single to multi-point excitation sources simulations, and the effectiveness of the proposed model on testing datasets can be referred from our prior work [26].

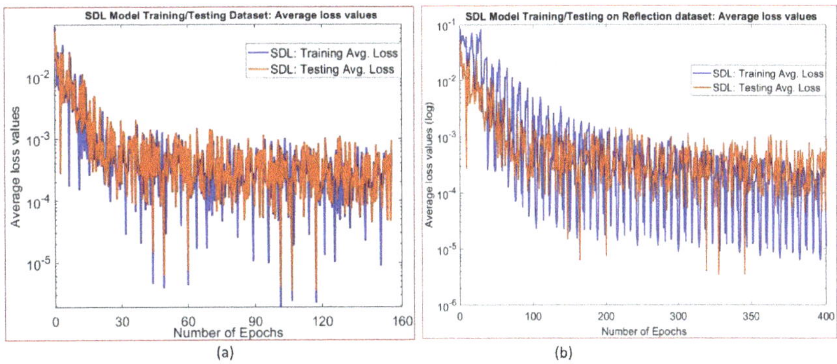

Figure 3. (**a**) Spatio-temporal deep learning (SDL) model, for forward wave propagation: training and testing average loss values over the number of epochs for the type-1 datasets. (**b**) Spatio-temporal deep learning (SDL) model, for reflected wave propagation: training and testing average loss values over the number of epochs for the type-2 datasets.

4.1.1. Evaluation Metric

To assess the efficiency of the SDL model prediction on wave propagation simulation, we have employed the Mean Absolute Error (MAE), Root Mean Square Error (RMSE), and Mean Absolute Percentage Error (MAPE), to compute using the equations below:

$$MAE = \frac{1}{n}\sum_{i=1}^{n}|Y_i - \bar{Y}_i| \tag{7}$$

$$RMSE = \sqrt{\left(\frac{1}{n}\right)\sum_{i=1}^{n}(Y_i - \bar{Y}_i)^2} \tag{8}$$

$$MAPE = \frac{100\%}{n}\sum_{i=1}^{n}\left|\frac{Y_i - \bar{Y}_i}{Y_i}\right| \tag{9}$$

here, Y_i is the ground truth from FE wave propagation simulation, $\bar{Y}_i$ is the predicted from SDL model generated wave propagation simulation, and n = number of instances.

4.1.2. Ultrasonic Wave Propagation on Different Geometrical Domain Modelling

This section presents the comparative study between the SDL model and FE, to simulate the forward wave propagation simulations for the different geometrical domains, shown in Figure 4. So, we have considered two scenarios of different geometrical domains (i.e., other than the trained domains) with point source excitation to evaluate the trained SDL model, which is not used during training or testing. The SDL model is, initially, trained with a single-step short-term prediction during the training process. Then, the model takes a 5-input image sequence from the training set and predicts the next following consecutive 10-output image sequence. These model-predicted images are fed back to the model's input, iteratively, to predict the complete long-term simulation. Then, the SDL model generated simulation is compared with the actual FE ultrasonic wave propagation simulation.

Figure 4 shows the FE and SDL model simulations over the total simulation time for geometry-1 and geometry-2. In particular, the top row in Figure 4a shows the FE datasets, and the bottom row shows the SDL model output datasets at identical time intervals at t = (0.9 μs, 1.5 μs, 2.2 μs, 2.2 μs, 3.6 μs), for the long-term simulation of geometry-1.

Similarly, Figure 4b depicts the wave propagation of geometry-2. The proposed SDL model could capture the constructive and destructive interference at the wavefront interaction and match the FE simulation. We can conclude that the SDL model-generated simulations are in good agreement with FE.

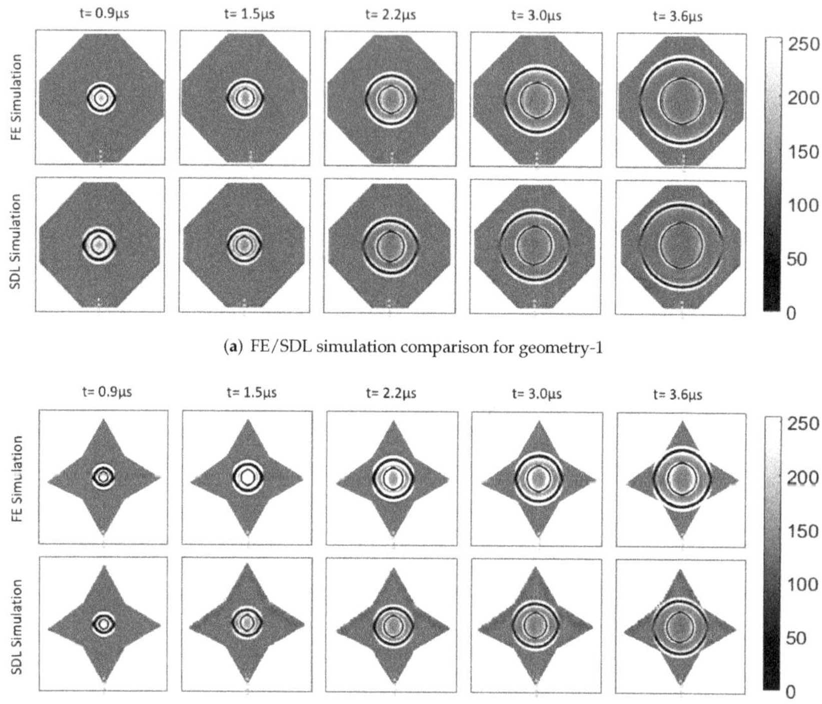

(a) FE/SDL simulation comparison for geometry-1

(b) FE/SDL simulation comparison for geometry-2

Figure 4. Forward wave propagation of different geometrical domain: The FE and SDL are modeled using 5 MHz central frequency and two cycles with a single-point excitation source on different geometrical domains. The SDL model predicted simulations are qualitatively compared with FE simulation. (**a**) Forward wave propagation simulation for geometry-1 and (**b**) forward wave propagation simulation for geometry-2, at exact time steps for different time instances.

For quantitative assessment, we have extracted the displacement values at x = 17.5 mm and y = 0 to 35 mm, in all the frames from Figure 4 (as shown by the vertically dotted yellow line). Figure 5a,b show the wavefront line scans for both geometry-1 and geometry-2. The line scans from the SDL model follow the same trend as the FE simulation line scans for all the frames, but some differences exist in amplitude and TOF. To find out the difference in amplitude and time of flight between the SDL and FE simulations, from the line scans in Figure 5 for each modelling scenario, we have used three methods (refer to Section 4.2) (1) the Mean Absolute Error (MAE), (2) Root Mean Square Error (RMSE), and (3) Mean Absolute Percentage Error (MAPE), to estimate the error in each frame. Table 1 shows the summary of the MAE, RMSE, and MAPE on amplitude and TOF error. The maximum MAE is in order of 10^{-2}, the maximum RMSE is in order of 10^{-1}, and MAPE is 115.9% on amplitude error, over all the time frames. The maximum MAE is in order of 10^{-1}, the maximum RMSE is 3.15, and MAPE is 3.34% on TOF error, over all the time frames. The evaluation finding shows that the proposed SDL model could, efficiently, simulate ultrasound wave propagation on the different geometrical domains, other than the trained domain, with good accuracy.

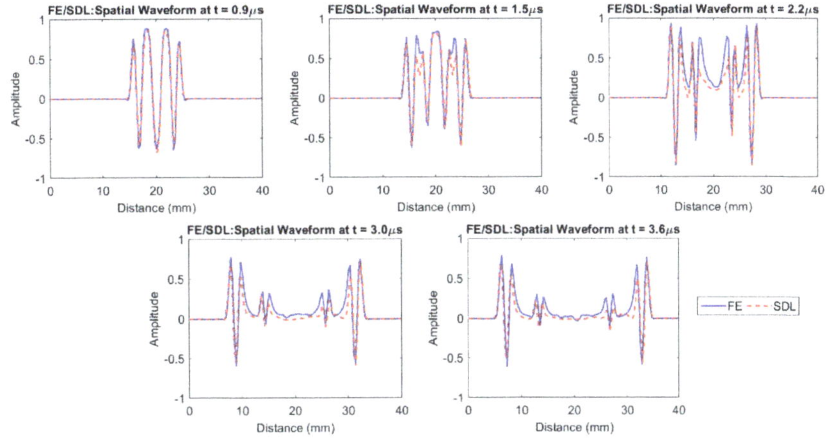

(**a**) Line scans between FE/SDL simulation, for geometry-1 at identical time steps

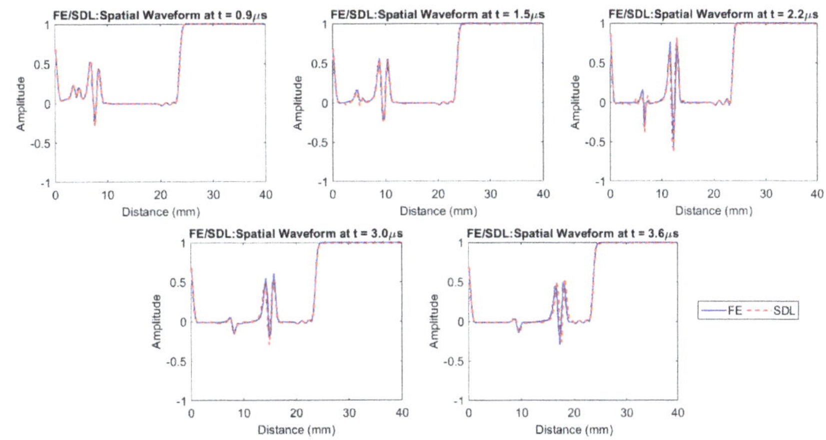

(**b**) Line scans between FE/SDL simulation, for geometry-2 at identical time steps

Figure 5. Line scans of different geometrical domains: the displacement values are extracted at x = 15 mm and y = 0 to 30 mm in each frame from the FE, and SDL-generated simulation of the different geometrical domains for different time instances in Figure 4. (**a**) The line scans were extracted from Figure 4a for geometry-1, and (**b**) the line scans were extracted from Figure 4b for geometry-2.

Table 1. Forward wave propagation on different geometrical domains datasets: the line scan amplitude and time of flight (TOF) differences between SDL and FE are computed using MAE, RMSE, and MAPE methods on the sequence of images at different time instances in Figure 5.

Line Scan	Datasets of Testing and Different Geometrical Domain	MAE					RMSE					MAPE (%)				
		Sequence of Images at					Sequence of Images at					Sequence of Images at				
		T-1	T-2	T-3	T-4	T-5	T-1	T-2	T-3	T-4	T-5	T-1	T-2	T-3	T-4	T-5
Amplitude	Geometry-1	0.00	0.02	0.05	0.04	0.04	0.05	0.12	0.14	0.20	0.24	9.50	25.2	30.6	75.6	92.7
	Geometry-2	0.00	0.01	0.06	0.07	0.06	0.07	0.14	0.20	0.21	0.19	115.9	16.4	22.5	34.8	39.6
Time of Flight	Geometry-1	0.00	0.17	0.17	0.83	1.42	0.00	0.41	0.41	1.63	3.15	0.00	0.28	0.24	1.62	3.34
	Geometry-2	0.29	0.21	0.21	0.21	0.21	0.53	0.46	0.46	0.46	0.46	0.46	0.39	0.46	0.48	0.67

4.2. SDL Model Implementation on Datasets Type-2: Reflection Wave Propagation Simulation

A similar approach is followed from Section 4.1 to train the SDL model using the type-2 datasets of reflection wave propagation simulation (from Section 2.3). The SDL architecture consists of six layers of an encoder structure and six layers of a decoder structure; these structures are stacked together. Each encoder or decoder structure contains 256 hidden dimensions, and during convolutional operation, a 5×5 kernel size and the same padding are used. The average loss value over the number of epochs is shown in Figure 3b. The average training loss is determined to be 1.0×10^{-5}, the testing loss is 4.0×10^{-5}, and the loss value becomes stabilized with an increase in the number of epochs.

4.2.1. Generalization to Different Geometrical Boundary Modelling: Reflected Wave Propagation from Boundaries

To generalize the SDL model, we have modeled the ultrasonic wave propagation in four different geometrical domains: curved, T-shaped, and triangular. The SDL model predicted simulation is compared with the FE. Figure 6a–c shows the reflected ultrasonic wave propagation simulation for curved, T-shaped, and triangular domains, respectively. In each figure, the top row simulations are from the FE, and the bottom row shows the predicted simulation from the SDL model. During the training process, the proposed SDL model is trained with straight edge boundaries, but it could generate the simulation for curved boundaries (Figure 6a) and diagonal boundaries (Figure 6c). The SDL model has learned the wave interactions at the domain sharp edge and could generate the reflection wave propagation from the sharp edges (Figure 6b). So, the SDL model accurately predicts reflection wave propagation from the side and back walls as well as matches with the FE.

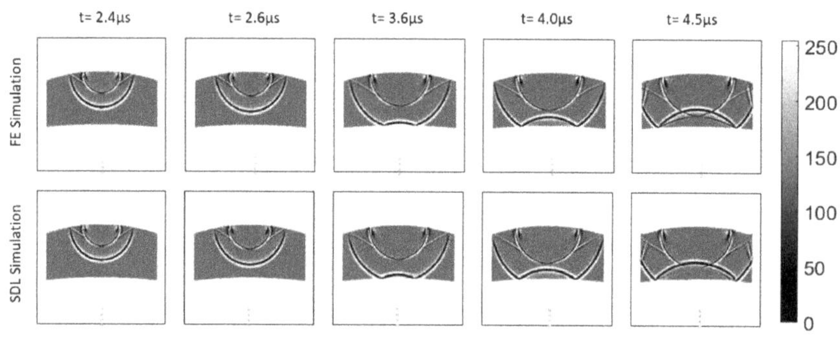

(**a**) FE/SDL simulation comparison for the curved boundary.

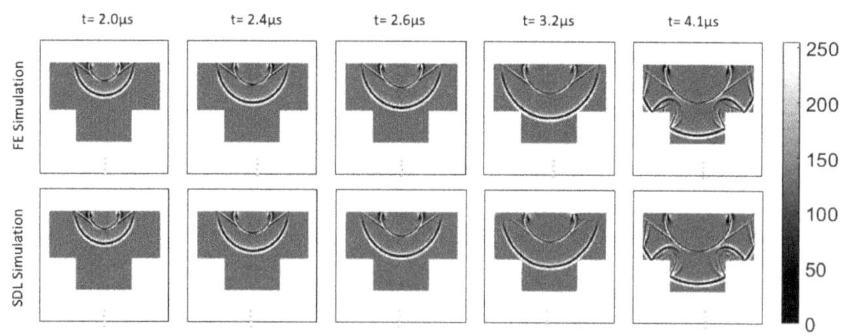

(**b**) FE/SDL simulation comparison for the T-shaped boundary.

Figure 6. *Cont.*

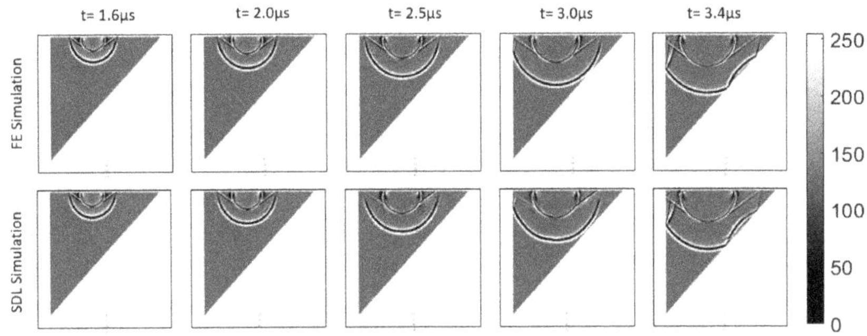

(c) FE/SDL simulation comparison for the triangular boundary.

Figure 6. Reflected wave propagation with different geometrical boundaries: FE and SDL models are used to simulate reflected wave propagation in different geometrical boundaries. The simulations generated using FE and SDL are compared to each other at identical time steps for different time instances. These simulations are modeled with an incident wave of 5 MHz frequency with two cycles, using a single-point excitation source on the top edge. (**a**) Represents the FE/SDL simulation for the curved domain, (**b**) illustrates the FE/SDL wave propagation simulation for the T-shaped domain, and (**c**) shows the wave propagation simulation for the triangular domain. For the illustration, a yellow dashed vertical line is created, manually, in all the frames, to extract the displacement values.

Quantitatively exploring the SDL model prediction on reflected wave propagation, we have extracted the displacement values along the line at x = 20 mm and y = 0 to 40 mm, from all the geometrical domain frames. Figure 7 shows the wavefront line scans between the SDL and FE model, for all the geometrical boundaries datasets. The SDL model-based line scans follow a similar trend to FE simulation-based line scans. To find out the amplitude and TOF difference between the two methods, we have employed MAE, RMSE, and MAPE, to calculate the error on each frame. A similar approach is adopted from Section 4.2. The comparison summary is shown in Table 2 for each frame for each domain. The maximum MAE is in order of 10^{-2}, the maximum RMSE is in order of 10^{-1}, and MAPE is 78.7% on amplitude error, over all the time frames. The maximum MAE is 1.86, the maximum RMSE is 2.56, and MAPE is 5.37% on an error in TOF, between all the geometrical domains. The SDL model could simulate curved domains (Figure 6a), sharp corner reflection (Figure 6b), and diagonal boundaries (Figure 6c), even though we have not used these geometries based datasets while the training process. The SDL model-generated simulations are reasonably in good agreement with FE simulation.

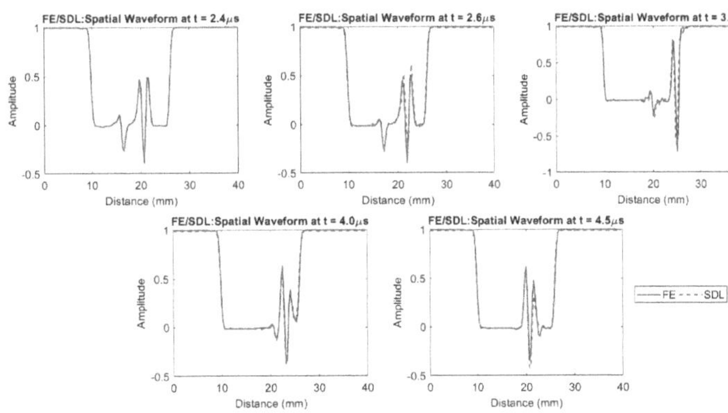

(**a**) Line scans between FE/SDL simulation for the curved boundary at identical time steps.

Figure 7. *Cont.*

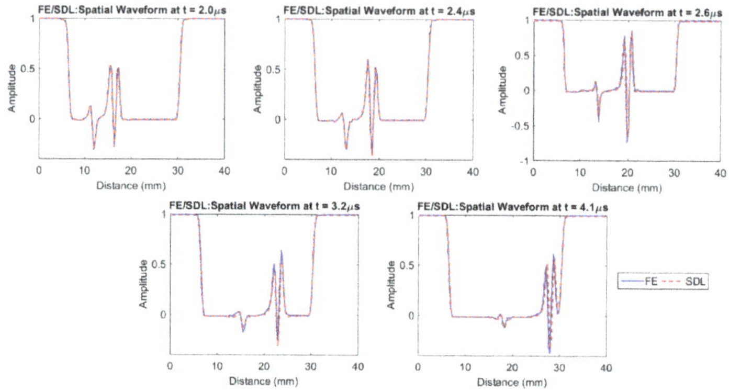

(**b**) Line scans between FE/SDL simulation for the T-shaped boundary at identical time steps.

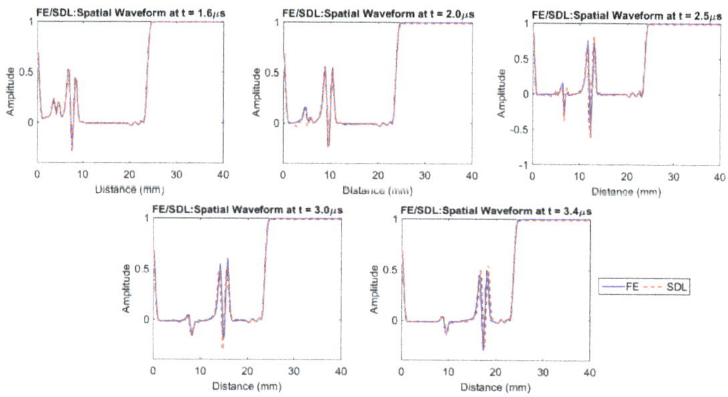

(**c**) Line scans between FE/SDL simulation for the triangular boundary at identical time steps.

Figure 7. Line scans of reflected wave propagation with different geometrical domains: the FE and SDL modeled simulations are compared, quantitatively, by extracting the displacement values in each frame at x = 15 mm and y = 0 to 40 mm for different time instances, shown in Figure 6. (**a**) The line scans are extracted from Figure 6a for the curved domain, (**b**) shows the line scans are extracted from Figure 6b for the T-shaped domain, and (**c**) represents the line scans extracted from Figure 6c for the triangular domain.

Table 2. Reflected wave propagation with different geometrical boundaries: the SDL model predicted simulation is compared with FE simulations. The MAE, RMSE, and MAPE methods are used for calculating the line-scan amplitude and time of flight (TOF) difference between SDL and FE, on the sequence of images at different time instances, shown in Figure 7.

Line Scan	Datasets of Different Geometrical Domain	MAE Sequence of Images at					RMSE Sequence of Images at					MAPE (%) Sequence of Images at				
		T-1	T-2	T-3	T-4	T-5	T-1	T-2	T-3	T-4	T-5	T-1	T-2	T-3	T-4	T-5
Amplitude	Curved	0.00	0.00	0.00	0.01	0.01	0.00	0.10	0.13	0.06	0.06	0.00	55.9	48.3	36.6	26.6
	T-shaped	0.00	0.00	0.00	0.01	0.00	0.00	0.00	0.06	0.08	0.18	0.00	0.00	78.7	73.0	74.0
	Triangular	0.00	0.01	0.01	0.01	0.00	0.00	0.06	0.07	0.07	0.19	0.00	57.8	67.4	131	71.4
Time of Flight	Curved	0.00	0.50	0.86	0.07	0.29	0.00	1.16	1.51	0.27	0.85	0.00	1.05	1.36	0.20	0.90
	T-shaped	0.00	0.00	0.43	0.79	1.86	0.00	0.00	0.76	1.16	2.54	0.00	0.00	1.16	2.10	4.32
	Triangular	0.00	1.14	0.23	1.64	0.64	0.00	1.75	0.48	2.56	1.04	0.00	5.37	1.19	5.34	1.98

Once trained, the SDL model takes approximately 180 s to generate the 1040 frames in one simulation sequence, for 13 μs total simulation time, with the time interval of 0.0125 μs. The conventional FE solver takes 3600 s, using the same computer processor. Hence, we can infer that the proposed SDL model could simulate reflected wave propagation simulation for different geometrical boundaries with reduced computational requirements than FE simulation.

4.2.2. SDL Model Simulation for Reflected Wave Propagation with Varying Frequencies and Cycles

To determine the effectiveness of the proposed SDL model, we have modeled the different scenarios of varying excitation frequencies and the varying number of cycles on rectangular domains with a single-point excitation source on the top edge, to simulate reflected wave propagation. These datasets are not used during the training or testing of the network. We have modeled two sets of FE simulations on the rectangular domain, first, by varying the excitation frequencies of 4 MHz, 5 MHz, and 7 MHz with two cycles of the incident wave, and, second, by modelling with three cycles using 5 MHz frequency. The FE simulations are performed using a similar approach to Section 2.3. Here, we compare the SDL modeled simulation with the FE simulation, for different scenarios. Figure 8 shows the comparison of reflected wave propagation simulation for different excitation frequencies with two-cycle signal width, and the corresponding displacement values extracted along the yellow dashed line from Figure 8 for all the frames are shown in Figure 9a,b. Similarly, Figure 10 shows the comparison of the reflection wave propagation simulation modeled using a varying number of cycles with a 5 MHz central frequency. The corresponding displacement values extracted along the yellow dashed line from Figure 10 for all the time frames are shown in Figure 11a,b. We can observe that the proposed SDL model could generate reflection wave propagation for varying frequencies and cycles. The SDL simulations are in good agreement with FE simulations. The SDL-model-based line scans follow a similar trend as the FE simulation; however, amplitude and TOF differences are present in the line scans as the future time step prediction increases. We have utilized the same metrics that we used to evaluate different geometrical boundaries predictions, and the results are illustrated in Table 3. In this case, we are comparing the MAE, RMSE, and MAPE values for two different scenarios as follows: (1) the simulation between the frequencies of 4 MHz, 5 MHz, and 7 MHz with two cycles and (2) simulation, with the number of cycles being two and three, with 5 MHz frequency. We can notice the following conclusion from Table 3: (1) with an increase in the excitation frequencies other than the training frequency, the MAE, RMSE, and MAPE values increase, but the overall magnitude is in the order of 10^{-2} for MAE and 10^{-1} for RMSE, while the maximum MAPE is 221% on amplitude. (2) Increasing the number of cycles, the MAE, RMSE, and MAPE values are increased, but, overall, are in order of 10^{-2}, for MAE and 10^{-1} for RMSE, while the maximum MAPE is 447 % on amplitude. (3) When increasing the number of cycles, the MAE, RMSE, and MAPE values are increased in amplitude and TOF.

The trend in MAE, RMSE, and MAPE values of amplitude as well as TOF is, closely, following the increasing function with respect to increasing in future time step prediction. The apparent trend from Table 3 is that the performance of the SDL model prediction on reflected wave propagation simulation deteriorates, as the number of future time steps to be predicted increases. Even if we can observe that from Figures 6, 8, and 10, we can infer that the results obtained from the SDL model are in reasonably good agreement with the FE simulation until the t = 4.0 μs frame, for all the domains. Still, there is some compounding errors that started accumulating time instances afterward. The generation of accurate waveform amplitude from back-wall reflection depends on the various parameters in the network architecture, such as the number of hidden layers, kernel size, input–output frames, etc. However, the best suitable learnable hyperparameters and network parameters need to be selected, which may require additional computational resources and time to train the model. The SDL models developed in this work are applicable for solving the forward- and reflected-wave propagation in different 2D geometrical domains. The same approach can be extended for modelling the 3D domain simulation and the wave phenomenon of scattering effect with defects.

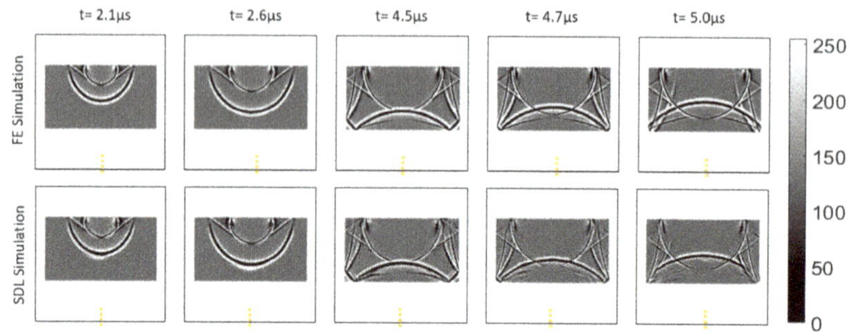

(**a**) FE/SDL simulation comparison for the 4 MHz central frequency, with two cycles.

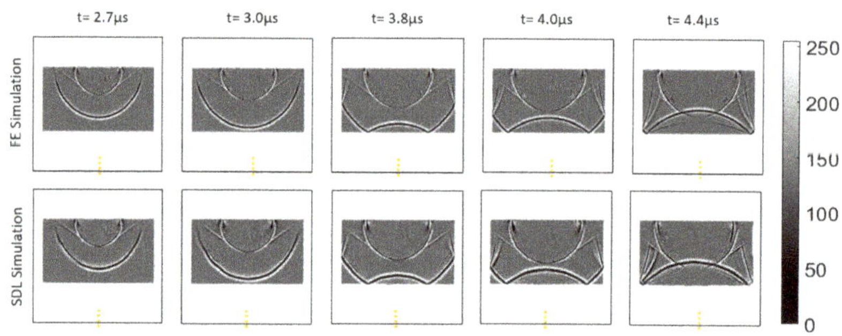

(**b**) FE/SDL simulation comparison for the 7 MHz central frequency, with two cycles.

Figure 8. Reflected wave propagation with varying frequency: the SDL model is used for modelling reflected wave propagation in a rectangular domain for varying frequencies and two cycles with a single-point excitation source on the top edge, compared with FE simulation. (**a**) Shows the wave propagation simulation for 4 MHz central frequency, and (**b**) represents the wave propagation simulation for 7 MHz.

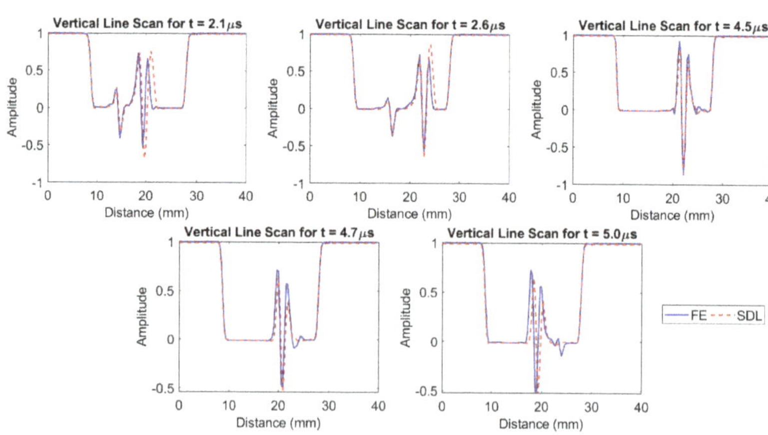

(**a**) Line scans between FE/SDL simulation with the 4 MHz central frequency, at identical time steps.

Figure 9. *Cont.*

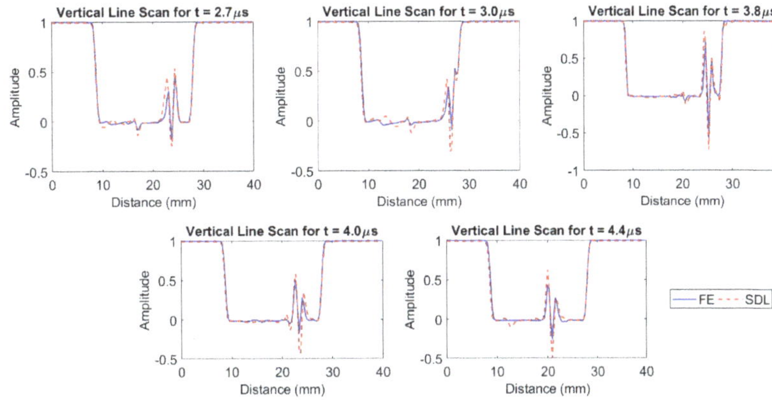

(**b**) Line scans between FE/SDL simulation with the 7 MHz central frequency, at identical time steps.

Figure 9. Line scans of reflected wave propagation with varying frequency: The FE- and SDL-model-generated simulations are compared, quantitatively, by extracting the displacement values, which are extracted at x = 15 mm and y = 0 to 40 mm for different time instances, from Figure 8. (**a**) Represents the line scans that are extracted from Figure 8a for 4 MHz, and (**b**) illustrates the line scans that are extracted from Figure 8b, for 7 MHz in the rectangular domain.

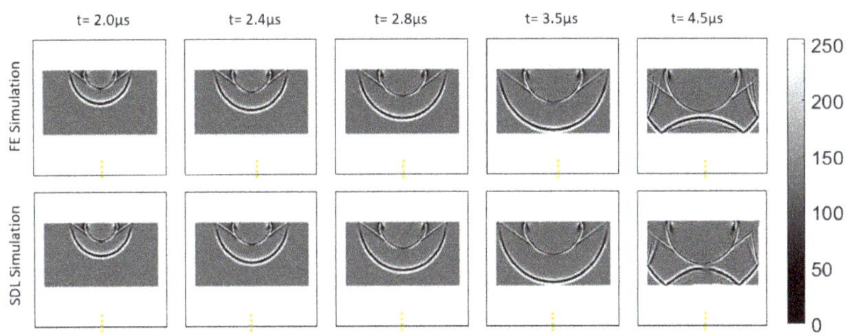

(**a**) FE/SDL simulation comparison for the 5 MHz central frequency with two cycles.

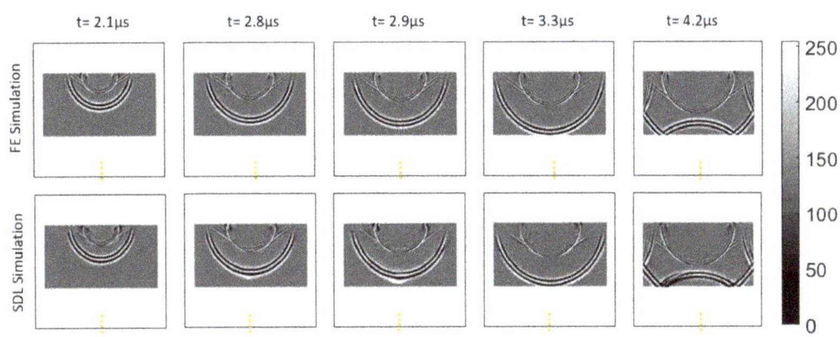

(**b**) FE/SDL simulation comparison for the 5 MHz central frequency with three cycles.

Figure 10. Reflected wave propagation with a varying number of cycles: A qualitative comparison between the FE and SDL simulations modeled in the rectangular domain, with the varying number of cycles with a 5 MHz central frequency with a single-point excitation source on the top edge. (**a**) represents the simulation of two cycles, and (**b**) illustrates the simulation of three cycles, at identical time instances.

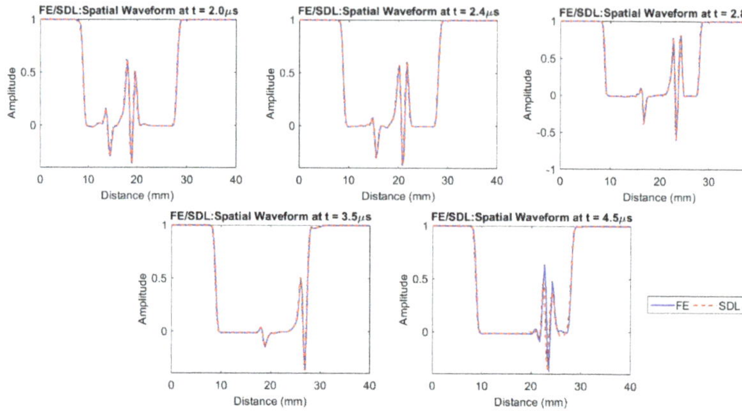

(**a**) Line scans between FE/SDL simulation with the 2-cycles at identical time steps

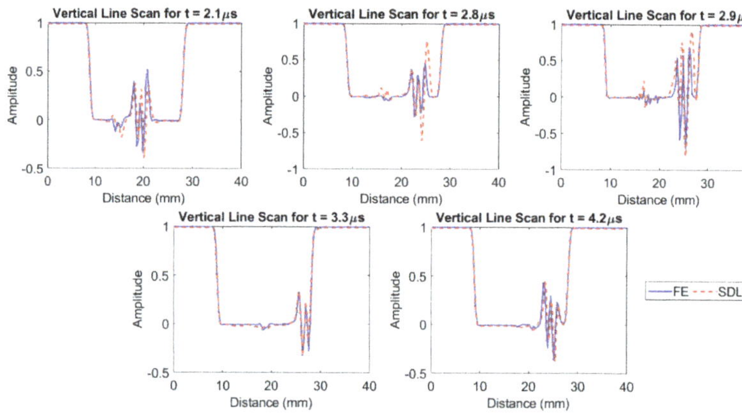

(**b**) Line scans between FE/SDL simulation with the 3-cycles at identical time steps

Figure 11. Line scans of reflected wave propagation with a varying number of cycles: The SDL model predicted simulation with FE simulation to examine quantitatively; the displacement values are extracted at x = 15 mm and y = 0 to 40 mm for different time instances from Figure 10. (**a**) Represents the line scans are extracted from Figure 10a for two cycles, and (**b**) illustrates the line scans are extracted from Figure 10b, for three cycles in the rectangular domain.

Table 3. Reflection wave propagation on the rectangular domain with varying frequencies and cycles: The SDL model line scan amplitude and time of flight (TOF) are compared with FE using MAE, RMSE, and MAPE methods at different time instances, as shown in Figures 9 and 11.

Line Scan	Datasets of Varying Frequencies and Cycles	MAE					RMSE					MAPE (%)				
		Sequence of Images at					Sequence of Images at					Sequence of Images at				
		T-1	T-2	T-3	T-4	T-5	T-1	T-2	T-3	T-4	T-5	T-1	T-2	T-3	T-4	T-5
Amplitude	5 MHz with 2 cycles	0.00	0.00	0.00	0.00	0.01	0.00	0.00	0.00	0.00	0.19	0.00	0.00	0.00	0.00	50.1
	5 MHz with 3 cycles	0.01	0.00	0.02	0.01	0.01	0.23	0.22	0.21	0.05	0.21	118	477	108	99.2	111
	4 MHz with 2 cycles	0.00	0.00	0.01	0.02	0.01	0.24	0.12	0.07	0.13	0.28	132	221	24.8	44.7	54.0
	7 MHz with 2 cycles	0.00	0.01	0.01	0.01	0.01	0.09	0.15	0.09	0.12	0.11	203	111	87.7	116	135
Time of Flight	5 MHz with 2 cycles	0.00	0.00	0.00	0.00	0.56	0.00	0.00	0.00	0.00	0.88	0.00	0.00	0.00	0.00	1.19
	5 MHz with 3 cycles	1.14	1.00	2.86	0.86	1.59	2.03	1.68	5.05	2.72	2.38	3.03	2.39	5.44	2.72	2.95
	4 MHz with 2 cycles	0.36	1.50	0.21	0.57	0.79	0.71	2.87	0.46	0.93	1.16	1.51	2.60	0.78	1.19	2.03
	7 MHz with 2 cycles	0.36	0.64	0.43	0.50	1.14	0.71	0.96	0.65	0.71	2.45	0.81	1.37	0.93	0.94	5.21

5. Conclusions

In this work, the data-driven spatio-temporal deep learning (SDL) model is used to rapidly compute ultrasonic wave propagation for modelling forward and reflected ultrasonic wave propagation in the different geometrical domains as well as varying frequencies and cycles. The SDL is trained with simulation-assisted FE simulation data. Here, we have used two types of training datasets to teach two individual SDL models, using multiple point sources of wave simulation in a single domain and reflected wave propagation from boundaries with different geometries. The SDL can learn representations of the time-domain ultrasonic wave propagation phenomenon from the training datasets, thus employing a data-driven approach to understand the underlying physics to build an AI predictive model to simulate ultrasonic wave propagation and reflection from boundaries. The SDL model can be used for modelling forward and reflected ultrasonic wave propagation simulations, in the different geometrical domains and varying incident wave parameters, such as frequencies and number of cycles with significantly reduced time (20×) and computation resources, which compare well with the FE model simulations. Furthermore, the SDL models have, traditionally, been successfully used for creating conventional 2D wave propagation simulations. Understanding the seismic wave propagation in the geological structure is complex, due to various geological and geometrical structures, to take into account that the various numerical methods are used to model seismic wave propagation in the geological structure but need huge numerical costs and memory storage for handling larger domains with higher frequencies. However, the proposed SDL model can be a direct replacement, where time and a high-processor computer are the constraints for seismic wave propagation simulation. Hence, the SDL model approach may be further extended for modelling the scattering effects of the defects and virtual source imaging in nondestructive evaluation and biomedical imaging.

Author Contributions: T.G. and K.B., together, conceptualized and developed the methodology. T.G. generated the training datasets from a numerical simulation. K.B. helped with the validation and interpretation of results. All authors have equally contributed to the theoretical and numerical foundation of the manuscript. All authors have read and agreed to the published version of the manuscript.

Funding: This research work was supported by SERB, DST, the government of India, and a CII partnernership with Baker Hughes, under the prime minister's Fellowship Scheme for Doctoral Research, 2020.

Institutional Review Board Statement: Not applicable.

Informed Consent Statement: Not applicable.

Data Availability Statement: The datasets generated during and/or analysed during the current study are available from the corresponding author on reasonable request.

Conflicts of Interest: The authors declare no conflict of interest.

References

1. Zhu, Y.C.; AlZoubi, A.; Jassim, S.; Jiang, Q.; Zhang, Y.; Wang, Y.B.; Ye, X.D.; Du, H. A generic deep learning framework to classify thyroid and breast lesions in ultrasound images. *Ultrasonics* **2021**, *110*, 106300. [CrossRef] [PubMed]
2. Liu, Y.; Liu, E.; Chen, Y.; Wang, X.; Sun, C.; Tan, J. Study on Propagation Depth of Ultrasonic Longitudinal Critically Refracted (LCR) Wave. *Sensors* **2020**, *20*, 5724. [CrossRef] [PubMed]
3. Gantala, T.; Balasubramaniam, K. Automated Defect Recognition for Welds Using Simulation Assisted TFM Imaging with Artificial Intelligence. *J. Nondestruct. Eval.* **2021**, *40*, 28. [CrossRef]
4. Chen, S.; Wang, H.; Xu, F.; Jin, Y.Q. Target Classification Using the Deep Convolutional Networks for SAR Images. *IEEE Trans. Geosci. Remote. Sens.* **2016**, *54*, 4806–4817. [CrossRef]
5. Raja, N.; Balasubramaniam, K. Experimental Study on Dispersion Effects of F (1,1) Wave Mode on Thin Waveguide When Embedded with Fluid. *Sensors* **2021**, *21*, 322. [CrossRef]
6. Nakahata, K.; Sugahara, H.; Barth, M.; Köhler, B.; Schubert, F. Three dimensional image-based simulation of ultrasonic wave propagation in polycrystalline metal using phase-field modeling. *Ultrasonics* **2016**, *67*, 18–29. [CrossRef]

7. Guha, A.; Aynardi, M.; Shokouhi, P.; Lissenden, C.J. Shear-Actuation and Vibrometer Reception of Penetrating Ultrasonic Guided Wave Modes in Human Tibia. *Appl. Sci.* **2020**, *10*, 8397. [CrossRef]
8. Moon, S.; Kang, T.; Han, S.; Kim, K.M.; Jin, H.H.; Kim, S.W.; Kim, M.; Seo, H. FEA-Based Ultrasonic Focusing Method in Anisotropic Media for Phased Array Systems. *Appl. Sci.* **2021**, *11*, 8888. [CrossRef]
9. Shivaprasad, S.; Pandala, A.; Krishnamurthy, C.V.; Balasubramaniam, K. Wave localized finite-difference-time-domain modelling of scattering of elastic waves within a polycrystalline material. *J. Acoust. Soc. Am.* **2018**, *144*, 3313–3326. [CrossRef]
10. Dutykh, D.; Katsaounis, T.; Mitsotakis, D. Finite volume schemes for dispersive wave propagation and runup. *J. Comput. Phys.* **2010**, *230*, 3035–3061. [CrossRef]
11. Manidipa, D.; Ghosh, S.K. Data-Driven Approaches for Spatio-Temporal Analysis: A Survey of the State-of-the-Arts. *J. Comput. Sci. Technol.* **2020**, *35*, 665–696. [CrossRef]
12. Hochreiter, S.; Schmidhuber, J. Long Short-Term Memory. *Neural Comput.* **1997**, *9*, 1735–1780. [CrossRef] [PubMed]
13. Schmidhuber, J. Deep learning in neural networks: An overview. *Neural Netw.* **2015**, *61*, 85–117. [CrossRef] [PubMed]
14. Connor, J.; Martin, R.; Atlas, L. Recurrent neural networks and robust time series prediction. *IEEE Trans. Neural Netw.* **1994**, *5*, 240–254. [CrossRef]
15. Rautela, M.; Senthilnath, J.; Moll, J.; Gopalakrishnan, S. Combined two-level damage identification strategy using ultrasonic guided waves and physical knowledge assisted machine learning. *Ultrasonics* **2021**, *115*, 106451. [CrossRef]
16. Srivastava, N.; Mansimov, E.; Salakhudinov, R. Unsupervised Learning of Video Representations using LSTMs. In Proceedings of the 32nd International Conference on Machine Learning, Lille, France, 6–11 July 2015; Bach, F.; Blei, D., Eds.; PMLR: Lille, France, 2015; Volume 37, pp. 843–852.
17. Liu, F., Shen, T.; Luo, Z.; Zhao, D., Guo, S. Underwater target recognition using convolutional recurrent neural networks with 3-D Mel-spectrogram and data augmentation. *Appl. Acoust.* **2021**, *178*, 107989. [CrossRef]
18. Liu, G.Y.; Kong, D.Y.; Hu, S.G.; Yu, Q.; Liu, Z.; Chen, T.P.; Yin, Y.; Hosaka, S.; Liu, Y. Smart electronic skin having gesture recognition function by LSTM neural network. *Appl. Phys. Lett.* **2018**, *113*, 084102. [CrossRef]
19. Sorteberg, W.E.; Garasto, S.; Cantwell, C.C.; Bharath, A.A. Approximating the solution of surface wave propagation using deep neural networks. In Proceedings of the INNS Big Data And Deep Learning Conference, Sestri Levante, Italy, 16–19 April 2019; Springer: Berlin/Heidelberg, Germany, 2019; pp. 246–256.
20. Pan, P.; Chen, H.; Li, Y.; Cai, N.; Cheng, L.; Wang, S. Tumor segmentation in automated whole breast ultrasound using bidirectional LSTM neural network and attention mechanism. *Ultrasonics* **2021**, *110*, 106271. [CrossRef]
21. Ahmed, S.; Kamal, U.; Hasan, M.K. DSWE-Net: A deep learning approach for shear wave elastography and lesion segmentation using single push acoustic radiation force. *Ultrasonics* **2021**, *110*, 106283. [CrossRef]
22. Hughes, T.W.; Williamson, I.A.D.; Minkov, M.; Fan, S. Wave physics as an analog recurrent neural network. *Sci. Adv.* **2019**, *5*, Eaay6946. [CrossRef]
23. Shi, X.; Chen, Z.; Wang, H.; Yeung, D.Y.; Wong, W.k.; Woo, W.c. Convolutional LSTM Network: A Machine Learning Approach for Precipitation Nowcasting. *Adv. Neural Inf. Process. Syst.* **2015**, *28*, 802–810.
24. Wang, Y.; Gao, Z.; Long, M.; Wang, J.; Philip, S.Y. Predrnn++: Towards a resolution of the deep-in-time dilemma in spatiotemporal predictive learning. In Proceedings of the International Conference on Machine Learning, Stockholm Sweden, 10–15 July 2018; pp. 5123–5132.
25. Su, J.; Byeon, W.; Kossaifi, J.; Huang, F.; Kautz, J.; Anandkumar, A. Convolutional Tensor-Train LSTM for Spatio-temporal Learning. *Adv. Neural Inf. Process. Syst.* **2020**, *33*, 13714–13726.
26. Gantala, T.; Balasubramaniam, K. DPAI: A Data-driven simulation-assisted-Physics learned AI model for transient ultrasonic wave propagation. *Ultrasonics* **2022**, *121*, 106671. [CrossRef] [PubMed]
27. Valencia, M.L.; Cantwell, C.D.; Fotiadis, S.; Pignatelli, E.; Bharath, A.A. Simulating Surface Wave Dynamics with Convolutional Networks. *arXiv* **2020**, arXiv:2012.00718
28. Fotiadis, S.; Pignatelli, E.; Valencia, M.L.; Cantwell, C.; Storkey, A.; Bharath, A.A. Comparing recurrent and convolutional neural networks for predicting wave propagation. *arXiv* **2020**, arXiv:2002.08981.
29. Raissi, M.; Perdikaris, P.; Karniadakis, G. Physics-informed neural networks: A deep learning framework for solving forward and inverse problems involving nonlinear partial differential equations. *J. Comput. Phys.* **2019**, *378*, 686–707. [CrossRef]
30. Shukla, K.; Di Leoni, P.C.; Blackshire, J.; Sparkman, D.; Karniadakis, G.E. Physics-informed neural network for ultrasound nondestructive quantification of surface breaking cracks. *J. Nondestruct. Eval.* **2020**, *39*, 1–20. [CrossRef]
31. Noakoasteen, O.; Wang, S.; Peng, Z.; Christodoulou, C. Physics-Informed Deep Neural Networks for Transient Electromagnetic Analysis. *IEEE Open J. Antennas Propag.* **2020**, *1*, 404–412. [CrossRef]
32. Rajagopal, P.; Drozdz, M.; Skelton, E.A.; Lowe, M.J.; Craster, R.V. On the use of absorbing layers to simulate the propagation of elastic waves in unbounded isotropic media using commercially available Finite Element packages. *NDT E Int.* **2012**, *51*, 30–40. [CrossRef]
33. Yu, B.; Tola, K.D.; Lee, C.; Park, S. Improving the Ability of a Laser Ultrasonic Wave-Based Detection of Damage on the Curved Surface of a Pipe Using a Deep Learning Technique. *Sensors* **2021**, *21*, 7105. [CrossRef]

applied sciences

MDPI

Article

An Experimental and Theoretical Comparison of 3D Models for Ultrasonic Non-Destructive Testing of Cracks: Part I, Embedded Cracks

Michel Darmon *, Gwenael Toullelan and Vincent Dorval

Université Paris-Saclay, CEA, List, F-91120 Palaiseau, France; gwenael.toullelan@cea.fr (G.T.); vincent.dorval@cea.fr (V.D.)
* Correspondence: michel.darmon@cea.fr

Featured Application: The proposed models' validation will enable the proper use of ultrasonic simulation for designing NDT methods for embedded crack detection and characterization.

Abstract: Ultrasonic Non-Destructive Testing (NDT) methods are broadly used for detection and characterization/imaging of cracks. Simulation is of great interest for designing such NDT methods. To model the ultrasonic 3D response of a crack, ultrasonic high frequency asymptotic (semi-analytical) models (such as the Physical Theory of Diffraction—PTD) are known to provide accurate predictions for most classical NDT configurations, and 3D numerical models have also emerged more recently. The aim of this paper is to carry out for the first time an experimental and theoretical comparison of 3D models for ultrasonic NDT of embedded cracks in 3D configurations. Semi-analytical models and a hybrid 3D FEM strategy—combining high-order spectral Finite Elements Method (FEM) for flaw scattering and an asymptotic ray model for beam propagation—have been compared. Both numerical validations and comparisons between simulation and experiments prove the effectiveness of PTD in numerous configurations but validate and demonstrate the improvement provided by the 3D hybrid code, notably for small flaws compared to the wavelength and for shear waves.

Keywords: Non-Destructive Testing (NDT); ultrasounds; crack; scattering; numerical models comparison; experimental validation

Citation: Darmon, M.; Toullelan, G.; Dorval, V. An Experimental and Theoretical Comparison of 3D Models for Ultrasonic Non-Destructive Testing of Cracks: Part I, Embedded Cracks. *Appl. Sci.* **2022**, *12*, 5078. https://doi.org/10.3390/app12105078

Academic Editor: Marco Scalerandi

Received: 17 March 2022
Accepted: 12 May 2022
Published: 18 May 2022

1. Introduction

Nowadays, simulation plays a key role in the design and demonstration of performances of Non-Destructive Testing (NDT) methods. Conventional ultrasonic inspection methods have been used for the NDT of cracks for several decades. The analysis of the echoes generated by cracks, generally specular or diffraction ones, can lead to their detection or imaging. To simulate the ultrasonic response of a crack during an ultrasonic inspection, system models [1,2] have been devised. Such models consist of modeling the propagated beam, its interaction with scatterers and the reception by a probe.

Historically, semi-analytical models were first studied to model flaw scattering. Two classical ones were revisited in elastodynamics from the end of 1970s: the Kirchhoff approximation (KA) [3] and the Geometrical Theory of Diffraction (GTD) [4], which have complementary areas of validity. KA enables handling reflections from planar or multi-faceted cracks, volumetric voids [5] and impedance interfaces [6,7]. GTD is preferred to KA for simulating scattering by crack edges (notably for TOFD configurations [8,9]) but fails in the near-incident and specular reflection directions (shadow boundaries). A GTD solution has also recently been proposed for wedge scattering [10–12]. Several system models based on KA [3] or GTD [13,14] were conceived first for 2D configurations and then developed in 3D for KA [1,5] and for GTD [15], then using an incremental Huygens model [16]. Strategies were then created to take into account both reflection and edge diffraction. Recently, an approach based on the Huygens–Fresnel diffraction theory has been proposed [17] to evaluate the signals diffracted by the edges, but it is equivalent [18] to KA and limited to 2D time-of-flight diffraction (TOFD) configurations; it also cannot describe head waves [19,20]. Uniform corrections of GTD which are able to predict a spatially uniform scattered field has led to recent developments for elastic waves. An elastodynamic version of the Uniform Asymptotic Theory (UAT) [21] was first proposed [4]; however, the UAT requires an artificial extension of the scattering surface and additional costly tracing of fictitious rays reflected on this extended surface. The elastodynamic Uniform Theory of Diffraction (UTD) [22] has also been established for the scattering from a stress-free half-plane and represents a good alternative even if leading to small differences with UAT. Finally, an ultrasonic system model accounting for both reflection and diffraction has been proposed for 3D crack-like flaws in 3D configurations [23]. It is based on the Physical Theory of Diffraction (PTD), initially developed in electromagnetism [24] and extended to elastodynamics [25]. For a half-plane, PTD has been shown to be identical to UAT at the used leading order. Moreover, the region of the model's validity has been extended [26] to cover some transition zones surrounding critical rays, where the shear diffracted waves and head waves interfere. Important efforts have been made to propose an analytical modelling of some head waves contributions, notably lateral waves in the TOFD configuration [27,28]: for planar surfaces, lateral waves are head waves propagating along such surfaces, and recent advances have been obtained in their modelling [29–33]; for complex or cylindrical surfaces, lateral waves can include bulk wave contributions [34]. The PTD-based system model [26] can be considered as the most sophisticated and valid semi-analytical approach existing at the present date.

Nevertheless, the ultrasonic high frequency asymptotic models, despite providing accurate predictions in a wide range of situations, have a limited domain of validity identified through various experimental validation campaigns [35–37]: they can fail at predicting responses from defects in some complex configurations when creeping and bulk head waves or caustics occur. In the same way, their precision drops for defects with characteristic dimension of the order of the wavelength. To overcome these limitations, the use of hybrid models such as the 2D CIVA ATHENA model [5] combining numerical and asymptotic methods has been demonstrated as an effective strategy. However, the computational cost of the numerical solver remains a crucial issue for 3D simulations. Different numerical codes [38–40] have been devised for simulating flaw scattering but have limited performances for 3D problems. In this paper, we use a 3D hybrid model [41,42], coupling FEM (Finite Element Method) for flaw scattering and a ray model for beam propagation. In this model, the numerical parameters are set automatically, and the use of high-order spectral elements [43,44] and techniques of domain decomposition ensures high numerical performance.

A comparison of analytical, semi-analytical and numerical techniques has been performed for 2D/3D ultrasonic field modeling [45]. For flaw scattering simulation, different analytical approximations (such as GTD, UAT, UTD, KA and PTD) have been compared for the wedge scattering problem in acoustics [46] or for the rigid half-plane problem in elastodynamics [25]. When modelling a complete inspection with both beam propagation and flaw scattering, a comparison between an analytical model (integral equation [47]) and two numerical (COMSOL FEM and finite differences) models has been recently carried out in some 2D ultrasonic NDT configurations [48]. A comparison between the semi-analytical PTD model and the hybrid 2D numerical model CIVA ATHENA model has also been studied in detail for 2D configurations [26]. Preliminary numerical comparisons between PTD and the 3D hybrid FEM model have been reported in two simple 2D configurations [42]. To our knowledge, no study has yet compared the most advanced semi-analytical (as PTD) and numerical 3D crack ultrasonic NDT models in 3D configurations and their results with experience. The models' comparison proposed here relate to embedded planar defects inspected in 3D configurations with a special interest in defect characteristic lengths of the order of the wavelength for which asymptotic models are no longer valid. The models involved are the semi-analytical (GTD, PTD) and the hybrid numerical FEM models. We also present comparisons between simulations and experiments for the models' validation. The analytical and numerical simulation methods are briefly recalled in Section 2 of this paper; then, numerical (Section 3) and experimental (Section 4) validations are described for 3D-embedded flaws.

2. Simulation Methods

All the simulation methods used are integrated in the CIVA NDT platform [49] for simulation [2] and analysis. The semi-analytical models (GTD, PTD) have been widely described in Refs. [15,26]. The Physical Theory of Diffraction (PTD) combines GTD (ray theory) and Kirchhoff Approximation (KA—integral method) and applies to arbitrary large scatterers. PTD [25,26] replaces the Kirchhoff edge diffraction contribution by GTD, the asymptotical approximation of the edge problem solution. The PTD scattered field is the sum of the Kirchhoff scattered field and a modified GTD field, whose diffraction coefficient is the difference between the GTD and Kirchhoff edge diffraction coefficients. Finally, this summation enables building a PTD total field which is spatially uniform, contrary to GTD, which leads to several singularities (shadow boundaries).

The 3D hybrid FEM model is a coupling method [42] between an FEM scattering model with a ray-based asymptotic field model (the CIVA pencil method for field computation [2]). This hybrid methods consequently consists of simulating the propagation field until the defect vicinity employing the CIVA pencil method [50] and then the scattering of this field from the defect thanks to a numerical approach. The flaw's echographic signal is finally obtained by the way of the Auld's reciprocity principle [51], which expresses it as a combination of fields from the emitter and receiver with and without a scattering defect. The numerical solver is built upon a "dedicated" high-order spectral finite element method [43,44], which enables us to address 3D configurations, and second, the constitution of the numerical parameters are automatically deduced from the inspection configuration. A theoretical description of the hybrid FEM method is given in Appendix A. Some additional details (meshing around the defect, numerical scheme, etc.) can be found in [41,52]. Different calculation strategies adopted in the CIVA 2017 release used in this paper depend on the defect's location (embedded or surface breaking) and on the sensibility area defined by the user to reduce the computation time. For breaking flaws, the coupling reciprocity integral [51] is carried out on both the flaw and component surfaces inside the sensibility area (in red in Figure 1b).

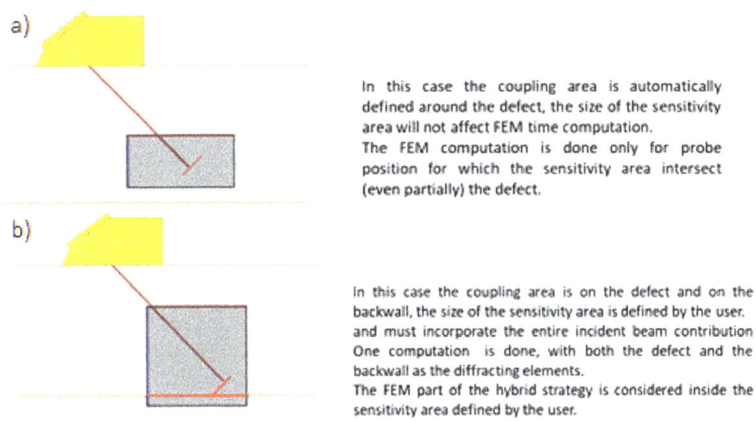

Figure 1. Different coupling strategies used in the CIVA hybrid FEM model for (**a**) embedded and (**b**) breaking cracks.

Finally, the simulation methods utilized here both use the same ray-based asymptotic field model (the CIVA pencil method for field computation [2]) to simulate wave propagation, but they differ in the method they employ to model flaw scattering. The model based on GTD or on the Physical Theory of Diffraction (PTD) is a semi-analytical high frequency model, whereas, for the hybrid CIVA-FEM model, spectral finite elements (using high order polynomials) are used.

For contact probes, only longitudinal waves are taken into account in the probe by the simulation, and since a coupling gel is used in experiments to make a bond between the probe and the specimen, a solid–solid sliding interface model is used for transmission coefficients at the specimen interface.

3. Numerical Validations of 3D Embedded Flaws Simulations

In a first step prior to this study, numerical validations were performed in a wide range of classical 2D NDT configurations by comparing simulations from the proposed hybrid FEM method, the asymptotic PTD model [10] and another coupling method (2D CIVA ATHENA). Such numerical validations showed that the PTD approach, which has a lower computation time, breaks downs for small flaw heights compared to the wavelength as shown in [10]. Preliminary comparisons between PTD and hybrid FEM have been reported by some of the present authors in two simple 2D configurations [42] and lead to the same conclusions. The hybrid FEM method appears thus be a solution to overcome the limitation of analytical models for small flaws.

The proposed 3D hybrid FEM model is evaluated here by comparing numerically it with existing analytical 3D models as the 3D PTD model. In these validations, crack inspections in ferritic steel components are simulated using several different configurations: TOFD [28] configurations, which are usually used to detect and characterize cracks from their edge diffraction echoes, and pulse echo configurations with various incidences to study both the reflection (normal incidence) and diffraction (oblique incidence) from the crack. The influence of the flaw height and extension is also evaluated notably for small sizes to determine the improvement provided by the FEM model in 3D configurations. For the simulations, the following mechanical parameters have been considered: longitudinal and shear sound velocities of 5900 and 3230 m/s and a density of 7.9 g/cm^3. All the input data for simulation are exactly the same for all the used models. These data describe the inspection configuration [1]: the characteristics of the component, the used probes, the inspection scanning, the inspected flaws and the simulation settings.

3.1. Longitudinal Waves

P waves are first considered. The first TOFD configuration is described in Figure 2a. The specimen includes a 5 mm (or 0.5 mm) high rectangular crack with a varying extension (perpendicular to the plane of Figure 2a). It is inspected using P45 waves in TOFD mode at 2.25 MHz.

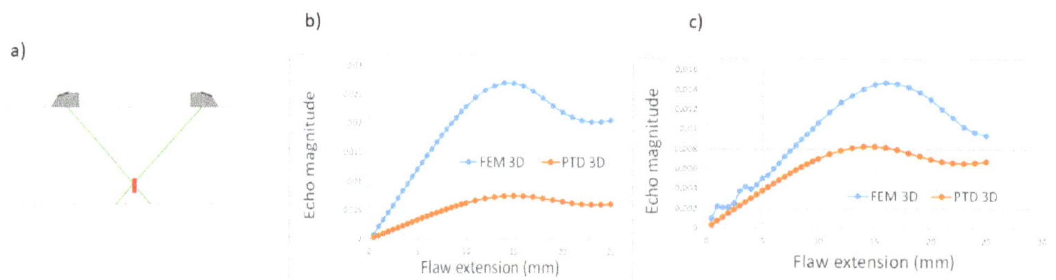

Figure 2. (**a**) Inspection of a 5 mm high (height represented by a red segment) and varied extent rectangular crack inspected using P45 waves in TOFD mode at 2.25 MHz; Echo magnitude versus the flaw extension (**b**) for the 0.5 mm high crack and (**c**) for the 5 mm high crack.

The first numerical comparison of models consists in studying the effect of the flaw extension on maximal flaw echo amplitude (Figure 2b). For both the 0.5 mm and 5 mm high cracks, very important differences are observed between the FEM and PTD models in terms of maximal echo amplitudes. Such differences are more important for the smallest height. Nonetheless, for the 5 mm height differences are not negligible even for large extensions for which the echo magnitude tends to stabilise (25 mm is close to the beam width). For this 5 mm height, the echoes from the top and bottom edge are well dissociated, and the amplitude noticeably oscillates as a function of the extension for the FEM model. These oscillations may be due to interference between waves scattered from each edge and their corners, the latter being better predicted by FEM than by the PTD model. The PTD model used here relies on the incremental Huygens model to account for the finite edge extension [16,26,53,54].

The second numerical comparison in the previous P45° TOFD configuration (Figure 2a) consists of studying the effect of the flaw height on maximal flaw echo amplitude for the 25 mm extent crack (extension large enough to include the beam width). The flaws' heights from 0.5 to 5 mm are studied, which correspond to adimensional wave number 0.6 < ka < 6. Again, significant differences are observed in Figure 3a between the FEM and PTD models in terms of maximal echo amplitudes, especially for the smallest heights. Whereas the echo amplitude is quite constant versus the height for the PTD, the FEM model predicts higher amplitudes when the scattered waves are mixed for small heights.

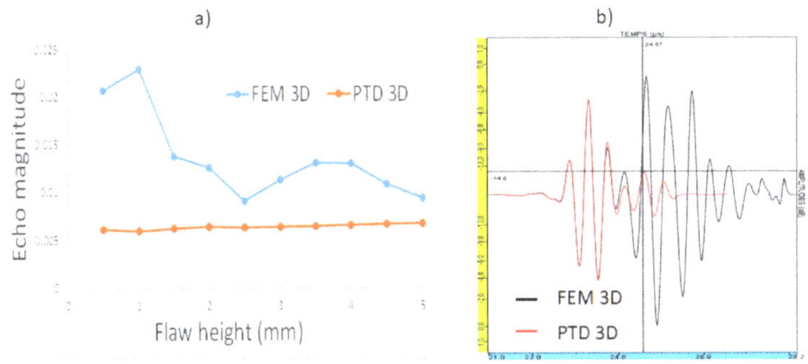

Figure 3. Inspection of a 25 mm extent rectangular crack inspected using P45 waves in TOFD mode at 2.25 MHz: (**a**) Echo magnitude versus the flaw height; (**b**) Ascans simulated both by the 3D FEM and PTD models for the 5 mm high crack.

By comparing the Ascans simulated by both the 3D FEM and PTD and generated by the 5 mm high and 25 mm extent rectangular crack (Figure 3b), the two models lead to a similar prediction from the top edge response, but PTD provides an underestimation for the bottom edge echo (which yields the maximal amplitude from the flaw). This is due to the use of P45° waves and the presence at the 38° inspection angle of a minimum predicted by GTD for the amplitude of the bottom edge echo versus the angle of observation in TOFD. As shown in [55], GTD and CIVA ATHENA (hybrid FEM model) both predict a minimum, but it is much more pronounced with GTD. Consequently, the difference of prediction of this minimum by the two models explains the gaps obtained in terms of echo magnitude even for large flaw heights in TOFD P45°. The FEM model also predicts later contributions, notably due to the Rayleigh waves propagating on the flaw surfaces.

The previous section concludes that the P45° waves TOFD configuration exhibits differences between PTD and FEM predictions due to the presence of a minimum in the directivity pattern for a particular observation angle. In the next P60° waves TOFD configuration, the PTD and FEM models are expected to lead to similar results for large flaws. When comparing the Ascans simulated both by the 3D FEM and PTD models using inspection at 60° still at 2.25 MHz (configuration shown in Figure 4a), the two models give rise to similar echoes for both the top and bottom edges (Figure 4b). The FEM model also predicts later contributions. The amplitude difference between the 3D FEM and PTD models observed using P45 waves in TOFD is not reproduced for P60 waves.

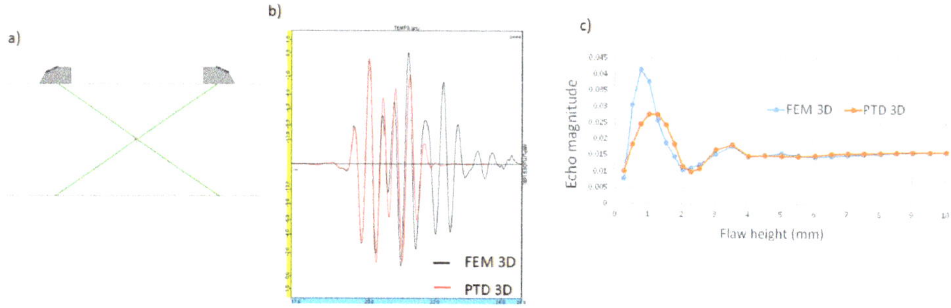

Figure 4. (**a**) Inspection of a 25 mm extent rectangular crack of varying height using P60 waves in TOFD mode at 2.25 MHz. (**b**) Superimposition of Ascans simulated both by the 3D FEM and PTD for the echoes generated for the 5 mm high crack. (**c**) Echo magnitude versus the flaw height.

In the previous configuration with P60 waves at 2.25 MHz, another validation consists of studying the effect of the flaw height on maximal flaw echo amplitude. The flaws' heights from 0.25 to 10 mm are studied, which corresponds to $0.3 < ka < 12$. For small heights (especially for heights less than 1 mm; $ka < 1.2$), significant differences in terms of echo magnitude are observed between the two models.

Comparisons of the Ascans simulated by 3D PTD and 3D FEM (Figure 5) show that according to finite elements, the two different echoes arising from the top and bottom edges are relatively well dissociated for heights higher than 5 mm ($ka > 6$). From this height, these two echoes are very well simulated by PTD, but the later Rayleigh waves are nevertheless not taken into account in the PTD model. For small heights, important differences are observed between the two models in terms of waveforms, which is most likely due to primary edge diffracted waves and secondary diffractions of multiply reflected Rayleigh waves (not modelled in the PTD model) having close time of flights.

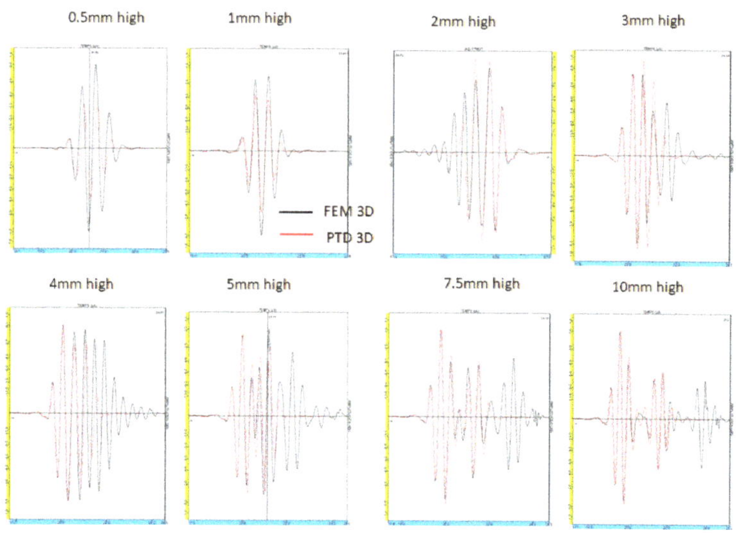

Figure 5. Inspection of a 25 mm extent rectangular crack of varied height inspected using P60 waves in TOFD mode at 2.25 MHz: Ascan versus the flaw height.

3.2. Shear Waves

3.2.1. Immersion Pulse Echo Mode: S45° Waves—Various Incidences on the Flaw

Figure 6a describes the first validation case in the pulse echo mode involving defects of various heights, a fixed 0.5 mm extension and of an arbitrary tilt angle α. A 12.7 mm diameter immersed transducer positioned 20 mm above the component generates transversal 45° waves at 5 MHz.

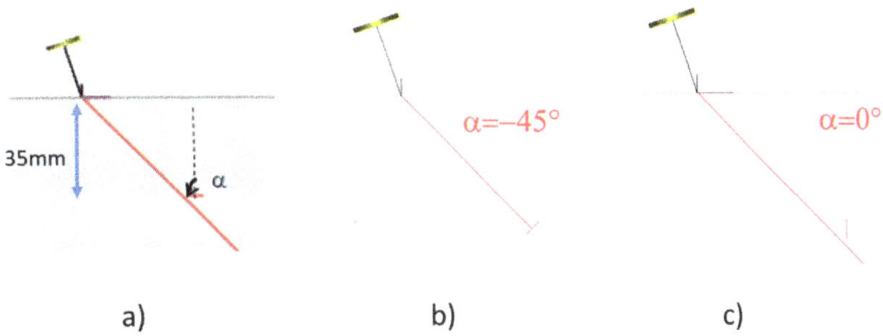

Figure 6. (**a**) pulse echo configuration with the SV45° wave at 5 MHz of a rectangular defect of varying height (5 mm here—red segment), 0.5 mm extent and various tilts; (**b**) α = −45°, specular reflection configuration; (**c**) α = 0°, classical configuration of a vertical flaw.

The probe is not very divergent, and the flaw is in the area of the maximal field amplitude (33–40 mm depths). The flaw extension is small (0.5 mm extent), which will enable us to study the 3D effects of the flaw size in configurations unfavourable to analytical models such as PTD.

In Figure 7, the 3D FEM and PTD models predict the flaw echo amplitude versus the tilt angle α for three different flaw heights: 5, 1 and 0.5 mm (ka ~ 2.4 both in height and extension in the latter case). An improvement consisting in a smoothing of PTD coefficients near critical angles has been devised and described in [26]: it is referred to as "PTD3D critical smoothing". The results referred to as "PTD3D" were obtained without this treatment, using a GTD code based on previous works [56]. Its response exhibits peaks around critical angles.

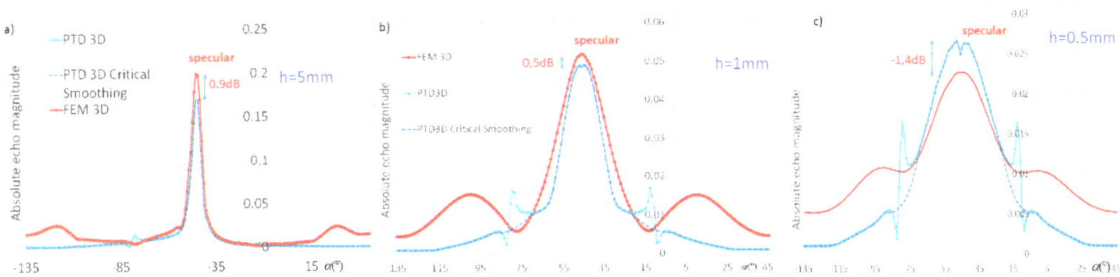

Figure 7. A planar component containing an embedded rectangular flaw of 0.5 mm extent and of height: (**a**) 5, (**b**) 1 and (**c**) 0.5 mm. Comparison of FEM and PTD (standard or with smoothing around the critical angle) models.

For specular reflection, PTD generally leads to a relatively good approximation and to simulated results close to that of FEM even for small heights. Nonetheless, for the smallest height (0.5 mm), the PTD response versus the tilt angle exhibits two local maxima, contrary to FEM. This suggests that some phenomena are better predicted by the numerical model than by PTD, possibly due to interferences between the waves reflected from the flaw surface and those diffracted from the contour.

The difference between PTD and FEM can be significant away from the specular direction, especially beyond the critical angle. This is due to head waves not being accounted for in the PTD model presented here. Consequently, the 3D extension of the flaw can be assumed to be a source of inaccuracies for the PTD model, possibly due to quantitative errors in the PTD prediction of corner diffractions.

PTD differs most significantly from the FEM reference—for specular reflection when the flaw height or extent is small—near and above the critical angle for S waves (head waves) and small heights.

3.2.2. Contact Pulse Echo Mode: S45° Waves—Vertical Flaw

The next numerical comparison case involves the inspection of a vertical rectangular crack inspected using SV45 waves in pulse echo mode at 2.25 MHz (Figure 8a). The used probe emitting at 2.25 MHz is a contact circular planar probe of 6.35 mm diameter. The first numerical comparison consists in studying the effect of the flaw extension on maximal flaw echo amplitude for the 5 mm high crack (Figure 8b). For small extensions (<4 mm), some amplitude oscillations occur, probably due to interferences between waves scattered from both the edges and the corners. Important differences are observed between the FEM and PTD models in terms of maximal echo amplitudes for all extensions. It is unexpected that differences are obtained even for large extensions (25 mm ~ beam width) in the case of a 5 mm high crack corresponding to a large dimensionless factor ka ~ 11.25. In order to analyze the previous differences, the Ascans simulated by 3D FEM and 3D PTD are plotted in Figure 9 for the 40 mm extent (the largest extension in Figure 8b) and 5 mm high rectangular crack both at 2.25 MHz (corresponding to the case studied in previous Figure 8) and also at 5 MHz. At 5 MHz, the Ascan simulated by 2 D FEM CIVA ATHENA is also provided. The first echo chronologically observed in Figure 9a,b is obviously the top edge diffracted wave. Echoes generated by bulk wave diffraction from the bottom edge (after 40.5 µs) and by later waves can be better distinguished at the higher frequency (5 MHz). A head wave arrives at almost the same time as a wave diffracted from the bottom edge: it corresponds to the diffraction at the top edge of the incident S wave path into a P creeping head wave, which is then rediffracted into a bulk S wave at the bottom tip (see Figure 9c), and its time of flight, which can be theoretically calculated by ray theory, is close to that of the bottom edge diffraction (as explained later using Table 1). Chronologically, the first wave arriving after the bottom edge diffracted wave is a bulk diffraction of a Rayleigh wave. When the beam hits the top tip, a Rayleigh wave (R_1) is generated. It propagates along each crack face towards the opposite tip. Upon reaching the bottom tip, R_1 sheds the bulk S_5 wave. The path of this Rayleigh wave then diffracted in the bulk corresponds to that of the so-called S_5 wave, which is due to a secondary diffraction (see its path drawn in Figure 6 of Ref. [26]).

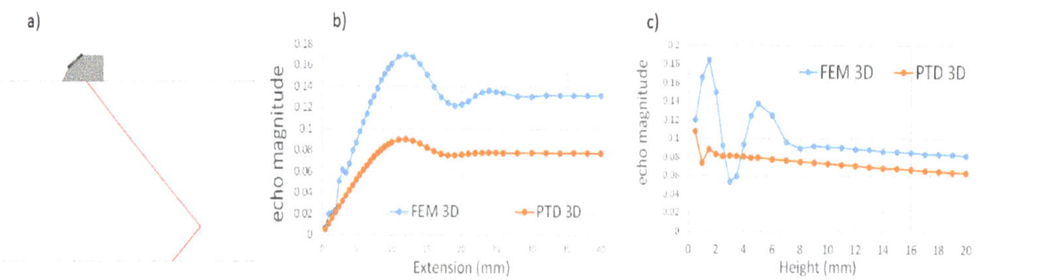

Figure 8. (**a**) Inspection of rectangular crack using SV45 waves in Pulse Echo mode at 2.25 MHz (circular planar probe of 6.35 mm diameter), (**b**) of 5 mm height and varying extension and (**c**) of a 34 mm extension and varying height.

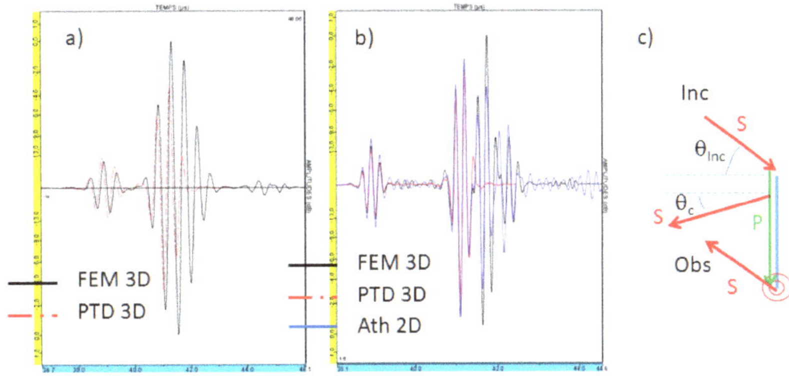

Figure 9. Inspection of a 40 mm extent and 5 mm high rectangular crack inspected using SV45 waves in Pulse Echo mode (circular plane probe of 6.35 mm diameter); Ascans simulated by 3D FEM and 3D PTD: (**a**) at 2.25 MHz (**b**) at 5 MHz with also the normalized Ascan modelled by 2D FEM Civa/Athena. (**c**) S head waves diffracted from a crack (in blue) under 45° incidence: the S wave shed by the bottom tip irradiated by the P creeping wave coming back to the probe at 45° direction and the S head wave radiated during the propagation of the P creeping wave along the crack surface.

Table 1. Analysis of Ascans simulated by 3D FEM and 3D PTD of the inspection of a 40 mm extent and 5 mm high rectangular crack using SV45 waves in Pulse Echo mode (circular planar probe of 6.35 mm diameter) at 5 MHz: time of flight of the different waves theoretically calculated using ray theory or obtained after 3D FEM simulation; PTD amplitude versus 3D FEM (dB).

	Top Edge Diffraction	Head Wave	Bottom Edge Diffraction	Secondary Rayleigh Wave Diffraction
Theoretical time of flight (μs)	0	1.94	2.19	2.81
FEM simulated time of flight (μs)	0	Lost in the bottom edge echo	2.12	2.70
PTD amplitude versus 3D FEM (dB)	−0.9	Not calculated by the PTD model	−1.0	Not calculated by the PTD model

When considering the time of flight of the top edge diffracted wave as the time origin, the time of flight of the different waves can be theoretically calculated using ray theory and compared to that obtained after 3D FEM simulation (see Table 1).

The comparison of the times of flight shows that the second echo chronologically observed corresponds to the bottom edge diffracted wave interfering slightly with the head wave. Such a head wave is of lower amplitude since it decreases quickly during its propagation along the flaw surface and since the incidence is not critical (in Figure 9c, $\theta_{inc} = 45°$ whereas $\theta_c = 33°$ for steel). The third echo corresponds to the secondary Rayleigh wave diffraction S_5. PTD predicts correctly the amplitudes of the primary bulk waves diffracted by the two edges but the PTD model does not simulate the secondary Rayleigh wave diffraction S_5. It is this latter wave which leads to the more important amplitude in the 3D FEM Ascan.

At the lower frequency of 2.25 MHz, the secondary Rayleigh wave diffraction echo coalesces with the bottom edge diffracted echo, which explains the difference of simulated maximal amplitudes between PTD and FEM shown in Figure 8b for large extensions.

After studying the effect of the flaw extension, a second numerical comparison (Figure 8c) consists of studying the effect of the flaw height on maximal flaw echo amplitude for the 34 mm extent crack (extension wider than the beam size). For rather small heights (<8 mm), the FEM model highlights oscillations probably due to interferences between waves scattered from the edges, and Rayleigh waves multiply reflected at edges and also to a lesser extent to multiply reflected head waves. For large heights, a difference of 2 dB is observed between the two models. The maximal echo amplitude slightly decreases with height due to the decrease in the field with the increasing distance from the bottom edge.

4. Experimental Validations

In a second step of validation, experimental measurements have been carried out on ferritic steel components. First, TOFD inspections of large flaws involving 3D effects (sometimes very important) have been considered.

4.1. Inspection of Large Flaws

Several notches and holes have been fabricated in a planar specimen (Figure 10a). In order to study the influence of the orientations of both probes and flaws, two 6.35 mm diameter probes emitting 45° P-waves at 2.25 MHz have been positioned in a TOFD configuration with a 60 mm Probe Center Separation and rotated from the 0° skew angle to the 34° skew angle. Figure 10b,c present a typical case of the 11° skew.

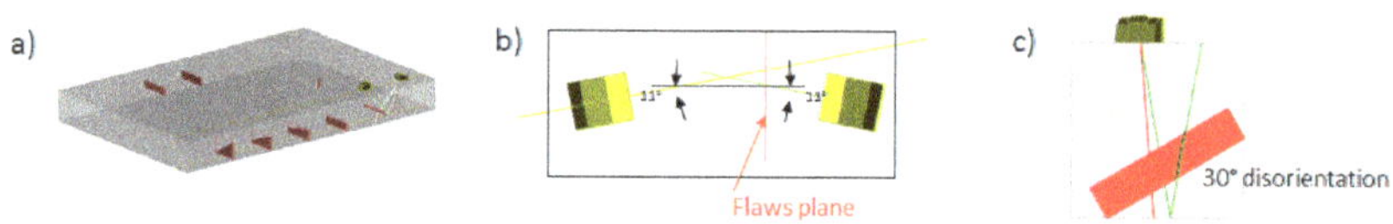

Figure 10. (**a**) A planar component containing disoriented backwall breaking flaws and a 2 mm diameter SDH; the TOFD configuration with the 11° probes' skew: (**b**) top view and (**c**) side view showing the flaw used for simulating the top edge of the 30° disoriented notch.

Measurements have been carried out on a rectangular 0° flaw and three defects (with vertical disorientation of 10, 20 and 30° for the top edge) and calibrated against a 2 mm diameter SDH. The resulting experimental B-Scan is shown in Figure 11a. The variation in the amplitude of the echo from the top tip (equivalent to that of an embedded flaw) with the vertical disorientation is displayed in Figure 11b. The results simulated by different models are compared in terms of echo amplitude: (1) an older model based on the so-called 2.5D GTD [5], which involves the projection of the incoming and scattered wave vectors on the plane normal to the flaw edge and 2D GTD coefficients related to these projections, (2) 3D PTD model and (3) the hybrid 3D FEM model.

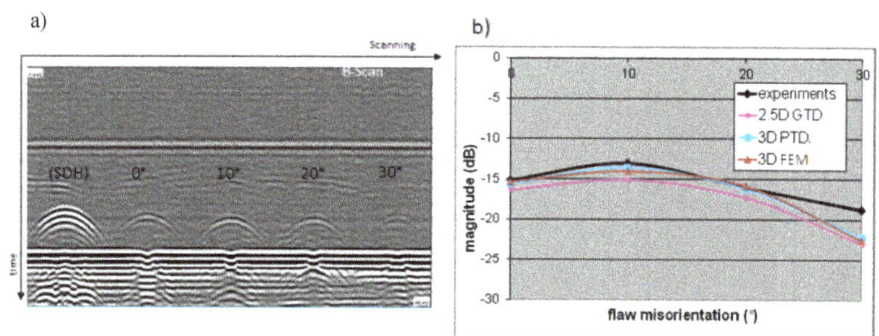

Figure 11. (**a**) Experimental B-scan obtained when scanning the side drilled hole (SDH) and the three defects (with vertical misorientation of 10° to 30° for the top edge); (**b**) validation of the 2.5D GTD, 3D PTD and the hybrid 3D FEM models against the measured echoes from the top tip of misoriented backwall breaking flaws.

The differences between the 3D PTD model, the hybrid 3D FEM model and experimental data are less than 1 dB, except for the 30° misoriented flaw, for which the signal-to-noise ratio is low (see Figure 11b). Note that the 3D PTD-based model provides a slight improvement over the 2.5D GTD model in that configuration.

In the view of studying more important 3D effects, a second experimental validation has been performed to evaluate the effect of skew flaw angle on edge diffraction amplitude. Tests have been carried out on a planar rectangular backwall breaking flaw (10 mm high and 40 mm wide) using a pair of transducers (2.25 MHz, 45° longitudinal waves, 6.35 mm diameter) in contact mode. The probes have been positioned in a TOFD configuration with a 60 mm Probe Space Center and with flaw skew angle varying from 0° to 70° (Figure 12a).

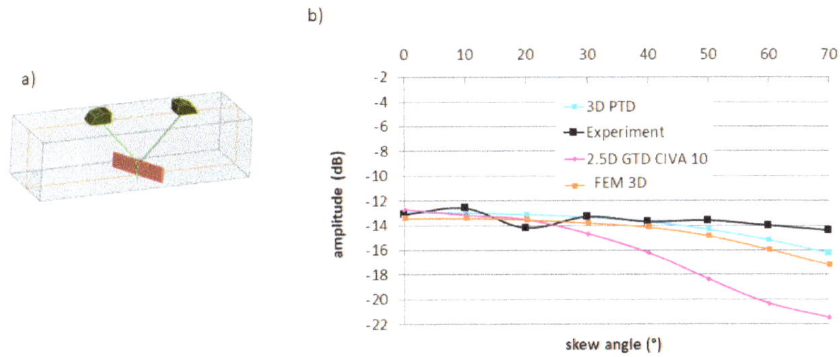

Figure 12. (**a**) TOFD configuration on a skewed backwall breaking crack; (**b**) top edge diffraction echoes amplitudes versus skew angle for measure, 2.5D GTD and 3D PTD and FEM simulations.

In Figure 12b, it can be seen that the influence of the skew angle on the measured diffraction echo amplitudes is minor. The 3D hybrid FEM model leads to a good prediction with a maximal disagreement of about 2.5 dB with measure; the 3D PTD also effectively surpasses the older 2.5D GTD model (available in a previous CIVA release), which breaks down for important skew (>40°). It can be noticed that the 3D PTD and hybrid 3D FEM models lead to close results for this configuration involving a flaw of both large height and extension.

The previous experimental validations have thus confirmed that the 3D hybrid method provides in the case of a large flaw a prediction in very good agreement with experiments and also with the other CIVA 3D model (PTD). The hybrid method correctly simulates larges flaws and 3D configurations.

4.2. Inspection of Small Flaws Compared to the Wavelength

Several measurements have also been carried out to evaluate the models validity on cracks with a small height with respect to the wavelength.

The component used for the following experimental validations is described in Figure 13. The thickness of the specimen, which is the depth to be inspected in experiments (see Figure 14b for an example of inspection configuration), is indicated in red in Figure 13.

The specimen contains the following:

- Four electro-eroded notches (5 mm height × 30 mm extension) and four side-drilled holes (2 mm diameter × 40 mm extension) which are embedded with respective bottom ligaments (distance between notch extremity and backwall) of 5, 10, 15 and 20 mm;
- One backwall breaking notch (not considered in the following validation study);
- Four flat bottom holes on the right side.

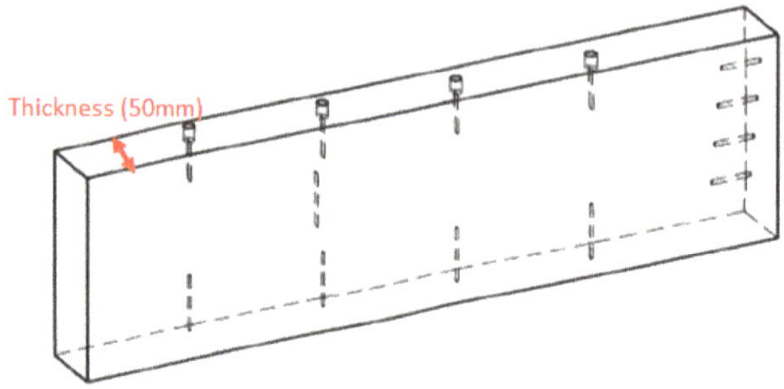

Figure 13. Specimen used for the following experimental validations.

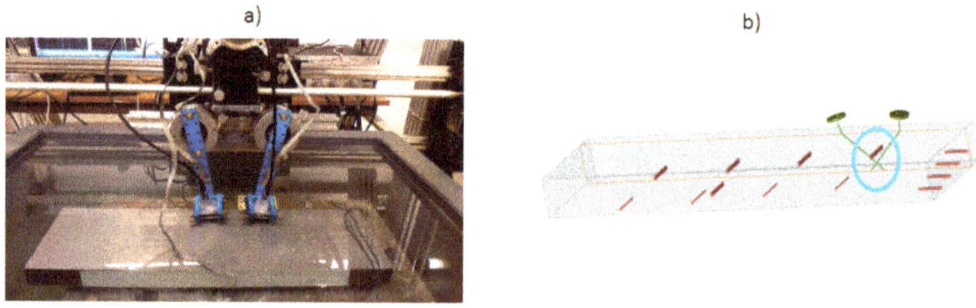

Figure 14. (**a**) Photograph of the setup; (**b**) TOFD inspection configuration using immersion probes: study of P45 echoes scattered by the flaw surrounded in blue.

The experimental validations presented here concern inspections in immersion of the notch of the highest ligament 20 mm (surrounded in blue in Figure 14b). We privilege validation on the flaw with the highest ligament for a better confidence in the measurements, even though measurements on the other flaws were satisfactory. A 10 mm extent starter hole was pre-drilled before producing the notch of 30 mm extension by electro erosion. To study small notch heights with respect to the wavelength, since the flaw height (5 mm) is relatively important, low frequency transducers (0.5 MHz and 1 MHz) have been employed in experiments. We controlled the measurements reproducibility with an obtained confidence interval of the order ±3 dB owing to the low time resolution of the echoes observed at the low frequencies used (0.5 MHz and 1 MHz). Measurements of side-drilled holes have been carried out to set the input signal in simulation [2].

4.2.1. Inspections with Compressional (P) Waves

The specimen has been first inspected in a TOFD configuration for which the probes radiate P45° waves. Figure 14a depicts the device, and the inspection configuration is described in Figure 14b.

The experimental and simulated Bscans obtained using the 1 MHz probes and the 3D hybrid FEM model in simulation for the notch are provided in Figure 15. The Bscans chronologically display the lateral wave, the echo from the defect and the echo due to the P waves backwall reflection. At this low frequency (1 MHz), we observe a single echo arising from the entire flaw. The experimental and simulated results are very similar: the observed relative amplitudes of all echoes are very well reproduced in simulation. Details on the simulation of lateral wave or backwall reflection in TOFD can be found in [13]. The comparison of maximal amplitude for the flaw echo lead to a simulation/experience gap of −1.6 dB for the 3D hybrid FEM model against −2.2 dB for the 3D PTD model.

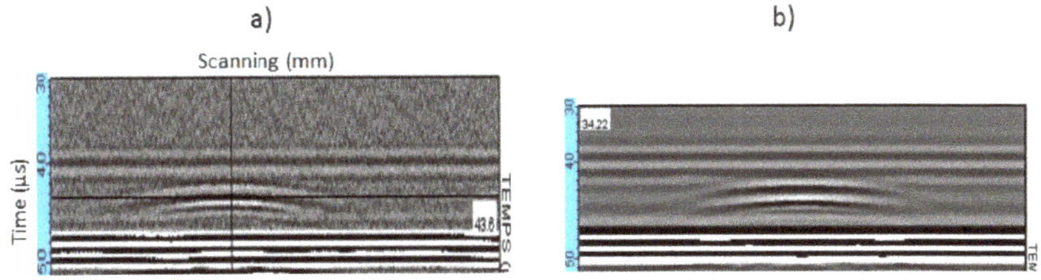

Figure 15. For the TOFD P45° waves inspection at 1 MHz: (**a**) experimental Bscan; (**b**) Bscan simulated using the 3D hybrid FEM model in simulation.

The 3D hybrid FEM model provides a slightly better prediction than the analytical model for small flaws and TOFD P waves at 1 MHz.

4.2.2. Inspections with Shear (S) Waves

Since the numerical validations have shown that differences are more important between the numerical and analytical models for shear waves, experimental acquisitions employing transversal waves have also been carried out in pulse-echo and TOFD configurations.

Figure 16a shows the configuration of ultrasonic pulse echo NDT of the embedded planar electro-eroded slots (depicted in Figure 13). Transducers acting in immersion and generating oblique transversal waves at 45° at 0.5 and 1 MHz have been used. The experimental BScan at 0.5 MHz shown in Figure 16b highlights the emission echo versus time in blue, the echo from the entry surface, the response of the four slots and the flat bottom holes. The echogenic signature of the slot include the direct, corner and indirect echoes. All these echoes can be mixed when the slot backwall ligament is low (two slots at left); in that case, the corner echo amplitude is higher.

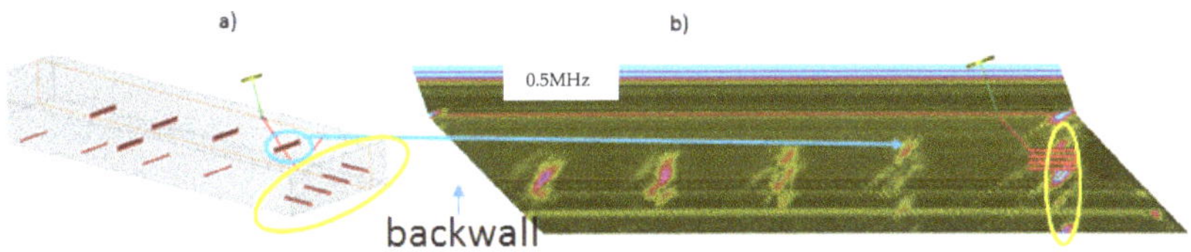

Figure 16. (**a**) Configuration of ultrasonic pulse echo NDT of the embedded planar electro-eroded slots (depicted in Figure 13); (**b**) experimental True BScan using the immersion probe at 0.5 MHz. The entry surface and backwall of the component are indicated in grey. The slot utilized for validation is shown thanks to the blue arrow, whereas the yellow ellipse indicates the component right corners and the flat bottom holes.

As seen in Figure 17a, the FEM model leads to the better prediction. The gap between PTD and FEM reaches 8 dB at 0.5 MHz, PTD losing its effectiveness for small flaw heights with respect to the wavelength. The prediction of the echo signal also seems better when using the FEM model (Figure 17b,c).

The previous validation confirms that the FEM-based 3D hybrid model could bring an interesting contribution to the simulation of small defects compared to the wavelength for shear waves.

The last experimental validation deals with the S45° TOFD inspection of the specimen at both 0.5 MHz and 1 MHz, whose configuration is schematized in Figure 18a. One can observe chronologically on the experimental BScan of Figure 18b for the 0.5 MHz case the lateral wave, the echo due to the specular reflection on the entry surface, the echo from the defect and the echo due to the S waves reflection on the backwall.

a)

Gap Simulation/Exp (dB)	0.5MHz	1MHz
3D FEM	+2.5	0.0
PTD 3D	-5.6	-4.0

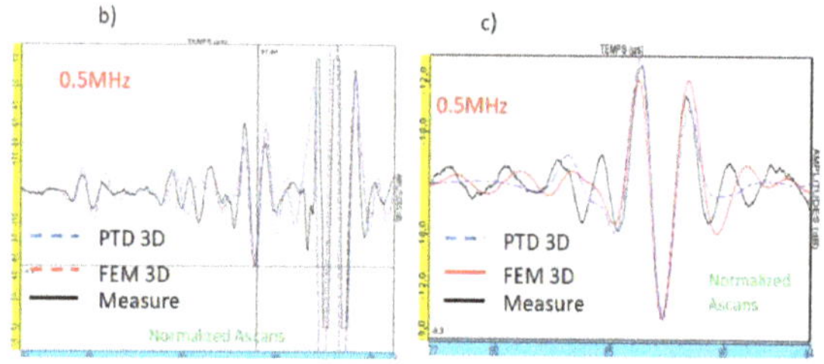

Figure 17. (**a**) Disagreements with measurements obtained with the 3D FEM and PTD simulations; (**b**) Predicted and measured signals with the 0.5 MHz probe; (**c**) the same as (**b**) with normalization on the maximal amplitude.

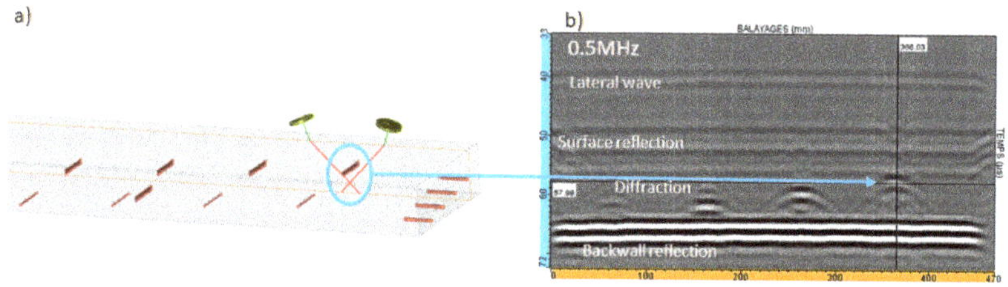

Figure 18. (**a**) S45° TOFD immersion inspection of several embedded planar electro eroded slots; (**b**) experimental BScan obtained when scanning the notches and the flat bottom holes at 0.5 MHz.

The table of Figure 19a presents the comparison with measure of both the 3D hybrid FEM method and the PTD model simulations of the maximal echo amplitude from the validation defect. The 3D hybrid FEM gives rise again to a better prediction in amplitude for the 0.5 MHz case corresponding to the smallest flaw height/wavelength ratio. The measured and simulated normalized Ascans are presented in Figure 19b at 0.5 MHz. One can notice the effectiveness of the simulation which models all these waves of very different natures both in terms of wave forms and of amplitude ratios (notably relatively to the lateral wave and entry surface echoes for the FEM model).

a)

Gap Simulation/Exp (dB)	0.5MHz	1MHz
3D FEM	-1.3	-1.8
3D PTD	-4.0	-1.5

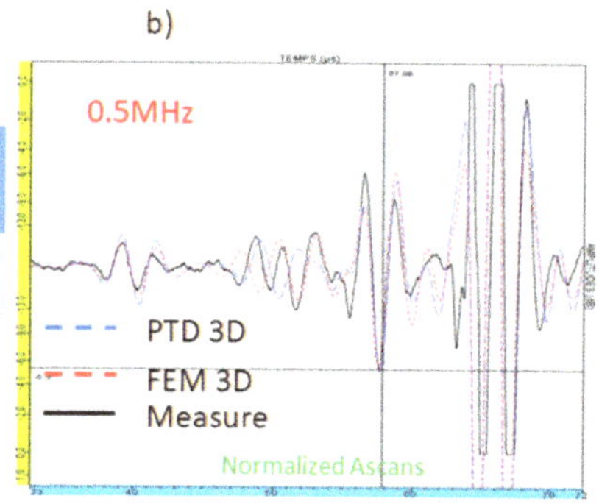

Figure 19. (**a**) Gap between simulation and experience for the 3D hybrid FEM model and for the 3D PTD model; (**b**) Superimposition of the normalized experimental and simulated AScans at 0.5 MHz.

The experimental validation has to be pursued on small flaws notably in 3D configurations, but the first numerical and experimental validations show that the hybrid method provides a much better prediction than analytical models for small flaws (especially with shear waves).

5. Conclusions

An efficient semi-analytical ultrasonic crack measurement model based on the Physical Theory of Diffraction (PTD) and an original hybrid asymptotic/numerical approach based on a high-order spectral-like finite element method were recently proposed in the literature. The numerical comparison of the new 3D hybrid model with semi-analytical models for 3D planar embedded defects has been carried out in this study. It highlights important differences in terms of echo amplitude and waveforms for small flaw heights or extensions and the FEM method seems to provide a more physical description in all configurations by simulating phenomena not or badly accounted for by analytical models as corner diffraction, Rayleigh and head waves. When the crack extension is large, the overall validity range of PTD on the flaw half height $ka > (ka)_{max}$—where ka is the dimensionless wave number—is wider for P waves ($(ka)_{max} \in [1,3]$ approximately) than for S waves ($(ka)_{max} \in [5,10]$ approximately). When the crack height is large, a similar criterion (around ka > 5–10) could be approximately found for the overall PTD validity range with the flaw extension. The PTD limitations are more noticeable when edge diffraction echoes are predominant with respect to specular echoes and when head waves occur; PTD can be valid for smaller ka than those given in the criterion in configurations of specular reflection. The existing PTD code could be improved by accounting for the secondary diffractions due to Rayleigh waves.

Experimental validations on notches have first confirmed the good prediction provided by PTD for simulating 3D top edge diffraction of large flaws. Measurements of embedded slots carried out at low frequency have shown that the proposed 3D hybrid method leads to a significant improvement for simulating small flaws compared to the wavelength, notably for shear waves. Similar comparisons have been made for surface-breaking cracks and should be the object of a complementary future publication. In-progress improvements and developments of the 3D hybrid method (notably for handling complex-shaped cracks) also have to be evaluated.

Author Contributions: Conceptualization, M.D.; methodology, M.D.; software, M.D. and V.D.; validation, M.D. and G.T.; formal analysis, M.D.; writing—original draft preparation, M.D.; writing—review and editing, M.D. and V.D.; funding acquisition, G.T. All authors have read and agreed to the published version of the manuscript.

Funding: This research received no external funding.

Conflicts of Interest: The authors declare no conflict of interest.

Appendix A. Hybrid Method Description

In this appendix, a quick description of the hybrid method is given; more theoretical details can be found in [42,52].

The adopted strategy consists of coupling a semi-analytical propagation model to a scattering numerical model. The main principle of such coupling is to use the computation of the incident field from CIVA by applying the beam computation module [50], and to compute the interaction with the flaw using a dedicated scattering model. The echo-response from the defect is then given by the Auld's reciprocity principle [51], which links healthy (incident field) and damaged (defect scattering) components. Indeed, in the time domain, the elastodynamic response of a defect is expressed as:

$$S_{\Gamma_F} = \int_{\partial B_{\Gamma_F}} \underline{y}^{tot,E} * (\underline{\underline{\sigma}}.\underline{n})^{inc,\,R} - (\underline{\underline{\sigma}}.\underline{n})^{tot,E} * \underline{y}^{inc,R} d\Gamma \quad \text{(A1)}$$

where the involved displacements $\underline{y}$ and stresses $\underline{\underline{\sigma}}$ are defined below, $\underline{n}$ refers to the normal integration surface (surrounding the flaw and depicted in dash blue in Figure A1) and indices E and R denote two states. E quantities correspond to the ultrasonic fields radiated by the emitting probe in the presence of the defect and R quantities are linked to the field radiated by the receiver (used as a fictitious emitter) in healthy components (without flaws).

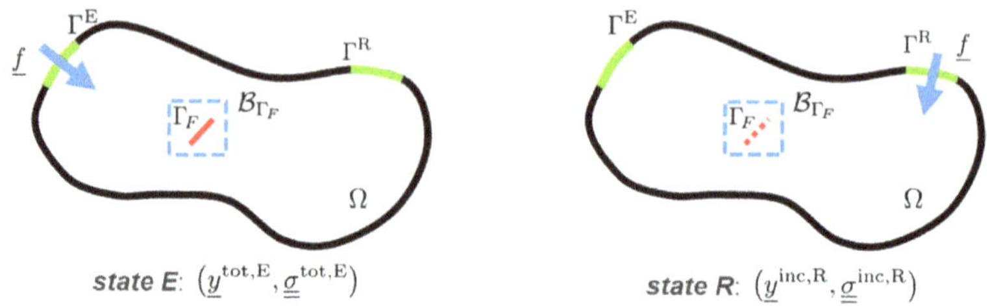

Figure A1. The two different states for reciprocity application. For state R, the dashed red line mimics the absence of the flaw.

The total displacement field is defined as follows for each state as the sum of the incident and diffracted fields (similar notations/decompositions are employed for the stresses also involved in Equation (A1)):

$$\underline{y}^{tot,\alpha} = \underline{y}^{inc,\alpha} + \underline{y}^{dif,\alpha}, \quad \alpha \in \{E, R\} \tag{A2}$$

In the case of cracks, a stress-free boundary condition applies:

$$\underline{\underline{\sigma}}^{tot,E}.\underline{n}\big|_{\Gamma} = 0, \tag{A3}$$

and Equation (A1) reduces to:

$$S_{\Gamma_F} = \int_{\Gamma_F} \left(\underline{y}^{inc,E} + \underline{y}^{dif,E}\right) * \left(\underline{\underline{\sigma}}.\underline{n}\right)^{inc,\,R} d\Gamma \tag{A4}$$

The two incident fields involved in the previous Equation (A4) are calculated by the ray-based asymptotic model [50], whereas the diffracted field in state E is obtained by a numerical model.

In the latter calculation, the input data of the scattering model is obtained in an area of interest associated with the flaw (around the defect) using the ray-based beam computation model in the healthy part. The response of the flaw to this incident field is then computed using the numerical dedicated model (high order spectral finite elements).

The advantages of such an approach are that the flaw response does not disturb the incident field calculated by the ray method. In practice, the defect should be at sufficient distance from the component edges to not interact with them. Otherwise, it is necessary to consider the defect and the component edges in the diffraction pattern.

Computational performances are significantly enhanced by the use of high-order spectral finite elements [43,44] defined on hexahedral meshes. This numerical method is widely spread in the community of numerical solutions of transient high-frequency wave propagation problems since they combine the flexibility of finite element methods and the performances of standard finite differences by allowing a fully explicit numerical scheme thanks to a diagonal mass matrix—in the literature, this technique is referred to as the mass-lumping technique. By using a second-order leap-frog time discretization, a fully explicit numerical scheme is finally obtained and transparent boundary conditions are performed using the so-called perfectly matched layer (PML) formulation. Moreover, by allowing high-order polynomials to represent the solutions, the spectral finite elements require less discretization points to reach a given precision, which is a major asset when simulating 3D configurations.

The assumption of embedded crack is utilized to derive an unbounded diffracted field formulation:

$$\left\{\begin{array}{c} \rho\partial_t^2\underline{y}^{dif,E} - \underline{\nabla}.\underline{\underline{\sigma}}^{dif,E} = 0 \text{ in } \mathcal{B}_{\Gamma_F}, \\ \underline{\underline{\sigma}}^{dif,E}.\underline{n} = -\underline{\underline{\sigma}}^{inc,E}.\underline{n} \text{ on } \Gamma_F, \\ \text{\& transparent boundary conditions on } \partial\mathcal{B}_{\Gamma_F}. \end{array}\right. \tag{A5}$$

The first equation is the fundamental equation of dynamics; the second one corresponds to the boundary conditions applied to the flaw surfaces and transparent boundary conditions are used to model an unbounded domain. Finally, the solution to the diffracted field formulation (A5) is approximated using the numerical model based on high-order spectral finite elements.

References

1. Schmerr, L.W.; Song, S.-J. *Ultrasonic Nondestructive Evaluation Systems: Models and Measurements*; Springer: Berlin/Heidelberg, Germany, 2007.
2. Darmon, M.; Chatillon, S. Main Features of a Complete Ultrasonic Measurement Model—Formal Aspects of Modeling of Both Transducers Radiation and Ultrasonic Flaws Responses. *Open J. Acoust.* **2013**, *3*, 43–45. [CrossRef]
3. Chapman, R.K. *Ultrasonic Scattering from Smooth Flat Cracks: An Elastodynamic Kirchhoff Diffraction Theory*; CEGB Report, NWR/SSD/82/0059/R; North Western Region NDT Applications Centre: Manchester, UK, 1982.
4. Achenbach, J.D.; Gautesen, A.K.; McMaken, H. *Rays Methods for Waves in Elastic Solids*; Pitman: Boston, MA, USA, 1982.
5. Darmon, M.; Leymarie, N.; Chatillon, S.; Mahaut, S. Modelling of Scattering of Ultrasounds by Flaws for NDT. In *Ultrasonic Wave Propagation in Non Homogeneous Media*; Springer: Berlin, Germany, 2009; Volume 128, pp. 61–71.
6. Dorval, V.; Chatillon, S.; Lu, B.; Darmon, M.; Mahaut, S. A General Kirchhoff Approximation for Echo Simulation in Ultrasonic NDT. In Proceedings of the 38th Annual Review of Progress in Quantitative Nondestructive Evaluation (QNDE), Burlington, VT, USA, 17–22 July 2011; Volume 1430, pp. 193–200.
7. Lu, B.; Darmon, M.; Potel, C.; Zernov, V. Models Comparison for the Scattering of an Acoustic Wave on Immersed Targets. In Proceedings of the 10th Anglo-French Physical Acoustics Conference (AFPAC), Villa Clythia, Fréjus, France, 19–21 January 2011; Volume 353, p. 12009.
8. Toullelan, G.; Raillon, R.; Chatillon, S.; Dorval, V.; Darmon, M.; Lonné, S.; Chimenti, D.E.; Bond, L.J. Results of the 2015 UT Modeling Benchmark Obtained with Models Implemented in CIVA. *AIP Conf. Proc.* **2016**, *1706*, 190001. [CrossRef]
9. Raillon-Picot, R.; Toullelan, G.; Darmon, M.; Lonné, S. Experimental Study for the Validation of CIVA Predictions in TOFD Inspections. In Proceedings of the 10th International Conference on NDE in Relation to Structural Integrity for Nuclear and Pressurized Components, Cannes, France, 1–3 October 2013.
10. Chehade, S.; Djakou, A.K.; Darmon, M.; Lebeau, G. The Spectral Functions Method for Acoustic Wave Diffraction by a Stress-Free Wedge: Theory and Validation. *J. Comput. Phys.* **2019**, *377*, 200–218. [CrossRef]
11. Chehade, S.; Darmon, M.; Lebeau, G. 2D Elastic Plane-Wave Diffraction by a Stress-Free Wedge of Arbitrary Angle. *J. Comput. Phys.* **2019**, *394*, 532. [CrossRef]
12. Chehade, S.; Darmon, M.; Lebeau, G. 3D Elastic Plane-Wave Diffraction by a Stress-Free Wedge for Incident Skew Angles below the Critical Angle in Diffraction. *J. Comput. Phys.* **2021**, *427*, 110062. [CrossRef]
13. Baskaran, G.; Balasubramaniam, K.; Krishnamurthy, C.V.; Rao, C.L. Ray Based Model for the Ultrasonic Time-of-Flight Diffraction Simulation of Thin Walled Structure Inspection. *J. Press. Vessel Technol.* **2005**, *127*, 262–268. [CrossRef]
14. Nath, S.K. Effect of Variation in Signal Amplitude and Transit Time on Reliability Analysis of Ultrasonic Time of Flight Diffraction Characterization of Vertical and Inclined Cracks. *Ultrasonics* **2014**, *54*, 938–952. [CrossRef]
15. Darmon, M.; Chatillon, S.; Mahaut, S.; Fradkin, L.; Gautesen, A. Simulation of Disoriented Flaws in a TOFD Technique Configuration Using GTD Approach. In Proceedings of the 34th Annual Review of Progress in Quantitative Nondestructive Evaluation, Golden, CO, USA, 22–27 July 2007; Volume 975, pp. 155–162.
16. Darmon, M.; Djakou, A.K.; Chehade, S.; Potel, C.; Fradkin, L. Two Elastodynamic Incremental Models: The Incremental Theory of Diffraction (ITD) and a Huygens Method. *IEEE Trans. Ultrason. Ferroelectr. Freq. Control* **2019**, *66*, 998–1005. [CrossRef]
17. Kolkoori, S.; Venkata, K.C.; Balasubramaniam, K. Quantitative Simulation of Ultrasonic Time of Flight Diffraction Technique in 2D Geometries Using Huygens–Fresnel Diffraction Model: Theory and Experimental Comparison. *Ultrasonics* **2015**, *55*, 33–41. [CrossRef]
18. Lhemery, A. *Modèles Théoriques et Études Pratiques En Imagerie Ultrasonore Large Bande; Applications Au Contrôle Non Destructif Des Matériaux Habilitation à Diriger des Recherches*; Université Paris VII Denis Diderot: Paris, France, 2000.
19. Gridin, D. The Radiating near Field of a Circular Normal Transducer of Arbitrary Apodization on an Elastic Half-Space. *J. Acoust. Soc. Am.* **1999**, *106*, 1237–1246. [CrossRef]
20. Gridin, D.; Fradkin, L.J. The High-Frequency Asymptotic Description of Pulses Radiated by a Circular Normal Transducer into an Elastic Half-Space. *J. Acoust. Soc. Am.* **1998**, *104*, 3190–3198. [CrossRef]
21. Lewis, R.M.; Boersma, J. Uniform Asymptotic Theory of Edge Diffraction. *J. Math. Phys.* **1969**, *10*, 2291–2305. [CrossRef]
22. Djakou, A.K.; Darmon, M.; Fradkin, L.; Potel, C. The Uniform Geometrical Theory of Diffraction for Elastodynamics: Plane Wave Scattering from a Half-Plane. *J. Acoust. Soc. Am.* **2015**, *138*, 3272–3281. [CrossRef]
23. Dorval, V.; Darmon, M.; Chatillon, S.; Fradkin, L. Simulation of the UT Inspection of Planar Defects Using a Generic GTD-Kirchhoff Approach. *AIP Conf. Proc.* **2015**, *1650*, 1750–1756.
24. Ufimtsev, P.Y. *Fundamentals of the Physical Theory of Diffraction*; John Wiley & Sons: Hoboken, NJ, USA, 2007.

25. Zernov, V.; Fradkin, L.; Darmon, M. A Refinement of the Kirchhoff Approximation to the Scattered Elastic Fields. *Ultrasonics* **2012**, *52*, 830–835. [CrossRef]
26. Darmon, M.; Dorval, V.; Djakou, A.K.; Fradkin, L.; Chatillon, S. A System Model for Ultrasonic NDT Based on the Physical Theory of Diffraction (PTD). *Ultrasonics* **2016**, *64*, 115–127. [CrossRef]
27. Chaffai, S.; Darmon, M.; Mahaut, S.; Menand, R. Simulations Tools for TOFD Inspection in Civa Software. In Proceedings of the ICNDE 2007, Istanbul, Turkey, January 2007.
28. Darmon, M.; Ferrand, A.; Dorval, V.; Chatillon, S.; Lonné, S. Recent Modelling Advances for Ultrasonic TOFD Inspections. *AIP Conf. Proc.* **2015**, *1650*, 1757–1765.
29. Zernov, V.; Gautesen, A.; Fradkin, L.J.; Darmon, M. Aspects of Diffraction of a Creeping Wave by a Back-Wall Crack. *J. Phys. Conf. Ser.* **2012**, *353*, 012017. [CrossRef]
30. Huet, G.; Darmon, M.; Lhemery, A.; Mahaut, S. Modelling of Corner Echo Ultrasonic Inspection with Bulk and Creeping Waves. In *5th Meeting of the Anglo-French-Research-Group*; Léger, A., Deschamps, M., Eds.; Springer: Berlin/Heidelberg, Germany, 2009; Volume 128, pp. 217–226.
31. Zernov, V.; Fradkin, L.; Gautesen, A.; Darmon, M.; Calmon, P. Wedge Diffraction of a Critically Incident Gaussian Beam. *Wave Motion* **2013**, *50*, 708–722. [CrossRef]
32. Fradkin, L.J.; Darmon, M.; Chatillon, S.; Calmon, P. A Semi-Numerical Model for near-Critical Angle Scattering. *J. Acoust. Soc. Am.* **2016**, *139*, 141–150. [CrossRef]
33. Fradkin, L.J.; Djakou, A.K.; Prior, C.; Darmon, M.; Chatillon, S.; Calmon, P.-F. The Alternative Kirchhoff Approximation in Elastodynamics with Applications in Ultrasonic Nondestructive Testing. *ANZIAM J.* **2020**, *62*, 406–422. [CrossRef]
34. Ferrand, A.; Darmon, M.; Chatillon, S.; Deschamps, M. Modeling of Ray Paths of Head Waves on Irregular Interfaces in TOFD Inspection for NDE. *Ultrasonics* **2014**, *54*, 1851–1860. [CrossRef] [PubMed]
35. Raillon, R.; Bey, S.; Dubois, A.; Mahaut, S.; Darmon, M. Results of the 2010 Ut Modeling Benchmark Obtained with Civa: Responses of Backwall and Surface Breaking Notches. *AIP Conf. Proc.* **2011**, *1335*, 1777–1784. [CrossRef]
36. Raillon, R.; Bey, S.; Dubois, A.; Mahaut, S.; Darmon, M. Results of the 2009 Ut Modeling Benchmark Obtained with Civa: Responses of Notches, Side-drilled Holes and Flat-bottom Holes of Various Sizes. *AIP Conf. Proc.* **2010**, *1211*, 2157–2164. [CrossRef]
37. Raillon-Picot, R.; Toullelan, G.; Darmon, M.; Calmon, P.; Lonné, S. Validation of CIVA Ultrasonic Simulation in Canonical Configurations. In Proceedings of the World Conference of Non Destructive Testing (WCNDT), Durban, South Africa, 16–20 April 2012.
38. Dhvani, A.I. Available online: https://www.dhvani-analytic-intelligence.com/simsonic.html (accessed on 4 February 2022).
39. OnScale Solve—The Cloud Engineering Simulation Platform. Available online: https://onscale.com/ (accessed on 4 February 2022).
40. Pogo—High Speed Wave Simulations. Available online: http://www.pogo.software/ (accessed on 15 April 2021).
41. Imperiale, A.; Chatillon, S.; Darmon, M.; Leymarie, N.; Demaldent, E. UT Simulation Using a Fully Automated 3D Hybrid Model: Application to Planar Backwall Breaking Defects Inspection. *AIP Conf. Proc.* **2018**, *1949*, 050004. [CrossRef]
42. Imperiale, A.; Leymarie, N.; Fortuna, T.; Demaldent, E. Coupling Strategies Between Asymptotic and Numerical Models with Application to Ultrasonic Non-Destructive Testing of Surface Flaws. *J. Theor. Comput. Acoust.* **2019**, *27*, 1850052. [CrossRef]
43. Cohen, G. *Higher-Order Numerical Methods for Transient Wave Equations*; Springer Science & Business Media: Berlin/Heidelberg, Germany, 2002; ISBN 978-3-540-41598-5.
44. Komatitsch, D.; Tromp, J. Introduction to the Spectral Element Method for Three-Dimensional Seismic Wave Propagation. *Geophys. J. Int.* **1999**, *139*, 806–822. [CrossRef]
45. Kundu, T.; Placko, D.; Rahani, E.; Yanagita, T.; Dao, C. Ultrasonic Field Modeling: A Comparison of Analytical, Semi-Analytical, and Numerical Techniques. *IEEE Trans. Ultrason. Ferroelectr. Freq. Control* **2010**, *57*, 2795–2807. [CrossRef]
46. Lü, B.; Darmon, M.; Fradkin, L.; Potel, C. Numerical Comparison of Acoustic Wedge Models, with Application to Ultrasonic Telemetry. *Ultrasonics* **2016**, *65*, 5–9. [CrossRef]
47. Bövik, P.; Boström, A. A Model of Ultrasonic Nondestructive Testing for Internal and Subsurface Cracks. *J. Acoust. Soc. Am.* **1998**, *102*, 2723. [CrossRef]
48. Semenova, M.; Wirdelius, H.; Persson, G. Comparison between Three Mathematical Models of Three Well Defined Ultrasonic NDT Cases. *IOP Conf. Ser. Mater. Sci. Eng.* **2020**, *747*, 012061. [CrossRef]
49. Civa Software Website. Available online: http://www.Extende.com/ (accessed on 11 May 2022).
50. Gengembre, N.; Lhémery, A. Pencil Method in Elastodynamics: Application to Ultrasonic Field Computation. *Ultrasonics* **2000**, *38*, 495–499. [CrossRef]
51. Auld, B.A. General Electromechanical Reciprocity Relations Applied to the Calculation of Elastic Wave Scattering Coefficients. *Wave Motion* **1979**, *1*, 3–10. [CrossRef]
52. Imperiale, A.; Demaldent, E. A Macro-Element Strategy Based upon Spectral Finite Elements and Mortar Elements for Transient Wave Propagation Modeling. Application to Ultrasonic Testing of Laminate Composite Materials. *Int. J. Numer. Methods Eng.* **2019**, *119*, 964–990. [CrossRef]
53. Djakou, A.K.; Darmon, M.; Potel, C. Elastodynamic Models for Extending GTD to Penumbra and Finite Size Flaws. *J. Phys. Conf. Ser.* **2016**, *684*, 012002. [CrossRef]
54. Djakou, A.K.; Darmon, M.; Potel, C. Elastodynamic Models for Extending GTD to Penumbra and Finite Size Scatterers. *Phys. Procedia* **2015**, *70*, 545–549. [CrossRef]

55. Raillon-Picot, R.; Toullelan, G. *Caractérisation Expérimentale de CIVA, Contrôle TOFD Par Ultrasons, Échos de Diffraction de Défauts Plans Verticaux et Échos de Trous Génératrice, Mode Longitudinal*, CEA/DISC/LMC/12 RT128. 2012.
56. Darmon, M.; Chatillon, S.; Mahaut, S.; Calmon, P.; Fradkin, L.J.; Zernov, V. Recent Advances in Semi-Analytical Scattering Models for NDT Simulation. *J. Phys. Conf. Ser.* **2011**, *269*, 012013. [CrossRef]

applied sciences

MDPI

Article

Numerical Simulation of Enhanced Photoacoustic Generation and Wavefront Shaping by a Distributed Laser Array

Ruijie Hou [1], Bin Xu [1], Zhiying Xia [1], Yang Zhang [1], Weiping Liu [1], Christ Glorieux [2], John H. Marsh [3], Lianping Hou [3], Xuefeng Liu [1,*] and Jichuan Xiong [1,*]

1 School of Electronic and Optical Engineering, Nanjing University of Science and Technology, Nanjing 210094, China; hrj@njust.edu.cn (R.H.); xubinhit@njust.edu.cn (B.X.); zhiying_Xia@njust.edu.cn (Z.X.); 218104010144@njust.edu.cn (Y.Z.); 18362961563@163.com (W.L.)
2 Laboratory of Acoustics—Soft Matter and Biophysics, Department of Physics and Astronomy, KU Leuven, Celestijnenlaan 200D, 3001 Leuven, Belgium; christ.glorieux@kuleuven.be
3 James Watt School of Engineering, University of Glasgow, Glasgow G12 8QQ, UK; john.marsh@glasgow.ac.uk (J.H.M.); Lianping.Hou@glasgow.ac.uk (L.H.)
* Correspondence: liuxf_1956@sina.com (X.L.); jichuan.xiong@njust.edu.cn (J.X.)

Abstract: In photoacoustic imaging, the use of arrayed laser sources brings several advantages. Acoustic waves can be generated with flexible control of wavefronts, bringing functionality such as ultrasonic beam steering and focusing. The use of arrays reduces the optical intensity while increasing the strength of the ultrasonic wave, bringing the advantages of improved signal-to-noise ratio (SNR) while avoiding laser-induced damage. In this paper, we report a numerical model for studying the generation and shaping of acoustic wavefronts with laser arrays. The propagation of mechanical waves, photoacoustically generated by thermal expansion, is simulated and discussed in detail. In addition, a partially delayed distributed array is studied both theoretically and quantitatively. The developed model for wavefront control through time-delayed laser pulses is shown to be highly suited for the optimization of laser array generation schemes.

Keywords: photoacoustic imaging; finite element method; laser array; wavefront superposition

Citation: Hou, R.; Xu, B.; Xia, Z.; Zhang, Y.; Liu, W.; Glorieux, C.; Marsh, J.H.; Hou, L.; Liu, X.; Xiong, J. Numerical Simulation of Enhanced Photoacoustic Generation and Wavefront Shaping by a Distributed Laser Array. *Appl. Sci.* **2021**, *11*, 9497. https://doi.org/10.3390/app11209497

Academic Editors: Giuseppe Lacidogna and Habil. Michel Darmon

Received: 30 April 2021
Accepted: 14 July 2021
Published: 13 October 2021

1. Introduction

Photoacoustic imaging as a non-invasive detection technology has been widely used in biomedical imaging [1], nondestructive testing [2], and other fields. Due to the thermoelastic effect [3], when a laser beam irradiates a sample, optical absorption in the material causes the light energy to initiate thermal expansion, resulting in deformation of the illuminated area and its surroundings. If the illumination is a fast transient pulse or if the light is modulated periodically, then the material and its surroundings deform correspondingly, thereby generating an acoustic wave.

In photoacoustic imaging, structural features or defects in the object can be characterized by detecting reflections or other perturbations in the ultrasonic waves. The spatiotemporal features of the ultrasonic wave are related to those of the incident laser light, but although an intense beam can generate strong acoustic waves, it may damage the material. However, laser intensities far below the damage threshold generate acoustic waves that are relatively weak. The signal-to-noise ratio (SNR) of the reconstructed image can be quantitatively expressed as

$$\mathrm{SNR} \sim \mathrm{S} \times \gamma \times \mathrm{A} \times \sqrt{\mathrm{P} \times \mathrm{M}}/\mathrm{N} \quad (1)$$

where S is the photo-generated ultrasound intensity, γ is the sensitivity of the detector, A is the imaging area, P is the number of projections, M is the number of signal averages, and N is the noise floor of the system [4]. Thus, if the excited ultrasound signal intensity, which is proportional to the optical energy, is reduced by a factor of K, to recover the SNR, averaging over K^2 times is needed. Therefore, in order to obtain a higher SNR, researchers tend to increase the magnitude of the generated photoacoustic signal. In current biomedical photoacoustic imaging research, to enhance the imaging quality or increase the imaging depth, the laser pulse energy density used in the experiment is usually higher than the ANSI specified threshold of 20 mJ/cm^2 [5], which can result in ablation of the tissue [6,7]. It is vital to explore methods for generating strong acoustic waves using energy densities below the laser ablation threshold. To achieve efficient photoacoustic wave generation, Murray et al. used a continuous wave laser with intensity modulation to excite ultrasound waves [8], which greatly improved the SNR and simultaneously kept the sample surface intact. However, this method resulted in the generation of multiple acoustic modes and reflected waves, and interpretation of the signals was complicated. Huang et al. proposed a photoacoustic method that uses a laser array instead of a single light source to excite acoustic waves [9], which increases the intensity of the ultrasonic wave without increasing the energy density illuminating the sample surface.

Research of Matt Clark's group demonstrated that waves can be focused and steered using a spatially distributed laser source, with a spatial light modulator or specially designed computer-generated holograms [10–13]. They also developed a method using laser-induced phased arrays (LIPAs) to detect defects in metals and provide in-process monitoring during additive manufacturing (AM) [13,14]. It has also been shown that temporal control of the elements in a linearly arrayed laser source is an efficient way to enhance the intensity of the generated acoustic wave or steer the propagation direction [15–21]. Although wave steering and signal enhancement by controlling the spatial or temporal distribution of the generated laser source have been demonstrated, the method for manipulating the wavefront of a laser-generated acoustic wave is still uninvestigated.

The use of acoustic metamaterials to manipulate the wavefront of propagating acoustic waves from various sources has been proposed and developed. However, it is difficult to apply these to laser-generated acoustic waves due to the operation bandwidth and the requirement of the wave to be transmitted through or reflected from the meta-surface.

Beamforming of acoustic waves through array sensors has been widely used to improve the contrast, spatial resolution, and signal-to-noise ratio of ultrasound images. Based on our understanding of previous research, two categories of techniques are conceptualized as the beamformer approaches, either at the excitation-side and the probing-side of acoustic waves.

The traditional probe-side beamforming method achieves the purpose of signal enhancement by processing the received signal from the perspective of signal acquisition rather than directly generating stronger acoustic signals from the perspective of excitation. This is a fatal flaw in extremely absorbent biological tissues.

Traditional probe-side beamformer methods include the delay and sum (DAS) method [22], the minimum variance (MV) method [23,24], etc. The principle of these methods is to align the target beams according to the delay time and add them sequentially to enhance the signal. DAS is independent of echo data, which leads to the main lobe width being too wide and the side lobes intensity being too high [24]. The minimum variance (MV) method, e.g., adaptive beam synthesis, calculates a dynamic weight value based on the echo data, makes full use of the characteristics of the echo signal, reduces the side lobe signal, and improves the resolution of the image. The basic principle is to minimize the array output by keeping the gain in the desired direction constant.

For traditional excitation-side forming, Von Ramm et al. achieved beam steering and enhancement by applying a time delay to each piezoelectric plate in an 8-element line array [25]. The distance between the line sources is about one wavelength. The delay time is controlled by the line source spacing and rotation angle.

Shi-Chang Wooh [26] proposed that a larger inter-element spacing can help improve the directivity of the steered wave, yet only to a certain extent; the grating lobes will be introduced (or the beam is not steerable) when the inter-element spacing exceeds its critical value, which is about half the wavelength of the sound wave and controlled by the desired maximum steering angle.

The traditional excitation-side beamformer can achieve controllable superposition between different sources, thereby achieving high-degree-of-freedom beam direction control. In principle, it can also achieve the wavefront formation and control discussed in this article. However, a systematic study of the process of exciting, shaping, and steering photoacoustic waves using laser array sources is therefore still essential. The purpose of this article is to explore whether it is feasible to realize the formation and control of the wavefront in the process of photoacoustic excitation, as well as the mechanism of its formation and propagation.

In this paper, the characteristics of surface and bulk waves excited by a laser array with partial time delay are numerically simulated using the finite element method (FEM). We verify whether, by applying a time delay between different elements of the laser array, the photoacoustic wave can be enhanced and whether this enhanced wave still propagates as a wavefront maintaining its complete shape.

2. Theoretical Model and Simulation

2.1. Numerical Simulation

Both the finite element method (FEM) and finite-difference time-domain (FDTD) method can be used to model the process of light-excited acoustic waves. In terms of computational speed, FDTD has a great advantage. However, considering the complexity of photoacoustic imaging, the FEM method was selected as it is more versatile and rigorous.

Three configurations for exciting photoacoustic waves were simulated, namely single pulse single-point excitation (Model 1), single-pulse laser array excitation (Model 2), and laser array with time-delayed excitation between spots (Model 3). Except for the number of excitation light sources and the time delay between the spots, other parameters in the simulation remained unchanged, including the sample size, material, energy density, and pulse duration.

The modeled sample was a homogeneous, isotropic cylindrical aluminum block with a height of 4 mm and a radius of 4 mm. The laser array was incident perpendicularly to the top surface of the cylinder. Material parameters of aluminum from reference [27] were used for the simulation, with a surface wave sound velocity of 2940 m/s, a transverse wave sound velocity of 3080 m/s, and a longitudinal wave sound velocity of 6320 m/s. A laser source was used to generate the ultrasound, with a pulse width of 20 ns, pulse energy of 160 μJ, and energy density of 510 mJ/cm^2. The laser spot radius on the sample surface was 100 μm, and it was assumed that 80% of the light was absorbed. Figure 1 shows the aluminum cylinder in the Cartesian coordinate system, with three laser spots A, B, and C focused on the top surface. The simulation used the solid heat transfer module and the solid mechanics module of COMSOL Multiphysics to simulate in the transient field

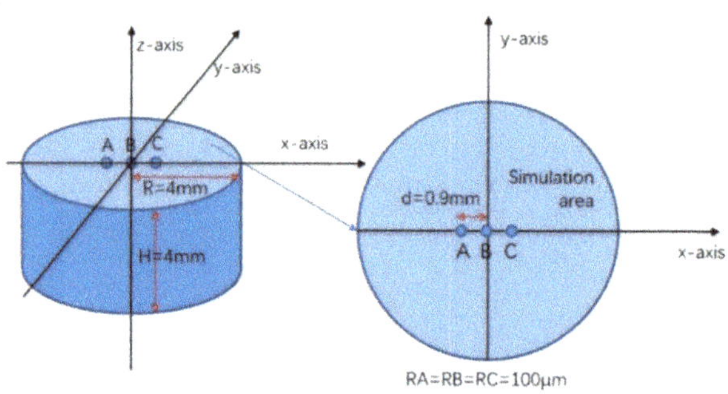

Figure 1. Cylindrical aluminum block and the three laser illumination spots A, B, and C on its top surface.

Model 1 and Model 2 were used as reference simulations to compare with Model 3. In Model 1, a single excitation pulse was used to illuminate the single spot "B". Model 2 was based on Model 1, with excitation spots "A" and "C" added on either side of the single laser spot B. The size and energy density of these three laser spots were the same as for Model 1. As shown in Figure 1, through pre-simulation results, the distance between adjacent laser sources was selected as 0.9 mm, which is about one wavelength of the excited acoustic wave. Model 3 was based on Model 2 with a time delay imposed on the central spot B. It can be seen from the geometric properties of the propagation of acoustic waves in Figure 2a that as the waves propagate, the difference between d_1 and d_2 becomes smaller. When the distance between adjacent sources is much smaller than d_1, and the wavefront has traveled sufficiently far, $\delta_d = d_2 - d_1 < \varepsilon \times v$. ε is defined as the allowable time error, which is chosen to be as same as the simulation time step, i.e., 0.01 μs, v is the velocity of the excited acoustic waves. During this time, the acoustic wave travels about one-tenth of the acoustic wavelength. The new merged wavefront due to superposition can be considered as a continuous wavefront that can propagate in an infinite medium.

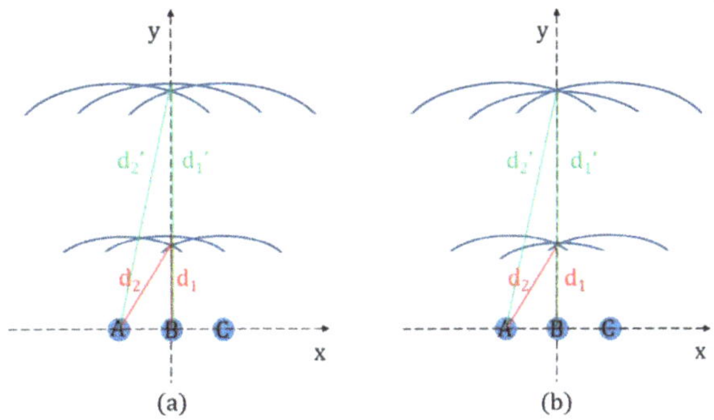

Figure 2. (**a**) Schematic diagram of laser array excitation with no time delay between pulses. (**b**) Schematic diagram of laser array excitation with a central pulse (B) delayed.

The distance Y at which the superposed acoustic waves add constructively on the y-axis can be calculated from

$$Y^2 + L^2 = (Y + \varepsilon \times v)^2 \quad (2)$$

where L is the distance between adjacent sources, and $\varepsilon \times v$ is the maximum acceptable error. In Model 2, the distance Y for acoustic waves without a time delay was calculated to be 13.5 mm, which is larger than the size of the sample. The corresponding traveling time is about 4.5 μs, calculated from the velocity of the surface acoustic wave. Furthermore, using this arrangement in practical applications would make it difficult to image features such as defects or impurities within the sample.

Figure 2b shows the propagation of acoustic waves when excitation of the central source B is delayed. If the delay in exciting the central source B is δ_t, the delayed acoustic wave from B adds constructively to the acoustic waves from the symmetrical sources A and C at a time t, which can be calculated from

$$[v \times (t + \delta_t)]^2 = (v \times t)^2 + L^2 \quad (3)$$

when the delay time is chosen to be 0.1 μs, constructive wave superposition occurs 0.5 μs after the first excitation at a distance on the y-axis of about 1.2 mm, which is less than 10% of the distance without any time delay. When the delay time is increased to 0.2 μs, the time for constructive superposition is shortened to 0.325 μs with a propagation distance of only 0.375 mm.

To ensure that the superposed propagating waves do not separate again, δ_d is required to be always less than the acceptable error value.

$$\delta_d = d_2 - d_1 = \sqrt{L^2 + [v \times (t + t_0 - \delta_t)]^2} - v \times (t + t_0 - \delta_t) < \varepsilon \times v \quad (4)$$

where t is the propagation time after constructive interference, t_0 is the time that constructive interference first occurs, and δ_t is the delay time. For all times $t > 0$, $\delta_d < \varepsilon \times v$, that is, there is no solution when $\delta_d \geq \varepsilon \times v$. We can see that when the delay time is 0.1 or 0.2 μs, the solution of the inequality is an empty set. Therefore, after constructive superposition occurs, the enhanced wavefront propagates as a complete new wavefront without separating again. In the results and discussions section, we will confirm that the enhanced wavefront propagates as a complete wavefront by fitting and reconstructing the wavefront.

2.2. Solid Heat Transfer

The energy distribution of the laser can be described as

$$I = I_0 \times f(r) \times g(t) \quad (5)$$

where I_0 is the maximum power density, and f(r) and g(t) are the spatial and temporal distributions of the laser pulse, respectively.

The thermal conduction equation can be written as

$$\rho C_p = \frac{\partial T(r,z,t)}{\partial t} \nabla k \nabla T(r, z, t) = Q \quad (6)$$

where T(r, z, t) represents the temperature distribution at time t; ρ is the material density; C_p is the thermal capacity of the material; and k is the thermal conductivity.

We consider the excited spot as a heat source loaded at the center of the cylinder surface. To simplify the simulation and reduce the computation time, we consider other boundaries of the cylinder as thermally insulated boundaries. The boundary conditions of the other surfaces can therefore be written as

$$-k\frac{\partial T(r,z,t)}{\partial z} = 0 \quad (7)$$

And the boundary conditions of the absorption layer can be written as

$$-k\frac{\partial T(r,z,t)}{\partial z} = IA(T) \quad (8)$$

where A(T) is the optical absorptivity of the focus plane [28].

2.3. Solid Mechanics

When the laser pulse is focused on the surface of the sample, a transient displacement field is excited due to thermoelastic expansion. In a homogeneous medium, the displacement satisfies:

$$\rho\frac{\partial^2 \mathbf{u}}{\partial t^2} = (\lambda + 2\mu)\nabla(\nabla \mathbf{u}) - \mu \nabla \times \nabla \times \mathbf{u} - \alpha(3\lambda + 2\mu)\nabla T(r, z, t) \quad (9)$$

where **u** is the transient displacement, ρ is the density of the material, α is the thermoelastic expansion coefficient, and λ and μ are the first-order and second-order Lamé constants, which are determined by the material-related quantities represented by the strain-stress relationship. The first-order Lamé constant λ represents the compressibility of the material, which is equivalent to the bulk elastic modulus or Young's modulus, and the second-order

Lamé constant μ represents the shear modulus of the material. The values of λ and μ for aluminum are 6.10 and 2.49, respectively [29].

3. Results and Discussions

Figure 3 shows the z-direction displacement due to the surface wave propagating in the xy-plane. Comparing Model 1, Model 2, and Model 3, we can see that a wavefront showing constructive superposition appears first in Model 3. The color bar used in each subfigure in Figure 3 is the same, ranging from -1×10^{-6} (dark blue) to 0 (light blue) to 3×10^{-6} mm (red). It can be seen that due to the time delay, the wavefront in Model 3 lags behind Model 1 and Model 2, but its intensity is significantly enhanced.

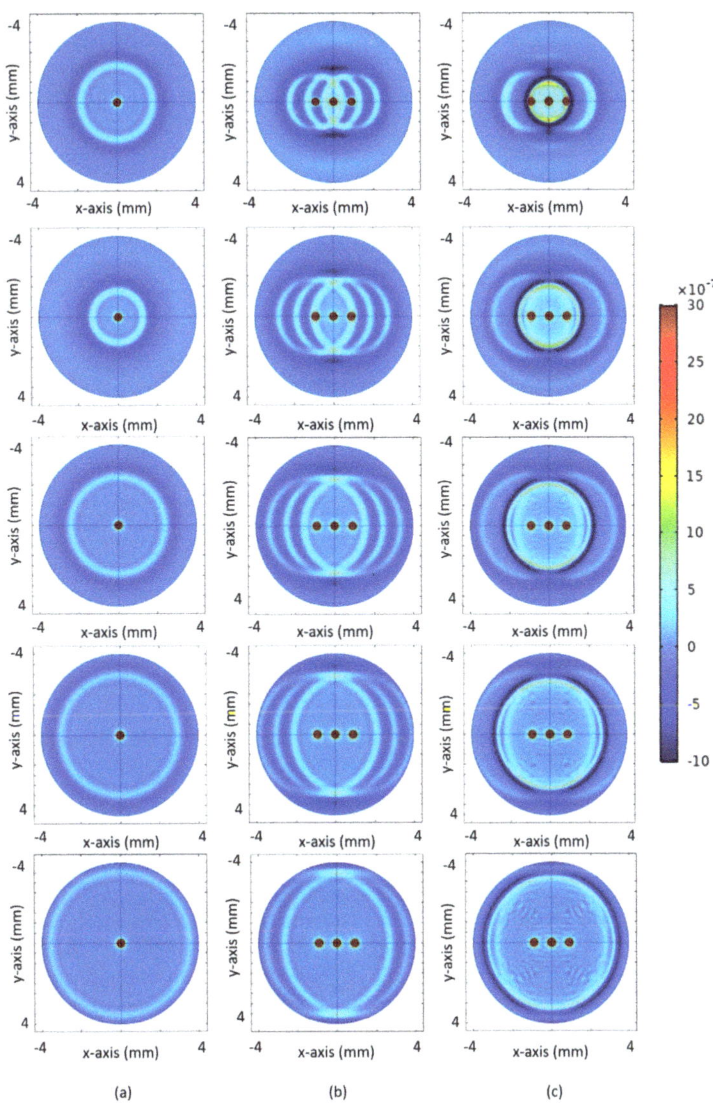

Figure 3. Displacement in the z-direction due to the surface wave propagating in the xy-plane at different times. (**a**) Model 1; (**b**) Model 2; (**c**) Model 3. The corresponding times are 0.6, 0.8, 1, 1.2, and 1.4 μs (from top to bottom).

In Figure 4, we plot the displacement for the three models at points at which the distance to the x-axis and y-axis are both 1, 1.5, 2, and 2.5 mm, referred to as point 1, point 2, point 3, and point 4, respectively. It can be seen that the peak displacement for Model 3 is much larger than that for Model 1 and Model 2 because of the constructive superposition of the acoustic wavefronts. The simulation results agree closely with the theoretical model, but the noise of Model 3 is larger than that of Model 1 and Model 2, and this problem needs to be considered in the subsequent data processing.

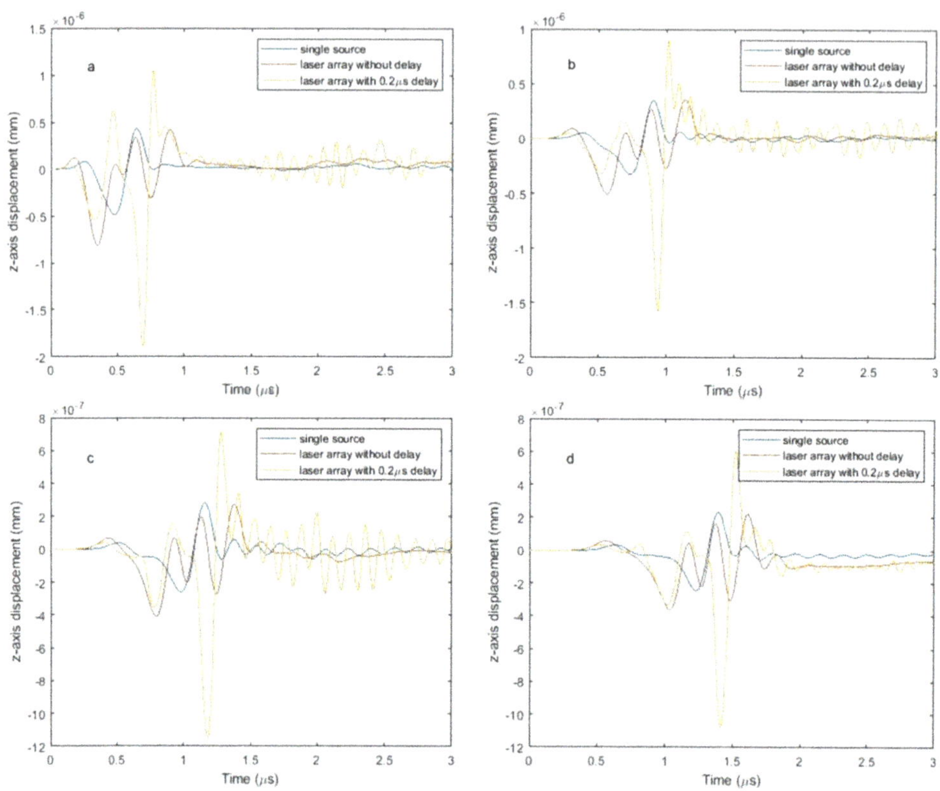

Figure 4. Displacement in the z-direction as a function of time of Model 1, Model 2, and Model 3 for (**a**) point 1, (**b**) point 2, (**c**) point 3, and (**d**) point 4.

MATLAB was used to fit the wavefront data to confirm that the enhanced wavefront continues to propagate in the form of a constructively superposed wave. A surface composed of points of equal amplitude is considered a wavefront surface. For the purpose of choosing a fitting method with a simple form and high accuracy, we used second to fifth-order polynomial fitting with the Zernike polynomial. Rayleigh waves propagate in the form of cylindrical wavefronts along the free surface of the medium [30], so only the first three orders were considered when using Zernike polynomials for fitting. The first three orders of the Zernike polynomial are shown in Table 1.

Due to symmetry, only the region with $x > 0$ was considered when fitting. Due to acoustic absorption, the surface wave decays exponentially inside the material. The surface wave on the z-axis far from the heat source is very weak and can barely be discriminated from noise, so only data within 0.1 mm of the heat source was used for fitting.

Figure 5 depicts the wavefronts of Model 1, Model 2, and Model 3 fitted by polynomials of different orders. The fitting error is displayed in Table 2. For higher-order polynomials, the fitting error is reduced. The results using third-order Zernike polynomial fitting are similar to third-order polynomial fitting. In both Model 1 and Model 3, fitting using fourth-order and fifth-order polynomials offers significantly better accuracy compared with second-order and third-order fitting, and it is recommended to adopt the fourth-order polynomial fitting as the difference between fourth- and fifth-order polynomial fitting is marginal. Regardless of the order of fitting for Model 2, the fitting error is larger than that for Model 1 and Model 3, and the fitted results are also not ideal. It can be seen from the original data of Model 2 in Figure 5 that weaker wavefronts appear distinct from the main wavefront. This implies that the use of multiple sources without time delay for simultaneous excitation not only cannot form a single enhanced wavefront, but the weaker wavefronts also lead to difficulties in interpreting ultrasonic images and extracting data.

Table 1. First 3 orders of the Zernike polynomial in Cartesian coordinate.

j	n	Z_j
1	1	y
2	1	x
3	2	2xy
4	2	$-1 + 2(x^2 + y^2)$
5	2	$x^2 - y^2$
6	3	$3x^2y - y^3$
7	3	$-2y + 3y(x^2 + y^2)$
8	3	$-2x + 3x(x^2 + y^2)$
9	3	$x^3 - 3xy^2$

For further verification, we conducted numerical calculations based on the parametric indirect microscopic imaging (PIMI) method [31], which is a method that can optically visualize the sound field wavefront by measuring the phase retardation and polarization ellipse orientation angle of the probe light. Linearly polarized light with rotation of its polarization angle was used as the probe beam incident on the surface of the sample. The polarization states of the reflected light were perturbed by the surface acoustic wave. By measuring the reflected light at four different polarization angles, I(0°), I(90°), I(45°), and I(135°), the Stokes parameters, the phase delay δ, and the polarization orientation angle ψ were calculated. The calculated polarization parametric images of the probe beam can be related to the ultrasonic wave field via the photo-elastic effect [31].

The visualization of the acoustic field is shown in Figure 6. The Stokes parameters S0, S1, S2, S3 phase delay δ and polarization orientation angle ψ can be obtained from I(0°), I(90°), I(45°), I(135°). I(0°) is the intensity of the linearly horizontal polarized light component, I(90°) is the intensity of the linearly vertical polarized light component, I(45°) is the intensity of the linearly +45° polarized (L+45P) light component (L for linearly and P for polarization), I(135°) is the intensity of the linear −45° polarized (L−45P) light components. From Figure 6, we found that S3 and ψ show high sensitivity to the laser-induced surface acoustic wave, while the wavefront can barely be seen in the phase retardation δ.

As a comparison, the propagation of acoustic waves in the xy-plane induced by three acoustic emitters at the same locations as the laser array source was simulated using MATLAB's k-wave toolbox [32]. The heat sources in Model 1, Model 2, and Model 3 were replaced by sound sources with the same size, shape, distributions, and delay time. The simulation area was a square with a side length of 8 mm, and an 800 × 800 mesh was used. The speed of sound in the medium was taken to be 2940 m/s, the same as the surface wave sound velocity of aluminum. The uniformly distributed source was defined by the acoustic pressure, which was assigned a value of −0.1 Pa because thermally induced SAWs have a negative initial acoustic pressure. The initial frequency of the acoustic signal was specified as 3.3 MHz, approximately equal to the frequency of the sound wave excited in the FEM simulation. The input signals of the acoustic sources A, B, and C in Model 3 are shown in Figure 7. The input signal for source B has a time delay of 200 ns relative to sources A and C.

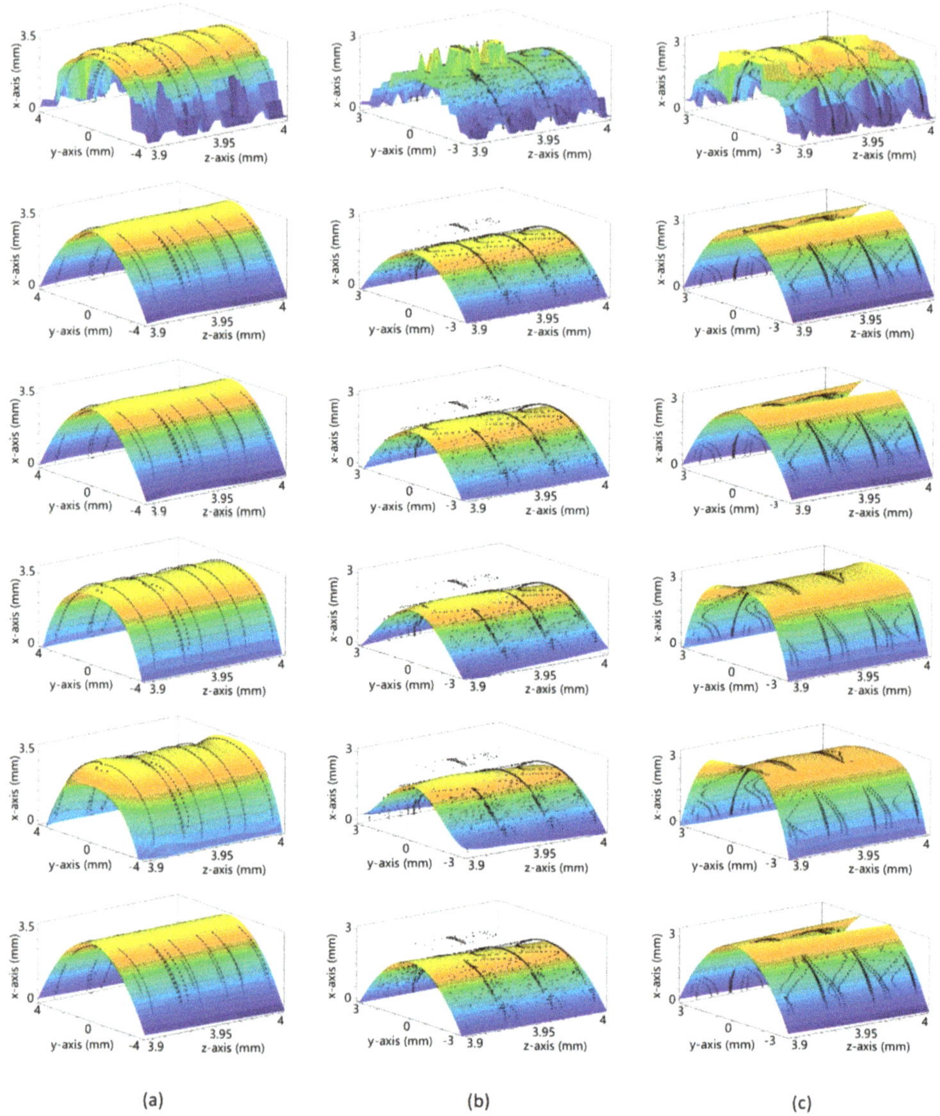

Figure 5. Processed data fitted with 2–5 order polynomials and Zernike polynomials at 1.2 μs. From top to bottom: original data interpolation results, second, third, fourth, and fifth-order polynomial fitting, Zernike polynomial fitting. (**a**) Model 1 (single heat source); (**b**) Model 2 (three heat sources without delay); (**c**) Model 3 (three heat sources with delay).

Table 2. Fitting error: SSE (sum of squares for error), R-square, and RMSE (root mean squared error).

Model	Fitting Method	SSE	R-Square	RMSE
1	2-order polynomial	99.88	0.9394	0.2551
	3-order polynomial	99.07	0.9399	0.2544
	4-order polynomial	57.67	0.965	0.1944
	5-order polynomial	55.95	0.9656	0.1919
	3-order Zernike polynomial	99.84	0.9395	0.2553
2	2-order polynomial	183.5	0.8763	0.2328
	3-order polynomial	177.8	0.8802	0.2293
	4-order polynomial	170.1	0.8854	0.2245
	5-order polynomial	166.2	0.888	0.2221
	3-order Zernike polynomial	180.5	0.8783	0.231
3	2-order polynomial	158.2	0.9337	0.2464
	3-order polynomial	156.2	0.9345	0.2451
	4-order polynomial	68.16	0.9714	0.162
	5-order polynomial	65.72	0.9724	0.1593
	3-order Zernike polynomial	156.6	0.9343	0.2454

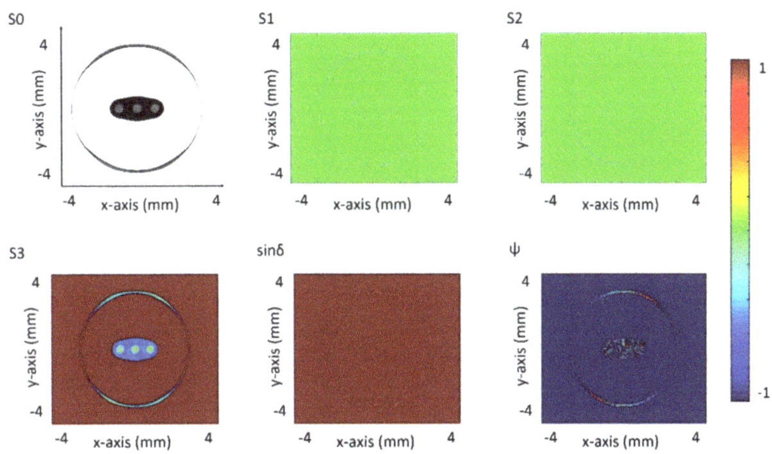

Figure 6. Modeled Stokes parameter images of S0, S1, S2, S3, sinδ, ψ.

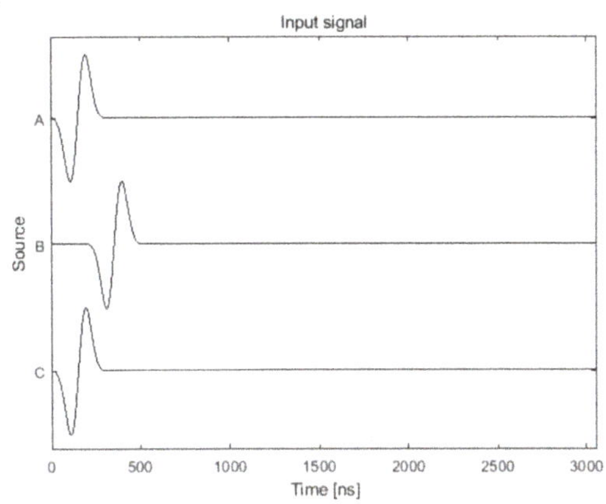

Figure 7. The input signals of the acoustic sources A, B, and C.

The wavefronts for Model 1, Model 2, and Model 3 at different times, shown in Figure 8, are generally consistent with the sound field distributions simulated by FEM. Four probes were placed at Points 1, 2, 3, and 4 (at 1, 1.5, 2, and 2.5 mm to both x-axis and y-axis), the same as in the FEM simulation. Data collected by these probes for Model 1, Model 2, and Model 3 are plotted in Figure 9. Sensor data of Model 1 (red curve) only has one peak, probe data of Model 2 (green curve) has three peaks with similar amplitude, while probe data of Model 3 (blue curve) has two peaks and the second peak is stronger than the first peak.

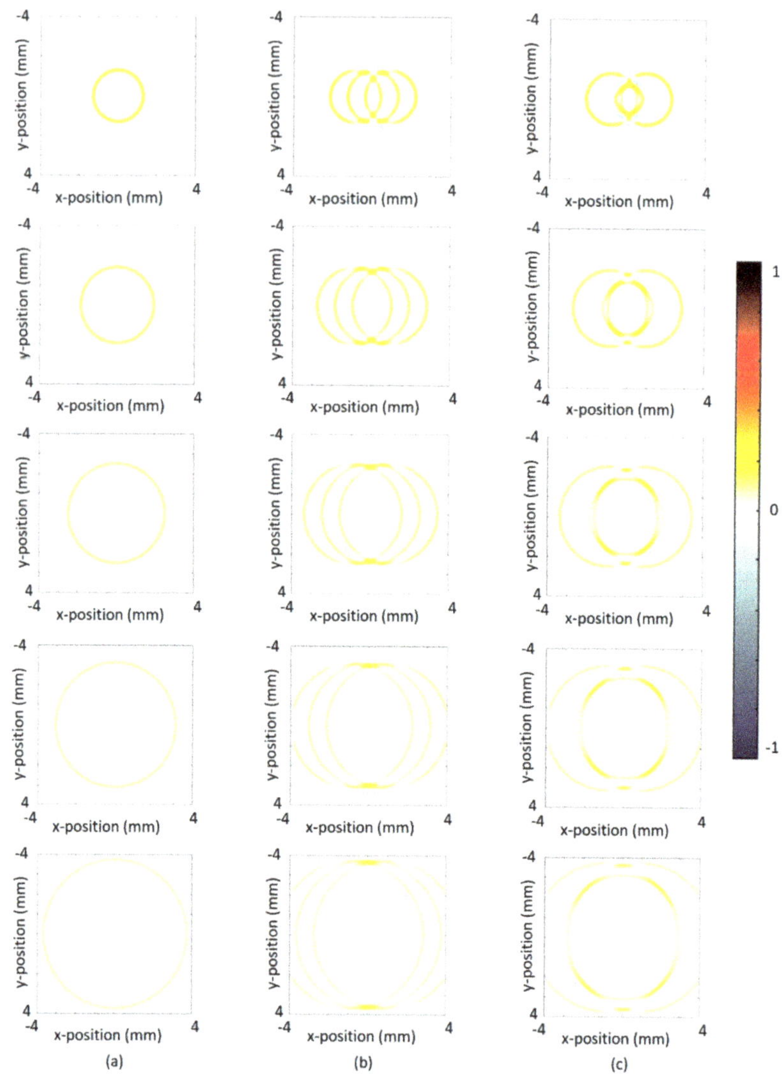

Figure 8. Wavefront at different time. (**a**) Model 1, (**b**) Model 2, and (**c**) Model 3 at 0.6, 0.8, 1, 1.2, and 1.4 μs (from top to bottom).

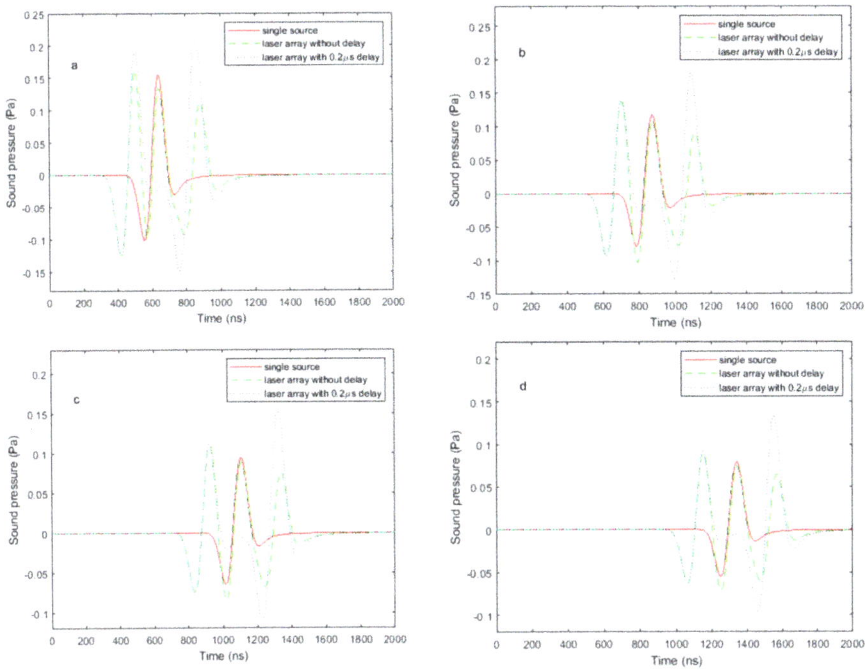

Figure 9. Sound pressure measured by sensors that were placed at (**a**) Point 1, (**b**) Point 2, (**c**) Point 3, and (**d**) Point 4.

Figure 10 shows normalized FEM simulation results and k-wave simulation results separately displayed on the same graph. The simulation results of the two methods have the same trend, but there are slight differences in details, k-wave simulations are simply based on a superposition of sound waves, while FEM simulates the complete process of sound waves generated by thermal expansion. There are also some differences between the different methods. In Model 1, Model 2, and Model 3, the arrival time of sound waves at detection points in FEM is earlier than that in k-wave. This is because the starting point of sound wave propagation in FEM is at the boundary of the heat source (strain caused by temperature gradient), while the starting point of sound wave propagation in the k-wave simulation is at the center of the sound source. The propagation distance of photo-generated acoustic waves in FEM is reduced by half the heat source radius compared to k-wave, which leads to the sound waves in the FEM simulation at the detection point arriving earlier.

In addition to the main wave, there are waves that are generated by multi-point excitation in Models 2 and 3, but these fail to fully participate in the superposition. The k-wave approach was only used as a comparative simulation to verify the suitability of the model. The actual photoacoustic process is closer to FEM simulation. Therefore, only the signal differentiation problem in the FEM model is considered here. Figure 10 shows that the excited sound wave has only a single peak in the negative direction, which can be used as the main peak of detection.

In the FEM simulation, there is slight vibration after the sound wave passes. This is because of the coarsening of edge mesh and the inhomogeneity of mesh division due to the compromise of calculation time. This kind of slight vibration can be eliminated by refining the mesh and decreasing the time step.

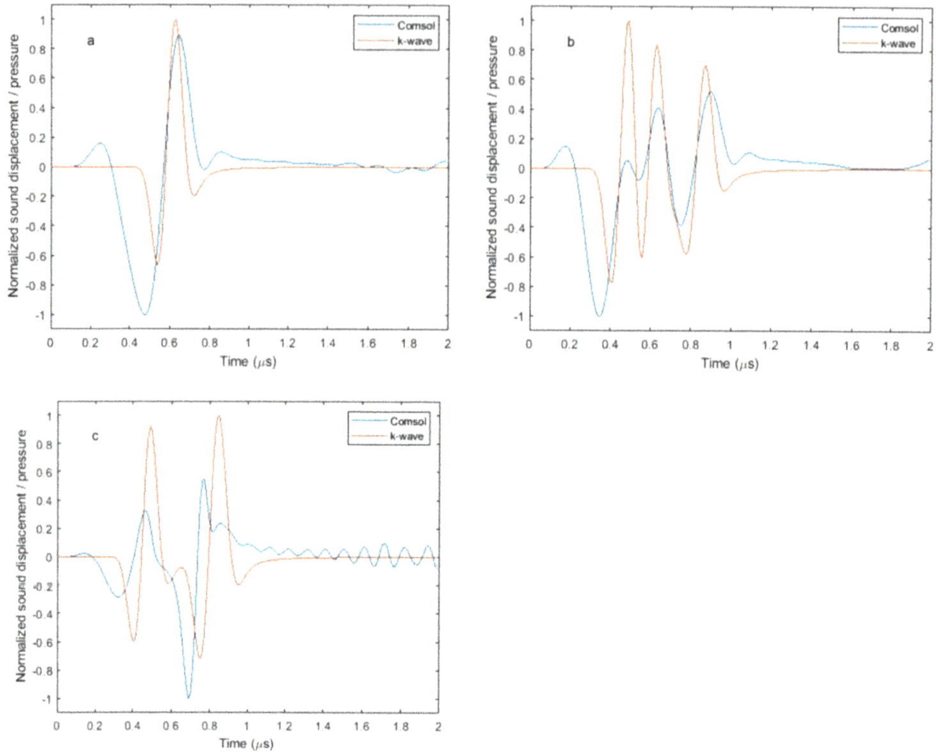

Figure 10. Normalized sound pressure field (k-wave) and displacement field (FEM) detected at Point 1 for (**a**) Model 1, (**b**) Model 2, and (**c**) Model 3.

Simulations of the bulk wave are shown in Figure 11. Clearly, the laser array enhances the intensity of the bulk wave. As expected, the amplitude of the acoustic wave generated by the laser array is larger than that generated by a single laser source with the same energy density. In Figure 11a–c, the difference between the arrival times of the peak of the bulk wave in Model 1 and Model 2 at Points 5, 6, and 7 (The points on the z-axis that are 1, 2, and 3 mm away from the central heat source B) are 0.22, 0.10, and 0.06 μs, respectively, after the excitation time. This is because, for detection points further from the excitation source, the propagation paths of bulk waves originating from spots A and C (symmetrically located on either side of spot B) gradually approach the propagation path of the wave originating from spot B. Because there are two sources, the dominant bulk wave propagating along the z-axis results from a superposition of the waves excited at spots A and C. In Figure 11a, the smaller peak is the acoustic wave excited by source B. In Figure 11b,c, this peak adds constructively to the waves excited by A and C, which increases the width of the wave. Comparing the peak amplitudes of waves at locations 1, 2, and 3 mm from spot B on the z-axis, it can be seen that the amplitudes of the bulk waves generated in Model 2 are, respectively, 1.1, 1.9, and 2.4 times of those excited by Model 1. In the far-field, the bulk wave excited by a distributed laser array is enhanced by a factor of more than 2 compared to a single spot. This phenomenon is analogous to the mechanism of the widely used synthetic aperture method for the enhancement of the detection SNR [33].

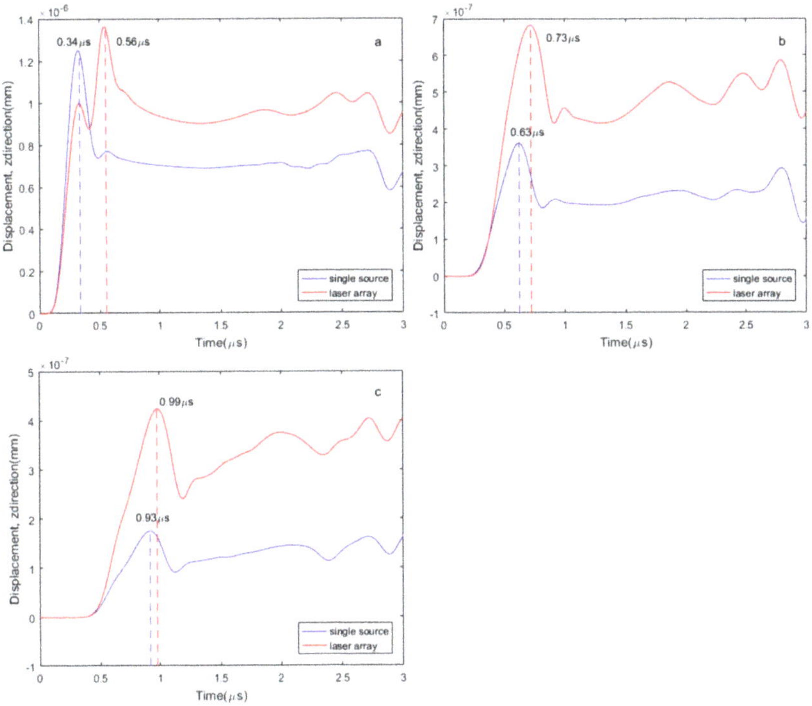

Figure 11. Displacement in the z-direction as a function of time for Model 1 (single laser source) and Model 2 (laser array) at (**a**) Point 5, (**b**) Point 6, and (**c**) Point 7.

Figure 12 shows the attenuation curve, obtaining the amplitude of the z-direction displacement at six points along the z-axis: 1, 1.5, 2, 2.5, 3, and 3.5 mm away from source B. It can be seen that the slope of the attenuation curve for Model 1 is steeper than that for Model 2. This phenomenon arises because the wave has a plane component.

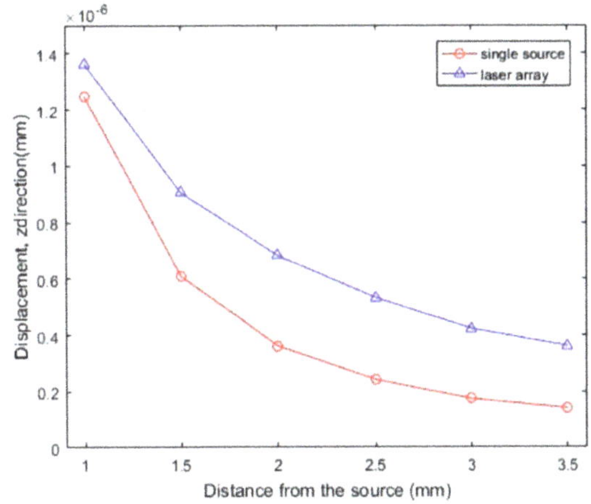

Figure 12. The peak z-direction displacement of the wave for different locations for the cases of single laser and laser array sources.

4. Conclusions

This study compares surface and bulk acoustic waves excited by a single laser source, a laser array composed of three sources without time delay, and a laser array of three sources with a time delay imposed on the central source. It is shown that for the same excitation energy density, arrayed photoacoustic excitation leads to surface acoustic waves that are around four times stronger in particular directions than that of a single source. Moreover, by applying an appropriate time delay to the center source, the time and distance required for constructive superposition to form a complete wavefront can be effectively shortened. The use of the laser array has a significant enhancement on the amplitude of bulk acoustic waves propagating into the object, analogous to the synthetic aperture method in ultrasonic detection. In a future study, we will systematically investigate the characteristics of wave generation by laser array sources with different energy spatial distributions, temporal profiles, and delays between pulses in order to obtain a better understanding of the mechanisms of wave generation using laser arrays. Using a laser array with an appropriate delay time can enhance the intensity of the excited acoustic signal without damaging the sample, and the excited acoustic wavefront still propagates in a regular shape with little clutter. This method provides a new way to manipulate the wavefront and improve the generation efficiency in nondestructive photoacoustic imaging applications.

Author Contributions: Conceptualization, J.X., X.L., J.H.M. and L.H.; software, R.H., B.X. and Y.Z.; writing—original draft preparation, R.H.; writing—review and editing, J.X., Z.X., W.L. and C.G.; supervision, J.X. and X.L. All authors have read and agreed to the published version of the manuscript.

Funding: This research was funded by the National Major Scientific Instruments and Equipment Development Project, grant no. 61827814, Beijing Natural Science Foundation, grant no. Z19J0018, the Fundamental Research Funds for the Central Universities, grant no. 30920010011, the Ministry of Education collaborative project (B17023). This research was supported by the U.K. Engineering and Physical Sciences Research Council (grant EP/R042578/1).

Institutional Review Board Statement: Not applicable.

Informed Consent Statement: Not applicable.

Conflicts of Interest: The authors declare no conflict of interest.

References

1. Xu, M.; Wang, L.V. Photoacoustic Imaging in Biomedicine. *Rev. Sci. Instrum.* **2006**, *77*, 41101. [CrossRef]
2. Shen, Z.; Xu, B.; Ni, X.; Lu, J. Numerical Simulation of Laser-Generated Ultrasonic Waves in Layered Plates. *J. Phys. D Appl. Phys.* **2004**, *37*, 2364–2370. [CrossRef]
3. Pitarresi, G.; Patterson, E.A. A Review of the General Theory of Thermoelastic Stress Analysis. *J. Strain Anal. Eng. Des.* **2003**, *38*, 405–417. [CrossRef]
4. Ma, R.; Taruttis, A.; Ntziachristos, V.; Razansky, D. Multispectral Optoacoustic Tomography (MSOT) Scanner for Whole-Body Small Animal Imaging. *Opt. Express* **2009**, *17*, 21414. [CrossRef] [PubMed]
5. Thomas, R.J.; Rockwell, B.A.; Marshall, W.J.; Aldrich, R.C.; Zimmerman, S.A.; Rockwell, R.J. A Procedure for Multiple-Pulse Maximum Permissible Exposure Determination under the Z136.1-2000 American National Standard for Safe Use of Lasers. *J. Laser Appl.* **2001**, *13*, 134–140. [CrossRef]
6. Hajireza, P.; Shi, W.; Bell, K.; Paproski, R.J.; Zemp, R.J. Non-Interferometric Photoacoustic Remote Sensing Microscopy. *Light Sci. Appl.* **2017**, *6*, e16278. [CrossRef]
7. Cao, Y.; Hui, J.; Kole, A.; Wang, P.; Yu, Q.; Chen, W.; Sturek, M.; Cheng, J.X. High-Sensitivity Intravascular Photoacoustic Imaging of Lipid-Laden Plaque with a Collinear Catheter Design. *Sci. Rep.* **2016**, *6*, 1–8. [CrossRef]
8. Murray, T.W.; Balogun, O. High-Sensitivity Laser-Based Acoustic Microscopy Using a Modulated Excitation Source. *Appl. Phys. Lett.* **2004**, *85*, 2974–2976. [CrossRef]
9. Huang, J.; Krishnaswamy, S.; Achenbach, J.D. Laser Generation of Narrow-band Surface Waves. *J. Acoust. Soc. Am.* **1992**, *92*, 2527–2531. [CrossRef]
10. Clark, M.; Sharples, S.D.; Somekh, M.G. Diffractive Acoustic Elements for Laser Ultrasonics. *J. Acoust. Soc. Am.* **2000**, *107*, 3179–3185. [CrossRef]

11. Sharples, S.D.; Clark, M.; Somekh, M.G. All-Optical Adaptive Scanning Acoustic Microscope. *Ultrasonics* **2003**, *41*, 295–299. [CrossRef]
12. Stratoudaki, T.; Clark, M.; Somekh, M.; Arca, A. CHeap Optical Transducers (CHOTs) for Generation and Detection of Longitudinal Waves. In Proceedings of the IEEE International Ultrasonics Symposium, IUS, Dresden, Germany, 7–10 October 2012; pp. 961–964.
13. Stratoudaki, T.; Clark, M.; Wilcox, P.D. Laser induced ultrasonic phased array using full matrix capture data acquisition and total focusing method. *Opt Express.* **2016**, *24*, 21921–21938. [CrossRef]
14. Pieris, D.; Stratoudaki, T.; Javadi, Y.; Lukacs, P.; Catchpole-Smith, S.; Wilcox, P.D.; Clare, A.; Clark, M. Laser Induced Phased Arrays (LIPA) to Detect Nested Features in Additively Manufactured Components. *Mater. Des.* **2020**, *187*, 108412. [CrossRef]
15. Jarzynski, J.; Berthelot, Y.H. The Use of Optical Fibers to Enhance the Laser Generation of Ultrasonic Waves. *J. Acoust. Soc. Am.* **1989**, *85*, 158–162. [CrossRef]
16. Noroy, M.H.; Royer, D.; Fink, M.A. Shear-Wave Focusing with a Laser-Ultrasound Phased-Array. *IEEE Trans. Ultrason. Ferroelectr. Freq. Control* **1995**, *42*, 981–988. [CrossRef]
17. Scott Steckenrider, J.; Murray, T.W.; Wagner, J.W.; Deaton, J.B. Sensitivity Enhancement in Laser Ultrasonics Using a Versatile Laser Array System. *J. Acoust. Soc. Am.* **1995**, *97*, 273–279. [CrossRef]
18. Murray, T.W.; Deaton, J.B.; Wagner, J.W. Experimental Evaluation of Enhanced Generation of Ultrasonic Waves Using an Array of Laser Sources. *Ultrasonics* **1996**, *34*, 69–77. [CrossRef]
19. Hopko, S.N.; Ume, I.C.; Erdahl, D.S. Development of a Flexible Laser Ultrasonic Probe. *J. Manuf. Sci. Eng. Trans. ASME* **2002**, *124*, 351–357. [CrossRef]
20. Mi, B.; Ume, C. Real-Time Weld Penetration Depth Monitoring with Laser Ultrasonic Sensing System. *J. Manuf. Sci. Eng. Trans. ASME* **2006**, *128*, 280–286. [CrossRef]
21. Pei, C.; Demachi, K.; Fukuchi, T.; Koyama, K.; Uesaka, M. Cracks Measurement Using Fiber-Phased Array Laser Ultrasound Generation. *J. Appl. Phys.* **2013**, *113*, 163101. [CrossRef]
22. Besson, A.; Carrillo, R.E.; Bernard, O.; Wiaux, Y.; Thiran, J.P. Compressed delay-and-sum beamforming for ultrafast ultrasound imaging. In Proceedings of the 2016 IEEE International Conference on Image Processing (ICIP), Phoenix, AZ, USA, 25–28 September 2016; pp. 2509–2513. [CrossRef]
23. Holfort, I.K.; Gran, F.; Jøensen, J.A. Broadband minimum variance beamforming for ultrasound imaging. *IEEE Trans Ultrason Ferroelectr Freq. Control* **2009**, *56*, 314–325. [CrossRef] [PubMed]
24. Capon, J. High-Resolution Frequency-Wavenumber Spectrum Analysis. *Proc. IEEE* **1969**, *57*, 1408–1418. [CrossRef]
25. Von Ramm, O.T.; Smith, S.W. Beam Steering with Linear Arrays. *IEEE Trans. Biomed. Eng.* **1983**, *BME-30*, 438–452. [CrossRef] [PubMed]
26. Wooh, S.C.; Shi, Y. Simulation study of the beam steering characteristics for linear phased arrays. *J. Nondestruct. Eval.* **1999**, *18*, 39–57. [CrossRef]
27. Guan, J.; Shen, Z.; Ni, X.; Wang, J.; Lu, J.; Xu, B. Numerical Simulation of the Reflected Acoustic Wave Components in the near Field of Surface Defects. *J. Phys. D Appl. Phys.* **2006**, *39*, 1237–1243. [CrossRef]
28. Kalenskii, A.V.; Zvekov, A.A.; Aduev, B.P. The Influence of Temperature on the Spectral Dependences of Aluminum's Optical Properties. *Opt. Spectrosc. (English Transl. Opt. I Spektrosk.)* **2018**, *124*, 501–508. [CrossRef]
29. Crecraft, D.I. The Measurement of Applied and Residual Stresses in Metals Using Ultrasonic Waves. *J. Sound Vib.* **1967**, *5*, 173–192. [CrossRef]
30. Al Wardany, R.; Rhazil, J.; Ballivy, G.; Gallias, J.L.; Saleh, K. Use of Rayleigh Wave Methods to Detect Near Surface Concrete Damage R. In Proceedings of the 16th World Conference on NDT, WCNDT, Montreal, QC, Canada, 30 August–3 September 2004. Available online: https://www.ndt.net/article/wcndt2004/pdf/civil_structures/729_wardany.pdf (accessed on 17 February 2021).
31. Liu, G.; Xiong, J.; Cao, Y.; Hou, R.; Zhi, L.; Xia, Z.; Liu, W.; Liu, X.; Glorieux, C.; Marsh, J.H.; et al. Visualization of Ultrasonic Wave Field by Stroboscopic Polarization Selective Imaging. *Opt. Express* **2020**, *28*, 27096. [CrossRef]
32. Treeby, B.E.; Jaros, J.; Rohrbach, D.; Cox, B.T. Modelling Elastic Wave Propagation Using the K-Wave MATLAB Toolbox. In Proceedings of the IEEE International Ultrasonics Symposium, IUS, Chicago, IL, USA, 3–6 September 2014; pp. 146–149.
33. Ylitalo, J. On the Signal-to-Noise Ratio of a Synthetic Aperture Ultrasound Imaging Method. *Eur. J. Ultrasound* **1996**, *3*, 277–281. [CrossRef]

applied sciences

MDPI

Article

The Spec-Radiation Method for Layered Fluid Media

Andreas Sebastian Schmelt * and Jens Twiefel

Institute of Dynamic and Vibration Research, Leibniz University of Hannover, 30823 Garbsen, Germany; twiefel@ids.uni-hannover.de
* Correspondence: schmelt@ids.uni-hannover.de

Abstract: The real-time evaluation for non-destructive air-coupled ultrasonic testing of panel materials is a big task for several industries. To make these tests more and more accurate, efficient and reliable calculation methods from ultrasonic holography are essential. In the past, we presented the spec-radiation method as a fast and accurate method for such tasks. The spec-radiation method calculates the sound field utilizing data from a measurement plane at another parallel or tilted plane, especially the sound field at the surface of a panel. This can be used to detect flaws. There is a limitation of the current method: using the data on the panel surface limits the accuracy of the detected flaws. A big step forward could be expected if the sound field in the material were known. As a first step, we developed the spec-radiation method forward to consider multiple material layers. For now, we made the major assumption that all layers have fluid-like properties. Hence, transversal waves were neglected. This extension of the spec-radiation method was validated utilizing an experiment. We present that flaws in the panel material can be detected with higher accuracy at a similar speed compared to our former approach.

Keywords: spec-radiation; acoustic holography; non-destructive testing; layered media; flaw detection

Citation: Schmelt, A.S.; Twiefel, J. The Spec-Radiation Method for Layered Fluid Media. *Appl. Sci.* **2022**, *12*, 1098. https://doi.org/10.3390/app12031098

Academic Editor: Habil. Michel Darmon

Received: 3 December 2021
Accepted: 17 January 2022
Published: 21 January 2022

1. Introduction

Today's panel materials industry is dependent on fast, reliable, accurate, and non-destructive testing to guarantee the materials' quality. This is because the demands on the panel material are increasing, while at the same time, material and thus costs must be saved. This affects a wide range of industries, such as wind turbine manufacturers (Jasiuniene et al. [1]), the construction industry (Conta et al. [2]), aeronautical applications (Fahr [3]), the wood-based panel industry (Fang et al. [4]), and many others. Sokolov [5] used ultrasound for the first time to test a material for defects. Shortly thereafter, Firestone developed the first echo material tester based on the reflection principle, and Trost developed an ultrasonic forceps based on ultrasonic transmission. After that, the further development of ultrasonic testing was inevitable (Deutsch et al. [6], Krautkrämer and Krautkrämer [7]).

Ultrasonic testing can be roughly divided into two groups: contact testing, in which the transmitter and receiver are in direct contact with the material to be tested, and non-contact testing, in which the ultrasonic waves are transmitted from a transmitter to the material via an ambient medium, and from there to the receiver. Ultrasonic testing with contact to the test object offers the advantage that sound energy can be transmitted easily with low losses. However, the applied contact force has an influence on the result (Gyekenyesi et al. [8]). Due to the high difference in impedance between the transmitter or receiver and the test object, a thin layer of a contact medium is often applied to further increase the sound energy. This contact medium consists in most cases of water-containing oils, gels, or fats (Willcox and Downes [9]). However, with some materials, this has the disadvantage that the contact oil leaves residues on the test object or is even absorbed by the test object (as is the case with wood-based materials, for example), thus changing the material properties of the test object (Schafer [10]). These disadvantages led to the development of non-contact ultrasonic testing (Fang et al. [4]). Non-contact ultrasonic testing is especially suitable for objects with higher temperatures in corrosive or other dangerous environments or areas that are difficult to access. It allows large distances from the structures to be tested (Green [11]), and due to the missing contact force, it cannot influence the results. Usually, water is used as the ambient medium for contactless ultrasonic testing. This is because water has good impedance properties with respect to most solid materials (Jasiuniene et al. [1], Zhang et al. [12], Mitri et al. [13]). For materials such as wood, however, testing under water is not possible because the material would absorb the water, change its properties, and/or even be destroyed. Air-coupled ultrasonic testing was developed for such materials. A major challenge is the large difference in the impedance of air compared to most solid materials. According to Stößel [14], in carbon-fiber-reinforced plastic (CFRP) plates, only 0.000004% of an incident wave is transmitted into the material; the rest is reflected, and only about 0.005% of the energy is transmitted by a piezoceramic to the air. To be able to use air-coupled ultrasonic testing successfully, the transmitters are provided with a matching layer. These are connected directly to the vibration generator (e.g., the piezoceramic) and usually have a characteristic length of a quarter of the wavelength ($\lambda/4$) of the respective material of the matching layer (Chimenti [15], Álvarez Arenas [16]). The application of modern materials allows the successful use of air-coupled ultrasound (Hillger et al. [17]). Air-coupled ultrasound is currently used in a wide range of applications, e.g., for the examination of turbine blades made of glass-fiber-reinforced plastics (GFRPs) (Raisutis et al. [18]) or wood-based materials (Sanabria et al. [19]), to name a few.

During the production process of wooden particleboard, delamination, air inclusions, or insufficiently glued areas can occur (Dunky and Niemz [20]). The first who detected such flaws in wood by ultrasound were Niemz [21], Bucur and Böhnke [22]. They assessed whether delamination was present based on the change in the propagation time of the signal. If the particleboard is to be inspected during production, it usually still has a very high temperature (>100 °C) and vapors escaping, that can lead to receiver damage (Dunky and Niemz [20]). This means that the receivers must be placed at a large distance (>100 mm) from the particleboard. These large distances from the wooden particleboard make a direct detection of flaws very difficult, if not impossible, since the sound waves diffract and interfere all the way to the receiver (Laybed and Huang [23]).

According to Döring [24], the three best-known methods of presenting the information from a non-destructive test with air-coupled ultrasound are the A-, B-, and C-scans. An A-scan is the time-dependent measurement of a single point; a B-scan is a line of A-scans; a C-scan is a set of multiple B-scans. The C-scan in particular enables good detection of flaws or delaminations in shape and position, but is also the most time consuming because a large number of points must be measured.

However, as already mentioned, at greater distances from the test object, it becomes increasingly difficult to find a clear indicator for flaws due to the diffraction and interference of the sound waves. If the transmitter is stationary during the measurements and only the receiver is moved, tomographic measurements can be performed (Chimenti [15]). If, in this way, the information of a plane is measured and known, methods of acoustic holography can be applied to calculate from that plane to another plane directly above the object under test. The acoustic holography originates from the holography in optics invented by Gabor [25]. Acoustic holography enables the calculation of sound fields starting from a known plane in the sound field. Due to the coherent properties of ultrasonic waves, diffraction-induced inaccuracies can be eliminated in this way (Singh [26]). An important method of acoustic holography for the detection of flaws is the re-radiation method (Sanabria et al. [27], Marhenke et al. [28], Marhenke et al. [29], Marhenke et al. [30], Schmelt et al. [31], Schmelt et al. [32], Schmelt et al. [33]). With the re-radiation method, flaws in the sub-wavelength range can be identified (Marhenke et al. [28]). Under the assumption that the solid test object is similar to a fluid and only longitudinal waves can propagate, a well-interpretable result has already been obtained on the side facing away from the receiver (Marhenke et al. [28]), and objects in a sound field have also been determined (Schmelt et al. [31], Tsysar and Sapozhnikov [34]). The re-radiation method is based on the Rayleigh–Sommerfeld diffraction integral and thus offers a well-interpretable description in the spatial domain. Due to the description in the spatial domain, the computation effort is not low, since several two-dimensional Fourier transformations (FFTs) and inverse Fourier transformations (iFFTs) have to be performed for the evaluation. The spec-radiation method (Schmelt and Twiefel [35]) offers the possibility to perform faster computations, since the data are processed in the spatial frequency domain, so fewer FFTs and iFFTs have to be performed. The smallest detectable flaws that can be detected theoretically are, according to Wolf [36], in the size of $\lambda/2$ with the re-radiation, as well as the spec-radiation method. The spec-radiation method is based on the angular spectrum method, which was first used by Booker and P. C. Clemmow [37] to describe light propagation. Ratcliffe [38], then it was possible to describe the diffraction of light in the ionosphere with this method. Boyer et al. [39] then showed that it is possible to apply this method to the propagation of sound waves, and he was able to identify holes in steel plates with a 5 MHz transmitter. To characterize the sound field of ultrasonic transmitters, Schafer and Lewin [40] and de Belleval and Messaoud-Nacer [41] proposed the use of the angular spectrum method. While Peng et al. [42] used it for object identification, Yan and Hamilton [43] used it for the analysis of tissue harmonic imaging. A finite-element model based on the angular spectrum method was built by Aanes et al. [44], and Liu and Waag [45] described the forward and backward propagation of sound waves using the method. Conta et al. [2] showed the influence of zero-padding. Matsushima [46] showed the application of the angular spectrum method in the field of computer holography. Schmelt and Twiefel [35] utilized it for the spec-radiation method to identify flaws in panel materials. In Schmelt and Twiefel [47], the spec-radiation method was extended to the calculation of tilted planes in sound fields, to identify flaws in tilted panels. However, all these applications assume that the sound waves propagate in one medium, so that there are no layered media considered.

In this publication, we present an extension to the spec-radiation method to calculate the sound field in layered fluid media. We show this in the first step including the determination of the particle deflection. The objective of the publication was to obtain an increased accuracy in the detection of flaws in thicker materials and on the side of the particleboard facing away from the receiver. For this purpose, it was necessary to calculate this in the material and thus take into account the occurring diffraction, as well as the reflections and transmissions. With the extension presented here and the assumptions we made, we then show, utilizing an experiment with a wooden particleboard, that a flaw on the side opposite the receiver can be detected with increased accuracy compared to calculation of the sound field on the top surface of the wooden particleboard.

2. Material and Methods

In this section, the material and methods are represented. In Section 2.1, the spec-radiation method based on the angular spectrum method is described. The spec-radiation method is extended to layered fluid media in Section 2.2. The practical implementation is described in Section 2.3. In Section 2.4, the experimental setup is introduced.

2.1. Spec-Radiation Method

The spec-radiation method is based on the angular spectrum method, and the derivation described here was mainly based on that of Goodman [48] and Matsushima [46]. Due to the nature of the spec-radiation method, only harmonic waves were used for the calculation.

Starting from a source wave field described by $p(x,y,z=0,\omega)$ in a plane, the FFT of it can be written as follows:

$$\begin{aligned} P(f_x,f_y,z=0,\omega) &= \mathfrak{F}_{x,y}(p(x,y,z=0,\omega)) \\ &= \iint_{-\infty}^{\infty} p(x,y,z=0,\omega)\mathrm{e}^{-\mathrm{i}2\pi(f_x x+f_y y)}\mathrm{d}x\mathrm{d}y. \end{aligned} \quad (1)$$

Here, x, y, and z are spatial coordinates. $\mathfrak{F}_{x,y}$ is the Fourier transform operator for the coordinates x and y. $p(x, y, z = 0, \omega)$ is the sound pressure wave field at $z = 0$. ω is the time-dependent angular frequency; i is the imaginary unit, and f_x, f_y, and f_z are the spatial frequencies with:

$$(f_x, f_y, f_z) = \left(\frac{k_x}{2\pi}, \frac{k_y}{2\pi}, \frac{k_z}{2\pi}\right). \tag{2}$$

k_x, k_y, and k_z are the wave numbers in the different spatial dimensions, and therefore, they are the components of the wave vector. The frequency f_z is not independent of the other spatial frequencies and the wavelength. For the determination of f_z, we have to take a deeper view into the wave vector. The components of the wave vector are dependent on the following condition:

$$|\mathbf{k}| = \frac{2\pi}{\lambda} \tag{3}$$

with **k** as the wave vector. This means that the spatial frequencies always satisfy:

$$f_x^2 + f_y^2 + f_z^2 = \left(\frac{1}{\lambda}\right)^2. \tag{4}$$

By changing Equation (4), it becomes obvious that f_z is dependent on the other two spatial frequencies:

$$f_z(f_x, f_y, \lambda) = \sqrt{\left(\frac{1}{\lambda}\right)^2 - f_x^2 - f_y^2}. \tag{5}$$

This equation represents the so-called Ewald sphere, and each plane wave is represented by a point on the surface of the sphere. Another important connection of spatial frequencies in the description of the propagation of plane waves is the relationship with the directional cosine of the wave propagation:

$$\begin{aligned} f_x &= \frac{\cos(\alpha)}{\lambda} \\ f_y &= \frac{\cos(\beta)}{\lambda} \\ f_z &= \frac{\cos(\gamma)}{\lambda} \end{aligned} \tag{6}$$

which means that a point on the Ewald sphere can also be described by three directional cosines of the propagation angles of the plane wave. To calculate the sound wave field of another plane, Equation (1) can be multiplied by the propagation function $H(f_z, z)$. $P(f_x, f_y, z, \omega)$ becomes:

$$\begin{aligned} P(f_x, f_y, z, \omega) &= H(f_z, z)\ \mathfrak{F}_{x,y}(p(x, y, z = 0, \omega) \\ &= \mathrm{e}^{-\mathrm{i}2\pi f_z z} \iint_{-\infty}^{\infty} p(x, y, z = 0, \omega)\mathrm{e}^{-\mathrm{i}2\pi(f_x x + f_y y)}\mathrm{d}x\mathrm{d}y. \end{aligned} \tag{7}$$

While computing the equation, it has to be considered that the square root of Equation (5) can also become complex. Thus, two cases have to be discussed:

$$\begin{aligned} f_x^2 + f_y^2 &\leq \left(\frac{1}{\lambda}\right)^2, \text{ leads to complex solutions} \\ f_x^2 + f_y^2 &> \left(\frac{1}{\lambda}\right)^2, \text{ leads to real solutions.} \end{aligned} \tag{8}$$

In the first case, when the square root of Equation (5) becomes complex, the information is outside the Ewald sphere and represents exponential decaying evanescent waves. For the through-transmission technique used in this publication to identify flaws, their contribution is rather small, and it was neglected in the further steps. The second case describes the information that is located directly on the surface of the Ewald sphere. With the propagation function $H(f_z, z)$, it is possible to calculate the sound propagation forward $(t \geq 0)$ and backward $(t \leq 0)$ in time. To transform the sound field back into the spatial domain and time domain, a 3D iFFT must be applied to Equation (7):

$$\begin{aligned} p(x,y,z,t) &= \mathfrak{F}^{-1}_{f_x,f_y,\omega}(P(f_x,f_y,z,\omega)) \\ &= \mathfrak{F}^{-1}_{f_x,f_y,\omega}(H(f_z,z)\,\mathfrak{F}_{x,y}(p(x,y,0,\omega))). \end{aligned} \tag{9}$$

With this equation, starting from a starting plane, the complete sound field in a homogeneous fluid volume can now be described.

2.2. Extended Spec-Radiation Method for Layered Fluid Media

The extension is derived for a rather simple case first and is generalized later. There are two different fluid media layers, as depicted in Figure 1. According to Brekhovskikh and Godin [49], the layer interface interaction is described by the part of the wave that has an influence at the interface. Hence, a corresponding impedance has to be defined; we call it here the specific impedance. The specific impedance for one single plane wave is calculated by:

$$Z = \frac{p(x,y,z)}{v_z(x,y,z)}, \tag{10}$$

which leads to the well-known formulation:

$$Z = \frac{c\rho}{\cos(\gamma)}, \tag{11}$$

with γ as the propagation direction angle, ρ as the density, and c as the speed of sound. With this knowledge, the reflection factor:

$$R = \frac{Z_2 - Z_1}{Z_2 + Z_1} \tag{12}$$

and the transmission factor:

$$T = \frac{2Z_2}{Z_2 + Z_1} = 1 + R \tag{13}$$

are determined. Indices 1 and 2 mark Medium 1 and Medium 2. With the reflection and transmission factor, the sound wave pressure in both media can be calculated.

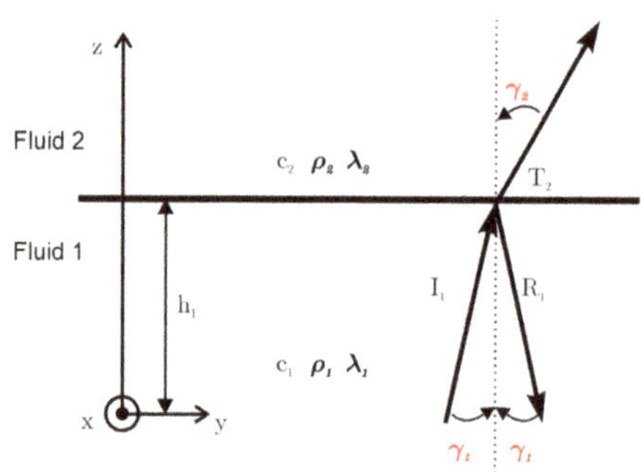

Figure 1. Schematic representation of the incident, reflected, and transmitted waves at a fluid–fluid interface in a semi-infinite volume. Included are the speed of sound c_1 and c_2, the densities ρ_1 and ρ_2, and the wavelengths λ_1 and λ_2 of the two fluid media.

If Equation (10) is set up for the spectrum, i.e., in the Fourier domain for a specific ω, it looks as follows:

$$Z_{\mathfrak{F}} = \frac{P(f_x, f_y, z)}{V(f_x, f_y, z)}. \tag{14}$$

Here, the index $\mathfrak{F}$ at the specific impedance indicates that this is the specific impedance for the Fourier domain In the Fourier domain, the velocity is described by the pressure with the following equation:

$$V(f_x, f_y, z) = \frac{P(f_x, f_y, z) k_z}{\omega \rho}, \tag{15}$$

as can be found in Möser [50]. This turns Equation (14) into:

$$Z_{\mathfrak{F}} = \frac{c\rho}{\lambda f_z}. \tag{16}$$

This is now an expression for the impedance that can be used in the spec-radiation method. When comparing Equations (11) and (16), the relation of Equation (6) is again recognizable. This means that if λ and f_z are known, the cosine of the angle of propagation is also known. With Equation (16), the reflection factor can be written as:

$$R_{\mathfrak{F}} = \frac{Z_{\mathfrak{F},2} - Z_{\mathfrak{F},1}}{Z_{\mathfrak{F},2} + Z_{\mathfrak{F},1}} = \frac{\frac{c_2\rho_2}{\lambda_2 f_{z,2}} - \frac{c_1\rho_1}{\lambda_1 f_{z,1}}}{\frac{c_2\rho_2}{\lambda_2 f_{z,2}} + \frac{c_1\rho_1}{\lambda_1 f_{z,1}}} \tag{17}$$

and the transmission factor as:

$$T_{\mathfrak{F}} = \frac{2Z_{\mathfrak{F},2}}{Z_{\mathfrak{F},2} + Z_{\mathfrak{F},1}} = \frac{2\frac{c_2\rho_2}{\lambda_2 f_{z,2}}}{\frac{c_2\rho_2}{\lambda_2 f_{z,2}} + \frac{c_1\rho_1}{\lambda_1 f_{z,1}}}. \tag{18}$$

The entire sound pressure in Fluid 1 according to Figure 1 is thus the sum of the incident and the reflected sound waves:

$$\begin{aligned} p_{f1}(x, y, 0 \le z < h_1, t) =& \mathfrak{F}^{-1}_{f_x, f_y, \omega}(H(f_{z,1}, 0 \le z^+ < h_1)\mathfrak{F}_{x,y}(p(x, y, z_0, \omega))) + \\ & \mathfrak{F}^{-1}_{f_x, f_y, \omega}(H(f_{z,1}, 0 \le z^- < h_1) R_{\mathfrak{F},12} \mathfrak{F}_{x,y}(p(x, y, z_{h1}, \omega))). \end{aligned} \tag{19}$$

Here, z^+ indicates that the sound propagation is calculated in the positive z-direction, and z^- indicates that the sound propagation is calculated in the negative z-direction. $R_{\mathfrak{F},12}$ is the reflection factor in the Fourier domain for the interface of Fluids 1 and 2. The sound pressure in Fluid 2 is:

$$p_{f2}(x, y, z \ge h_1, t) = \mathfrak{F}^{-1}_{f_x, f_y, \omega}(H(f_{z,2}, z^+ \ge h_1) T_{\mathfrak{F},12} \mathfrak{F}_{x,y}(p(x, y, z_{h1}, \omega))). \tag{20}$$

Here, $T_{\mathfrak{F},12}$ is the transmission factor at the interface between Fluid 1 and Fluid 2. To determine the vectorial particle deflection from the scalar sound pressure, the sound pressure can be converted into the scalar deflection potential:

$$\chi(x, y, z, \omega) = \frac{p(x, y, z, \omega)}{\rho \omega^2}. \tag{21}$$

From the deflection potential, the vectorial particle deflection can be calculated in the Fourier domain through:

$$\mathbf{s} = \text{grad}(\chi(x,y,z,\omega)) = \mathfrak{F}^{-1}_{f_x,f_y}\begin{pmatrix} i2\pi f_x \mathfrak{F}_{x,y}(\chi(x,y,z,\omega)) \\ i2\pi f_y \mathfrak{F}_{x,y}(\chi(x,y,z,\omega)) \\ Vi2\pi f_{z,j} \mathfrak{F}_{x,y}(\chi(x,y,z,\omega)) \end{pmatrix}. \tag{22}$$

Here, V is a propagation sign. $V = -1$ for waves that propagate in the direction of $z > z_0$ and $V = 1$ for waves that propagate in the direction of $z < z_0$. $f_{z,j}$ is the spatial frequency in the propagation direction for Fluid 1 ($j = 1$) or 2 ($j = 2$).

In the next section, the practical implementation for the calculation of layered fluid media is discussed.

2.3. Practical Implementation

To describe the calculation from one interface to the next, the flowchart in Figure 2 is used. In the first step, the constant known parameters such as the speed of sound, density, the z-position of the interfaces, and the measured sound pressure must be set.

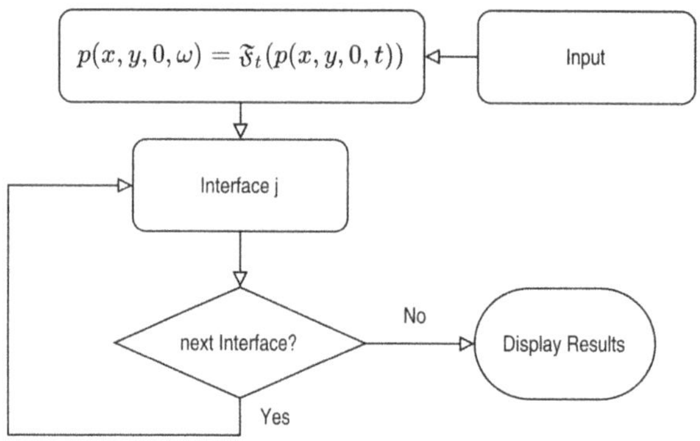

Figure 2. Flowchart for the practical realization of the spec-radiation method for layered fluid media.

Then, the sound pressure is transformed in the frequency domain. The sound field generated by the incident wave at the first interface layer is determined by Equation (9) (spec-radiation method). In the block "Interface j" (with the interface number j = 1, 2, 3,. . .), the reflected and transmitted part is calculated. A flowchart of the block "Interface j" is depicted in Figure 3.

In the block "Interface j", the first decision is among which frequencies the calculation has to be performed. For the representation of a harmonic signal, the calculation of a single frequency can be sufficient, while for the calculation of a single sinusoidal wave, several frequencies are needed. Here, all frequencies from 1.2 kHz up to 120 kHz were used. This means that 164 frequency step points were taken into account with the conjugate complex parts as well. All other frequencies were set to 0. This procedure saves much computing time. Then, the different wavelengths have to be calculated. Next, the sound pressure was transformed with a 2D FFT over the spatial coordinates. Any value that was not on the Ewald sphere was set to 0. Then, the two different spatial frequencies for the two media at the interface in the propagation direction had to be calculated. In the next step, the specific impedances and, with the specific impedances, the reflection and transmission coefficient were determined in the Fourier domain. The propagation function H was then calculated with the spatial frequency for the medium from the incident wave. In the following step, the sound pressure of the reflected and transmitted part at the interface can be calculated. If one is only interested in the sound pressure values, then the block "Interface j" could be left out. Here also, the deflection potential can be calculated, if needed, for the reflected and transmitted part. At last, in the block, the question is: Is one interested in the particle deflection? If so, then the particle deflection could be calculated from the deflection potentials in the Fourier domain as in Equation (22). Now, the block can be left alone. It has to be decided whether another interface should be determined or not. If not, then the generated results can be displayed. If so, then the block "Interface j" must be executed again. The transmitted wave up to the second interface results from the transmitted part at the first interface at $z = h_1$. The calculation has to be considered, now that the transmitted and reflected part starts again at a local $z = 0$. The incident wave at Interface 2 is the transmitted part of Interface 1. As at Interface 1, the incident wave is the start for the transmitted and reflected part, and they start again at a local $z = 0$. After the calculation, it must be decided again whether there is another interface layer or not. It must also be decided whether the reflected sound field should be calculated back to the first interface. If this is the case, then the procedure from Figure 3 is carried out again, whereby the incident sound wave is now the reflected part of Interface 2. It was again assumed that the calculation starts at a local $z = 0$. For the calculation, the indices at the impedances were now simply exchanged. Similarly, for the transmission factor H, the spatial frequency of the second medium was now used, due to the fact that the incident sound wave comes from the second medium. Depending on how many reflections and transmissions through different boundary layers are to be performed, this procedure must be repeated.

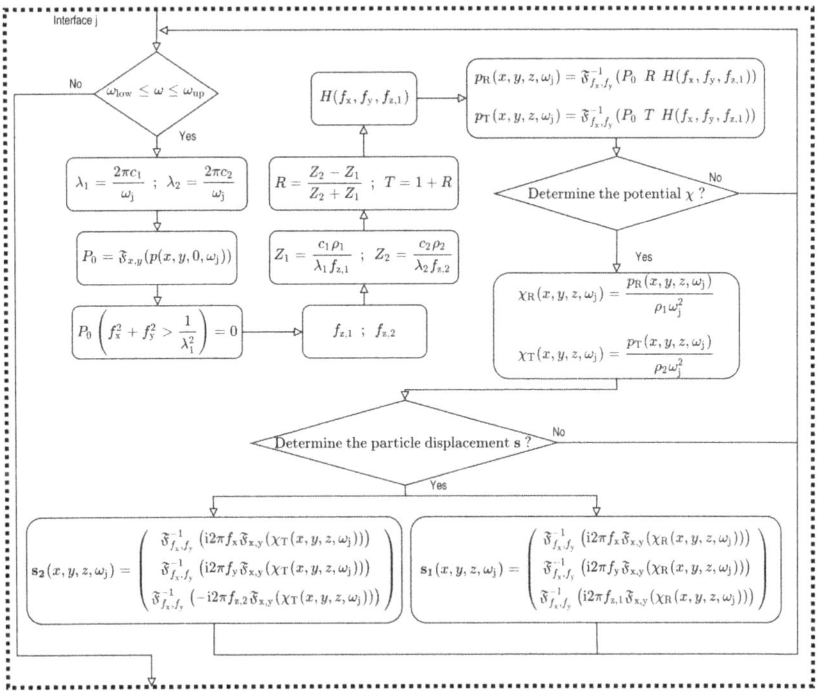

Figure 3. Flowchart of the block "Interface j" for the practical realization of the spec-radiation method for layered fluid media.

2.4. Experimental Setup

This experimental setup (see Figure 4) allows the measurement of a sound field emitted by an ultrasonic transducer. The original purpose is the detection of flaws in a particleboard; hence, such a board can be placed in between the sound source and the sensor. Air was used as the coupling medium between the particleboard and the receiver/transmitter. An ultrasonic transducer called AT50 from the company Airmar was used as a transmitter. It has an active diameter of 45 mm and operates at 50 kHz. The small frequency allows large distances between the particleboard and the receiver /transmitter. However, the accuracy is lower than with higher frequencies. In air the wavelength was at 50 kHz approximately $\lambda = 6.9$ mm. Thus, due to the theory for the spec-radiation method, the smallest detectable size in air was $\lambda/2 = 3.45$ mm. An amplitude of 100 V operates the transmitter to send out a burst of 10 sinusoidal periods with a repeating frequency of 1 Hz. With a distance of 280 mm from the transmitter to the particleboard, the distance was always larger than the near-field length (approximately 72 mm according to Krautkrämer and Krautkrämer [7]) of the transmitter in air at this frequency. The transmitter was stationary during the whole measurement. As a sample, a commercially available wooden medium-density fiberboard (MDF) was used.

Figure 4. Experimental setup.

The particleboard had a thickness of 25 mm. To determine the longitudinal speed of sound in the particleboard, a contact through transmission test was carried out with 50 kHz and was determined to be 450 m/s. The density of the particleboard was 730 kg/m^3. As flaw imitations, pieces of paper were used (Schmelt et al. [32], Schmelt et al. [33], Schmelt and Twiefel [35], Schmelt and Twiefel [47], Marhenke et al. [30], Marhenke et al. [28]) due to the similar material properties compared to wooden particleboard. Other studies often use Teflon tape for studying CFRP or GFRP material for the same reason (Fahr [3]). When the piece of paper was placed with tape underneath the particleboard, a thin air film was always present, and this film served as the flaw imitation. It caused a similar impedance change as real air inclusion in the material. The two pieces of paper used as the flaw imitation were circular with a diameter of ø10 mm and ø6 mm, which is larger and smaller than the wavelength of 9 mm in the particleboard, but larger than the half-wavelength. The particleboard was also stationary. As a receiver, a MEMS microphone (SPU0410LR5H-QB Knowles) was used. Its upper frequency limit is 80 kHz. The receiver had a constant distance of 177 mm from the particleboard. The receiver could be moved to obtain the information for C-scans. Therefore, the XYZ-traversing stage from RoboCylinder with the traverse path of x = 0–400 mm, y = 0–200 mm, and z = 0–200 mm was used. To control the entire system, a standard computer was used. Via USB, the XYZ stage was connected to the computer and controlled by a LabView program. With a National Instruments (NI) system, the transmitter was controlled and the signal of the microphone recorded. The NI system consisted of a chassis PXIe-1085, with an inserted card for the communication with the computer (PXIe-8398), a card to record the measurement data (PXIe-5171R with a maximum sampling frequency of 250 MHz and 8 channels), and a card (PXIe-5423) as a frequency generator to control the transmitter. For the amplification of the control signal of the transmitter, the power amplifier HSA 4052 from the company NF was used to generate the required amplitude of 100 V. The measured area had a size of 152 mm × 152 mm with a measurement point distance of 2 mm in x and y. Hence, 76 × 76 points were measured, and the time step was 2×10^{-7} s with a time length of 3.2768×10^{-3} s.

3. Results

3.1. Example with a Three-Layer Model

A three-layer model was selected as an example in this section. This three-layer model is shown schematically in Figure 5. The excitation was an academic piston transducer with a diameter of ø40 mm and an amplitude of one. The tilted orientation of 5° to the normal axis was computed with the procedure of Schmelt and Twiefel [47]. It emits pulse-like sinusoidal waves with a frequency of 50 kHz. The amplitude in the source plane and the time signal is displayed in Figure 6. The time step of the signal was $\Delta t = 2 \times 10^{-7}$ s, and the signal was 0.000819 s long.

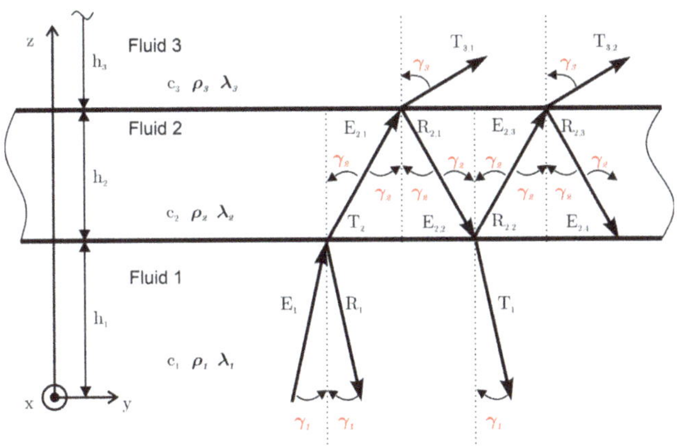

Figure 5. Schematic representation of the 3-layer fluid model. Depicted are the incident E_1, reflected $R_{n,m}$, and transmitted $T_{n,m}$ waves at fluid–fluid interfaces. n is the number of interfaces, and m is the number of interactions (example: $R_{2,3}$ is the third reflection in the fluid at Interface 2). h_n is the height of the fluid layer; ρ_n is the density; λ_n is the wavelength.

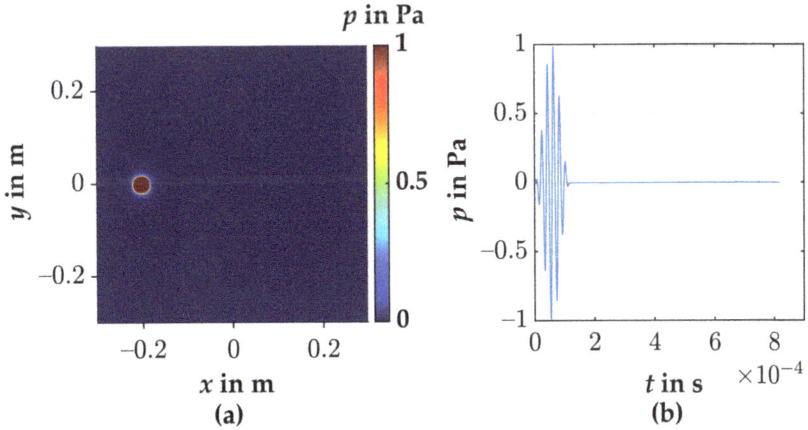

Figure 6. (**a**) Sound pressure amplitude of the tilted academic piston transducer in the plane $z = 0$ m; (**b**) transient pressure signal of the academic piston transducer.

Fluids 1 and 3 were air with a speed of sound of 343 m/s and a density of 1.2041 kg/m^3. Fluid 2 was water with a speed of sound of 1484 m/s and a density of 997 kg/m^3. $\Delta x = \Delta y = 2$ mm; $\Delta f_x = \frac{1}{x_{max}}$; $\Delta f_y = \frac{1}{y_{max}}$. The simulation space had a size of 0.6 m $\times$ 0.6 m $\times$ 0.3 m (length $\times$ width $\times$ height). A step size in the z-direction of $\Delta z = 1$ mm was used in this example. The first interface was at $h_1 = 0.1$ m, and the second interface was at $h_1 + h_2 = 0.2$ m. To calculate the sound pressure between the interfaces, the spec-radiation method without the extension as in Section 2.1 can be used. Here, only the reflected and transmitted parts, as depicted in Figure 5, were calculated. More reflections and transmission can be calculated, but to demonstrate how the procedure works, only these parts were determined. The individual layers between the interfaces are shown as cross-sections in Figure 7, as well as the superpositioned result of the half-3D volume.

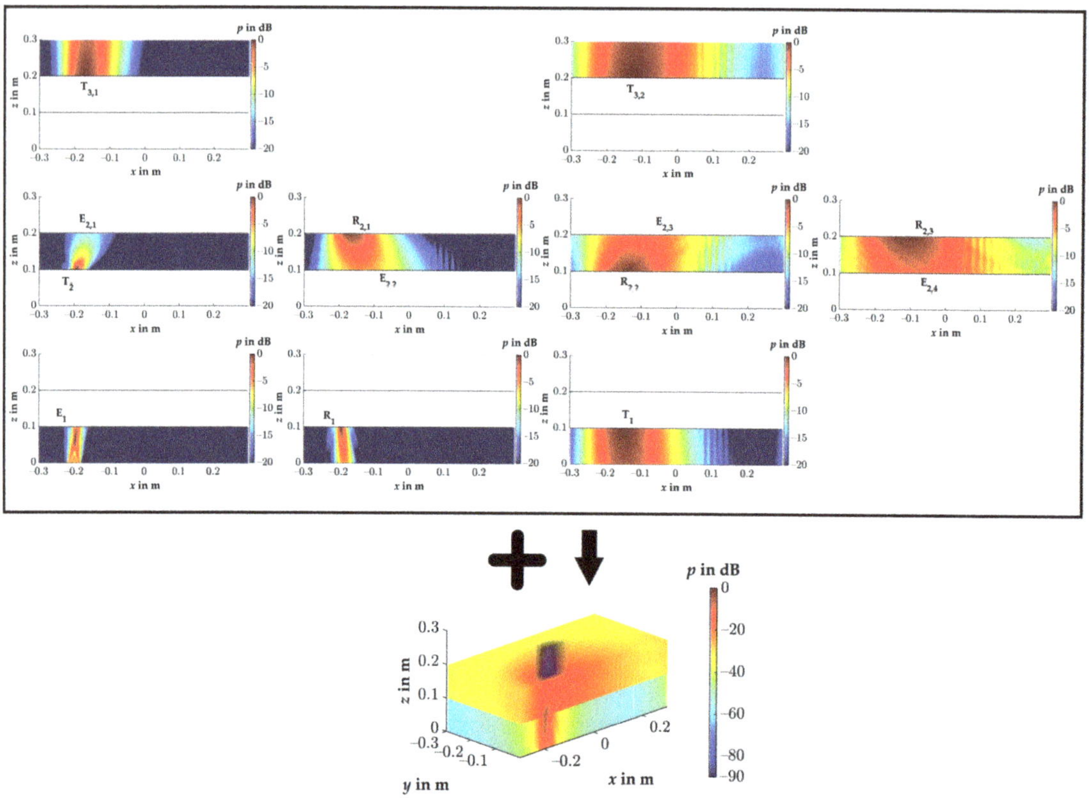

Figure 7. Calculated sound fields in the layered fluids and the superposition of the sound fields.

Here, each image was scaled in color to the maximum in the respective image to make the representation easier to understand. It can be seen that when all the results are superimposed, only with a color scaling down to -90 dB related to the maximum of the total volume, after the second interface, the transmittance becomes visible. The influences of reflection and refraction are also clearly visible. In Medium 2, it can be seen from the much wider sound field that there was a much larger wavelength present. It can be seen that from the angle of incidence of 5°, an angle of approximately 22° arose in Medium 2, as defined by Snell's law of refraction. Due to Snell's law, the refraction angle is here:

$$\gamma_2 = \arcsin\left(\frac{c_2}{c_1}\sin(5^\circ)\right) \approx 22^\circ. \tag{23}$$

Figure 8 depicts the particle deflection at time $t = 0.4$ ms in a cross-section of the calculated volume. In Figure 8, a region between the plane $z = 0$ m and the second interface at $z = 0.2$ m is presented. It can be seen that at this time, the sound waves already reflected at the first interface. In Medium 2, the much longer wavelength, due to the higher speed of sound, was also evident. In Figure 8 is an enlarged section of the area depicted. The oscillation direction of the particles is displayed with vectors. The representation of the vectors was generated with the MATLAB function "quiver". The vectors clearly show that the waves were longitudinal, since the direction of oscillation was in the direction of propagation.

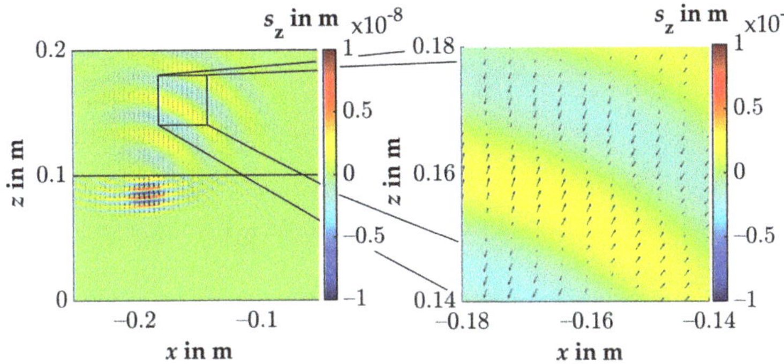

Figure 8. Particle deflection in a cross-section of the calculated volume of Figure 7 at time 0.4 ms with an enlarged section to show with vectors that here, there are longitudinal waves propagating.

We demonstrated with this example how the method works, that it is possible to evaluate the individual components of reflection and transmission, and that by superposition of all components, a complete sound field can be obtained. We also showed that it is possible to determine the particle deflection. We also created a video of another example of transient sound pressure propagation and included it in the Supplementary Files (Video S1). Here, using a three-layer model, as shown in Figure 3, the sound propagation can be observed. The transmitter is in the center of the volume, and only half of the volume is shown. Medium 1 is only a quarter and Medium 3 is only half as high as Medium 2. The sound pressure above in the third medium was increased by a factor of 100 because, otherwise, it would not be visible in this example. Such videos require much memory and time (some days for this example), because the time history of every point in space has to be stored and displayed.

3.2. Example with a Three-Layer Model: Perfect Impedance Match

Another academic example is the perfect impedance matching of the individual layers to each other. With a perfect impedance match, no reflections occur at an interface layer. However, this requires that the condition:

$$c_1\rho_1 = c_2\rho_2 \tag{24}$$

be fulfilled.

We present here the three-layer model (see Figure 5) again to demonstrate that the method presented here also included this special case. The simulation space had the same geometric dimensions as in the previous Section 3.1. The step sizes in the x-, y-, and z-directions remained as in the previous Section 3.1, and the same number of reflections and transmissions were calculated as depicted in Figure 5. The medium up to the first interface was air with a speed of sound of $c_1 = 343$ m/s and a density of $\rho_1 = 1.2041$ kg/m^3. The second medium had a speed of sound $c_2 = 2c_1$ and a density of $\rho_2 = \frac{\rho_1}{2}$. The third medium had a speed of sound $c_3 = \frac{c_1}{2}$ and a density of $\rho_3 = 2\rho_1$. Thus, the conditions for the three media for perfect impedance matching were fulfilled. The same excitation signal as in Figure 6b was used, and the transmitter also had the same size and position as in Figure 6a, with the difference that the transmitter was now tilted by 10°. As a result of the calculation and the superposition of all components, Figure 9 was obtained as a cross-section of the volume for the maximum sound pressure in dB related to the maximum in the entire volume.

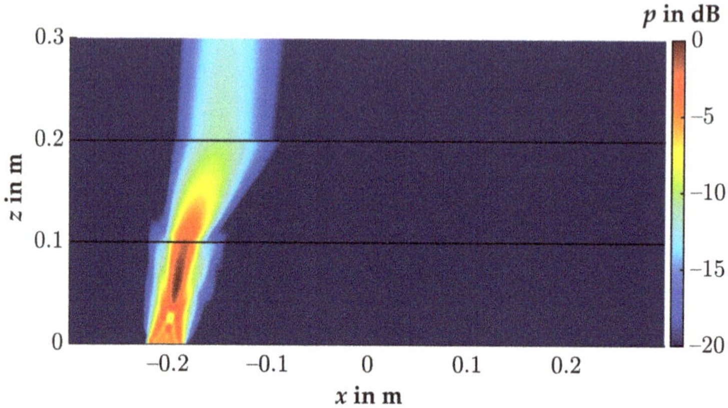

Figure 9. Maximum sound pressure in dB related to the maximum in the entire volume. Perfect impedance matching of a 3-layer model. The results were evaluated for frequencies up to 250 MHz.

Up to the first interface at $z = 0.1$ m, the tilted transmitter could be detected well. Likewise, the near-field could be identified well. A reflection from the interface was, as expected, not recognizable. After the first interface, the speed of sound in the medium was twice as high as in the first medium, which can also be recognized very well by the propagation angle. Here, as it was to be expected according to Snell's law of refraction, it was approximately twice as large as the angle of incidence. However, no reflection can be seen even up to the second interface at $z = 0.2$ m, from any interface. The sound passed from Medium 2 to Medium 3 without reflection. Here, due to the fact that the speed of sound was only a quarter compared to Medium 2, the angle of propagation was also only about a quarter of what is present in Medium 2.

With this special case, we once again proved the functionality of the spec-radiation method

3.3. Experimental Application

For the evaluation, the measured area was increased symmetrically by zero-padding to a size of 256 × 256 points with the measured area inside. With the spec-radiation method, the sound field on the board surface can be computed. The important frequency is the exciting frequency. Therefore, the evaluation was performed for a frequency of 50 kHz.

The result of this evaluation is depicted in Figure 10. In both results, the flaw may be detected, but not accurate in shape and size. This was due to the fact that the sound field behind the flaw had diffraction and interference patterns that affected the sound field on the board surface. In (a) is the result of the 10 mm flaw and in (b) the result of the 6 mm flaw. In the next step, we evaluated the result with the proposed extension, so that the result was evaluated at the z-position of the flaw in the particleboard.

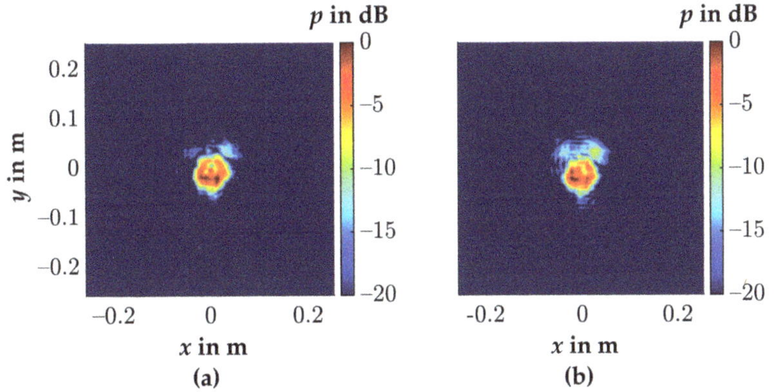

Figure 10. Evaluation result on the top surface of the wooden particleboard, with (**a**) 10 mm and (**b**) 6 mm under the particleboard. The results were evaluated at a frequency of 50 kHz.

We therefore calculated this through the particleboard. In contrast to the calculation of the forward propagation, it has to be considered that the start plane was the measuring plane. This means that the start sound pressure was the transmitted sound pressure $P_{\mathrm{T}}(f_x, f_y, z, \omega_{\mathrm{j}})$. To obtain the incidence sound pressure at the interface, the transmitted sound pressure has to be divided by $T_{\mathfrak{F}}$, so that:

$$P(f_x, f_y, z, \omega_{\mathrm{j}}) = \frac{1}{T_{\mathfrak{F}}} P_{\mathrm{T}}(f_x, f_y, z, \omega_{\mathrm{j}}). \tag{25}$$

The result is depicted in Figure 11. In (a), the 10 mm flaw is clearly visible as a circle with a diameter of 10 mm and the pressure drops stronger than −20 dB. In (b), the flaw with a diameter of 6 mm is visible more clearly, but is still not perfect. The area is larger than a circle of 6 mm. This was due to the fact that the 6 mm circle was close to the detection limit of $\lambda/2$ which was, in the wooden particleboard, approximately 4.5 mm. The pressure drop was not as strong, so that some air between the tape and the particleboard around the flaw could have raised the area here. The flaw was easier to detect in comparison with the result at the top surface of the particleboard. The evaluation time took approximately 14 s per interface layer with the corresponding dataset.

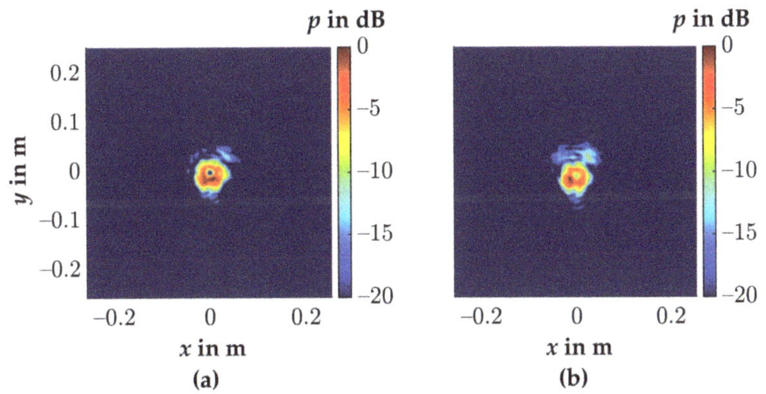

Figure 11. Evaluation result on the lower surface of the wooden particleboard, with (**a**) 10 mm and (**b**) 6 mm under the particleboard. The results were evaluated at a frequency of 50 kHz.

By calculating many planes from the microphone plane to the flaw plane, the sound field can also be represented as a cross-section at $x = 0$ m. This is depicted in Figure 12 for the ø10 mm flaw. Here, two different scalings of the results were chosen. Everything above the particleboard ($z = 0$ m to -0.177 m) was scaled in dB related to the maximum in the air volume above the particleboard. Everything inside the particleboard ($z = -0.178$ m to -0.202 m) was scaled in dB with respect to the maximum inside the particleboard. This resulted in a matching color scaling of the two areas. If everything would be scaled to the maximum inside the particleboard, it would no longer be possible to display the entire sound field in a well-recognizable color, because the sound field above the board would then be at approximately -55 dB.

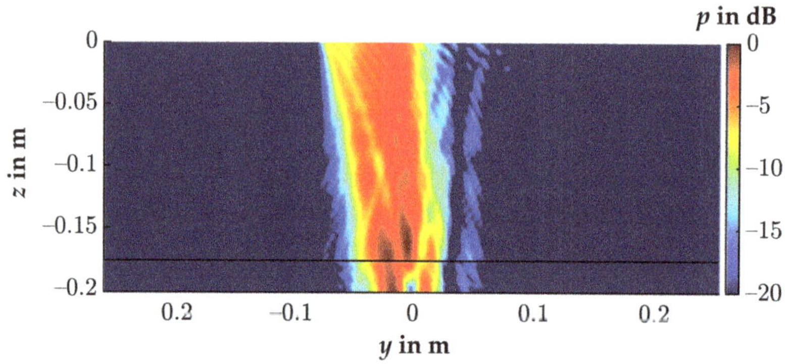

Figure 12. Cross-section plane at $x = 0$ m of the calculated volume with the ø10 mm flaw. The sound pressure is shown in dB. For the color scaling in the air, the maximum of the sound pressure in the air volume is taken up to the surface of the particleboard ($z = 0$ m to -0.177 m). For the color scaling in the particleboard, the maximum sound pressure in the particleboard ($z = -0.178$ m to -0.202 m) is used. If both are now displayed up to -20 dB, the result is a matching color scale. The results were evaluated at a frequency of 50 kHz.

From this cross-sectional view, it is clear that on the surface of the particleboard ($z = -0.177$ m), the flaw could not be clearly detected. What could only be detected there was a minimum of the sound pressure due to the diffraction of the sound around the flaw, as can also be seen in Figure 10a. The flaw itself, however, was completely masked by the diffraction. Only by calculating into the particleboard, the flaw became recognizable by a sound pressure drop (sound shadow) at $y = 0$ m. Here, the shadow extended to about half the thickness of the particleboard and had its largest diameter and strongest sound pressure drop in the plane of the flaw.

In Figure 13, the cross-sectional plane at $x = 0$ m is shown for the measurement with the ø6 mm flaw. There, at the position of approximately $y = 0.01$ m and $z = -0.202$ m to -0.187 m, the sound shadow of the ø6 mm flaw can be seen. Here, it is noticeable that the sound pressure drop was lower than for the ø10 mm flaw. In this figure, however, it is also clear that the supposed flaw on the surface of the particleboard in Figure 10b is deceptive. This is because the position of the flaw in the xy-plane could not be correctly determined with the evaluation only up to the upper surface of the particleboard, which means that an evaluation through the particleboard was necessary here.

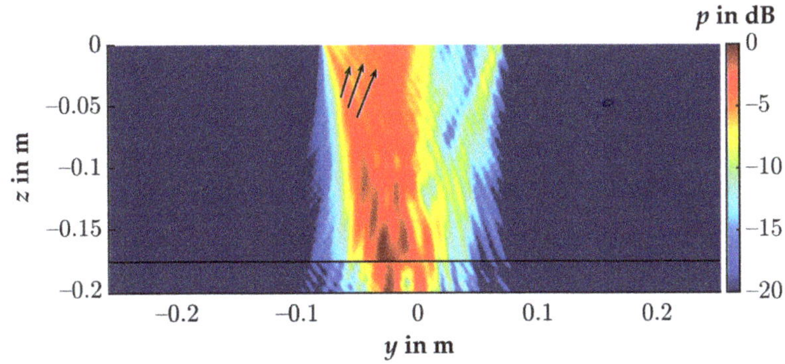

Figure 13. Cross-section plane at $x = 0$ m of the calculated volume with the ø6 mm flaw. The sound pressure is shown in dB. For the color scaling in the air, the maximum of the sound pressure in the air volume is taken up to the surface of the particleboard ($z = 0$ m to -0.177 m). For the color scaling in the particleboard, the maximum sound pressure in the particleboard ($z = -0.178$ m to -0.202 m) is used. If both are now displayed up to -20 dB, the result is a matching color scale. The results were evaluated at a frequency of 50 kHz.

In Figures 12 and 13, there are lines to be seen, which one might think do not actually belong there. We highlighted these by way of example in Figure 13 with three arrows. These were artifacts caused by the calculation due to reflections in the measurement. Because of the narrow space and the long recording time of the measurement data, sound waves could reach the XYZ-traversing unit and were reflected from there, thus reaching the measurement area of the microphone again. In this example shown here, however, the influence on the identification of the flaws was negligible. However, such phenomena must always be taken into account in experiments of this kind.

4. Discussion

A new innovative extension for the spec-radiation method was introduced in this publication. With the spec-radiation method, we showed in the past that the detection of flaws in wooden particleboard could be performed very fast and reliably. We showed that the spec-radiation method is faster than comparable methods to detect flaws by calculating back to the top surface of a panel material (Schmelt and Twiefel [35]). We also showed that it is possible to calculate directly to the top of a tilted panel material (Schmelt and Twiefel [47]). Nevertheless, in thicker material, it is not enough to calculate the sound distribution on the top of the panel material; due to interferences and diffraction, flaws may be masked. Therefore, the next logical step was to calculate through the material. With this extension, it was now possible to calculate through different layered fluid media with the spec-radiation method. A computer with the operating system Windows 10 Pro for Workstations and 128 GB RAM and 24 cores of the type Intel(R) Xeon(R) Silver 4116 CPU @ 2.1 GHz was used for the evaluations made here. We showed on different examples how the spec-radiation method can be used to simulate and visualize sound propagation. While the representation of the maximum sound pressure in a three-dimensional volume is no longer a big challenge, the creation of three-dimensional videos is a challenge for today's systems. For the creation of such videos, the transient information of each calculated point in space must be stored. Subsequently, all information must be called up and displayed at one point in time in order to save a frame of the video. For the three-layer model shown in Figure 7, this would be with (time × length × width × height) $4096 \times 300 \times 300 = 110{,}592{,}000{,}000$ data points. With that said, there are still many possibilities for optimization since not every data point is needed for the visualization. For the detection of flaws, we showed that we can identify flaws smaller than the wavelength also in the particleboard. Here, however, it becomes clear once again that reflections from the environment must always be taken into account, as they can possibly have a negative influence on the detection of flaws.

5. Conclusions

In conclusion, with this publication, we developed a new innovative extension to the spec-radiation method. In Section 2, we described in detail both the analytical description and the practical implementation and presented an experimental setup. We showed how to calculate the particle deflection potential and the particle deflection with the spec-radiation method in the Fourier domain. We presented a flowchart model that was used to explain the practical implementation. In Section 3.1, we first showed on a three-layer model how the extension of the spec-radiation method presented here can be applied to simulate the sound propagation from a transmitter. We showed that the particle deflection can also be represented by vectors and that each reflection or transmission can be considered individually. By superposing the reflections and transmissions, we showed how the complete sound field of a 3D volume can be obtained. With a video, we showed that with this extension, it is also possible to represent the transient three-dimensional sound pressure field. With the special case of perfect impedance matching, we proved the results of the method in Section 3.2. In Section 3.3, we showed with an experiment a wooden particleboard and flaw imitations at the transmitter side that the procedure to detect flaws with the extended version of the spec-radiation method is possible. We showed that the flaws were better detectable and characterized in geometry and location by calculating the sound field through the particleboard in comparison with the result at the top surface of the particleboard. Therefore, we made the assumption that the particleboard behaves as a fluid and that only longitudinal waves propagate. We presented the result and explained the backward propagation through an interface layer. We are confident that this type of evaluation will help optimize production and quality assurance processes due to the accurate knowledge that can be gained from the flaws. Due to the very fast evaluation time, the method has great potential.

Supplementary Materials: The following are available online at https://www.mdpi.com/article/10.3390/app12031098/s1, Video S1: Transient sound propagation evaluated with the spec-radiation method.

Author Contributions: Conceptualization, A.S.S. and J.T.; methodology, A.S.S.; software, A.S.S.; validation, A.S.S. and J.T.; formal analysis, A.S.S.; investigation, A.S.S.; resources, A.S.S.; data curation, A.S.S.; writing—original draft preparation, A.S.S.; writing—review and editing, A.S.S. and J.T.; visualization, A.S.S.; supervision, J.T.; project administration, J.T.; funding acquisition, J.T. All authors have read and agreed to the published version of the manuscript.

Funding: The publication of this article was funded by the Open Access Fund of Leibniz Universität Hannover.

Institutional Review Board Statement: Not applicable.

Informed Consent Statement: Not applicable.

Acknowledgments: This research was supported by J. Wallaschek.

Conflicts of Interest: The authors declare no conflict of interest.

Abbreviations

The following abbreviations are used in this manuscript:

CFRP	carbon-fiber-reinforced plastic
GFRP	glass-fiber-reinforced plastic
FFT	Fourier transformation
iFFT	inverse Fourier transformation
MDF	medium-density fiberboard
MEMS	micro-electro-mechanical systems

References

1. Jasiuniene, E.; Raisutis, R.; Sliteris, R.; Voleiis, A.; Jakas, M. Ultrasonic NDT of wind turbine blades using contact pulse-echo immersion testing with moving water container. *Ultragarsas J.* **2008**, *63*, 28–32.
2. Conta, S.; Santoni, A.; Homb, A. Benchmarking the vibration velocity-based measurement methods to determine the radiated sound power from floor elements under impact excitation. *Appl. Acoust.* **2020**, *169*, 107457. [CrossRef]
3. Fahr, A. *Aeronautical Applications of Non-Destructive Testing*; DEStech Publications, Inc.: Lancaster, PA, USA, 2014.
4. Fang, Y.; Lin, L.; Feng, H.; Lu, Z.; Emms, G.W. Review of the use of air-coupled ultrasonic technologies for nondestructive testing of wood and wood products. *Comput. Electron. Agric.* **2017**, *137*, 79–87. [CrossRef]
5. Sokolov, S.Y. On the problem of the propagation of ultrasonic oscillations in various bodies. *Elek. Nachr. Tech.* **1929**, *6*, 454–460.
6. Deutsch, V.; Platte, M.; Vogt, M. *Ultraschallprüfungen*; Springer: Berlin/Heidelberg, Germany, 1997. [CrossRef]
7. Krautkrämer, J.; Krautkrämer, H. *Werkstoffprüfung mit Ultraschall*; Springer: Berlin/Heidelberg, Germany, 1980. [CrossRef]
8. Gyekenyesi, A.L.; Harmon, L.M.; Kautz, H.E. The Effect of Experimental Conditions on Acousto-Ultrasonic Reproducibility. *Proc. SPIE* **2002**, *4704*, 177–186. [CrossRef]
9. Willcox, M.; Downes, G. *A Brief Description of NDT Techniques*; NDT Equipment Limited: Toronto, ON, Canada, 2003.
10. Schafer, M. The Effect of Experimental Conditions on Acousto-Ultrasonic Reproducibility. In Proceedings of the IEEE Ultrasonics Symposium—An International Symposium (Cat. No.00CH37121), San Juan, PR, USA, 22–25 October 2000; pp. 771–778. [CrossRef]
11. Green, R.E. Non-contact ultrasonic techniques. *Ultrasonics* **2004**, *42*, 9–16. [PubMed]
12. Zhang, Y.; Sidibé, Y.; Maze, G.; Leon, F.; Druaux, F.; Lefebvre, D. Detection of damages in underwater metal plate using acoustic inverse scattering and image processing methods. *Appl. Acoust.* **2016**, *103*, 110–121. [CrossRef]
13. Mitri, F.G.; Greenleaf, J.F.; Fatemi, M. Comparison of continuous-wave (CW) and tone-burst (TB) excitation modes in vibro-acoustography: Application for the non-destructive imaging of flaws. *Appl. Acoust.* **2009**, *70*, 333–336. [CrossRef]
14. Stößel, R. Air-Coupled Ultrasound Inspection as a New Non-Destructive Testing Tool for Quality Assurance. Ph.D. Thesis, Fakultät für Maschinenbau, Universität Stuttgart, Stuttgart, Germany, 2004. [CrossRef]
15. Chimenti, D.E. Review of air-coupled ultrasonic materials characterization. *Ultrasonics* **2014**, *54*, 1804–1816. [CrossRef]
16. Álvarez Arenas, T.E.G. Acoustic Impedance Matching of Piezoelectric Transducers to the Air. *IEEE Trans. Ultrason. Ferroelectr. Freq. Control* **2014**, *51*, 624–633. [CrossRef]
17. Hillger, W.; Bühling, L.; Ilse, D. Review of 30 years ultrasonic systems and developments for the future. In Proceedings of the 11th European Conference on Non-Destructive Testing (ECNDT2014), Prague, Czech Republic, 6–10 October 2014.
18. Raisutis, R.; Jasiuniene, E.; Sliteris, R.; Vladisauskas, A. The review of non-destructive testing techniques suitable for inspection of the wind turbine blades. *Ultragarsas* **2008**, *63*, 26–30.
19. Sanabria, S.; Mueller, C.; Neuenschwander, J.; Niemz, P.; Sennhauser, U. Air-coupled ultrasound as an accurate and reproducible method for bonding assessment of glued timber. *Wood Sci. Technol.* **2011**, *45*, 645–659. [CrossRef]
20. Dunky, D.; Niemz, P. *Holzwerkstoffe und Leime*; Springer: Berlin/Heidelberg, Germany, 2002. [CrossRef]
21. Niemz, P. Bestimmung von Fehlverklebungen mittels Schallaufzeitmessung. *Holz als Roh-und Werkst.* **1995**, *53*, 236. [CrossRef]
22. Bucur, V.; Böhnke, I. Factors affecting ultrasonic measurements in solid wood. *Ultrasonics* **1994**, *32*, 385–390. [CrossRef]
23. Laybed, Y.; Huang, L. Ultrasound time-reversal MUSIC imaging with diffraction and attenuation compensation. *IEEE Trans. Ultrason. Ferroelectr. Freq. Control* **2012**, *59*, 2186–2200. [CrossRef]
24. Döring, D. Air-Coupled Ultrasound and Guided Acoustic Waves for Application in Non-Destructive Material Testing. Ph.D. Thesis, Fakultät Luft-und Raumfahrttechnik der Universität Stuttgart, Stuttgart, Germany, 2011. [CrossRef]
25. Gabor, D. Holography. *Science* **1972**, *177*, 299–313. [CrossRef]
26. Singh, V. Acoustical imaging techniques for bone studies. *Appl. Acoust.* **1989**, *27*, 119–128. [CrossRef]
27. Sanabria, S.; Marhenke, T.; Furrer, R.; Neuenschwander, J. Calculation of volumetric sound field of pulsed air-coupled ultrasound transducers based on single-plane measurements. *IEEE Trans. Ultrason. Ferroelectr. Freq. Control* **2018**, *65*, 72–84. [CrossRef] [PubMed]
28. Marhenke, T.; Neuenschwander, J.; Furrer, R.; Zolliker, P.; Twiefel, J.; Hasener, J.; Wallaschek, J.; Sanabria, S. Air-coupled ultrasound time reversal (ACU-TR) for subwavelength non-destructive imaging. *IEEE Trans. Ultrason. Ferroelectr. Freq. Control* **2020**, *67*, 651–663. [CrossRef] [PubMed]
29. Marhenke, T.; Sanabria, S.; Chintada, B.; Furrer, R.; Neuenschwander, J.; Goksel, O. Acoustic field characterization of medical array transducers based on unfocused transmits and single-plane hydrophone measurements. *Sensors* **2019**, *19*, 863. [CrossRef] [PubMed]
30. Marhenke, T.; Sanabria, S.; Twiefel, J.; Furrer, R.; Neuenschwander, J.; Wallaschek, J. Three dimensional sound field computation and optimization of the delamination detection based on the re-radiation. In Proceedings of the 12th European Conference on Non-Destructive Testing (ECNDT 2018), Gothenburg, Sweden, 11–15 June 2018.
31. Schmelt, A.; Marhenke, T.; Twiefel, J. Identifying objects in a 2D-space utilizing a novel combination of a re-radiation based method and of a difference-image-method. In Proceedings of the 23rd International Congress on Acoustics, Aachen, Germany, 9–13 September 2019.
32. Schmelt, A.; Marhenke, T.; Hasener, J.; Twiefel, J. Investigation and Enhancement of the Detectability of Flaws with a Coarse Measuring Grid and Air Coupled Ultrasound for NDT of Panel Materials Using the Re-Radiation Method. *Appl. Sci.* **2020**, *10*, 1155. [CrossRef]

33. Schmelt, A.; Li, Z.; Marhenke, T.; Twiefel, J. Aussagefähigkeit von Fehlstellenimitaten in der ZfP. In Proceedings of the DAGA 2020—46. Jahrestagung für Akustik, Hannover, Germany, 16–19 March 2020; pp. 1133–1136. ISBN: 978-3-939296-17-1.
34. Tsysar, S.; Sapozhnikov, O. Ultrasonic holography of 3D objects. In Proceedings of the IEEE International Ultrasonics Symposium, Rome, Italy, 20–23 September 2009; pp. 737–740. [CrossRef]
35. Schmelt, A.; Twiefel, J. The Spec-Radiation Method as a Fast Alternative to the Re-Radiation Method for the Detection of Flaws in Wooden Particleboards. *Appl. Sci.* **2020**, *10*, 6663. [CrossRef]
36. Wolf, E. Three-dimensional structure determination of semi-transparent objects from holographic data. *Opt. Commun.* **1969**, *1*, 153–156. [CrossRef]
37. Booker, H.G.; Clemmow, P.C. The concept of an angular spectrum of plane waves, and its relation to that of polar diagram and aperture distribution. *IEEE-Part III Radio Commun. Eng.* **1950**, *97*, 11–17. [CrossRef]
38. Ratcliffe, J.A. Some Aspects of Diffraction Theory and their Application to the Ionosphere. *Rep. Prog. Phys.* **1956**, *19*, 188–267. [CrossRef]
39. Boyer, A.L.; Hirsch, P.M.; Jordan, J.A.; Lesem, L.B.; Van Rooy, D.L. Reconstruction of tultrasonic images by backward propagation. *Proc. Acoust. Hologr.* **1970**, *3*, 333–348. [CrossRef]
40. Schafer, M.E.; Lewin, P.A. Transducer characterization using the angular spectrum method. *J. Acoust. Soc. Am.* **1989**, *85*, 2202–2214. [CrossRef]
41. de Belleval, J.F.; Messaoud-Nacer, N. Ultrasonic transducer beams model, using transient angular spectrum. *Rev. Prog. Quant. Nondestruct. Eval.* **1999**, *18*, 1101–1106. [CrossRef]
42. Peng, H.; Lu, J.; Han, X. High frame rate ultrasonic imaging system based on the angular spectrum principle. *Ultrasonics* **2006**, *44*, e97–e99. [CrossRef]
43. Yan, X.; Hamilton, M.F. Angular spectrum decomposition analysis of second harmonic ultrasound propagation and its relation to tissue harmonic imaging. In *Ultrasonic and Advanced Methods for Nondestructive Testing and Material Characterization*; e-Journal of Nondestructive Testing (NDT); World Scientific Publishing: Singapore, 2007; pp. 155–168; ISSN 1435-4934. [CrossRef]
44. Aanes, M.; Lohne, K.D.; Lunde, P.; Vestrheim, M. Ultrasonic beam transmission through a water-immersed plate at oblique incidence using a piezoelectric source transducer. Finite element angular spectrum modeling and measurements. In Proceedings of the 2012 IEEE International Ultrasonics Symposium, Dresden, Germany, 7–10 October 2012; pp. 1972–1977. [CrossRef]
45. Liu, D.L.; Waag, R.C. Propagation and backpropagation for ultrasonic wavefront design. *IEEE Trans. Ultrason. Ferroelectr. Freq. Control* **1997**, *44*, 1–13. [CrossRef]
46. Matsushima, K. *Introduction to Computer Holography*; Springer: Cham, Switzerland, 2020. [CrossRef]
47. Schmelt, A.; Twiefel, J. Flaw Detection on a Tilted Particleboard by Use of the Spec-Radiation Method. *Appl. Sci.* **2020**, *10*, 8513. [CrossRef]
48. Goodman, J.W. *Introduction to Fourier Optics*; W. H. Freeman and Company: New York, NY, USA, 2017.
49. Brekhovskikh, L.M.; Godin, O.A. *Acoustics of Layered Media I*; Springer: Heidelberg, Germany, 2020. [CrossRef]
50. Möser, M. *Analyse und Synthese Akustischer Spektren*; Springer: Berlin/Heidelberg, Germany, 1988.

applied sciences

MDPI

Article

Simulation of Fluid Dynamics Monitoring Using Ultrasonic Measurements

Masaru Nagaso [1], Joseph Moysan [1,*], Christian Lhuillier [2] and Jean-Philippe Jeannot [2]

1 Aix Marseille Univ, CNRS, Centrale Marseille, LMA UMR 7031, 13453 Marseille, France; nagaso@lma.cnrs-mrs.fr
2 CEA/DES/IRESNE/DTN/STCP/LISM, 13108 St Paul lez Durance, France; christian.lhuillier@cea.fr (C.L.); jean-philippe.jeannot@cea.fr (J.-P.J.)
* Correspondence: joseph.moysan@univ-amu.fr

Featured Application: Monitoring of liquid metal flows by ultrasounds.

Abstract: The simulation of the propagation of ultrasonic waves in a moving fluid will improve the efficiency of the ultrasonic flow monitoring and that of the in-service monitoring for various reactors in several industries. The most recent simulations are mostly limited to 3D representations of the insonified volume but without really considering the temporal aspect of the flow. The advent of high-performance computing (HPC) now makes it possible to propose the first 4D simulations, with the representation of the inspected medium evolving over time. This work is based on a highly accurate double simulation. A first computational fluid dynamics (CFD) simulation, performed in previous work, described the fluid medium resulting from the mixing of hot jets in a cold opaque fluid. There have been many sensor developments over the years in this domain, as ultrasounds are the only method able to give information in an opaque medium. The correct design of these sensors, as well as the precise and confident analysis of their measurements, will progress with the development of the modeling of wave propagation in such a medium. An important parameter to consider is the flow temperature description, as a temperature gradient in the medium deflects the wave path and may sometimes cause its division. We develop a 4D wave propagation simulation in a very realistic, temporally fluctuating medium. A high-performance simulation is proposed in this work to include an ultrasonic source within the medium and to calculate the wave propagation between a transmitter and a receiver. The analysis of the wave variations shows that this through-transmission setup can track the jet mixing time variations. The steps needed to achieve these results are described using the spectral-element-based numerical tool SPECFEM3D. It is shown that the low-frequency fluctuation of the liquid metal flow can be observed using ultrasonic measurements.

Keywords: ultrasounds; fluid dynamics; liquid metal; monitoring; nondestructive testing; high-performance computing

Citation: Nagaso, M.; Moysan, J.; Lhuillier, C.; Jeannot, J.-P. Simulation of Fluid Dynamics Monitoring Using Ultrasonic Measurements. *Appl. Sci.* **2021**, *11*, 7065. https://doi.org/10.3390/app11157065

Academic Editors: Habil. Michel Darmon and Marco Scalerandi

Received: 15 June 2021
Accepted: 24 July 2021
Published: 30 July 2021

1. Introduction

Flow monitoring is of importance for many applications involving various liquids and gases. There exist some methods to characterize flows using global measurements effective in many industrial situations [1]. Many technologies use pressure variations to deduce flow rate. When the fluid has conductive properties, other technologies use electromagnetic measurements. In many cases, a precise analysis of the velocity distribution in the flow is required. When optical paths are possible, a classical solution consists of making Doppler measurements with particles floating in the fluid. Major developments in three-dimensional velocity field measurements using the tomographic particle image velocimetry (PIV) technique have taken place in past years [2]. When the flow is opaque or when it cannot be optically observed directly, then ultrasonic solutions are often implemented. In the case of liquid metals, the floating particles may be impurities (such as metallic inclusions or oxide films), and ultrasonic measurements are also used to analyze the purity of liquid metal [3,4]. There are also studies in magnetohydrodynamics (MHDs) where flows, in metallic melts, can be driven by magnetic fields. The study of turbulent melt flows requires 2D flow mapping at a frame rate of several hertz. Thieme et al. developed an ultrasound array doppler velocimeter (UADV) using linear arrays of large piezoelectric elements. They demonstrated they can detect the transition from laminar flow to turbulent flow [5]. In such media, simultaneously monitoring temperatures in the flow and the fluid dynamics presents the most difficulties. Zürner et al. proposed to combine multiple crossing ultrasound beam lines and an array of thermocouples to realize a 3D characterization of the complex dynamics in a cylindrical convection cell that is filled with the liquid metal alloy GaInSn [6]. This method makes it possible to study complex fluid dynamics (Rayleigh–Bénard convection) to understand the transport processes in several turbulent convection flows such as the geodynamo in the core of the earth [7] or liquid metal batteries for renewable energy storage [8]. There is also a great interest in liquid metal plasma-facing components (LM-PFCs) within the framework of research aiming at developing the next generation of fusion reactors. They are expected to enhance plasma confinement and withstand large heat and particle fluxes better than solid components [9,10].

In this work, we are mainly concerned with fast-breeder reactors for which a liquid metal alloy is used as a coolant fluid. A number of countries, including Japan, India, Lithuania/Belgium, France, the United States of America and Germany have sought to develop ultrasonic ranging or imaging devices for liquid-metal-cooled fast reactors [11]. In his report, Griffin provided an extensive review of these devices. These systems are designed to inspect the reactor core region for the presence of foreign objects prior to refueling and to detect possible component degradation or damage due to normal operation. Along with imaging, there are wide-ranging investigations on the use of acoustic or ultrasonic sensors for liquid-metal-cooled reactors that include active or passive inspections using acoustic emission and boiling detection methods [11]. In her review of ultrasonic viewing systems for liquid metal reactors published in 2007, Jasiuniene explained there are no other physical means but ultrasound that would make it possible to inspect inner reactor parts submerged in an opaque hot liquid metal [12]. The harsh operating conditions in liquid metal significantly restrict the possible architecture of the visualization system and materials. In their review, Tarpara and al. developed solutions for undersodium ultrasonic viewing for fast-breeder reactors [13]. A major difficulty is obtaining a good wetting of the transducer to avoid the trapping of gas bubbles, which reduce transmitted energy [14,15]. To improve ultrasonic systems, various designs have been studied using waveguides [16], arrays of transducers [17], arrays of electromagnetic acoustic transducers (EMATs) [18] or dedicated acoustic mirrors [19].

In addition to the development of efficient sensors, the future of inspection requires an increasing knowledge of wave propagation in such media. The flow above a reactor core is turbulent, with local speeds of several meters per second and velocity gradients of about several meters per second per centimeter [20]. Therefore, the flow profile and the fluid turbulence create uncertainties in transit-time measurements for flowmeters [21]. In the case of acoustic thermometry, this could lead to errors in the measurement of the temperature. The development of all these applications requires modeling efforts to propose more accurate measurements or new applications. Nondestructive examinations are usually planned when the flow is steady, but it is also interesting to develop in-service inspections. In such cases, the liquid metal flow is no longer steady, the flow regime becomes turbulent and temperature gradients occur. Ultrasonic devices could provide reactor monitoring and structural health monitoring functions, but signal analysis should take into account wave propagation in a heterogeneous flow. Modeling wave propagation should include an accurate flow model. In their work, Massacret and al. [20] investigated the importance of temperature gradients using a ray-tracing code. They demonstrated that the beam deviates under the effect of the temperature gradient and may even split when the ultrasonic wave travels through several areas with different thermal gradients [22]. For a steady fluid, temperature heterogeneity could be described using stochastic methods, which randomly generate a fluctuating temperature field using a Gaussian random process. This method has the great advantage of proposing a simplified model and of not requiring huge computer resources, but it does not model the real flow well [23]. The signals used for monitoring often include backward echoes from geometrical parts. The influence of the edges has been discussed in several studies, including a comparison of diffraction theories [32] and a comparison with experiments [24].

In this work, a qualitative modeling leap is sought, which must include precise modeling of the fluid dynamics and 4D modeling of the acoustic wave propagation so that time is included in the acoustic simulation. The use of high-performance computing (HPC) is mandatory for such a project. For the flow modeling, we selected the PLAJEST setup, with the mixing of two hot jets with a cold one in a liquid metal medium [25]. This setup was completely simulated using computational fluid dynamics tools [26,27]. Thanks to these CFD data provided by the DES/IRESNE/DM2S/STMF laboratory of the French Alternative Energies and Atomic Energy Commission (CEA), we can develop a 4D wave propagation simulation in a very realistic, temporally fluctuating medium, which is the original objective described in this work.

For the ultrasonic wave simulation, we used the spectral-element numerical tool SPECFEM3D [28]. In the second section, we present the numerical models for the wave equation, and we indicate our motivations for choosing the SPECFEM3D code. In Section 2, the CFD data used to reproduce the PLAJEST experiment are detailed, and the frozen-flow hypothesis is justified. The fourth section describes a huge task of implementing the modeling of the wave propagation in a heterogeneous medium resulting from the CFD modeling. It is a very new collaborative work combining CFD simulation and wave propagation simulation. In the last section, we demonstrate that the temporal temperature fluctuations could be monitored by ultrasounds. The study is focused on the measurement of the frequency of these fluctuations, as it would prove the potential of such complete modeling for giving a 3D description of a heterogeneous flow and for monitoring it in time, thereby adding a fourth dimension to the simulation.

2. Numerical Models for Wave Equation

There are several ways to calculate wave propagation. For example, the ray-based method—or ray-tracing method—is one that requires a much smaller amount of computational resources than other full-wave-based methods. We therefore applied this method in our former studies [29–32]. Initially, this method considered only the deflection from Snell's law; however, nowadays, with the so-called "fat rays" or "pencils" technique, it may perform ray tracing in inhomogeneous and anisotropic media [33]. However, it is a trade-off relation between the lightness of numerical resources that the ray-based method requires and its accuracy, because the accuracy is degraded by the high-frequency or infinite-frequency approximation that the ray-based method applies [34].

The finite-difference time-domain method (FDTD) is a type of full-wave method [35]. In a finite-difference scheme, the spatial domain is discretized as grid points, which are usually evenly spaced, and all physical properties are defined at these fixed grid points or halfway between these grid points. These schemes correspond to spatially staggered schemes. This becomes a restriction on the placement of physical values and the use of symmetric differentiation operators, which require fictitious grid points located beyond the boundary. This causes difficulties for the accurate expressions of material boundaries and of boundary conditions, especially if the shape has a complex geometry. Several studies have been conducted on the FDTD methods for various media that include curved boundaries [36–39], which used an interpolation method based on the distance between a discretized point and a potentially curved boundary. Unfortunately, they are only partially satisfactory in terms of accuracy and also make FDTD techniques more complex, thus partially losing one of their main advantages, which is their simplicity.

The finite element time-domain method (FETD) is also a full-wave method that applies the finite element method to spatial decomposition. This method may easily define a complex geometry in the target computation domain thanks to the matured utility tools of computer-aided engineering (CAE). In addition, a curved boundary may be included when using a second-order finite element [40,41]. Compared with the ray-based method and FDTD, FETD requires larger computation resources, which is usually limiting when calculating the propagation of a wave in a large domain and with a high-frequency source wave.

The spectral-element method (SEM) [42–44] is a technique that is similar to FETD but reduces the limitation of FETD. The main difference is the type of basis function used for each method. While FETD uses Gauss points for its spatial discretization, SEM uses higher-order Gauss–Lobatto–Legendre points. SEM uses a modified formulation that leads to a perfectly diagonal mass matrix [45], and thus it is much more numerically efficient than FETD.

We therefore selected SEM as the numerical method to model wave propagation in our target, i.e., liquid sodium flow. We selected a numerical code SPECFEM3D, one of the most efficiently implemented SEMs for solving wave propagation in many types of material (elastic, viscoelastic, poroelastic, etc.). SPECFEM3D has already been applied for many cases and used with several HPC environments, e.g., [46–48].

As we planned to use the frozen-flow hypothesis, i.e., the temperature and density values are temporally constant while the wave propagates in the calculation domain, we started from an acoustic wave propagation equation in a static and heterogeneous medium [49].

$$\nabla^2 p - \frac{1}{c^2}\frac{\partial^2 p}{\partial t^2} = 0. \tag{1}$$

where ∇ is the Del or nabla operator, t is time, p is pressure and c is sound speed. Instead of solving this equation directly, SPECFEM3D solves an equation with a second-order time derivative of scalar potential χ:

$$p = -\ddot{\chi}. \tag{2}$$

The advantage of this implementation is that it automatically suppresses numerical artifacts that are known for appearing in the displacement formulation [50]. By combining Equations (2) and (1) and using the acoustic bulk modulus $\lambda = \rho c^2$ with a density ρ:

$$\frac{1}{\lambda}\ddot{\chi} = \frac{1}{\rho}\nabla \cdot \nabla \chi, \tag{3}$$

which is the acoustic wave equation that SPECFEM3D actually solves.

Rather than solving wave Equation (3) directly, the "weak form" equation is solved. It is derived from Equation (3) by multiplying a test function ω, integrating by parts and using Green's first identity:

$$-\int_\Omega \frac{1}{\rho}\nabla\chi \cdot \nabla\omega d\Omega + \int_\Gamma (\nabla\chi \cdot \boldsymbol{n})\omega d\Gamma = \int_\Omega \frac{1}{\lambda}\ddot{\chi}\omega d\Omega, \tag{4}$$

where Γ is the boundary of the domain Ω, and $\boldsymbol{n}$ is an outward normal vector unit along Γ.

In SPECFEM3D, this weak formulation is solved in a discretized form with the continuous Galerkin method. For a further explanation of the discretization, refer to [45,51].

This type of Gaussian integral is well known in the context of finite elements, but instead, SEM uses a generalized Gaussian integration, the so-called Gauss–Lobatto–Legendre integration, which has the supplemental collocation points at both ends of the discretization interval. Detailed explanations of such Gauss-type polynomials can be found in [52,53]. Functions on a given spectral element are approximated using the Lagrange polynomial with Gauss–Lobatto–Legendre (GLL) points. Figure 1 shows an example of a 2D first-order finite element and a fourth-order spectral element with a total of 25 GLL points.

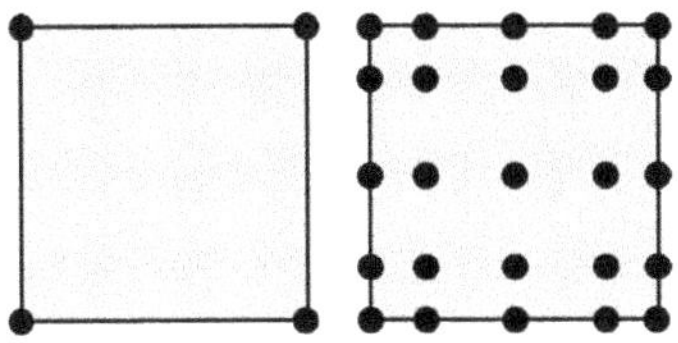

Figure 1. An example of 2D first-order finite element (**left**) and a fourth-order spectral element (**right**).

3. Fluid Dynamics Modeling in PLAJEST Experiment and Frozen-Flow Hypothesis

A large benchmark exercise was performed under the auspices of an international collaboration on thermal hydraulics for sodium-cooled fast-reactor development with participation from the Japan Atomic Energy Agency (JAEA), the U.S. Department of Energy (DOE) and the French Commissariat à l'Énergie Atomique et aux Énergies Alternatives (CEA). It was based on experiments performed to study the effects of thermal striping, where three differentially heated jets mix inside a cavity [25]. These experiments were carried out either in water or in liquid sodium, and velocity and temperature data were provided. The object of the benchmark was to numerically predict the results of these experiments and make comparisons with available measured data. For the French part, the numerical simulations were led using the TrioCFD simulation code (known as Trio_U by 2015). The large eddy simulation (LES) model was applied for the analyses. A computational domain reproducing the test sections was created. The discretization of the equations was based on unstructured staggered meshes and the resolution on a finite-volume approach. The numerical results were in good agreement with the experiments. Good results were obtained for the time-averaged fields and the power spectral densities of temperature fluctuations [27,54,55].

A schematic description of the three jets is drawn in the upper part of Figure 2. Three jets are mixed, with a cold jet between two hot jets. The three jets have a mean flow speed of 0.51 m/s. Two CFD maps, separated by 1 ms, are represented in this figure. The insonified area is represented, superimposed on the left map by a rectangular box. The box is centered at Z = 0.09 m, which corresponds to the most turbulent area. Letters E and R, respectively, represent the position of the virtual ultrasonic emitter and that of the virtual receiver. This area is 230 mm long, and the total time for the wave to propagate along this distance is less than 100 µs, a tenth of the time between the two CFD maps. The emitter and the receiver correspond to a 1 MHz transducer with a diameter of 0.0254 m (i.e., 1 inch). A complete study of wave propagation is realized, and pressure time-series data are recorded on several planes in this volume, but not only in the plane of the virtual receiver (see Figure 3 thereafter).

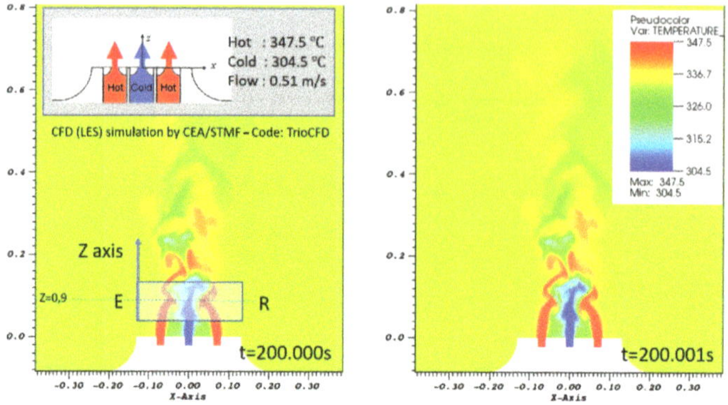

Figure 2. Two maps from the PLAJEST CFD simulation and representation of the insonified volume.

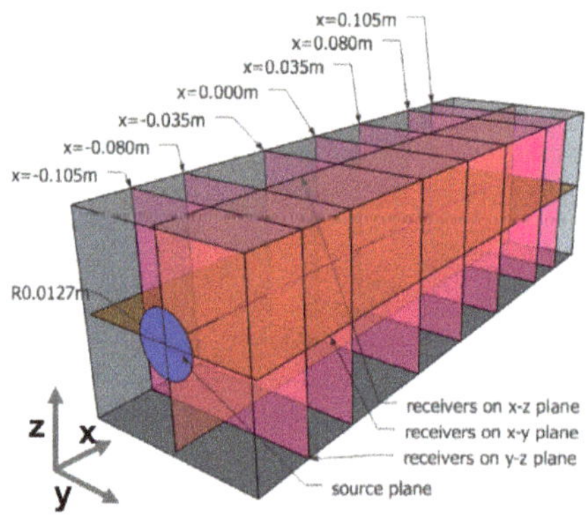

Figure 3. Definition of recorded planes in the insonified area.

When considering how the temperature field modifies the wave propagation (Step 10 in Figure 4), it implies taking into account the temperature-dependent properties of liquid sodium reported in [56]. Sobolev et al. demonstrated that the density of liquid sodium varies linearly with the temperature. Within the temperature range between the normal melting point and the normal boiling point, the density is calculated with :

$$\rho = 1014 - 0.235 \cdot T \tag{5}$$

at normal atmospheric pressure, where ρ is the density of sodium, and T is the sodium temperature in Kelvin degrees. On the other hand, the sound velocity in liquid sodium decreases monotonically with temperature, because of the decrease in the number of interatomic interactions. In the normal melting–boiling point (371–1155 °K) range, the sound velocity in pure liquid sodium may be described based on this linear relation:

$$c_p = 2723 - 0.531 \cdot T \tag{6}$$

where c_p is the celerity of ultrasonic waves in meters per second.

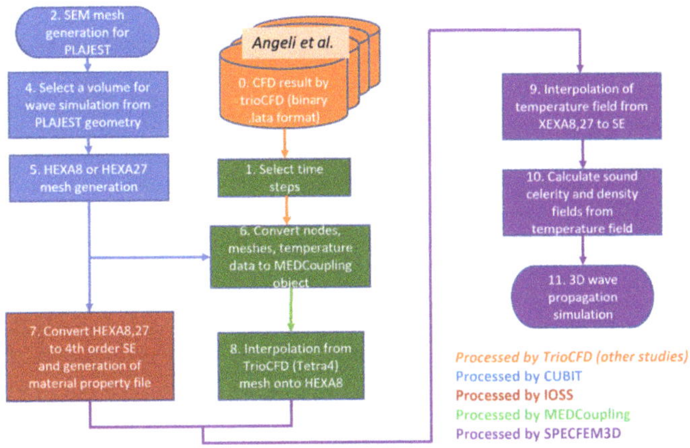

Figure 4. Data processing steps for mesh generation and preparation of the heterogeneous medium.

In Figure 5, the difference between the two temperature maps is represented. The maximum temperature difference is lower than 2 °K in 1 ms. Using a simple proportional rule, it is then possible to consider that the maximum temperature difference during wave propagation is about 0.2 °K. Considering the velocity relation with temperature in (Equation (6)), it is estimated that there is a very little variation of the liquid metal velocity and of the local temperature gradient. Furthermore, this little variation occurs in a very small area in the jets. This leads to the conclusion that wave propagation is very similar between the two time steps separated by 0.1 ms. Thanks to this complete CFD simulation, the frozen-flow hypothesis is clearly confirmed in this study.

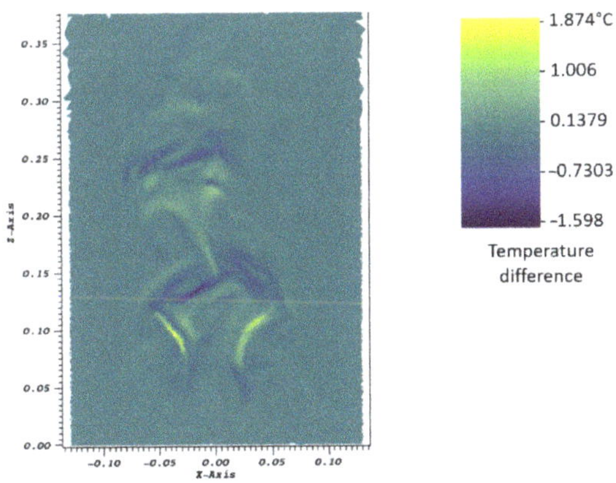

Figure 5. Temperature differences during 1 ms.

4. Implementation of Wave Propagation Modelling

This thermal–hydraulic data obtained from a calculation with the CFD code "TrioCFD" was used to completely describe the medium in which ultrasounds propagate. The type of finite element of the meshes used in the two codes TrioCFD and SPECFEM3D is not the same, so we needed to perform preprocessing to import the result of CFD into our acoustic simulation. Figure 4 shows the entire workflow of this process.

Tetrahedral elements with four nodes were used in the CFD calculation with TrioCFD. The total number of elements was 5,582,706, and the characteristic mesh length was set to 1.40 mm. The first 200 s of their calculation were removed so as to obtain a good stabilization of the flow state in this CFD calculation. Thus, this calculation eventually provided us with

200 to 210 s of data of fluctuating flow status with a time step of 1 ms. The total number of time steps was 10,000. In the result data, temperature field values are defined at the center of each TrioCFD tetrahedral mesh element, and flow velocity values are defined on the vertex nodes. These values were to be transferred to our hexahedral mesh for SPECFEM3D. Interpolation onto each node of the SPECFEM3D hexahedral mesh was performed using the simulation data management tool called MEDCoupling. MEDCoupling is part of the pre-/postprocessing platform SALOME (https://www.salome-platform.org/downloads/current-version) and is also available as a library.

In order to prepare a mesh for SPECFEM3D calculation, we firstly selected a partial region for wave propagation simulation from the entire PLAJEST simulation model (Step 4 in Figure 4). This extraction of the domain was performed in order to eliminate acoustically uninteresting parts from the PLAJEST geometry, and this enabled us to reduce the required amount of computer memory, which is one of the limitations of wave simulations in large 3D models.

After selecting the domain, we built the first-/second-order hexahedral mesh for SPECFEM3D using the meshing software CUBIT developed by Sandia National Laboratories (USA) (Step 5).

The actual mesh type that SPECFEM3D uses is not a first- nor second-order hexahedral mesh but a fourth-order spectral element. Thus, we then performed this conversion as the final step of mesh preparation (Step 7). In order to speed up this conversion process, we used the IOSS (IO Systems) library. This library is included in the finite element analysis supporting software called SEACAS (Sandia National Laboratories).

After finishing the preparation of mesh data, the temperature field transfer was carried out, i.e., interpolation of temperature values defined at the barycentre of each tetrahedral finite element to corner nodes of the hexahedral spectral elements (Step 8) using MEDCoupling. The MEDCoupling mesh-to-node transfer function does not support HEXA27 (second-order hexahedral finite elements); consequently, if the interpolation target mesh is of the HEXA27 type, we first split HEXA27 into eight parts of the HEXA8 type (first-order hexahedral finite elements) and then use the MEDCoupling interpolation function for those HEXA8 elements.

Then, SPECFEM3D performed the interpolation of these values onto each Gauss–Lobatto–Legendre point in Step 9.

The flow velocity data were not used for our simulation, as the temperature gradient has a greater influence on the wave path than the velocity vector field [22].

To determine the element size for the mesh, we used the two conditions: the Courant–Friedrich–Lewy condition (CFL), described in Equation (7) , and the number of elements per wavelength:

$$C_p \frac{\Delta t}{\Delta x_{gll}} \leq \alpha , \quad (7)$$

where Δt is the time step, and Δx_{gll} is the minimum interval between two GLL grid points. An averaged Courant number $\alpha = 0.4$ was selected. and the wave celerity was defined as the highest one $C_p = 2416.268\,\mathrm{m\,s^{-1}}$. This celerity was calculated with the lowest temperature value 274.5 °C in the CFD simulation. In Equation (7), Δx_{gll} is not the mesh size itself; rather, it is the distance interval between GLL grid points within the spectral elements. This led us to calculate a mesh size $\Delta x = 8.05 \times 10^{-4}$ m and a time step of 2.3×10^{-8} s. A total of 5.000 steps were simulated in order to have a sufficient total physical duration for the waves to travel through the entire simulation domain. The meshing used in the simulation thus comprises 3,250,000 spectral elements and a total number of GLL nodes of 215,320,764. This time step corresponds to a 43.5 MHz sampling frequency, which corresponds to a fine sampling for the ultrasonic simulated signal.

Figure 6 represents some examples of the preprocesses of temperature field transfer and mesh generation for SPECEFM3D. The temperature field in the tetrahedral elements of

TrioCFD is indicated with transparent color on the interpolated field of the hexahedral mesh for SPECFEM3D. The image on the left side is a close-up on one of these three examples.

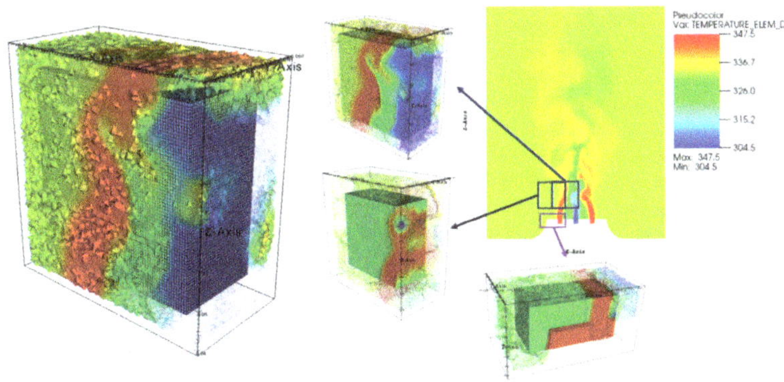

Figure 6. Examples of some interpolations of temperature fields from a tetrahedral mesh to a hexahedral mesh using the MEDCoupling pre-/postprocessing library.

5. Temporal Temperature Fluctuation Measurement by Ultrasounds

The analysis of the temperature fluctuation was performed over a total duration of approximately 10 s, from 200.000 s to 210.197 s. Because the computation time allocated on the supercomputers that we used was limited, it was not possible to run the temperature field interpolation and wave propagation calculation processes for all of these CFD time steps. Instead, we had to extract several time steps from the temperature field with a wider time interval from the CFD results. In order to select the minimum time interval to be used for our acoustic simulation, we exploited the power spectral density (PSD) curve calculated in Figure 7 [27]. This curve allows knowing the frequency of the temperature fluctuation at $x = -0.015$ m (between the left and the central jets), $y = 0.09$ m (middle point on the y axis) and $z = 0.1$ m. This is a normalized PSD calculated by dividing the original PSD with the maximum PSD value. The peak of this PSD curve is found around 3 Hz and is in accordance with experimental results, as mentioned by Angeli in [27]. This peak is lower than 5 Hz. This peak is followed by a slope (Kolmogorov slope).

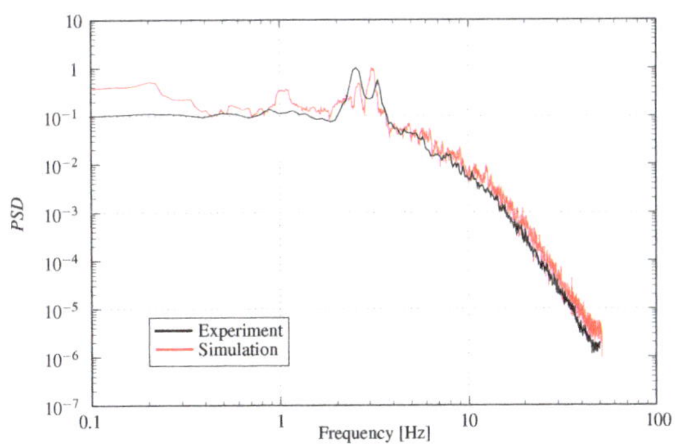

Figure 7. Normalized PSD curves of temperature history in the CFD calculation (blue line) at $x = -0.015$ m, $y = 0.09$ m and $z = 0.1$ m (from STMF laboratory).

In this work, all our simulations were performed on two of the largest supercomputers in Europe: CURIE (CEA TGCC) and OCCIGEN (CINES), both part of Grand Équipement National de Calcul Intensif (GENCI). Despite the use of these supercomputers, we knew that it would be difficult to perform calculations with a fine temporal step. In order to be able to follow the fluid dynamics, we wanted to confirm fluid fluctuations at a 3 Hz frequency as seen in PSD curves. We fixed the frequency limit to 5 Hz, then the Shannon sampling criterion imposed us to use a minimum sampling frequency of 10 Hz). Temperature fields were thus extracted from the complete volume with a 0.1 s interval.

The computation domain was divided into 256 parts, and parallel calculations were carried out. The duration of the interpolation of a temperature field from a TrioCFD result to SPECFEM3D was approximately 20 h using a single CPU at one chosen altitude and one time step (the insonified area represented in Figure 2). The average duration for an acoustic simulation was about 26 min, excluding the mesh generation and the interpolation processes of the temperature fields, which were performed once and for all. Complete results, signals at each point in the volume, cannot be fully stored on a physical disc because of the high number of calculations. Thus the data resulting from SPECFEM3D calculations were recorded for several planes, as described in Figure 3.

To analyze temporal temperature fluctuations within our heterogeneous medium and then to make the correlation with ultrasonic data, a PSD curve was calculated at three different points. These three points have the same y and z position ($y = 0.09$ m and $z = 0.1$ m) and a different x position ($x = -0.035$ m, 0.0 m and 0.035 m). The position $y = 0.09$ m corresponds to the middle of the jets along the y axis, and $z = 0.1$ m corresponds to an altitude with high turbulence. $x = 0.0$ m is in the middle of the central jet. $x = -0.035$ m and 0.035 m are two positions between the cold jet and the hot jet. These values are imposed by the initial choice of recorded planes (Figure 3). In these positions, temperature fluctuations are high but probably less than for positions $x = -0.015$ m and 0.015 m chosen by Angeli et al. Figure 8A,B shows the temperature PSD curve of the original CFD results (every millisecond) at three selected points and the same kind of curves but with only extracted time steps (every 0.1 s), respectively.

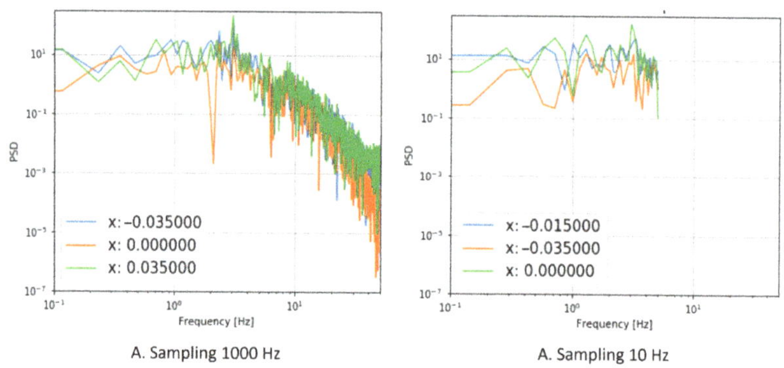

Figure 8. Temperature PSD curves calculated from the CFD data at three different x positions (y = 0.09 m; z = 0.1 m), with two sampling rates.

Using the finer sampling, the peak of the experimental PSD curve is confirmed at 3 Hz for each position. Using data with a coarser time step gives merely the same results for the central point, as the peak frequency at 3 Hz is clearly visible. From these comparisons, we thus verify that the peak frequency of the temperature fluctuation is possibly detected. In the next step, the idea is to search for this flow indicator using ultrasonic measurements. The ultrasonic parameters which are selected are the maximum amplitude of the wave at one point and the corresponding time of flight (TOF). Two examples are shown in Figure 9. The two points have the same altitude: one point is close to the jet, and the second one is after the three jets (from the emitter position). The position $x = 0.035$ m is close to the jets, and the temperature behavior exhibits clear periodic fluctuation. The position $x = 0.105$ m is far from the jets, and the temperature remains constant over the 10 s of the

CFD simulation, so the corresponding PSD curve gives no frequency information. At the position $x = 0.035$ m, among the two PSD calculated with ultrasonic parameters, only those calculated with the time of flight exhibit a clear 3 Hz peak as for the temperature PSD. At the position $x = 0.105$ m, the same observation is performed: the time-of-flight PSD exhibits a clear 3 Hz peak as for the temperature PSD. This result is very interesting, as it demonstrates that the history of temperature fluctuations is transported by the ultrasonic wave and a receiver positioned after the jets is able to analyze the temperature fluctuation.

The idea is further deepened using measurements of time-of-flight PSDs at various positions along the x-axis. These curves are compared with PSD curves calculated on CFD data. The resulting curves are presented in Figure 10.

The calculation of the PSDs with the low sampling CFD data (sampling frequency of 10 Hz) shows that only the PSDs on the positions $x = -0.035$ m and 0.035 m exhibit a frequency peak at 3 Hz. On the other hand, as expected, following the hypothesis of the transport of the fluctuation information by the ultrasonic wave, for positions beyond $x = -0.035$ m, all the time-of-flight PSDs contain this information. We also cumulated (added) all the PSDs calculated at the different points, the PSD curve being defined as a cumulated temperature PSD curve in Figure 10. The 3 Hz peak emerges clearly with a secondary peak at 1 Hz. The comparison with the PSD of the TOF at $x = 0.105$ m shows that the ultrasonic measurement clearly indicates this peak but does not find the secondary peak. There is therefore a limit in the sensitivity of the detection of these fluctuations by an ultrasonic measurement technique, but simulations and results demonstrate that a pair of transducers positioned in a through-transmission setup would allow the temperature fluctuation frequency to be registered.

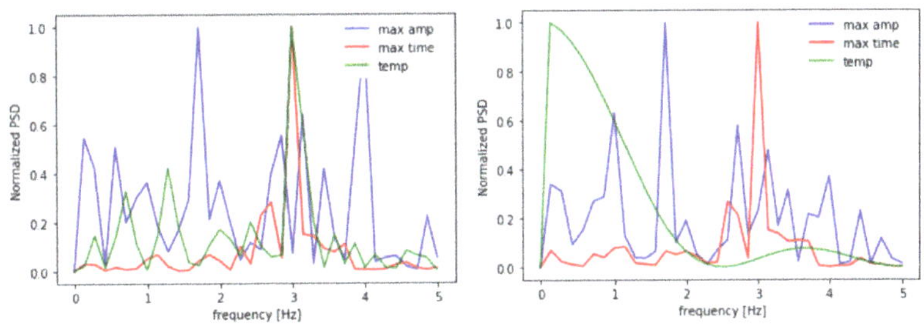

A. z alt 0.09 x_pos 0.035 B. z alt 0.09 x_pos 0.105

Figure 9. History and PSD analysis for temperature, time of flight and wave amplitude.

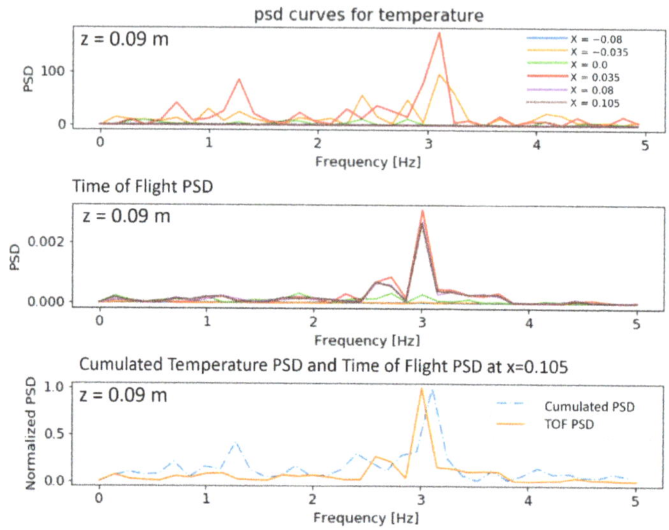

Figure 10. Temperature PSD vs. time-of-flight PSD.

6. Conclusions

We designed an extensive physical and numerical simulation strategy to study the potential of ultrasonic measurements to monitor turbulent flow. The study was focused on liquid metal flows for which ultrasonic measurements are the best candidate to monitor flows. In the case of sodium flow monitoring inside SFR nuclear reactors, it is planned that transducers will be positioned in the flows. There are not many experiments being conducted on liquid sodium flows, as they are still very difficult to perform, and there are not many SFR nuclear reactors in operation worldwide.

One of the main difficulties, when trying to assess measurement performance without using real experiments, lies in simulating ultrasonic propagation in a representative flow. Our simulation strategy consisted of selecting a high-level CFD simulation of a real experiment of liquid metal jet mixing. We also drew on previous studies concerning ultrasonic wave propagation in liquid sodium to define all physical parameters for the wave simulation. The novelty of this study consisted in making a wave propagation simulation for several CFD time steps to validate the potential of ultrasonic measurements to follow flow fluctuations. Our analysis of flow fluctuation enabled us to prove that the frozen-flow hypothesis was completely valid in this study. It should be underlined that such a hypothesis is rarely numerically studied. The study required the development of dedicated tools to implement the wave propagation modeling on high-performance computers. The SPECFEM3D code was chosen as the numerical calculation tool for its numerical efficiency because of matured development for massively parallelized computing on HPCs and the application of the spectral-element method.

Despite significant computer resources, we had to limit the maximum frequency observed to 5 Hz in order to be able to complete the study. This limit was enough to demonstrate that the main flow fluctuation frequency of 3 Hz can be monitored using a pair of transducers in a through-transmission setup. This demonstration opens the way for developing more simulations combining CFD simulations and wave propagation for several time steps so that better ways to monitor fluctuating flows could be developed. One limitation of these studies is the huge volume of data to be stored to study wave propagation in detail. This volume could be drastically reduced when transducer positions are not a parameter of investigation. The advantage of a pair of transducers is that it makes it possible to monitor flow between the two transducers. The wave propagates from the emitter in the flow and, to some extent, transfers the flow history to the receiver.

In this study, the flow history is the fluctuation of the flow temperature. This study proves that it is now possible to conduct simulations of wave propagation in a more realistic flow geometry. There is of course still a step to be taken to be able to conduct simulations in large and long flows.

Author Contributions: Conceptualization, M.N., J.M. and C.L.; methodology, M.N. and J.M.; software, M.N.; validation, J.M. and C.L.; formal analysis, M.N.; investigation, M.N.; resources, M.N.; data curation, M.N.; writing—original draft preparation, M.N. and J.M.; writing—review and editing, M.N, J.M. and C.L.; visualization, M.N.; supervision, J.M. and C.L.; project administration, J.-P.J.; funding acquisition, J.-P.J. All authors have read and agreed to the published version of the manuscript.

Funding: This study was funded by The French Alternative Energies and Atomic Energy Commission, Cadarache.

Institutional Review Board Statement: Not applicable.

Informed Consent Statement: Not applicable.

Acknowledgments: This study was granted access to the French HPC resources of TGCC/CINES under allocations 2016107626, 2016047165 and 2016047748 made by GENCI. We also thank Antoine Gershenfeld and Pierre Emmanuel Angéli, from the laboratory of CEA/DES/ IRESNE/DM2S/STMF, for sharing their thermal–hydraulic data of PLAJEST and their advice on how to analyze these data. They provided elements for the Figure 7. This work was developed within the framework of the MISTRAL joint research laboratory between Aix-Marseille University, CNRS, Centrale Marseille and CEA.

Conflicts of Interest: The authors declare no conflict of interest.

References

1. Baker, R. *Flow Measurement Handbook: Industrial Designs, Operating Principles, Performance, and Applications*; Cambridge University Press: Cambridge, UK, 2000.
2. Scarano, P. Tomographic PIV: Principles and practice. *Meas. Sci. Technol.* **2012**, *24*, 012001. [CrossRef]
3. Achard, J.; Ramel, N.; Beretta, G.; Menet, P.; Prigent, J.; Le Brun, P. Batscan™, Constellium In-melt Ultrasonic Inclusion Detector: Industrial Performance. In *Light Metals 2020*; Springer: Cham, Switzerland, 2020; pp. 936–943._126. [CrossRef]
4. Fergus, J. Sensors for monitoring the quality of molten aluminum during casting. *J. Mater. Eng Perform* **2005**, *4*, 267–275. [CrossRef]
5. Thieme, N.; Nauber, R.; Büttner, L.; Czarske, J.; Büchner, K.; Pätzold, O. Ultrasound flow mapping for 3D turbulent liquid metal flows. In Proceedings of the IEEE International Ultrasonics Symposium (IUS), Washington, DC, USA, 6–9 September 2017; pp. 1–4. [CrossRef]
6. Zürner, T.; Vogt, T.; Resagk, C.; Eckert, S.; Schumacher, J. Local Lorentz force and ultrasound Doppler velocimetry in a vertical convection liquid metal flow. *Exp. Fluids* **2018**, *59*, 3. [CrossRef]
7. Christensen, U.R.; Aubert, J. Scaling properties of convection-driven dynamos in rotating spherical shells and application to planetary magnetic fields. *Geophys. J. Int.* **2006**, *166*, 97–114. [CrossRef]
8. Kelley, D.H.; Weier, T. Fluid Mechanics of Liquid Metal Batteries. *Appl. Mech. Rev.* **2018**, *70*, 23. [CrossRef]
9. Hvasta, M.; Bruhaug, G.; Fisher, A.E.; Dudt, D.; Kolemen, E. Liquid Metal Diagnostics. *Fusion Sci. Technol.* **2020**, *76*, 62–69. [CrossRef]
10. Salavy, J.F.; Boccaccini, L.V.; Lasser, R.L.; Meyder, R.; Neuberger, H.; Poitevin, Y.; Rampal, G.; Rigal, E.; Zmitko, M.; Aiello, A. Overview of the last progresses for the European Test Blanket Modules projects. *Fusion Eng. Des.* **2019**, *82*, 2105–2112. [CrossRef]
11. Griffin, J.; Bond, L.; Peters, T.; Denslow, K.; Posakony, G.; Sheen, S.; Chien, H.; Raptis, A. *Under-Sodium Viewing: A Review of Ultrasonic Imaging Technology for Liquid Metal Fast Reactors*; Technical Report PNNL-18292; Pacific Northwest National Laboratory: Richland, WA, USA, 2009; doi:10.2172/1010482. [CrossRef]
12. Jasiuniene, E. Ultrasonic imaging techniques for non-destructive testing of nuclear reactors, cooled by liquid metals: Review. *Ultragarsas (Ultrasound)* **2007**, *62*, 39–43.
13. Tarpara, E.; Patankar, V.; Vijayan Varier, N. *Under-Sodium Ultrasonic Viewing for Fast Breeder Reactors: A Review*; Technical Report; BARC: Mumbai, India, 2016.
14. Paumel, K.; Descombin, O.; Moysan, J.; Corneloup, G.; Aguem, J.M. Acoustic coupling of ultrasonic transducers for in-service inspection of sodium fast reactors. In Proceedings of the 1st International Conference on Advancements in Nuclear Instrumentation, Measurement Methods and their Applications, Marseille, France, 7–10 June 2009; pp. 1–6. [CrossRef]

15. Paumel, K.; Moysan, J.; Chatain, D.; Corneloup, G.; Baqué, F. Modeling of ultrasound transmission through a solid-liquid interface comprising a network of gas pockets. *J. Appl. Phys.* **2011**, *110*, 044910. [CrossRef]
16. Kim, H.-W.; Joo, Y.-S.; Park, C.-G.; Kim, J.-B.; Bae, J.-H. Ultrasonic Imaging in Hot Liquid Sodium Using a Plate-Type Ultrasonic Waveguide Sensor. *J. Nondestruct. Eval.* **2014**, *33*, 676–683. [CrossRef]
17. Imbert, C.; Berton, J.; Gimenez, G. Realization of ultrasonic images of immersed metallic structures using a digital beamforming system. Experimental study. In Proceedings of the IEEE International Ultrasonics Symposium (IUS), San Antonio, TX, USA, 3–6 November 1996; pp. 765–770.
18. Le Bourdais, F.; Marchand, B. Development of electromagnetic acoustic transducer (EMAT) phased arrays for SFR inspection. *AIP Conf. Proc.* **2014**, *1581*, 1022–1029. [CrossRef]
19. Saillant, J.F.; Marlier, R.; Baqué, F.; Navacchia, F. Design and testing of an ultrasonic projector for operation in liquid sodium. In Proceedings of the IEEE International Ultrasonics Symposium (IUS), Glasgow, UK, 6–9 October 2019; pp. 96–99. [CrossRef]
20. Massacret, N.; Moysan, J.; Ploix, M.A.; Jeannot, J.P.; Corneloup, G. Modelling of ultrasonic propagation in turbulent liquid sodium with temperature gradient. *J. Appl. Phys.* **2014**, *115*, 204905. [CrossRef]
21. Iooss, B.; Lhuillier, C.; Jeanneau, H. Numerical simulation of transit-time ultrasonic flowmeters: Uncertainties due to flow profile and fluid turbulence. *Ultrasonics* **2002**, *40*, 1009–1015. [CrossRef]
22. Massacret, N.; Moysan, J.; Ploix, M.A.; Jeannot, J.P.; Corneloup, G. Simplified modeling of liquid sodium medium with temperature and velocity gradient using real thermal-hydraulic data: Application to ultrasonic thermometry in Sodium Fast Reactor. *AIP Conf. Proc.* **2013**, *1511*, 1693–1700. [CrossRef]
23. Nagaso, M.; Moysan, J.; Benjeddou, S.; Massacret, N.; Ploix, M.; Komatitsch, D.; Lhuillier, C. Ultrasonic thermometry simulation in a random fluctuating medium: Evidence of the acoustic signature of a one-percent temperature difference. *Ultrasonics* **2016**, *68*, 61–70. [CrossRef] [PubMed]
24. Ploix, M.A.; Corneloup, G.; Moysan, J.; Jeannot, J. Investigation of ultrasonic backward energy from various edges as a function of their 2D/3D geometry and of the incidence angle, for application to ultrasonic thermometry at the outlet of a tube. *Case Stud. Nondestruct. Test. Eval.* **2016**, *6*, 38–44. [CrossRef]
25. Kimura, N.; Miyakoshi, H.; Kamide, H. Experimental investigation on transfer characteristics of temperature fluctuation from liquid sodium to wall in parallel triple-jet. *Int. J. Heat Mass Transf.* **2007**, *50*, 2024–2036. [CrossRef]
26. Brillant, G.; Bataille, F.; Ducros, F. Large-eddy simulation of a turbulent boundary layer with blowing. *Theor. Comput. Fluid Dyn.* **2004**, *17*, 433–443. [CrossRef]
27. Angeli, P.E. Large-Eddy Simulation of thermal striping in WAJECO and PLAJEST experiments with TrioCFD. In Proceedings of the 16th International Topical Meeting on Nuclear Reactor Thermal Hydraulics (NURETH-16), Chicago, IL, USA, 30 August–4 September 2015; pp. 1223–1236.
28. Komatitsch, D. Méthodes Spectrales et Éléments Spectraux pour L'équation de L'élastodynamique 2D et 3D en Milieu Hétérogène (Spectral and Spectral-Element Methods for the 2D and 3D Elastodynamics Equations in Heterogeneous Media). Ph.D. Thesis, Institut de Physique du Globe, Paris, France, 1997; 187p.
29. Massacret, N. Étude d'une Méthode Ultrasonore D'estimation des Températures Locales du Sodium Liquide en Sortie Coeur RNR-Na. Ph.D. Thesis, Aix-Marseille University, Marseille, France, 2014.
30. Lü, B. Modélisation de la Propagation et de L'interaction d'une onde Acoustique pour la Télémétrie de Structures Complexes. Ph.D. Thesis, Université du Maine, Le Mans, France, 2011.
31. Lü, B.; Darmon, M.; Potel, C. Stochastic simulation of the high-frequency wave propagation in a random medium. *J. Appl. Phys.* **2012**, *112*, 054902. [CrossRef]
32. Lü, B.; Darmon, M.; Fradkin, L.; Potel, C. Numerical comparison of acoustic wedge models, with application to ultrasonic telemetry. *Ultrasonics* **2016**, *65*, 5–9. [CrossRef]
33. Gengembre, N.; Lhémery, A. Pencil method in elastodynamics: Application to ultrasonic field computation. *Ultrasonics* **2000**, *38*, 495–499. [CrossRef]
34. Jensen, F.B.; Kuperman, W.; Porter, M.; Schmidt, H. *Computational Ocean Acoustics*, 2nd ed.; Springer: Berlin, Germany, 2011; 794p.
35. Yee, K.S. Numerical Solution of Initial Boundary Value Problems involving Maxwell's Equations. *IEEE Trans. Antennas Propag.* **1966**, *14*, 302–307.
36. Jurgens, T.G.; Taflove, A.; Umashankar, K.; Moore, T.G. Finite-difference time-domain modeling of curved surfaces (EM scattering). *IEEE Trans. Antennas Propag.* **1992**, *40*, 357–366. [CrossRef]
37. Moczo, P.; Bystrický, E.; Kristek, J.; Carcione, J.M.; Bouchon, M. Hybrid modeling of *P-SV* Seism. Motion Inhomogeneous Viscoelastic Topographic Structures. *Bull. Seism. Soc. Am.* **1997**, *87*, 1305–1323.
38. Kristek, J.; Moczo, P. Seismic-Wave Propagation in Viscoelastic Media with Material Discontinuities: A 3D Fourth-Order Staggered-Grid Finite-Difference Modeling. *Bull. Seism. Soc. Am.* **2003**, *93*, 2273–2280. [CrossRef]
39. Tarrass, I.; Giraud, L.; Thore, P. New curvilinear scheme for elastic wave propagation in presence of curved topography. *Geophys. Prospect.* **2011**, *59*, 889–906. [CrossRef]
40. Kuhlemeyer, R.L.; Lysmer, J. Finite element method accuracy for wave propagation problems. *J. Soil Mech. Found. Div.* **1973**, *99*, 421–427. [CrossRef]
41. Moser, F.; Jacobs, L.J.; Qu, J. Modeling elastic wave propagation in waveguides with the finite element method. *Ndt E Int.* **1999**, *32*, 225–234. [CrossRef]

42. Gottlieb, D.; Orszag, S.A. *Numerical analysis of spectral methods: Theory and applications*; Society for Industrial and Applied Mathematics (SIAM): Philadelphia, PA, USA, 1977.
43. Patera, A.T. A spectral element method for fluid dynamics: Laminar flow in a channel expansion. *J. Comput. Phys.* **1984**, *54*, 468–488. [CrossRef]
44. Komatitsch, D.; Tromp, J. Introduction to the spectral-element method for 3-D seismic wave propagation. *Geophys. J. Int.* **1999**, *139*, 806–822. [CrossRef]
45. Komatitsch, D.; Tromp, J. Spectral-element simulations of global seismic wave propagation-I. Validation. *Geophys. J. Int.* **2002**, *149*, 390–412. [CrossRef]
46. Komatitsch, D.; Tsuboi, S.; Ji, C.; Tromp, J. A 14.6 billion degrees of freedom, 5 teraflops, 2.5 terabyte earthquake simulation on the Earth Simulator. In Proceedings of the SC'03 ACM/IEEE Conference on Supercomputing, Phoenix, AZ, USA, 15–21 November 2003; ACM: New York, NY, USA, 2003; Gordon Bell Prize Winner Article; pp. 4–11. [CrossRef]
47. Carrington, L.; Komatitsch, D.; Laurenzano, M.; Tikir, M.; Michéa, D.; Le Goff, N.; Snavely, A.; Tromp, J. High-frequency simulations of global seismic wave propagation using SPECFEM3D_GLOBE on 62 thousand processor cores. In Proceedings of the SC'08 ACM/IEEE Conference on Supercomputing, Austin, TX, USA, 15–21 November 2008; IEEE Press: New York, NY, USA, 2008; Article #60, Gordon Bell Prize Finalist Article; pp. 1–11. [CrossRef]
48. Tsuboi, S.; Ando, K.; Miyoshi, T.; Peter, D.; Komatitsch, D.; Tromp, J. A 1.8 trillion degrees-of-freedom, 1.24 petaflops global seismic wave simulation on the K computer. *Int. J. High Perform. Comput. Appl.* **2016**, *30*, 411–422. [CrossRef]
49. Landau, L.D.; Lifshitz, E.M. *Fluid Mechanics*; Pergamon Press: New York, NY, USA, 1959; 536p.
50. Hamdi, M.A.; Ousset, Y.; Verchery, G. A displacement method for the analysis of vibrations of coupled fluid-structure systems. *Int. J. Numer. Methods Eng.* **1978**, *13*, 139–150. [CrossRef]
51. Nagaso, M. Study of Ultrasound Wave Propagation in a Heterogeneous Fluid Medium for the Monitoring of an Operating Sodium-Based Nuclear Reactor. Ph.D. Dissertation, Université d'Aix-Marseille (AMU), ED 353, Marseille, France, 2018; A Full PHDTHESIS Entry.
52. Davis, P.J.; Rabinowitz, P. *Methods of Numerical Integration (Computer Science and Applied Mathematics)*; Academic Press: New York, NY, USA, 1984.
53. Canuto, C.; Hussaini, M.Y.; Quarteroni, A.; Zang, T.A. *Spectral Methods*; Springer, Berlin, Germany, 2011.
54. Angeli, P.E.; Puscas, A.; Fauchet, G.; Cartalade. FVCA8 Benchmark for the Stokes and Navier-Stokes Equations with the TrioCFD Code—Benchmark Session. In *Finite Volumes for Complex Applications VIII—Methods and Theoretical Aspects*; Springer: Cham, Switzerland, 2019; pp. 181–203.
55. Angeli, P.E. Verification and validation of LES of a triple parallel jet flow in the context of a thermal striping study. *Nucl. Eng. Des.* **2019**, *353*, 110210. [CrossRef]
56. Sobolev, V. *Database of Thermophysical Properties of Liquid Metal Coolants for GEN-IV*; Technical Report; BLG-1069; SCK-CEN: Mol, Belgium, 2011.

applied sciences

MDPI

Article

An Efficient Semi-Analytical Scheme for Determining the Reflection of Lamb Waves in a Semi-Infinite Elastic Waveguide

Robert C. Davey [1], Raphaël C. Assier [1,*] and I. David Abrahams [2]

[1] Department of Mathematics, The University of Manchester, Manchester M13 9PL, UK; robertcdavey@googlemail.com
[2] Department of Applied Mathematics and Theoretical Physics, University of Cambridge, Cambridge CB3 0WA, UK; ida20@cam.ac.uk
* Correspondence: raphael.assier@manchester.ac.uk

Abstract: The classical problem of reflection of Lamb waves from a free edge perpendicular to the centre line of an elastodynamic plate is studied. It is known that Lamb wave expansions for the displacement and stress fields poorly represent the irregular behaviour near corners, leading to the slow convergence of a series of such waves. The form of the irregularity for an elastodynamic corner is derived asymptotically, and a new solution method, which incorporates this corner behaviour analytically, is then implemented. Results are presented showing that this new approach represents the near-field and far-field behaviour very accurately, requiring very modest numbers of Lamb wave and corner modes. Further, it is revealed that the method can recover the trapped-mode phenomenon encountered in this configuration at the Lamé frequency and a specific Poisson's ratio that we find to be approximately 0.224798.

Keywords: elasticity; plates; Lamb waves; corner behaviour; trapped-mode

Citation: Davey, R.C.; Assier, R.C.; Abrahams, I.D. An Efficient Semi-Analytical Scheme for Determining the Reflection of Lamb Waves in a Semi-Infinite Elastic Waveguide. *Appl. Sci.* **2022**, *12*, 6468. https://doi.org/10.3390/app12136468

Academic Editor: Michel Darmon

Received: 14 April 2022
Accepted: 22 June 2022
Published: 25 June 2022

1. Introduction

We are interested in the use of guided wave techniques in non-destructive testing, applied specifically to the problem of hidden tamper detection in an arms control context. In arms control it is essential to ensure that all parties are in compliance with the terms of a treaty. A problem arising from this necessity is tamper detection [1]. The competent authorities wish to ensure that items of interest cannot be accessed for long periods of time and to this end they are sealed in freight-like containers. However, with sufficient time and effort any container can be breached, hence we will focus on the problem of making these containers 'tamper indicating'. This means that if a container is breached, it should be impossible to hide the evidence of this from a verification process, even if the breach has been deliberately concealed. Such a deliberately concealed breach is known as a 'hidden tamper'.

We wish to apply a well known, non-destructive testing method to tamper detection, i.e., that of guided elastic waves. This has been motivated by experimental evidence [2] that tampers can be detected easily, even by non-specialist technicians, by scanning for mode conversion of elastic waves. Refraction and reflection of the guided wave, or excitation of additional modes, may occur due to welds [3], cracks [4], or other inhomogenous imperfections at the boundary of the tamper; however, we are mainly interested in mode conversion due to asymmetry introduced by the tamper itself.

In elastic plates, there are two types of modes that propagate, symmetric and antisymmetric, and these modes travel at different speeds. For a plate of given thickness and material properties, the number of cut on (i.e., propagating) symmetric and antisymmetric modes depends on the frequency, with the lowest antisymmetric mode, A_0, moving more slowly than its symmetric counterpart, S_0. When elastic waves are initially generated from a known source location, the region where the slower modes are present at any given time is bounded. Any slower modes found outside this region must therefore be caused by mode conversion from the faster waves interacting with a hidden tamper. Due to the symmetric and antisymmetric nature of the two types of modes, converting between the two requires some asymmetry around the mid-plane of the plate. An example of how a tamper could introduce asymmetry is as follows: a section of the container could be removed to gain internal access, and later replaced by a plate of slightly greater thickness; it could then be polished flat on the outside wall to hide it from detection. However, it cannot be polished on the inside wall as the container is sealed, hence there will exist an indentation or protrusion on the inaccessible inner side; this feature is asymmetric and induces mode conversion. A schematic diagram of such a hidden tamper is given in Figure 1, which shows that the tamper is asymmetric about the mid-plane.

Figure 1. An exaggerated diagram showing a 'hidden tamper'. The tamper is not symmetrically placed about the mid-plane and will therefore induce mode conversion between the two types of guided elastic waves.

It has been shown [5] that the stress field near traction-free elastic corners with an internal angle greater than π are locally singular; in Figure 1, two such corners are present. This leads us to believe that the corners could have a large effect on the solution of the problem and their asymmetric positioning about the mid-plane implies that this effect will include mode conversion. The local stress fields near a traction-free corner of internal angle less than π are not singular, but are irregular. This means that they are bounded in the vicinity of the corner, but behave locally as a non-integer power of the distance from the corner. For sufficiently low frequencies, the wavelengths of the propagating modes are much longer than the corner lengthscales, therefore the local behaviour around the corner is unimportant away from it; however, even in such low-frequency regimes, the behaviour near sharp corners strongly affects the accuracy of methods for determining the global solution [6]; we shall see that this is true even for irregular, but non-singular, corner behaviour.

In this article, as a first step towards the tamper model, and as a useful way to demonstrate the novel solution method, we examine a simpler problem that incorporates corner behaviour. We will consider the reflection of an incoming time-harmonic, symmetric wave in a semi-infinite elastic waveguide, as shown in Figure 2. In this problem we have symmetry about the mid-plane and hence no asymmetric mode conversion; however, it will allow us to study how sharp corners affect the reflection problem. The corners in this problem each have an internal angle less than π, which means that the stress field around them is bounded. This problem may be solved by traditional methods, such as spectral collocation [7], allowing us to generate solutions to compare against; hence, this is a good test case to implement and validate our new method to deal with irregular corner behaviour. In addition to this goal, this canonical problem is also associated with the important question of the existence of a trapped mode, which in this case occurs at the Lamé frequency and at a particular value of Poisson's ratio, which we will discuss in Section 3.2 [7–9].

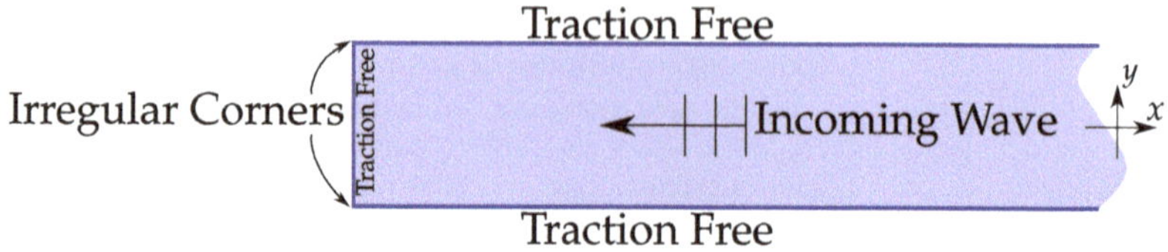

Figure 2. A simple canonical problem, containing two corners with locally irregular behaviour, at the points of intersection of the traction-free surfaces. A propagating symmetric Lamb wave is incoming from the right, and we wish to determine the resulting scattered field. The method we introduce will also work for an incoming asymmetric Lamb wave.

We wish to find the solution to this problem in terms of the in-plane natural modes of elastic waveguides, as these are the waves that are of experimental interest. These were first studied by Lamb over one hundred years ago [10] and are still an active area of research due to their very wide range of applications in areas such as nondestructive testing, seismology, and vibration control. The two sets of modes (in this article, we may ignore the third set of modes, the out-of-plane polarized shear waves, which are decoupled from the other two in the problem of interest) are the faster symmetric and the slower anti-symmetric modes, the wavenumbers of which satify different dispersion relations. Finding the roots of these relations is non-trivial, but methods relying on finite-product formulae [11], asympototic analysis, or projection onto Chebyshev polynomials [7] can be used efficiently to this end. The solutions to each dispersion relation consist of a single propagating mode that always exists (the A_0 or S_0 mode), a finite number of higher propagating modes that cut on depending on frequency, and an infinite number of evanescent modes.

It has been known for some time [6,12] that the solution to the present canonical problem in terms of a modal expansion of Lamb waves is slowly convergent near the free end, and poorly represents the behaviour of the field near the corners. This is essentially due to the Gibbs phenomenon. There are a variety of numerical treatments that can be applied in this and other waveguide problems to aid convergence, including the Lanczos σ-approximation (see, e.g., [13,14]).

Note that difficulties arising from geometric singularities in waveguides when using a naive modal expansion do not only occur in the elastodynamic case considered presently. In fact, the corner mode expansion method presented in this paper was first conceived in the context of easier, scalar, acoustic problems [15]. Moreover, other authors have considered such problems in, for instance, acoustic waves [16,17], water waves [18,19], horizontal shear waves [20], and electromagnetic waves [21,22]. Most of these methods take an ingenious numerical approach relying on the use of an expansion in terms of a carefully chosen family of orthogonal polynomials (Chebyshev, Gegenbauer, Jacobi, etc.), which can effectively be used to tackle certain corner singularities and 'fix' the naive modal expansions.

In the present work, rather than trying (in vain) to find known special functions with the required singular behaviour, we let them emerge naturally from the physical problem, giving rise to what we call the corner modes. This method has the advantage of not requiring an a priori knowledge of the singularities at hand and can accommodate rather complicated singular behaviours.

As hinted above, the aim of this paper is to apply a new approach tailored to the specific physical nature of the irregularity. We will introduce new modes that accurately represent the irregular behaviour near the corners and add them to the Lamb wave terms, which well represent the smooth part of the field elsewhere. This paper offers an application of the new method in the context of elastodynamics, whilst contemporaneous applications are presented in [15], for acoustic problems, and [23], for a model problem in elastostatics. However, the present work also requires an interesting additional step. That is, it is convenient for mathematical purposes to imagine that the plate shown in Figure 2 is infinite in extent, and has specified traction boundary forcing on the top and bottom faces in $x < 0$ to ensure that the prescribed boundary conditions on $x = 0$, $|y| \leq 1/2$ are satisfied (i.e., that the total normal and shear stresses on $x = 0$ are zero.) This geometry is illustrated in Figure 3, noting that the extension of the plate is a purely virtual construct; however, from Green's theorem, it can easily be shown that there will always exist forcing functions, $p(x), q(x), r(x), s(x)$ that ensure the correct boundary conditions on the actual plate edge, $x = 0$.

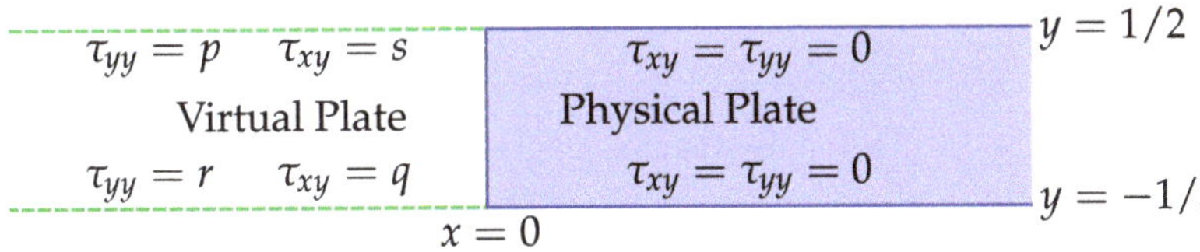

Figure 3. The construction we use to implement corner modes. The shaded physical plate has traction free conditions on top and bottom, but the condition on the face $x = 0$ will be applied later. The dashed lines denote an extended 'virtual' plate where we are imposing tractions on the surfaces which capture the correct local form of the stresses in the corners.

The remainder of this paper is organised as follows. In Section 2.1, we recast the problem as an infinite plate, with (as yet unknown) boundary tractions on the faces of the semi-infinite virtual extension of the plate. We then write the solution to this new problem in Fourier space. In Section 2.2, we find an asymptotic representation of the new unknown boundary conditions in terms of the behaviour associated with the corners. Next, we summarise the known similarity solutions of the stress field associated with a static elastic corner before going on to implement an asymptotic technique to recover the similarity solutions for the stresses in an elastodynamic corner. In Section 2.3, we discuss how the corner behaviour, as determined in Section 2.2, can be expressed as a sum of corner modes. These new corner modes will enable us to write the field as an expansion of the form

$$\text{Field} = \sum \text{Lamb Modes} + \sum \text{Corner Modes}.$$

In this representation, the Lamb modes capture well the far field behaviour, which is of interest experimentally, while the corner modes accurately describe the near field behaviour, which is important when applying the edge boundary conditions to determine the reflection coefficients. We then highlight the excellent convergence properties of such an expansion in Section 3. We present some results obtained by including both corner and Lamb modes, and illustrate the superiority of our method (compared to the pure Lamb mode expansion) when it comes to the reconstruction of the stresses on the free end of the plate and the prediction of the propagating waves reflection coefficients. We conclude by showing that our new method can predict the trapped-mode appearing at the Lamé frequency and find the specific value of the Poisson ratio at which it occurs as approximately $\nu = 0.224798$.

2. Materials and Methods

2.1. A Useful 'Equivalent' Problem: A Forced Infinite Elastic Plate

Virtual Plates

As mentioned above, we know from Green's theorem that the problem shown in Figure 3, with certain prescribed tractions $s(x)$, $p(x)$, $q(x)$, and $r(x)$, is equivalent, in $x \geq 0$, to the posed semi-infinite free-end model.

We have introduced this auxiliary problem because it has two useful properties that the original problem does not possess. First, it easily admits solutions in terms of the unknown tractions by way of Fourier transform, as we will demonstrate. Second, by approximating the prescribed tractions, it will allow us to find a convenient way of inputting the desired near-corner behaviour into a modal expansion.

The first step is to solve the equivalent problem in terms of the prescribed tractions. We begin with the time-harmonic (angular frequency ω) potential forms of the Navier–Lamé equations, and we non-dimensionalize spatial variables using the thickness of the plate, h, as a suitable lengthscale. The dimensionless potentials ϕ and ψ are related to the dimensionless displacement in the x-direction, u, and in the y-direction, v, by

$$u = \frac{\partial \phi}{\partial x} + \frac{\partial \psi}{\partial y}, \tag{1}$$

$$v = \frac{\partial \phi}{\partial y} - \frac{\partial \psi}{\partial x}. \tag{2}$$

the dimensionless potentials are governed by the Helmholtz equations

$$(\nabla^2 + k_l^2)\phi = 0, \tag{3}$$

$$(\nabla^2 + k_t^2)\psi = 0, \tag{4}$$

where k_l and k_t are, respectively, the dimensionless free space longitudinal and transverse wavenumbers for the material, given in terms of the plate density ρ and the Lamé constants λ and μ, as $k_l = h\tilde{k}_l = h\sqrt{\rho\omega^2/(\lambda+2\mu)}$ and $k_t = h\tilde{k}_t = h\sqrt{\rho\omega^2/\mu}$. One of these wavenumbers, along with Poisson's ratio $\nu = \lambda/(2\lambda+2\mu)$, say, may be used to fully define our elastic material for a given frequency and plate thickness. However, it is slightly more convenient to introduce the constant K as the ratio of the transverse wavenumber to the longitudinal wavenumber, so that $K^2 = k_t^2/k_l^2 = (2-2\nu)/(1-2\nu)$. The stresses are then related to the potentials in the usual way by

$$\tau_{xx} = K^2\frac{\partial^2\phi}{\partial x^2} + (K^2-2)\frac{\partial^2\phi}{\partial y^2} + 2\frac{\partial^2\psi}{\partial x\partial y}, \tag{5}$$

$$\tau_{xy} = 2\frac{\partial^2\phi}{\partial x\partial y} + \frac{\partial^2\psi}{\partial y^2} - \frac{\partial^2\psi}{\partial x^2}, \tag{6}$$

$$\tau_{yy} = (K^2-2)\frac{\partial^2\phi}{\partial x^2} + K^2\frac{\partial^2\phi}{\partial y^2} - 2\frac{\partial^2\psi}{\partial x\partial y}. \tag{7}$$

We seek solutions to the boundary value problem shown in Figure 3 by use of a Fourier transform defined by

$$F(\alpha) = \int_{-\infty}^{\infty} f(x)\mathrm{e}^{-\mathrm{i}\alpha x}\mathrm{d}x, \tag{8}$$

where α is the spectral variable. The Fourier transformed Helmholtz equations are given by

$$\left(\frac{\partial^2}{\partial y^2} + (k_l^2-\alpha^2)\right)\Phi = 0, \tag{9}$$

$$\left(\frac{\partial^2}{\partial y^2} + (k_t^2-\alpha^2)\right)\Psi = 0, \tag{10}$$

where the capital letters Φ and Ψ are the Fourier transforms of ϕ and ψ, respectively. The general solutions to these equations are given by

$$\Phi(\alpha, y) = A\cos(\beta_l y) + B\sin(\beta_l y), \tag{11}$$

$$\Psi(\alpha, y) = C\cos(\beta_t y) + D\sin(\beta_t y), \tag{12}$$

where we have defined

$$\beta_l^2 = (k_l^2 - \alpha^2), \tag{13}$$

$$\beta_t^2 = (k_t^2 - \alpha^2). \tag{14}$$

the stresses in Fourier space are denoted by T, and are given by

$$T_{xx}(\alpha,y) = -\alpha^2 K^2\Phi + (K^2-2)\frac{\partial^2\Phi}{\partial y^2} + 2i\alpha\frac{\partial\Psi}{\partial y}, \tag{15a}$$

$$T_{xy}(\alpha,y) = 2i\alpha\frac{\partial\Phi}{\partial y} + \frac{\partial^2\Psi}{\partial y^2} + \alpha^2\Psi, \tag{15b}$$

$$T_{yy}(\alpha,y) = -\alpha^2(K^2-2)\Phi + K^2\frac{\partial^2\Phi}{\partial y^2} - 2i\alpha\frac{\partial\Psi}{\partial y}. \tag{15c}$$

the Fourier transformed boundary conditions are $T_{yy}(\alpha,1/2) = P^+(\alpha)$, $T_{xy}(\alpha,1/2) = S^+(\alpha)$, $T_{yy}(\alpha,-1/2) = R^+(\alpha)$, and $T_{xy}(\alpha,-1/2) = Q^+(\alpha)$, where

$$T_{xy}(\alpha,1/2) = S^+(\alpha) = \int_{-\infty}^{0} s(x)e^{-i\alpha x}dx, \tag{16}$$

etc., and the superscript '+' ('−') notation denotes a 'plus' ('minus') function, that is, a function that is analytic in the upper (lower) half of the complex α-plane. Note that, from the theory of half-range Fourier transforms [24], we know that such well-behaved functions are analytic in the upper or lower half space of the α-plane.

We may use the symmetry of the general solutions to simplify the boundary conditions. Using the linear combination of boundary conditions given by $T_{xy}(\alpha,1/2) - T_{xy}(\alpha,-1/2) = S^+(\alpha) - Q^+(\alpha)$ and $T_{yy}(\alpha,1/2) + T_{yy}(\alpha,-1/2) = P^+(\alpha) + R^+(\alpha)$, we find that the unknown coefficients A and D can be found by solving a system decoupled from the system governing B and C. The solutions for A and D are known as the symmetric part and must satisfy the linear system given by

$$\begin{bmatrix} (\alpha^2-\beta_t^2)\cos\left(\frac{\beta_l}{2}\right) & -2i\alpha\beta_t\cos\left(\frac{\beta_t}{2}\right) \\ -2i\alpha\beta_l\sin\left(\frac{\beta_l}{2}\right) & (\alpha^2-\beta_t^2)\sin\left(\frac{\beta_t}{2}\right) \end{bmatrix} \begin{bmatrix} A \\ D \end{bmatrix} = \frac{1}{2}\begin{bmatrix} P^+(\alpha)+R^+(\alpha) \\ S^+(\alpha)-Q^+(\alpha) \end{bmatrix}. \tag{17}$$

for the antisymmetric part, we use the combination of conditions given by $T_{xy}(\alpha,1/2) + T_{xy}(\alpha,-1/2) = S^+(\alpha) + Q^+(\alpha)$ and $T_{yy}(\alpha,1/2) - T_{yy}(\alpha,-1/2) = P^+(\alpha) - R^+(\alpha)$. This generates the linear system given by

$$\begin{bmatrix} (\alpha^2-\beta_t^2)\sin\left(\frac{\beta_l}{2}\right) & 2i\alpha\beta_t\sin\left(\frac{\beta_t}{2}\right) \\ 2i\alpha\beta_l\cos\left(\frac{\beta_l}{2}\right) & (\alpha^2-\beta_t^2)\cos\left(\frac{\beta_t}{2}\right) \end{bmatrix} \begin{bmatrix} B \\ C \end{bmatrix} = \frac{1}{2}\begin{bmatrix} P^+(\alpha)-R^+(\alpha) \\ S^+(\alpha)+Q^+(\alpha) \end{bmatrix}. \tag{18}$$

In linear theory, the symmetric and antisymmetric solutions do not interact at a symmetric boundary. Therefore, if we assume a symmetric incoming wave, the contributions from the antisymmetric Lamb waves must be zero. This implies that $Q^+(\alpha) = -S^+(\alpha)$ and $R^+(\alpha) = P^+(\alpha)$. Similarly for the antisymmetric case, the symmetric waves must be zero, which requires $Q^+(\alpha) = S^+(\alpha)$ and $R^+(\alpha) = -P^+(\alpha)$.

We will present the method for symmetric waves here; the antisymmetric case follows similarly. By taking the inverse of the first matrix in (17), we find that A and D are functions of α given by

$$A(\alpha) = \frac{(\alpha^2-\beta_t^2)\sin\left(\frac{\beta_t}{2}\right)P^+(\alpha) + 2\alpha i\beta_t\cos\left(\frac{\beta_t}{2}\right)S^+(\alpha)}{\mathcal{D}_s(\alpha)}, \tag{19}$$

$$D(\alpha) = \frac{2\alpha i\beta_l\sin\left(\frac{\beta_l}{2}\right)P^+(\alpha) + (\alpha^2-\beta_t^2)\cos\left(\frac{\beta_l}{2}\right)S^+(\alpha)}{\mathcal{D}_s(\alpha)}, \tag{20}$$

where

$$\mathcal{D}_s(\alpha) = (\alpha^2-\beta_t^2)^2\cos\left(\frac{\beta_l}{2}\right)\sin\left(\frac{\beta_t}{2}\right) + 4\alpha^2\beta_l\beta_t\cos\left(\frac{\beta_t}{2}\right)\sin\left(\frac{\beta_l}{2}\right) \tag{21}$$

is the determinant of the matrix to be inverted in (17). The relation $\mathcal{D}_s(\alpha) = 0$ is known as the dispersion relation for symmetric Lamb waves. We may now use the expressions for A and D to find the transformed potentials and stresses. The symmetric part of the transformed potentials are given by

$$\Phi(\alpha, y) = \frac{(\alpha^2 - \beta_t^2)\sin\left(\frac{\beta_t}{2}\right)P^+(\alpha) + 2i\alpha\beta_t\cos\left(\frac{\beta_t}{2}\right)S^+(\alpha)}{\mathcal{D}_s(\alpha)}\cos(\beta_l y), \tag{22}$$

$$\Psi(\alpha, y) = \frac{2i\alpha\beta_l\sin\left(\frac{\beta_l}{2}\right)P^+(\alpha) + (\alpha^2 - \beta_t^2)\cos\left(\frac{\beta_l}{2}\right)S^+(\alpha)}{\mathcal{D}_s(\alpha)}\sin(\beta_t y). \tag{23}$$

From the potentials, we find the transformed stresses to be

$$T_{xx}(\alpha, y) = \frac{F_1(\alpha, y)S^+(\alpha) + F_2(\alpha, y)P^+(\alpha)}{\mathcal{D}_s(\alpha)}, \tag{24}$$

$$T_{xy}(\alpha, y) = \frac{G_1(\alpha, y)S^+(\alpha) + G_2(\alpha, y)P^+(\alpha)}{\mathcal{D}_s(\alpha)}, \tag{25}$$

where $\mathcal{D}_s$, F_1, F_2, G_1, and G_2 are analytic functions given by (21),

$$F_1(\alpha, y) = 2i\alpha\beta_t\left(\left(\alpha^2 - \beta_t^2\right)\cos\left(\frac{\beta_l}{2}\right)\cos(\beta_t y) - \cos\left(\frac{\beta_t}{2}\right)\left(K^2\alpha^2 + (K^2 - 2)\beta_l^2\right)\cos(\beta_l y)\right), \tag{26}$$

$$F_2(\alpha, y) = -\left(\alpha^2 - \beta_t^2\right)\sin\left(\frac{\beta_t}{2}\right)\left(K^2\alpha^2 + (K^2 - 2)\beta_l^2\right)\cos(\beta_l y) - 4\alpha^2\beta_l\beta_t\sin\left(\frac{\beta_l}{2}\right)\cos(\beta_t y), \tag{27}$$

$$G_1(\alpha, y) = \left(\alpha^2 - \beta_t^2\right)^2\cos\left(\frac{\beta_l}{2}\right)\sin(\beta_t y) + 4\alpha^2\beta_l\beta_t\cos\left(\frac{\beta_t}{2}\right)\sin(\beta_l y) \tag{28}$$

and

$$G_2(\alpha, y) = \left(\alpha^2 - \beta_t^2\right)\left(2i\alpha\beta_l\sin\left(\frac{\beta_l}{2}\right)\sin(\beta_t y) - 2i\alpha\beta_l\sin\left(\frac{\beta_t}{2}\right)\sin(\beta_l y)\right). \tag{29}$$

The solution for the Potentials (22) and (23), or stresses (24) and (25), fully describes the problem in terms of the Fourier transforms of the unknown functions s, p, q, and r. Whilst we do not know these functions exactly, we can approximate them from knowledge of their behaviour near a corner, which we will examine in Section 2.2.

2.2. Similarity Solutions in the Corner of an Elastic Body

We now seek to establish the form of the unknown boundary conditions on the virtual extension of the plate. To do this, let us consider the physical problem local to a corner of an elastic wedge. We know that, near a corner, these conditions will be formed of regular contributions from the Lamb waves and irregular contributions from the similarity solutions associated with this corner. As the Lamb waves have zero traction on the virtual boundaries, the conditions will be dominated by the similarity solutions, the derivation of which will be the subject of this section. In doing so, we will directly obtain the behaviour that the Lamb wave representation has trouble capturing.

The corners in this model problem both have the same internal angle, $\pi/2$; however, there is little extra complexity in presenting the solutions for arbitrary internal angle; hence this is what will be presented here. Figure 4 shows an infinite arbitrary-angled traction-free corner. We consider the two traction-free edges to meet at an angle ξ (where ξ is the interior angle). We introduce a planar polar coordinate system with $\rho = 0$ at the corner and with the line $\theta = 0$ as the bisector of the angle ξ. Choosing the bisector as the $\theta = 0$ coordinate curve allows us to consider two families of solutions (symmetric and antisymmetric in θ) separately, which simplifies the algebra. The normal vector to each edge lies in the θ direction, therefore the traction-free conditions are $\tau_{\theta\theta}|_{\pm\xi/2} = 0$ and $\tau_{\rho\theta}|_{\pm\xi/2} = 0$.

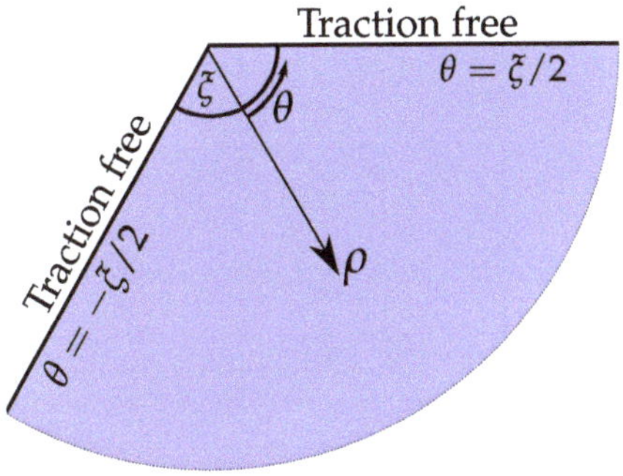

Figure 4. A general traction-free elastic corner. The elastic body is the shaded area extending to infinity.

2.2.1. Elastostatic Expansion

It has been known for over 50 years that static deformations (eigensolutions) in a two-dimensional corner of an elastic material can be found by use of an Airy stress function, an approach that is well understood for similar problems (see, e.g., [25]). We shall examine this briefly here, and then move on to the more-relevant elastodynamic case afterwards. Following [5], we pose an Airy stress function of the form $\chi(\rho, \theta) = \rho^{\gamma_m+1}\Theta(\theta)$, where $\gamma_m \in \mathbb{C}$, from which we can recover the stresses as

$$\tau_{\rho\theta} = -\frac{\partial}{\partial\rho}\left(\frac{1}{\rho}\frac{\partial\chi}{\partial\theta}\right), \qquad \tau_{\theta\theta} = \frac{\partial^2\chi}{\partial\rho^2}. \tag{30}$$

Then, Navier's equation is satisfied, and compatibility yields $\nabla^4\chi = 0$, which requires that the θ dependence is of the form

$$\Theta(\theta) = c_1\cos((\gamma_m+1)\theta) + c_2\sin((\gamma_m+1)\theta) + c_3\cos((\gamma_m-1)\theta) + c_4\sin((\gamma_m-1)\theta). \tag{31}$$

We now apply the boundary conditions, and use the symmetry properties of the θ dependence as given in (31) to simplify the problem. We do this by considering the linear combinations of conditions given by $\tau_{\rho\theta}|_{\xi/2} \pm \tau_{\rho\theta}|_{-\xi/2} = 0$ and $\tau_{\theta\theta}|_{\xi/2} \pm \tau_{\theta\theta}|_{-\xi/2} = 0$, which yield two decoupled linear systems of equations given by

$$\begin{bmatrix} \cos((\gamma_m+1)\xi/2) & \cos((\gamma_m-1)\xi/2) \\ (\gamma_m+1)\sin((\gamma_m+1)\xi/2) & (\gamma_m-1)\sin((\gamma_m-1)\xi/2) \end{bmatrix} \begin{bmatrix} c_1 \\ c_3 \end{bmatrix} = \begin{bmatrix} 0 \\ 0 \end{bmatrix}, \tag{32}$$

$$\begin{bmatrix} \sin((\gamma_m+1)\xi/2) & \sin((\gamma_m-1)\xi/2) \\ (\gamma_m+1)\cos((\gamma_m+1)\xi/2) & (\gamma_m-1)\cos((\gamma_m-1)\xi/2) \end{bmatrix} \begin{bmatrix} c_2 \\ c_4 \end{bmatrix} = \begin{bmatrix} 0 \\ 0 \end{bmatrix}. \tag{33}$$

In order for these conditions to possess non-trivial solutions to the homogeneous problem, the determinants must be equal to zero, which yields constraints on the corner exponents:

$$\sin(\gamma_m\xi) \pm \gamma_m\sin(\xi) = 0. \tag{34}$$

In (34), a positive sign corresponds to the determinant of the first system (32) and an Airy stress function symmetric around $\theta = 0$, while the negative sign corresponds to the determinant of the second system (33) and an Airy stress function antisymmetric about $\theta = 0$. In order for the solutions to be physical, we must also have finite displacements, which limits the solutions to those where $\Re[\gamma_m] > 0$.

The roots of (34), which can easily be found numerically, are labelled γ_m and ordered according to the size of the real parts, and also by imaginary parts when two roots have the same real part. For each solution of the compatibility condition (34) we find a non-trivial solution to the homogeneous problem with a single unknown constant; we can therefore write a general similarity solution as the series

$$\tau_{\rho\theta} = \sum_{m=1}^{M} \eta_m f_0(\gamma_m, \theta)\rho^{\gamma_m - 1}, \tag{35}$$

$$\tau_{\theta\theta} = \sum_{m=1}^{M} \eta_m g_0(\gamma_m, \theta)\rho^{\gamma_m - 1}, \tag{36}$$

for arbitrary number of terms M. Here f_0 and g_0 are known functions and η_m are arbitrary unknown constants, each associated with the mth similarity solution.

2.2.2. Elastodynamic Expansion

We now wish to consider the local problem relevant to this study, i.e., the local dynamic (irregular) behaviour in the vicinity of a corner of the form shown in Figure 4. That is, we want to find (eigen-)solutions to the dynamic Navier-Lamé equation in this corner domain with traction-free boundary conditions. We begin by noting that in the limit where the distance to the corner becomes small, $\rho \to 0$, the dynamic terms become small. This implies that the leading-order contributions in the elastodynamic case should correspond to the similarity solutions obtained in the static case.

In this section, we will show that for any given $L > 0$, the asymptotic expansions of the elastodynamic stresses $\tau_{\rho\theta}$ and $\tau_{\theta\theta}$ as $\rho \to 0$ are given by

$$\tau_{\rho\theta} = \sum_{m=1}^{M(L)} \eta_m \mathfrak{f}_m(\rho, \theta; L) + o(\rho^L) \quad \text{and} \quad \tau_{\theta\theta} = \sum_{m=1}^{M(L)} \eta_m \mathfrak{g}_m(\rho, \theta; L) + o(\rho^L), \tag{37}$$

where $M(L)$ is the largest positive integer m such that $\Re[\gamma_m - 1] < L$, and

$$\mathfrak{f}_m(\rho, \theta; L) = \sum_{\ell=0}^{\mathcal{L}_m(L)} f_\ell(\gamma_m, \theta)\rho^{\gamma_m - 1 + 2\ell} \quad \text{and} \quad \mathfrak{g}_m(\rho, \theta; L) = \sum_{\ell=0}^{\mathcal{L}_m(L)} g_\ell(\gamma_m, \theta)\rho^{\gamma_m - 1 + 2\ell}, \tag{38}$$

where $\mathcal{L}_m(L)$ is the largest integer ℓ such that $\Re[\gamma_m - 1 + 2\ell] < L$ and the functions f_ℓ and g_ℓ are known exactly. Hence, the only unknowns in the expansions (37) are the coefficients η_m. Note that for this expansion to be formally correct, we need to choose a $L > 0$ that is not equal to the real part of any of the static roots γ_m.

In order to justify this expansion, we again use the Helmholtz decomposition, but this time non-dimensionalised on the longitudinal dimensional wavenumber, $\tilde{k}_l = \sqrt{\rho\omega^2/(\lambda + 2\mu)}$, as there is no natural lengthscale for an infinite wedge. We again write the displacements in terms of potentials, as in (1) and (2), but here the Navier–Lamé equations reduce to

$$\nabla^2\phi + \phi = 0 \quad \text{and} \quad \nabla^2\psi + K^2\psi = 0, \tag{39}$$

where, as before, $K^2 = \tilde{k}_t^2/\tilde{k}_l^2 = (2 - 2\nu)/(1 - 2\nu)$.

Using separation of variables, we find that the non-dimensional potentials ϕ_m and ψ_m corresponding to a given static root γ_m, which solve (39), are given by

$$\phi_m(\rho, \theta) = \sum_{\ell \geqslant 0} J_{\ell_m(\ell)}(\rho)[a_{\ell_m(\ell)} \cos(\ell_m(\ell)\theta) + b_{\ell_m(\ell)} \sin(\ell_m(\ell)\theta)], \tag{40}$$

$$\psi_m(\rho, \theta) = \sum_{\ell \geqslant 0} J_{\ell_m(\ell)}(K\rho)[c_{\ell_m(\ell)} \cos(\ell_m(\ell)\theta) + d_{\ell_m(\ell)} \sin(\ell_m(\ell)\theta)], \tag{41}$$

where, for convenience, we have introduced the reduced index

$$\ell_m(\ell) = \gamma_m - 1 + 2\ell, \tag{42}$$

and $J_{\ell_m(\ell)}$ is the Bessel function of the first kind of order $\ell_m(\ell)$. The unknown coefficients $a_{\ell_m(\ell)}$, $b_{\ell_m(\ell)}$, $c_{\ell_m(\ell)}$, and $d_{\ell_m(\ell)}$ will be determined by imposing traction-free boundary conditions.

The normal and shear stresses in polar coordinates, $\tau^m_{\rho\theta}$ and $\tau^m_{\theta\theta}$, may now be written in terms of these potentials as

$$\tau^m_{\rho\theta}(\rho,\theta) = 2\left(-\frac{1}{\rho^2}\frac{\partial\phi_m}{\partial\theta} + \frac{1}{\rho}\frac{\partial^2\phi_m}{\partial\rho\partial\theta}\right) - 2\frac{\partial^2\psi_m}{\partial\rho^2} - K^2\psi_m$$

$$\tau^m_{\theta\theta}(\rho,\theta) = -2\frac{\partial^2\phi_m}{\partial\rho^2} - K^2\phi_m - 2\left(-\frac{1}{\rho^2}\frac{\partial\psi_m}{\partial\theta} + \frac{1}{\rho}\frac{\partial^2\psi_m}{\partial\rho\partial\theta}\right),$$

and traction-free conditions for the corner under consideration are $\tau^m_{\rho\theta}(\rho,\pm\xi/2) = \tau^m_{\theta\theta}(\rho,\pm\xi/2) = 0$. It is convenient to rewrite these conditions as

$$\tau^m_{\rho\theta}(\rho,\xi/2) - \tau^m_{\rho\theta}(\rho,-\xi/2) = 0, \quad \tau^m_{\theta\theta}(\rho,\xi/2) + \tau^m_{\theta\theta}(\rho,-\xi/2) = 0, \tag{43}$$

$$\tau^m_{\rho\theta}(\rho,\xi/2) + \tau^m_{\rho\theta}(\rho,-\xi/2) = 0, \quad \tau^m_{\theta\theta}(\rho,\xi/2) - \tau^m_{\theta\theta}(\rho,-\xi/2) = 0. \tag{44}$$

Note that the conditions (43) only involve the unknowns $a_{\ell_m(\ell)}$ and $d_{\ell_m(\ell)}$, while the conditions (44) only involve the unknowns $b_{\ell_m(\ell)}$ and $c_{\ell_m(\ell)}$.

Let us now, for brevity, consider a root γ_m that is associated with a symmetric Airy stress functions, i.e., a root of (34) with the + sign. The case of an anti-symmetric root can be dealt with similarly. It will be helpful to introduce the quantity C^m_ℓ, defined for $\ell \geqslant 0$ by

$$C^m_\ell = a_{\ell_m(\ell)} + K^{\ell_m(\ell)} d_{\ell_m(\ell)}. \tag{45}$$

Rewriting the Bessel functions as series expansions about $\rho = 0$, and collecting the terms of order $\rho^{\gamma_m - 3}$ in (43) (to do so, one only needs to consider the terms with $\ell = 0$ in (40) and (41)), we obtain

$$\sin(\ell_m(0)\xi/2)C^m_0 = 0 \quad \text{and} \quad \cos(\ell_m(0)\xi/2)C^m_0 = 0, \tag{46}$$

implying that $C^m_0 = 0$, and giving an explicit link between $a_{\ell_m(0)}$ and $d_{\ell_m(0)}$. Note that this implies that, as expected from the static case, our expansion does not actually have any terms of order $\rho^{\gamma_m - 3}$.

Using the relationship $C^m_0 = 0$, and collecting this time the terms of order $\rho^{\gamma_m - 1}$ in (43) (to do so, one only needs to consider the terms with $\ell = 0$ and 1 in (40) and (41)), we obtain

$$-d_{\ell_m(0)} K^{\ell_m(0)} (K^2 - 1)\ell_m(0)\cos(\ell_m(0)\xi/2) - C^m_1 \sin(\ell_m(1)\xi/2) = 0, \tag{47}$$

$$d_{\ell_m(0)} K^{\ell_m(0)} (K^2 - 1)\ell_m(1)\sin(\ell_m(0)\xi/2) - C^m_1 \cos(\ell_m(1)\xi/2) = 0. \tag{48}$$

This is a linear system of two equations for the two unknowns $d_{\ell_m(0)}$ and C^m_1, whose associated determinant can be shown to be $\sin(\gamma_m\xi) + \gamma_m \sin(\xi)$, which is known to be equal to zero given our choice of γ_m. Hence, this system admits non trivial solutions that take the form of an explicit expression for $d_{\ell_m(0)}$ in terms of C^m_1. Note that we recover here the static behaviour of Section 2.2.1.

Let us now collect the terms of order $\rho^{\gamma_m + 1}$ in (43) (to do so, one only needs to consider the terms with $\ell = 0, 1$ and 2 in (40) and (41)). This leads to two independent linear equations for the three unknowns $a_{\ell_m(1)}$, $d_{\ell_m(1)}$ and C^m_2. This system is always solvable in terms of the single constant C^m_2.

The process of increasing the order is now algorithmic. In a consistent expansion of the stresses up to and including the order $\rho^{\ell_m(\ell)}$ for $\ell \geqslant 1$, all the unknown coefficients $a_{\ell_m(1,\ldots,\ell)}$, and $b_{\ell_m(1,\ldots,\ell)}$ are known exactly in terms of a single unknown constant $C^m_{\ell+1}$. If we now follow exactly the same process, but for the boundary conditions (44), we will find that the coefficients $b_{\ell_m(\ell)}$ and $c_{\ell_m(\ell)}$ are all equal to zero. Hence, whatever the asymptotic order we want to reach (determined by a choice of $L > 0$), the consistent expansion obtained for the potentials corresponding to each static root γ_m only involves a single unknown coefficient. This is why we can reconstruct the stresses as in (37) and (38), where the f_ℓ and g_ℓ functions are known exactly. The exact form of these functions is found by an automatic implementation of the asymptotic procedure described above, which was implemented by the authors using the mathematical software package Mathematica.

2.3. *Expressing the Virtual Forcing in Terms of Corner Modes*

In Section 2.1, we found a Fourier-space solution in terms of functions describing the stress fields on the surfaces of the virtual extension to the plate. These functions can now be expressed in terms of the similarity solutions found in Section 2.2. Further, in this section we will use this information to introduce new modes that will prove extremely helpful in efficiently solving the present boundary value problem illustrated in Figure 2.

2.3.1. Inverse Transformation of the Forced Plate

We can now approximate the unknown functions from Section 2.1, s, p, q, and r, as a series of the similarity solutions associated with a corner of angle $\xi = \pi/2$, found in Section 2.2, which we can extend mathematically beyond the physical corner. We use polar coordinates consistent with that defined in Figure 4, i.e., angles are measured from the respective corner bisectors as shown in Figure 5. Hence, the virtual upper boundary is defined by $\theta = -3\pi/4$, whereas $\theta = 3\pi/4$ corresponds to the virtual lower boundary on the lines $y = \pm 1/2$, $x < 0$, and $\rho = -x$, so we can express the forcing terms in Cartesian form, which yield approximations, correct up to order $o((-x)^L)$, as

$$\tau_{xy}\big|_{y=\frac{1}{2}\ \&\ x<0} = s(x) \approx \sum_{m=1}^{M(L)} \eta_m \mathfrak{f}_m\left(-x, -\frac{3\pi}{4}; L\right), \tag{49}$$

$$\tau_{yy}\big|_{y=\frac{1}{2}\ \&\ x<0} = p(x) \approx \sum_{m=1}^{M(L)} \eta_m \mathfrak{g}_m\left(-x, -\frac{3\pi}{4}; L\right), \tag{50}$$

$$\tau_{xy}\big|_{y=-\frac{1}{2}\ \&\ x<0} = q(x) \approx \sum_{m=1}^{M(L)} \eta_m \mathfrak{f}_m\left(-x, \frac{3\pi}{4}; L\right), \tag{51}$$

$$\tau_{yy}\big|_{y=-\frac{1}{2}\ \&\ x<0} = r(x) \approx \sum_{m=1}^{M(L)} \eta_m \mathfrak{g}_m\left(-x, \frac{3\pi}{4}; L\right). \tag{52}$$

The solutions in Section 2.1 were found in terms of the Fourier transform of these virtual conditions, which we write as

$$S^+(\alpha) = \int_{-\infty}^{0} s(x)\mathrm{e}^{-\mathrm{i}\alpha x}\mathrm{d}x \approx \sum_{m=1}^{M(L)} \int_{-\infty}^{0} \eta_m \mathfrak{f}_m\left(-x, -\frac{3\pi}{4}; L\right)\mathrm{e}^{-\mathrm{i}\alpha x}\mathrm{d}x = \sum_{m=1}^{M(L)} \eta_m \mathfrak{F}_m^+\left(\alpha, \frac{-3\pi}{4}; L\right), \tag{53}$$

$$P^+(\alpha) = \int_{-\infty}^{0} p(x)\mathrm{e}^{-\mathrm{i}\alpha x}\mathrm{d}x \approx \sum_{m=1}^{M(L)} \int_{-\infty}^{0} \eta_m \mathfrak{g}_m\left(-x, -\frac{3\pi}{4}; L\right)\mathrm{e}^{-\mathrm{i}\alpha x}\mathrm{d}x = \sum_{m=1}^{M(L)} \eta_m \mathfrak{G}_m^+\left(\alpha, \frac{-3\pi}{4}; L\right), \tag{54}$$

$$Q^+(\alpha) = \int_{-\infty}^{0} q(x)\mathrm{e}^{-\mathrm{i}\alpha x}\mathrm{d}x \approx \sum_{m=1}^{M(L)} \int_{-\infty}^{0} \eta_m \mathfrak{f}_m\left(-x, \frac{3\pi}{4}; L\right)\mathrm{e}^{-\mathrm{i}\alpha x}\mathrm{d}x = \sum_{m=1}^{M(L)} \eta_m \mathfrak{F}_m^+\left(\alpha, \frac{3\pi}{4}; L\right), \tag{55}$$

$$R^+(\alpha) = \int_{-\infty}^{0} r(x)\mathrm{e}^{-\mathrm{i}\alpha x}\mathrm{d}x \approx \sum_{m=1}^{M(L)} \int_{-\infty}^{0} \eta_m \mathfrak{g}_m\left(-x, \frac{3\pi}{4}; L\right)\mathrm{e}^{-\mathrm{i}\alpha x}\mathrm{d}x = \sum_{m=1}^{M(L)} \eta_m \mathfrak{G}_m^+\left(\alpha, \frac{3\pi}{4}; L\right), \tag{56}$$

where the transforms can be determined analytically.

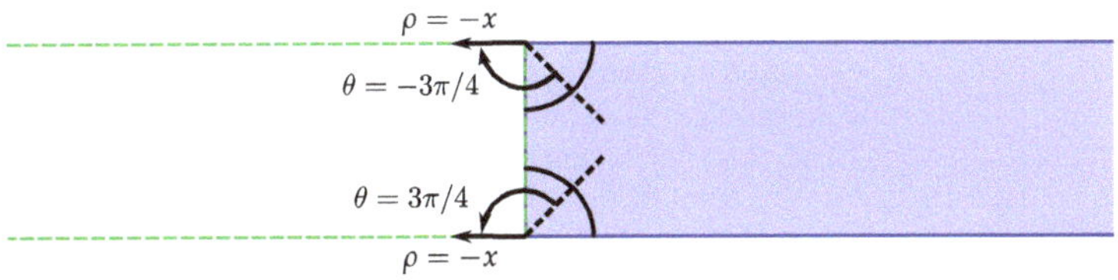

Figure 5. Location of the virtual surfaces in terms of the local corner coordinate system for each corner.

Combining the transforms of the virtual boundary conditions (53)–(56) and expressions for the stresses (24) and (25), we now have a complete solution in Fourier space, given in terms of an infinite series of similarity solutions, which have, built-in, the irregular corner behaviour. To find the expressions for the potentials and stresses in physical space, we use the inverse Fourier transform defined by

$$f(x) = \frac{1}{2\pi}\int_{-\infty}^{\infty} F(\alpha)\mathrm{e}^{\mathrm{i}\alpha x}\mathrm{d}\alpha. \tag{57}$$

We know that the integrands of the solution in this form will only contains pole-type singularities, not branch-cuts, and so we can evaluate these in $x \geqslant 0$ using Cauchy's residue theorem, by deforming the contour up to a finite imaginary part of α, along the contour $\mathcal{C}_1$, say, as shown in Figure 6. Note that it is straightforward to show that the contributions from the contours at $\pm\infty$ are zero. So, for example, we evaluate τ_{xy} as

$$\tau_{xy}(x,y) = \frac{1}{2\pi}\int_{\mathcal{C}} T_{xy}(\alpha,y)\mathrm{e}^{\mathrm{i}\alpha x}\mathrm{d}\alpha = \mathrm{i}\sum_{n=1}^{N}\operatorname*{Res}_{\alpha=\alpha_n}\left(T_{xy}(\alpha,y)\mathrm{e}^{\mathrm{i}\alpha x}\right) + \frac{1}{2\pi}\int_{\mathcal{C}_1} T_{xy}(\alpha,y)\mathrm{e}^{\mathrm{i}\alpha x}\mathrm{d}\alpha, \tag{58}$$

where α_n is the nth pole that lies between the real line and the contour $\mathcal{C}_1$. From the definition of T_{xy} in (25), these poles are only those associated with the Lamb wave dispersion relation $\mathcal{D}_s(\alpha) = 0$, as $S^+(\alpha)$ and $P^+(\alpha)$ are, by definition, known to be analytic in this region. Note that the original contour $\mathcal{C}$ is indented as shown in Figure 6 to include only the modes that obey the outgoing radiation condition for $x > 0$.

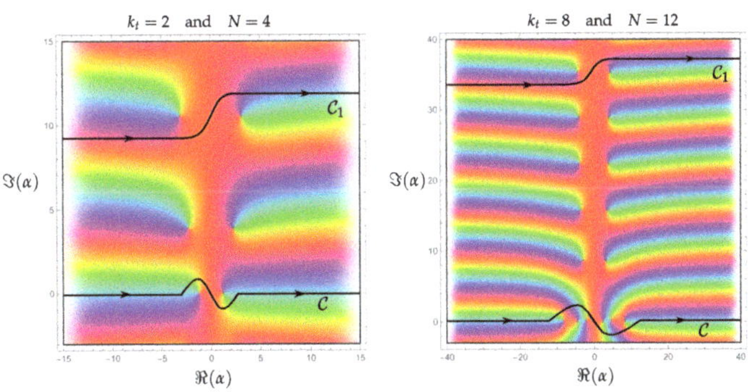

Figure 6. Phase portrait of the function $\beta_t(\alpha)/\mathcal{D}_s(\alpha)$ for $\nu = 0.3$, $k_t = 2$ (**left**) and $k_t = 8$ (**right**). It shows the location of the poles coinciding with the symmetric modes arising from the dispersion relation (21). The contours $\mathcal{C}$ and $\mathcal{C}_1$ extend horizontally to infinity. It can be shown that the functions decay for large $|\alpha|$ and so the contribution of the contours connecting $\mathcal{C}$ and $\mathcal{C}_1$ at infinity can be neglected. The contours are chosen to take only the rightward propagating and/or decaying modes. The total number of poles between the two contours is what we call N. On the left plot, $N = 4$, and 1 of the 4 modes is propagating, while on the right plot $N = 12$, and 3 of the 12 modes are propagating.

From (58), we observe that to obtain the stress field we may take the residue contributions from a discrete set of poles and add an integral contribution evaluated along the uplifted contour $\mathcal{C}_1$. We begin by examining the residues. We determine the poles, located at α_n, of the Lamb wave dispersion relation using the polynomial approximant method proposed by Chapman and Sorokin [11], and we also note that except when modes are cutting on or off all the poles are simple. We therefore find that the residues correspond to Lamb wave contributions, which can be evaluated as

$$\operatorname*{Res}_{\alpha=\alpha_n}\left(T_{xy}(\alpha,y)\right) = \frac{G_1(\alpha_n,y)S^+(\alpha_n) + G_2(\alpha_n,y)P^+(\alpha_n)}{\mathcal{D}_s'(\alpha_n)}, \tag{59}$$

where the $'$ denotes a derivative with respect to the argument, and we know that $\mathcal{D}_s'(\alpha_n)$ is non zero. Whilst this expression is easy to evaluate, we only know the behaviours of the imposed functions asymptotically for large $|\alpha|$ i.e., small $|x|$. We specified the behaviour on the virtual plates via the functions $s(x)$ and $p(x)$; their forms were chosen so that they gave the requisite behaviour near the corner using the asymptotic method set out in Section 2.2. From the properties of the Fourier transform we therefore know the forms of $S^+(\alpha)$ and $P^+(\alpha)$ for large $|\alpha|$. Hence, as the location of the poles (zeros of the Lamb wave dispersion relation) can lie relatively close to the origin, the values of $S^+(\alpha_n)$ and $P^+(\alpha_n)$ found in terms of corner modes will not be accurate. Hence, instead of evaluating them in this way, we will leave them as unknown constants for each α_n. Putting these into the expressions for the stresses, (24) and (25), allows us to write the nth Lamb mode term as

$$\tau_{xx}^n = \zeta_n t_x^l(\alpha_n, x, y), \tag{60}$$

$$\tau_{xy}^n = \zeta_n t_y^l(\alpha_n, x, y), \tag{61}$$

defined up to the as-yet unknown constant ζ_n. Note that there is only one constant for each Lamb mode because $G_1(\alpha_n, y)$ and $G_2(\alpha_n, y)$ reduce to the same function in y at any root of $\mathcal{D}_s(\alpha_n) = 0$.

We now evaluate the integral along the contour C_1. In contrast to the evaluation of the residue terms, we must choose this contour such that $|\alpha|$ is large and hence $S^+(\alpha)$ and $P^+(\alpha)$ will be known accurately in terms of a series expansion that captures the local behaviour of the fields near the top corner. The stresses associated with the corner modes can be expressed by substituting (53) and (54) into (22) and (23), and taking the inverse Fourier transform (57). This yields for τ_{xy}

$$\begin{aligned}\int_{C_1} T_{xy}(\alpha,y)\mathrm{e}^{\mathrm{i}\alpha x}\mathrm{d}\alpha &\approx \int_{C_1} \frac{G_1(\alpha,y)\sum_{m=1}^{M(L)}\eta_m\mathfrak{F}_m^+(\alpha,\frac{-3\pi}{4};L) + G_2(\alpha,y)\sum_{m=1}^{M(L)}\eta_m\mathfrak{G}_m^+(\alpha,\frac{-3\pi}{4};L)}{\mathcal{D}_s(\alpha)}\mathrm{e}^{\mathrm{i}\alpha x}\mathrm{d}\alpha \\ &= \sum_{m=1}^{M(L)}\eta_m\int_{C_1}\frac{G_1(\alpha,y)\mathfrak{F}_m^+(\alpha,\frac{-3\pi}{4};L) + G_2(\alpha,y)\mathfrak{G}_m^+(\alpha,\frac{-3\pi}{4};L)}{\mathcal{D}_s(\alpha)}\mathrm{e}^{\mathrm{i}\alpha x}\mathrm{d}\alpha. \end{aligned} \tag{62}$$

Everything inside each integral is known and, hence, for a given x and y, these can be evaluated numerically to give a mode associated with the asymptotic behaviour of the mth similarity solution of the corner.

We can write the mth corner mode stress terms as

$$\tau_{xx}^m = \eta_m\int_{C_1}\frac{F_1(\alpha,y)\mathfrak{F}_m^+(\alpha,\frac{-3\pi}{4};L) + F_2(\alpha,y)\mathfrak{G}_m^+(\alpha,\frac{-3\pi}{4};L)}{\mathcal{D}_s(\alpha)}\mathrm{e}^{\mathrm{i}\alpha x}\mathrm{d}\alpha = \eta_m t_x^c(C_1,\gamma_m,x,y;L), \tag{63}$$

$$\tau_{xy}^m = \eta_m\int_{C_1}\frac{G_1(\alpha,y)\mathfrak{F}_m^+(\alpha,\frac{-3\pi}{4};L) + G_2(\alpha,y)\mathfrak{G}_m^+(\alpha,\frac{-3\pi}{4};L)}{\mathcal{D}_s(\alpha)}\mathrm{e}^{\mathrm{i}\alpha x}\mathrm{d}\alpha = \eta_m t_y^c(C_1,\gamma_m,x,y;L), \tag{64}$$

so that the outgoing scattered stress field, expressed as a modal expansion in two types of modes, Lamb modes and corner modes, is

$$\tau_{xx}^{\text{sca}}(x,y) = \sum_{n=1}^{N} \zeta_n t_x^l(\alpha_n, x, y) + \sum_{m=1}^{M} \eta_m t_x^c(\mathcal{C}_1(N), \gamma_m, x, y; L), \tag{65}$$

$$\tau_{xy}^{\text{sca}}(x,y) = \sum_{n=1}^{N} \zeta_n t_y^l(\alpha_n, x, y) + \sum_{m=1}^{M} \eta_m t_y^c(\mathcal{C}_1(N), \gamma_m, x, y; L). \tag{66}$$

Here, $\mathcal{C}_1(N)$ is a contour that passes above the first N roots of the dispersion relation and below all others in the upper half plane. We note that both the evanescent Lamb modes and all the corner modes have exponentially decaying behaviour for large $x > 0$ and hence far away from the edge only the propagating Lamb modes are important; however, these decaying terms are required in order to accurately satisfy the boundary conditions on $x = 0$, as we will discuss in the next section.

2.3.2. Satisfaction of the Free-End Boundary Conditions by Collocation

We now wish to determine the unknown coefficients in the modal expansion, i.e., the Lamb mode constants ζ_n and the corner mode constants η_m. To do so, we need to satisfy the boundary conditions on $x = 0$, where the tractions of the total field must be zero; this implies that our scattered field must satisfy

$$\tau_{xx}^{\text{sca}}(0,y) = \sum_{n=1}^{N} \zeta_n t_x^l(\alpha_n, 0, y) + \sum_{m=1}^{M} \eta_m t_x^c(\mathcal{C}_1(N), \gamma_m, 0, y; L) = -\,\tau_{xx}^{\text{inc}}(0,y), \tag{67a}$$

$$\tau_{xy}^{\text{sca}}(0,y) = \sum_{n=1}^{N} \zeta_n t_y^l(\alpha_n, 0, y) + \sum_{m=1}^{M} \eta_m t_y^c(\mathcal{C}_1(N), \gamma_m, 0, y; L) = -\,\tau_{xy}^{\text{inc}}(0,y), \tag{67b}$$

where a superscript ^{inc} denotes the incoming wave. We need $(N+M)/2$ collocation points, on which we will enforce the two boundary conditions (67a) and (67b) exactly. This yields a linear system of $(N+M)$ equations that can be solved to find the $(N+M)$ unknown coefficients ζ_n and η_m.

The locations of the collocation points were found by trial and error, determining the best by inspection of the reconstructions. The optimally-positioned collocation points were found to vary, depending on which modal expansion we take.

For expansions involving just Lamb modes $(M = 0)$, the ideal points lie away from the corner, as these modes poorly represent the field near these points of irregularity. Therefore, we chose to use linear spacing for the $N/2$ collocation points in this case. If, instead, we chose a Chebyshev distribution of collocation points for the pure Lamb expansion, the convergence of the method would become very poor indeed.

When using an expansion including corner modes, the best points to take are those nearer to the corner; hence the optimal results were found by employing Chebyshev points. With corner modes included, we also have a choice of the number of Lamb modes N and the number of corner modes M, to use; we generally found that the best results are obtained when M is large and N is small, although N must always be large enough to include all the propagating modes.

It should be noted that the corner expansion method, unlike the pure Lamb mode expansion, seems to be very stable with respect to the choice of collocation points. Indeed, if, instead of a Chebyshev distribution, we used a linear distribution as in the pure Lamb case, then the results are only marginally worse. This may be expected as, effectively, once we have removed the corner singularities thanks to our corner modes, the remaining part of the field to be approximated is very regular indeed. We should also note at this stage that this novel expansion also worked well when choosing more collocation points than actually needed and solving the overdetermined system via a least squares approach.

A discussion on how the two different approaches (pure Lamb and with the corner expansion) behave and compare as N and M increase will be given in the following section;

we will show that, as expected, the corner mode expansion method has much better convergence properties than the pure Lamb approach.

3. Results

From the work presented in Section 2.1, we have posed a combined modal expansion solution: we have the usual Lamb modes, which accurately represent wave propagation within the plate, and we have introduced a new set of modes that accurately represent the local corner behaviour; the latter were found in Section 2.2. In this section, we will present the results obtained by numerically solving for the coefficients in the boundary conditions, (67a) and (67b), using collocation methods.

3.1. Reconstructions

The figures presented in this section show the tractions along the end face of the plate, as we wish to compare how well the different modal expansions recover the specified boundary conditions. The scattered stress field (65) and (66) on $x = 0$ should be equal to the negative of that for the incoming wave, as there is zero overall traction on the end face (see (67a) and (67b)). Note that we only need to consider the vertical end face, as the conditions at the top and bottom of the plate are automatically satisfied for each and every mode.

In Figure 7, we choose the wavenumber $k_t = 4$ and the Poisson ratio $\nu = 0.3$, and vary only the number of modes. The plots show the scattered field stresses, i.e., Equations (65) and (66), where the coefficients were found by collocation forced by the incident propagating symmetric Lamb wave (this is the only propagating mode for this value of k_t). This is plotted against the negative of the incoming wave in Figure 7a,b; for an exact solution, these would be identical. Also plotted, in Figure 7c,d, are the total normal and shear stress fields (i.e., due to both the incoming and outgoing waves) in order to demonstrate how well the results approximate the zero traction conditions on the face.

The results in Figure 7 compare the behaviour of four modal expansions; one is a standard Lamb wave expansion without corner modes that we deem to have converged and gives acceptable results in terms of conservation of energy and errors in the boundary conditions. The other results are with varying numbers of corner modes. It should be noted that, in the pure Lamb wave expansion, we need a considerable number of modes to ensure convergence; this means that the condition number of the collocation matrix is large in that case, demanding that high precision calculations be used to obtain good results. We note that the results show that we need significantly fewer total modes with our new expansion, i.e., even with a small number of corner modes, we can represent the near field behaviour with extremely small errors. We can therefore deduce that these modes are correctly capturing the key behaviour of the irregular fields near the corners.

We also plot the error in traction reconstructions associated with a variety of wavenumbers, which is equivalent to changing the forcing frequency; higher k_t corresponds to higher frequencies where there are more propagating modes. In practice, it is found that higher frequency problems require greater numbers of Lamb modes to ensure convergence with the same accuracy. The plots in Figure 8 show, again, the tractions along the face $x = 0$, this time considering three different values of k_t.

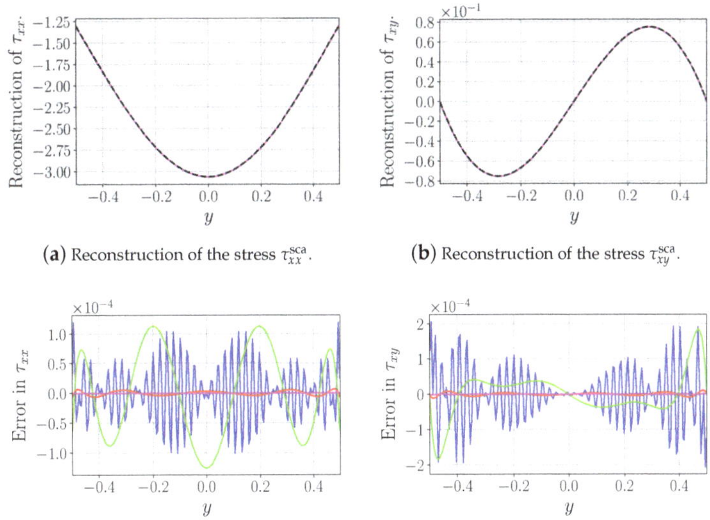

(**a**) Reconstruction of the stress τ_{xx}^{sca}.

(**b**) Reconstruction of the stress τ_{xy}^{sca}.

(**c**) Real part of the error of τ_{xx}.

(**d**) Imaginary part of the error of τ_{xy}.

Figure 7. The reconstructions of the stress fields using N symmetric Lamb modes and M corner modes. The curves correspond to: an expansion containing only Lamb waves, with $N = 80$ (blue line); a mixed expansion with $N = 2$ and $M = 4$ (green line); an expansion with $N = 2$ and $M = 8$ (red line); and an expansion with $N = 2$ and $M = 12$ (pink line). Subfigures (**a**,**b**) are the reconstructions of the scattered stress fields along the traction-free face. The negative of the incoming stress fields, which should lie on the same line, are plotted as (▪ ▪ ▪) for comparison. Subfigures (**c**,**d**) show the errors of the reconstructions of the total stress fields (sum of the incoming field and the outgoing field). In this example, $k_t = 4$ and $\nu = 0.3$.

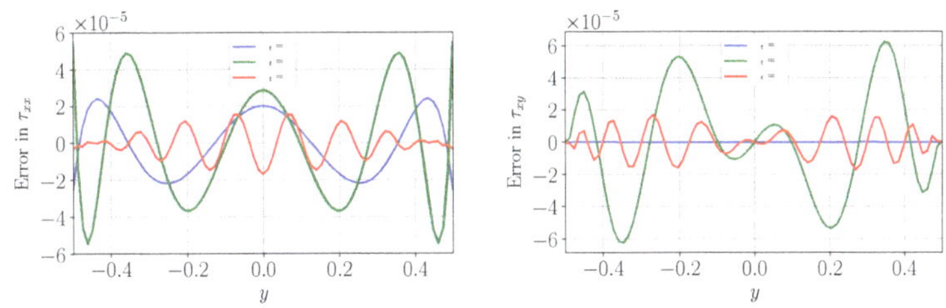

(**a**) Real part of the error of τ_{xx}.

(**b**) Imaginary part of the error of τ_{xy}.

Figure 8. A set of results showing the accuracy of the zero traction boundary condition for a variety of frequencies k_t for fixed Poisson ratio $\nu = 0.3$. The maximum values of the incoming field for both tractions are order one. There is one propagating mode apiece for $k_t = 2$ and 4, and three propagating modes when $k_t = 8$. The number of modes in each plot was chosen to maintain the same order of magnitude accuracy: for $k_t = 2$, $N = 2$ and $M = 2$; for $k_t = 4$, $N = 4$ and $M = 4$; and for $k_t = 8$, $N = 10$ and $M = 12$.

As an additional illustration of the superiority of our method compared to the pure Lamb mode approach, we study the convergence of the reflection coefficients (and the energy) in the challenging case of $k_t = 8$, for which there are three propagating modes. The results are summarised in Figure 9. We show that, even with only $M = 4$ corner modes, our method displays results that are more than two orders of magnitude better than the pure Lamb mode approach with the maximum number of Lamb modes considered ($N = 80$).

We also show that increasing the number of corner modes to the modest value of $M = 10$, we gain an accuracy of $O(10^{-8})$, which would be extremely challenging to obtain by the pure Lamb mode approach, even with a very large number N of Lamb modes.

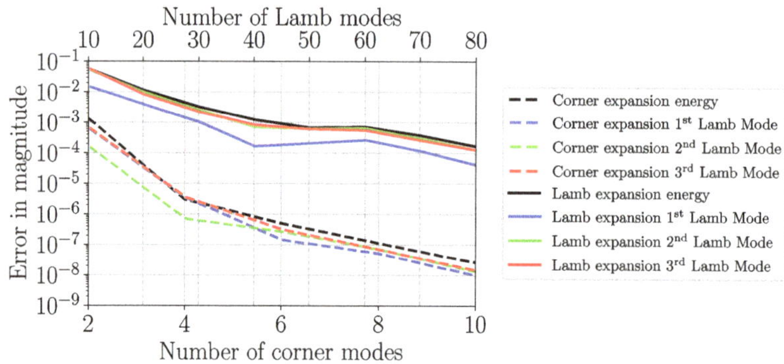

Figure 9. For $k_t = 8$ and $\nu = 0.3$ (the most challenging case of Figure 8), we illustrate the convergence of the energy and of the reflection coefficients of the three propagating modes. We display the absolute error between the computed reflection coefficients and a reference value, i.e., that obtained for $N = 10$ and $M = 12$, and the reference value for the energy is 1. The continuous lines correspond to the pure Lamb mode approach ($M = 0$) as the number N of Lamb modes increases. The dashed lines correspond to our new approach for a fixed number $N = 10$ of Lamb modes as the number M of corner modes increases.

3.2. Trapped Modes

The works of [7–9] discuss the existence of trapped modes in the context described here. At the Lamé frequency $k_t = \sqrt{2}\pi$, the propagating mode becomes orthogonal to the evanescent Lamb modes. In this case, the single propagating mode travelling leftwards can be trivially combined with the corresponding rightwards travelling mode to satisfy the traction-free condition, and hence the coefficients on all other modes are trivially zero. However, it has been observed that for two particular values of the Poisson ratio, $\nu^* = 0$ and $\nu^* \approx 0.2248$, there also exist a combination of evanescent Lamb modes that sum to satisfy the traction-free boundary conditions. In the case of $\nu^* = 0$, this trapped mode has formally been proven to exist; however, for $\nu^* \approx 0.2248$, its existence has only been established numerically.

Our modal expansion, which includes the new corner modes, can easily be employed to accurately determine the value of Poisson's ratio for the latter trapped mode solution. We search for a homogeneous solution of the problem solely in terms of the modes which decay as $x \to \infty$. In order to find the homogeneous solution, we construct a collocation matrix as before, with the propagating mode term removed, for a given ν at the Lamé frequency. We then vary ν and observe where the value of the determinant nearly vanishes. Figure 10 shows the resulting absolute values of the determinant, normalised on the magnitude at $\nu = 0.22479$ for ease of comparison, found when using this process for a range of numbers of corner modes. This clearly demonstrates that the modal expansion presented in this paper does find this trapped mode and that, for $M > 6$, its value has converged to six significant figures, at $\nu \approx 0.224798$.

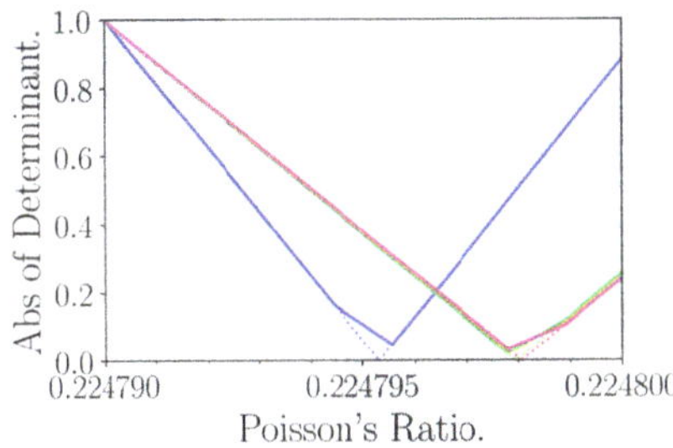

Figure 10. The absolute value of the determinant of the collocation matrix, excluding the propagating mode, for different numbers of corner modes. The respective lines correspond to four Lamb modes and $M = 4$ (—), $M = 6$ (—), $M = 8$ (—), and $M = 10$ (—) corner modes. The dotted lines are the projected values of the determinant, found by extending the curves on either side of zero; this aids the determination of location of zero. The value of Poisson's ratio at which the determinant vanishes can be seen to have converged, once $M = 6$ or larger, to $\nu \approx 0.224798$.

4. Discussion

In this article, we have asymptotically determined the form of the similarity solutions that are associated with the behaviour close to a corner in a semi-infinite elastodynamic thick plate. We recognized that the irregular corner behaviour causes difficulties for Lamb mode expansion techniques and so, to overcome this, we have introduced a new method for constructing corner modes that not only accurately capture the behaviour near the corner, but that also satisfy the plate boundary and radiation conditions. Additionally, because these modes are found using the same Fourier transform approach as for the Lamb waves, adding them does not result in an over-complete representation of the field.

As found in the results Section 3, our numerical work shows that including the corner modes into the expansion allows us to accurately capture the near-field behaviour, and as a consequence we need include far fewer modes to achieve comparable or superior results. Therefore, we believe that these corner mode solutions can be utilised in a wide variety of Lamb wave or other, scattering problems for which standard plate or duct wave modal expansion techniques converge slowly [15,23]. We have also demonstrated that, by considering these extra modes, we can more accurately capture the trapped-mode behaviour associated with this configuration; this allows us to quantify the Poisson's ratio with more precision than other extant methods, yielding the new result $\nu \approx 0.224798$ at which this trapped-mode occurs. This result has recently been confirmed in a private communication with Lawrie [26], who has approached the localised mode problem in a quite distinct way to the present work or to that described in [7–9].

Author Contributions: Conceptualization, R.C.A. and I.D.A.; methodology, R.C.D., R.C.A., and I.D.A.; software, R.C.D. and R.C.A.; validation, R.C.D., R.C.A., and I.D.A.; formal analysis, R.C.D., R.C.A., and I.D.A.; investigation, R.C.D., R.C.A., and I.D.A.; resources, R.C.A. and I.D.A.; writing—original draft preparation, R.C.D.; writing—review and editing, R.C.A. and I.D.A.; supervision, R.C.A. and I.D.A.; project administration, R.C.A. and I.D.A.; and funding acquisition, R.C.A. and I.D.A. All authors have read and agreed to the published version of the manuscript.

Funding: R.C.D. was in receipt of funding from AWE and EPSRC to support his postgraduate studentship, R.C.A. undertook part of the work under EPSRC grants EP/N013719/1 and EP/W018381/1, and I.D.A. undertook part of the work under EPSRC grants EP/K032208/1 and EP/R014604/1.

Acknowledgments: The authors are grateful to Richard E. Hewitt for his comments on early drafts of this manuscript and for facilitating the relationship with the industrial partner through his William Penney Fellowship.

Conflicts of Interest: The authors declare no conflict of interest. The funders had no role in the design of the study; in the collection, analyses, or interpretation of data; or in the writing of the manuscript. The funders approved the manuscript for publication.

References

1. AWE. *Private Correspondence*; Technical Report; AWE: Aldermastion, UK, 2016.
2. ACVR. *Experimental Data*; Technical Report; ACVR: Reno, NV, USA, 2017.
3. Abrahams, I.D.; Wickham, G.R. The propagation of elastic waves in a certain class of inhomogeneous anisotropic materials. I. The refraction of a horizontally polarized shear wave source. *Proc. R. Soc. Lond. Ser. Math. Phys. Sci.* **1992**, *436*, 449–478. [CrossRef]
4. Abrahams, I.D.; Wickham, G.R. Scattering of elastic waves by a small inclined surface-breaking crack. *J. Mech. Phys. Solids* **1992**, *40*, 1707–1733. [CrossRef]
5. Williams, M.L. Surface singularities resulting from various boundary conditions in angular corners of plates in extension. *J. Appl. Mech. Asme* **1952**, *19*, 526–528. [CrossRef]
6. Gregory, R.D.; Gladwell, I. The reflection of a symmetric Rayleigh-Lamb wave at the fixed or free edge of a plate. *J. Elast.* **1983**, *13*, 185–206. [CrossRef]
7. Pagneux, V. Revisiting the edge resonance for Lamb waves in a semi-infinite plate. *J. Acoust. Soc. Am.* **2006**, *120*, 649. [CrossRef]
8. Zernov, V.; Pichugin, A.; Kaplunov, J. Eigenvalue of a semi-infinite elastic strip. *Proc. R. Soc. Math. Phys. Eng. Sci.* **2006**, *462*, 1255–1270. [CrossRef]
9. Lawrie, J.B.; Kaplunov, J. Edge waves and resonance on elastic structures: An overview. *Math. Mech. Solids* **2012**, *17*, 4–16. [CrossRef]
10. Lamb, H. On waves in an Elastic Plate. *Proc. R. Soc. Lond. Ser. A Contain. Pap. Math. Phys. Character* **1917**, *93*, 114–128. [CrossRef]
11. Chapman, C.J.; Sorokin, S.V. The finite-product method in the theory of waves and stability. *Proc. R. Soc. Math. Phys. Eng. Sci.* **2010**, *466*, 471–491. [CrossRef]
12. Gregory, R.D.; Gladwell, I. The Cantilever Beam Under Tension, Bending or Flexure at Infinity. *J. Elast.* **1982**, *12*, 317–343. [CrossRef]
13. Nawaz, R.; Lawrie, J.B. Scattering of a fluid-structure coupled wave at a flanged junction between two flexible waveguides. *J. Acoust. Soc. Am.* **2013**, *134*, 1939–1949. [CrossRef] [PubMed]
14. Lawrie, J.B.; Afzal, M. Acoustic scattering in a waveguide with a height discontinuity bridged by a membrane: A tailored Galerkin approach. *J. Eng. Math.* **2017**, *105*, 99–115. [CrossRef]
15. Abrahams, I.D.; Assier, R.C.; Cotterill, P.A. A new method for improving convergence of modal expansions for discontinuous waveguide problems. 2022; *to be submitted*.
16. Linton, C. Accurate solution to scattering by a semi-circular groove. *Wave Motion* **2009**, *46*, 200–209. [CrossRef]
17. Homentcovschi, D.; Miles, R.N. A re-expansion method for determining the acoustical impedance and the scattering matrix for the waveguide discontinuity problem. *J. Acoust. Soc. Am.* **2010**, *128*, 628–638. [CrossRef] [PubMed]
18. Porter, R.; Evans, D.V. Complementary approximations to wave scattering by vertical barriers. *J. Fluid Mech.* **1995**, *294*, 155–180. [CrossRef]
19. Fernyhough, M.; Evans, D.V. Scattering by periodic array of rectangular blocks. *J. Fluid Mech.* **1995**, *305*, 263–279. [CrossRef]
20. Chang, K.H.; Tsaur, D.H.; Wang, J.H. Scattering of SH waves by a circular sectorial canyon. *Geophys. J. Int.* **2013**, *195*, 532–543. [CrossRef]
21. Lyapin, V.P.; Manuilov, M.B.; Sinyavsky, G.P. Quasi-analytical method for analysis of multisection waveguide structures with step discontinuities. *Radio Sci.* **1996**, *31*, 1761–1772. [CrossRef]
22. Lucido, M.; Panariello, G.; Schettino, F. Analysis of the Electromagnetic Scattering by Perfectly Conducting Convex Polygonal Cylinders. *IEEE Trans. Antennas Propag.* **2006**, *54*, 1223–1231. [CrossRef]
23. Cotterill, P.A.; Abrahams, I.D. Acoustic transmission through gaps in sound-reduction coatings. 2022; *to be submitted*.
24. Carrier, G.F.; Krook, M.; Pearson, C.E. *Functions of a Complex Variable*; McGraw-Hill: New York, NY, USA, 1966.
25. Fung, Y.C. *Foundations of Solid Mechanics*; Prentice-Hall: Hoboken, NJ, USA, 1965.
26. Lawrie, J.B. (Brunel University London, London, UK). Personal communication, 2022.

applied sciences

Article

Numerical Investigation of Excitation of Various Lamb Waves Modes in Thin Plastic Films

Rymantas Jonas Kazys, Justina Sestoke * and Egidijus Zukauskas

Prof. K. Barsauskas Ultrasound Research Institute, Kaunas University of Technology, 51423 Kaunas, Lithuania; rymantas.kazys@ktu.lt (R.J.K.); e.zukauskas@ktu.lt (E.Z.)
* Correspondence: justina.sestoke@ktu.lt; Tel.: +370-(37)-351-162

Featured Application: The developed new method will enable us to perform contactless air-coupled NDT of composite structures possessing a complex geometry.

Abstract: Ultrasonic-guided waves are widely used for the non-destructive testing and material characterization of plates and thin films. In the case of thin plastic polyvinyl chloride (PVC), films up to 3.2 MHz with only two Lamb wave modes, antisymmetrical A_0 and symmetrical S_0, may propagate. At frequencies lower that 240 kHz, the velocity of the A_0 mode becomes slower than the ultrasonic velocity in air which makes excitation and reception of such mode complicated. For excitation of both modes, we propose instead a single air-coupled ultrasonic transducer to use linear air-coupled arrays, which can be electronically readjusted to optimally excite and receive the A_0 and S_0 guided wave modes. The objective of this article was the numerical investigation of feasibility to excite different types of ultrasonic-guided waves, such as S_0 and A_0 modes in thin plastic films with the same electronically readjusted linear phased array. Three-dimensional and two-dimensional simulations of A_0 and S_0 Lamb wave modes using a single ultrasonic transducer and a linear phased array were performed. The obtained results clearly demonstrate feasibility to excite efficiently different guided wave modes in thin plastic films with readjusted phased array.

Keywords: air-coupled ultrasonic; Lamb waves; finite element modeling; plastic films

Citation: Kazys, R.J.; Sestoke, J.; Zukauskas, E. Numerical Investigation of Excitation of Various Lamb Waves Modes in Thin Plastic Films. *Appl. Sci.* **2022**, *12*, 849. https://doi.org/10.3390/app12020849

Academic Editors: Cem Selcuk and Habil. Michel Darmon

Received: 9 December 2021
Accepted: 12 January 2022
Published: 14 January 2022

1. Introduction

Ultrasonic-guided wave-based inspection techniques are very promising for non-destructed testing (NDT) and structural health monitoring (SHM) and are the primary techniques for long-range damage detection and characterization of plate-like structures. The guided Lamb waves propagate within the interior of the objects under investigation. The Lamb wave's fundamental A_0 and S_0 modes are widely used as they can be relatively easy to excite and are sensitive to various types of defects, which are found especially in composite materials.

Contemporary technologies include the employing of thin films and composite plate materials. There is a special class of widely used, very thin (~0.1 mm) materials—plastic tapes and films such as clear polyvinyl chloride films [1]. Polyvinyl, also known as poly (vinyl chloride) or PVC, is the third-most widely produced synthetic plastic. The production of polyvinyl chloride films in Europe reaches five million tons annually [2]. During the manufacturing of PVC films, various defects such as wrinkles, holes, rough surfaces, and thickness variations can arise.

Ultrasonic methods using guided waves enable the investigation of key elastic properties of materials that cannot be assessed by merely using other methods, for example excitation by lasers [3,4], electrostatic excitation methods [3], or electromagnetic acoustic transducers [5,6]. These guided waves are currently used for the non-destructive testing and evaluation (NDT) of thin film type materials [7]. Usually, they are excited when the vibrations of an acoustic transducer are transferred to the investigated item by direct contact via a coupling liquid. The contact liquids are impossible to use in many cases, as the investigated item may be contaminated, damaged or otherwise negatively affected [8,9]. This is relevant in the cases of thin films or composite structures. Some defects are detected by optical methods, but some defects such as holes and especially thickness variations can be better found by ultrasonic methods using guided waves [10–12]. Therefore, air-coupled ultrasonic investigation using guided waves can be a very attractive technique for the investigation of such type of materials.

The attenuation of ultrasonic waves in air and in the object under investigation increases with the frequency. The loss of ultrasonic signals may be reduced by generating guided waves while using lower frequencies [13–15].

At lower frequencies, the velocity of the ultrasonic-guided waves in thin films may become lower than the ultrasound velocity in air, which complicates excitation and reception of ultrasonic signals by air-coupled methods [16,17]. Therefore, it is appropriate to develop novel air-coupled ultrasonic measurement methods based on the excitation and reception of guided waves for a study of such objects. Special problems arise in the case of excitation and reception of symmetrically guided wave modes, which require fundamentally new methods of excitation and reception.

The guided waves in thin film type materials may be excited via air gap in two different ways depending on the ratio of propagation velocities of ultrasonic waves in the material and air. When the phase velocity of the guided waves in the object is bigger than the ultrasound velocity in air, then the optimal incident angle of the ultrasonic wave incident from air onto the plate exists at which the biggest amplitude of guided wave is obtained. The propagating guided wave in a thin film then radiates a leaky wave into air, which may be used for air-coupled reception. This mode is commonly applied for detecting defects inside thin plates and film type materials as the leakage increases at the defective zone [18–20].

In our previous investigations, we analyzed application of air-coupled linear phased ultrasonic arrays for excitation and reception of only the slow fundamental antisymmetric A_0 mode, in the frequency range in which the velocity of this mode is slower than the ultrasound velocity in air [16,17]. For the analysis of propagation of the ultrasonic wave through the air gap, the impulse response method was used [16]. Propagation of the A_0 mode in a plastic film was calculated by the analytic time harmonic solution method [14,16,17].

However, exploitation of the symmetric S_0 mode enables not only detection of various defects, but also evaluation of elastic properties of the item under a test [20]. For example, simultaneously excited S_0 and A_0 guided wave modes enabled the measurement of thickness and Young's modulus of thin PVC films [21]. For practical applications, it would be very attractive to use it for the same ultrasonic air-coupled transducer or array. To our knowledge, a detailed analysis of the performance of such methods is still missing [21]. For this purpose, we propose instead a single ultrasonic transducer to use linear air-coupled arrays, which can be electronically readjusted to optimally excite and receive the required guided wave modes. For the solution of this problem, we used a numerical simulation based on the application of finite element modelling.

The objective of this article was a numerical investigation of feasibility to excite different types of ultrasonic-guided waves, such as symmetrical S_0 and asymmetrical A_0 modes in thin plastic films with the same electronically readjusted linear phased array.

First, we simulated propagation of two fundamental A_0 and S_0 Lamb wave modes in thin plastic films using a single ultrasonic transducer. Second, to improve the excitation of the S_0 mode, a linear phased array was proposed for use. Calculations were performed using finite element methods (FEM). FEM has many advantages such as [22]: it allows solving acoustic contact problems, to model bodies made of different materials; a curvilinear region can be approximated by means of finite elements of smaller dimensions, which enables the increase of modelling accuracy [23]. FEA's main advantage is that it produces a much more detailed set of results than experimental investigations and is often quicker and less expensive. Fundamentally, instead of tackling a big problem directly, we divided it into smaller and more easily solvable problems to attain a unique result for the system as a whole. The discrete model approximates the behavior of the real physical structure. However, provided that the discretization mesh is dense enough, the approximation is sufficient to accurately model reality [24–26].

The numerical modelling was performed using the commercially available finite element software, Abaqus. Two numerous calculation methods, 3D and 2D, were used. Numerous calculations showed that the 2D technique significantly improves the computational efficiency compared to the 3D technique [27–29].

The paper consists of four sections. In Section 2, the numerical simulation of propagation of A_0 and S_0 guided wave modes in thin PVC plastic film is presented. The finite elements modeling results of A_0 and S_0 guided wave modes are described in Section 3. In Section 4, conclusions and discussion of the obtained results are given.

2. Finite Elements Models

Propagation of the guided A_0 and S_0 mode in the PVC film sample was simulated by the ABAQUS software package (Dassault Systemes, Johnston, Rhode Island, United States of America). The ABAQUS program was chosen for its high level of detail, manual design of the work piece, possibility of configuring the materials, and the very fine control of mesh; however, the ABAQUS program has no support for any materials and takes time to "setup" simulations as the user must manually set many simulation parameters. For the modelling of propagation of A_0 and S_0 modes in thin plastic films, two 2D and 3D calculation methods were chosen [30–33].

Modeling of elastic wave propagation was carried out by solving the following equation:

$$[\boldsymbol{M}]\left\{\ddot{\boldsymbol{U}}\right\} + [C]\left\{\dot{\boldsymbol{U}}\right\} + [K]\{\boldsymbol{U}\} = \{F(t)\}, \tag{1}$$

where $[\boldsymbol{M}]$ is the structural mass matrix, $[\boldsymbol{C}]$ is the element damping matrix, $[\boldsymbol{K}]$ is the structural stiffness matrix, $\{\boldsymbol{U}\}$ is the displacement vector, and $\{\boldsymbol{F}\}$ is the load vector. The Abaqus Explicit software uses a central difference method to integrate the equation of motion in time [25,34]:

$$\{\boldsymbol{U}_{t+\Delta t}\} = \left[\hat{\boldsymbol{M}}\right]^{-1}\left[\{\boldsymbol{F}(t)\} - \left([K] - \frac{2}{\Delta t^2}[\boldsymbol{M}]\right)\{\boldsymbol{U}_t\} - \left[\tilde{\boldsymbol{M}}\right]\{\boldsymbol{U}_{t-\Delta t}\}\right], \tag{2}$$

where Δt is the time step and

$$\left[\hat{\boldsymbol{M}}\right] = \frac{1}{\Delta t^2}[\boldsymbol{M}] + \frac{1}{2\Delta t}[C], \tag{3}$$

$$\left[\tilde{\boldsymbol{M}}\right] = \frac{1}{\Delta t^2}[\boldsymbol{M}] - \frac{1}{2\Delta t}[C]. \tag{4}$$

Three-dimensional models are in practice limited to small size models and low frequencies due to the required computational power to numerically solve the wave's propagation tasks. They also require relatively time-consuming post processing before the results can be obtained. However, there is a significant difference in the computational time between two-dimensional (2D) FE models and more accurate 3D models.

First of all, a 3D method was used to simulate the propagation of guided waves excited by one element with the excitation zone of 5 × 1 mm, the center of which is at x, y = 0. The calculations were performed for polyvinyl chloride (PVC, London, UK). Dimensions of the PVC sample selected for simulation were 240 × 240 mm. For reduction of the computational time, only a quarter of the film with symmetry boundary conditions was modeled by the 3D method. The modeled structure of one excitation zone is shown in Figure 1a. The simulation step in the time domain was set dt = 1 μs. The strip-like piezoelectric element radiates an ultrasonic wave, which propagates via the small air gap and excites a guided wave in the thin plastic film. The excitation was introduced as the uniformly distributed

1 N 150 kHz force impulse in the transmitter area. The frequency of 150 kHz was selected because at such a frequency only the only two modes, symmetrical S_0 and antisymmetrical A_0, may propagate.

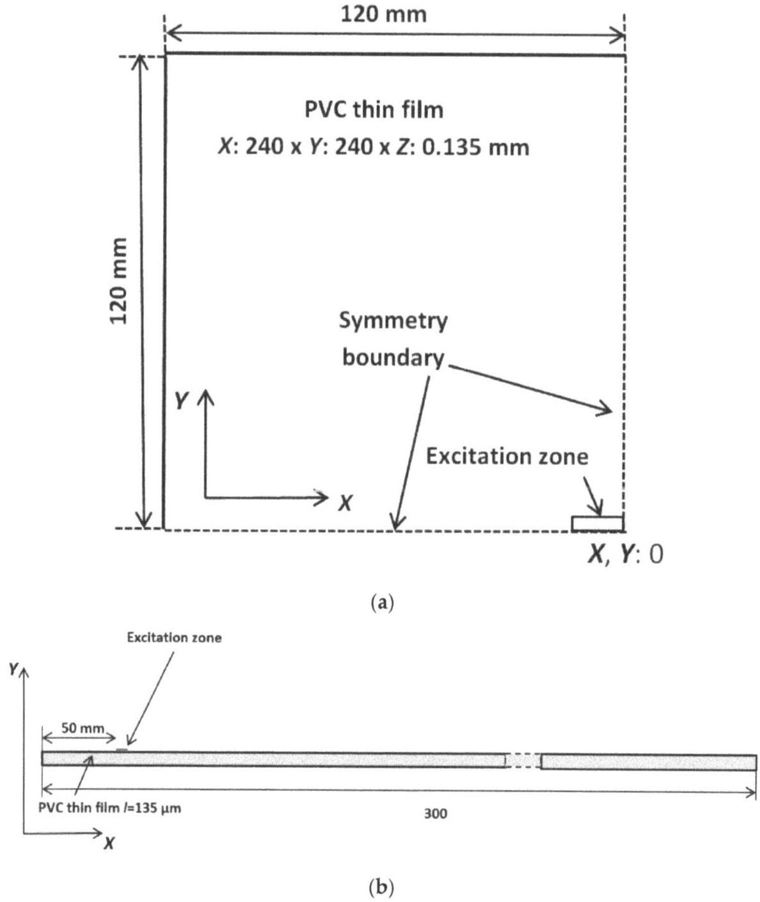

Figure 1. A model of the PVC sample used for the simulation: (**a**) for the 3D simulation; (**b**) for the 2D simulation.

Second, a 2D method was performed simulating an A_0 and S_0 guided wave modes excitation zone of 1 mm. This schematic diagram with one excitation zone is presented in Figure 1b. The excitation zone is placed $x = 50$ mm, $y = 0$ mm from the coordinate origin t at $x, y = 0$.

In Table 1, some properties corresponding to the computational efficiency of the 3D and 2D methods are compared, which show that the 2D simulation method has a high speed and needs a relatively smaller memory than the 3D method.

Table 1. Advantages and disadvantages of the simulation methods.

	Advantages and Disadvantages of Simulation Methods	
	2D Simulation	**3D Simulation**
The length of the simulated PVC film	300 mm	quarter of the film 120 mm × 120 mm
Finite element size	15 μm	65 μm
The simulation step in the time domain	40 ns	1 μs
Numbers of elements in the model	180,000	6,816,600
Calculation time	~20 min	~5 h
Data size	~15 GB	~120 GB

Calculation computer recourses:
Processor—Intel (R) Core (TM) i7—2700K CPU @ 3.50 GHz 3.50 GHz;
Installed RAM—32.0 GB;
System type—64-bit operating system, ×64-based processor.

The 3D model is very complicated and has some disadvantages: a large amount of data is required as input for the mesh used in terms of nodal connectivity and other parameters depending on the problem, and requires longer execution time. Our 3D model data is about 120 GB, with numbers of elements in the model being over 6 million, a calculation time of about 5 h, and where it is impossible to see the S_0 guided wave mode. For this reason, other simulation results were obtained by the 2D modeling method. In the 2D simulation method, we used a finite element size of 15 μm in the model, which is ~4.2 times smaller than in the 3D model and the length of the PVC film is 2.5 times longer than the 3D model.

The values of the phase and group velocities of the A_0 and S_0 guided waves modes in PVC film were calculated using the Semi Analytical Finite Element (SAFE) method. The parameters of the clear PVC thin film used for calculations are presented in Table 2 [35].

Table 2. Parameters of PVC thin film.

Parameter	Value
Density	$\rho = 1400\ kg/m^3$
Young's modulus	$E = 2.937$ GPa
Poisson's coefficient	$\nu = 0.42$

The calculated phase and group velocities versus frequency are shown in Figure 2. From the simulation results it follows that even in the frequency range up to 3.2 MHz, only two modes, symmetrical S_0 and antisymmetrical A_0, may propagate. For numerical simulations the frequency of the excitation signal 150 kHz was selected because at such a frequency only those two modes may propagate. The ultrasound phase velocity of the A_0 mode at the frequency is 232 m/s and the group velocity is 448 m/s are presented by red and blue colors in Figure 2. The ultrasound phase and group velocities of the S_0 mode do not depend on a frequency and are 1595 m/s. In Figure 2 they are indicated by a solid red horizontal line. The ultrasound velocity in air of 343 m/s is presented by a dotted line in Figure 2. Please note that the phase velocity of the A_0 mode is slower than the ultrasound velocity in air up to 240 kHz, therefore its air-coupled excitation by a classical method using deflected ultrasonic transducer looks impossible. However, we shall demonstrate that it is possible using a linear air-coupled array. The waveform and the spectrum of the excitation force are shown in Figures 3 and 4. In the Figures 3 and 4 n.u. represents normalized units. The excitation frequency is $f = 150$ kHz, and the numbers of periods is $n = 3$.

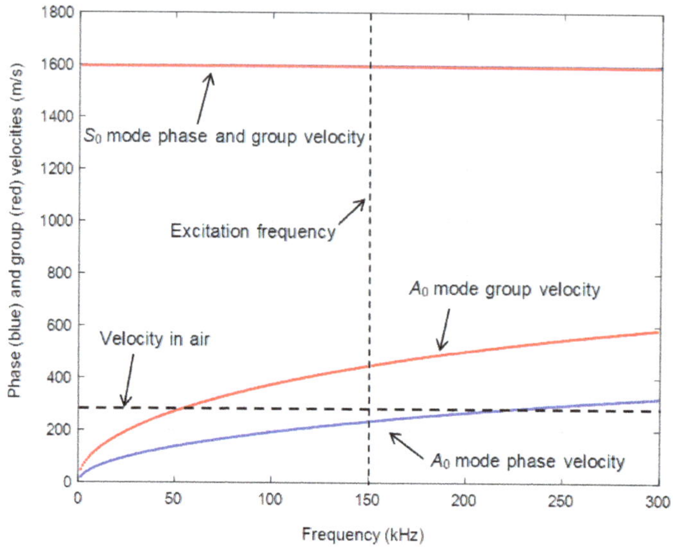

Figure 2. Calculated dispersion curves of PVC thin film.

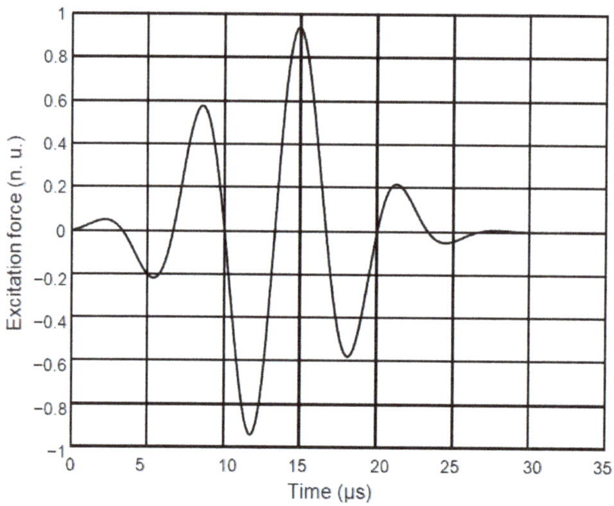

Figure 3. Excitation signal.

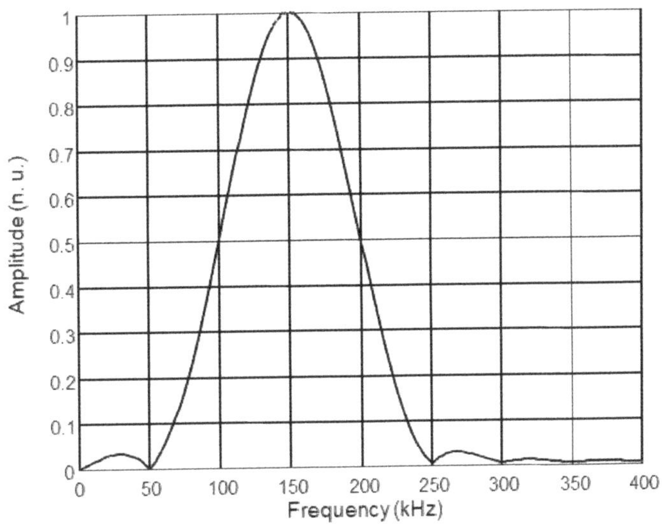

Figure 4. Spectrum of the excitation signal.

The simulation results obtained by the 2D and 3D FEM methods are presented in the following Section 3.

3. Finite Elements Modeling Results

3.1. Single Transducer Model

The main idea of the performed numerical investigation was to analyze possibilities to excite not one but a few guided wave modes possessing very different propagation velocities and distributions of displacements with the same linear phased array. For such purpose symmetrical S_0 and antisymmetrical A_0 modes fit very well. The biggest challenge is the excitation of the S_0 mode, as in this mode, the in-plane displacement component is much bigger than the off-plane component, e.g., normal to the surface of the film. For excitation of those guided waves we shall use a linear air-coupled array consisting of eight strip-like elements with radiating apertures of 5 × 1 mm.

The model of the thin PVC film was meshed using shell elements and the propagation of guided waves was first simulated by the 3D method. The thickness of the PVC film was l = 135 μm. The dimension of the finite elements was 62.5 μm, the numbers of elements in the model were 6 816 600 and only a quarter of the film with symmetry boundary conditions was modelled. Such a dimension is close to 1/20th of the S_0 wavelength λ = 10.6 mm at the analyzed 150 kHz frequency.

In order to evaluate the influence of ultrasonic field spreading due to a diffraction in the plane of the film, calculations by the 3D method were performed for a single array element with a rectangular aperture with dimensions 5 × 1 mm. The air-coupled ultrasonic transducer usually is placed very close to the film surface, for example 1 mm. In such case the simulation of the propagation of ultrasonic wave through the air gap can be replaced by the force acting on the surface of the film. This force is uniformly distributed in the rectangular area corresponding to the radiating aperture 5 × 1 mm of the array single element.

The simulated spatial distributions of the particle velocity y component at the time instant t = 200 μs is shown in Figure 5a. The B scan of this in-plane particle velocity component along y axis is presented in Figure 5b. The waves are more strongly radiated along the y axis direction due to directivity properties of the excitation zone.

In the B scan (Figure 5b) we have the signals measured at many positions along the same direction, the phase velocity can also be estimated by taking two signals measured at neighboring positions and estimating the propagation time of a particular phase point between those positions.

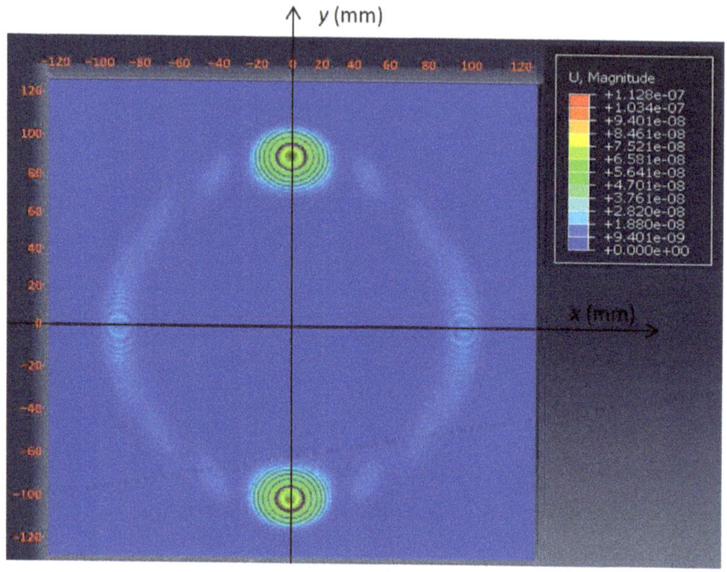

(**a**)

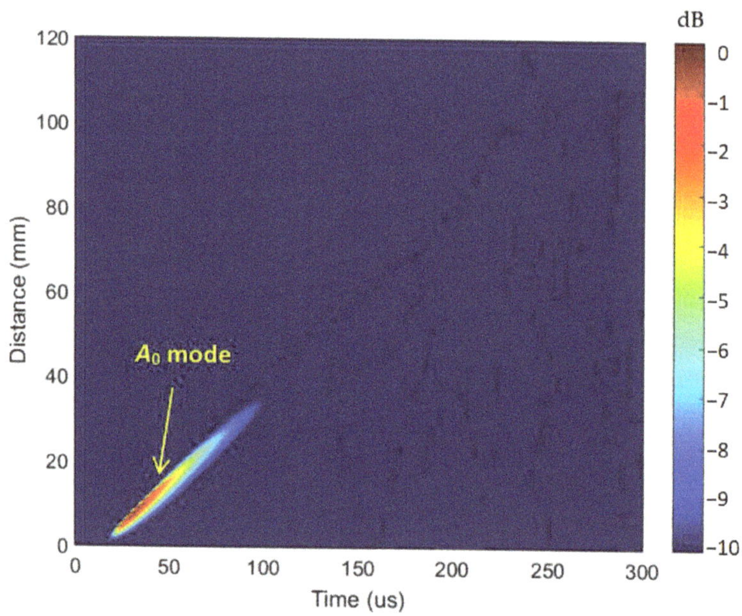

(**b**)

Figure 5. Simulation of guided waves in the thin plastic film excited by a single element by the 3D method: (**a**) spatial distributions of the particle velocity of magnitude (*X*, *Y* and *Z* components) at the time instant t = 200 μs; (**b**) the B scan of the y component in-plane particle velocity.

Such an estimation gives the group velocity 424 m/s, what allows concluding that a single array element excites mainly a strong A_0 mode in the thin film. There are very weak traces of the S_0 mode propagating with the velocity 1540 m/s, however they look so weak that are not suitable for any measurements. This 3D simulation was performed using shell elements; therefore, we decided to check if the simulation by 2D FEM exploiting conventional finite elements allows for the revealing of a better S_0 mode. On the other hand, the 2D FEM requires a smaller number of elements and much shorter simulation time (Table 1).

The schematic diagram with one excitation zone is shown in Figure 1b. The length of the simulated PVC film is 300 mm. The excitation zone was placed at the 50 mm from the origin of the coordinates. The B scan of the in-plane particle velocity component in the PVC film obtained by 2D simulation is presented in Figure 6. The ultrasonic pulses of the in-plane particle velocity at two different distances from the excitation zone (x = 50 mm, y = 0 mm) are shown in Figure 7a,b.

In order to identify what mode (or modes) is actually excited by a single array element the propagation velocity of the waves shown in Figure 5a should be evaluated. The phase velocities are usually found from the distances Δd_{ph} and Δd_{gr}, which are covered by a particular phase point during the time intervals Δt_{ph} and Δt_{gr}:

$$c_{ph} = \frac{\Delta d_{ph}}{\Delta t_{ph}}. \tag{5}$$

$$c_{gr} = \frac{\Delta d_{gr}}{\Delta t_{gr}}. \tag{6}$$

In the presented B scan we can see better the excited fast S_0 mode. The S_0 and A_0 modes reflected from the boundaries of the sample are also seen.

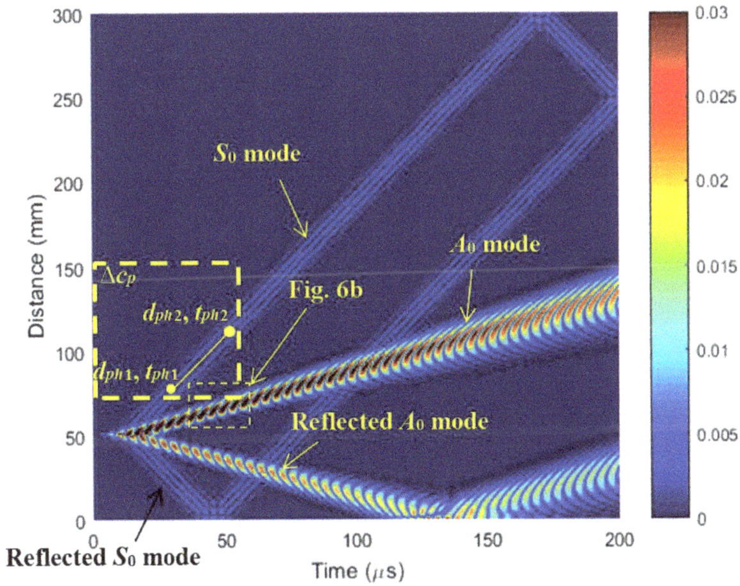

(a)

Figure 6. *Cont.*

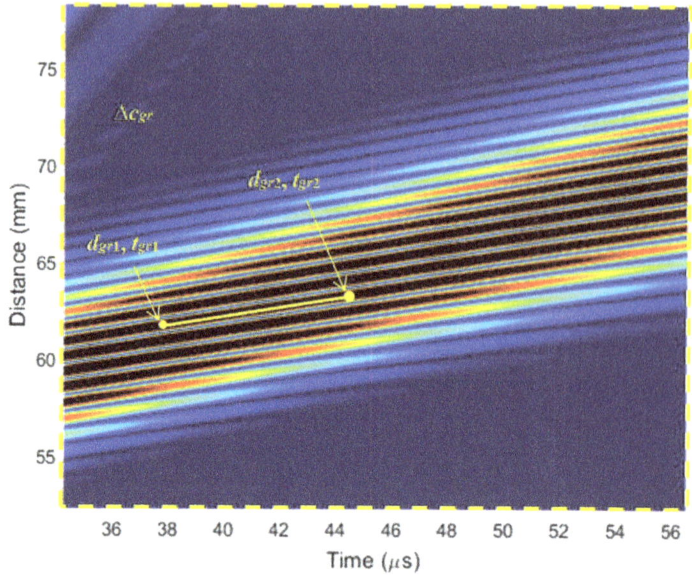

(**b**)

Figure 6. (**a**) The B scan of the in-plane x component particle velocity; (**b**) zoom of the B scan in Figure **6a** (yellow rectangule).

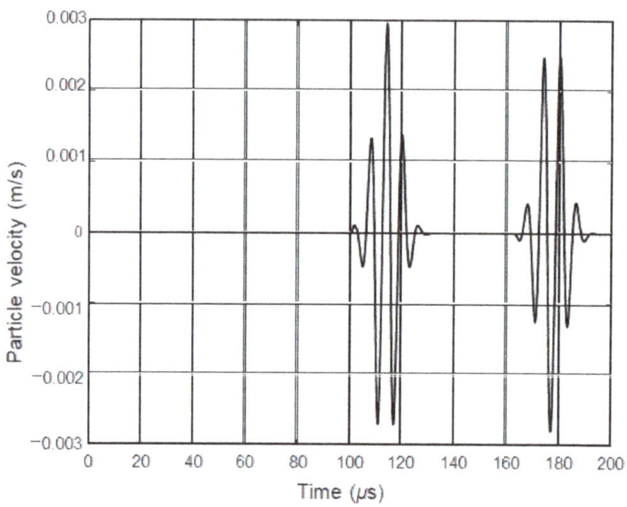

(**a**)

Figure 7. *Cont.*

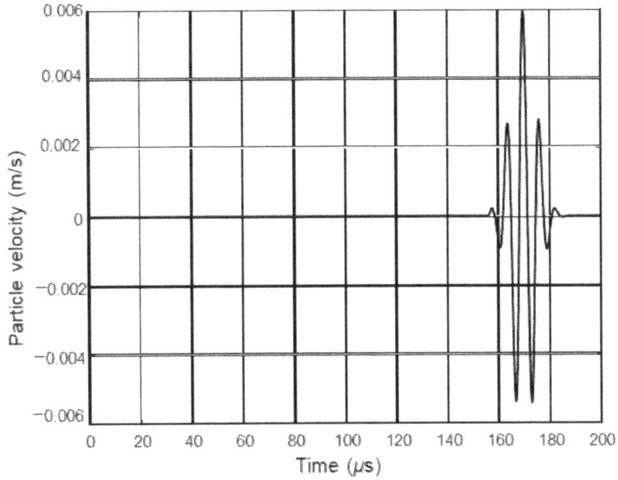

(**b**)

Figure 7. Ultrasonic pulses of the in-plane x component of the S_0 mode at two different distances from the excitation zone: (**a**) 210 mm; (**b**) 300 mm.

The ultrasound velocities of the waves shown in the B scan were calculated by Equations (5) and (6) and compared with the velocities of the guided waves modes obtained by the Semi Analytical Finite Element method (Figure 2). It is possible to conclude that there is a strong A_0 mode propagating with the group velocity c_{gr} = 464 m/s and the phase velocity c_{ph} = 233 m/s. The zoomed B scan of A_0 phase mode is presented in Figure 6b. The S_0 wave mode propagating at the phase velocity of c_{ph} = 1584 m/s can also be seen in Figure 6a.

The ultrasonic pulses of the particle velocities at two different distances x = 210 mm and x = 300 mm are shown in Figure 7a,b. At the distance 210 mm (Figure 7b), two pulses are seen. The first pulse is the directly propagating pulse and the second is the pulse reflected from the edge of the PVC sample. Please note a lower amplitude of the S_0 pulse at the distance x = 210 mm. From the results obtained it follows that a single element with the width 1 mm excites not only A_0 but also S_0 mode.

3.2. Linear Array Model

Therefore, we decided to check if it was feasible to excite S_0 mode by a properly phased linear array placed close to the film surface. For this purpose, we proposed a single ultrasonic transducer to use linear air-coupled arrays, which can be electronically readjusted to optimally excite and receive the required guided wave modes. This linear air-coupled array can improve the excitation of S_0 mode. Numerical investigation was performed simulating the excitation of the S_0 guided wave mode by a planar linear array. The schematic diagram is presented in Figure 8. The array consists of eight strip-like elements of 1 mm width. All spacings between elements should theoretically be equal to $\lambda_{S_0}/2$ = 5.3 mm. For the excitation of the S_0 modes, the linear air-coupled array elements are excited successively with the time delay necessary for this mode to propagate the distance between the adjacent elements. The delay time between elements is $\Delta\tau$ = 3.3 μs and calculated by the following Equation (7):

$$\Delta\tau = \frac{\frac{\lambda_{S_0}}{2}}{c_{Sph}(f,l)}. \quad (7)$$

where λ_{S_0} is the wavelength of the S_0 mode, l is the thickness of the thin PVC film, and f is the frequency. In such case each strip-like piezoelectric element of the linear array excites the S_0 mode in phase, thus increasing the amplitude of the propagating guided wave. The B scan of the x component in-plane particle velocity of the array is presented in Figure 9.

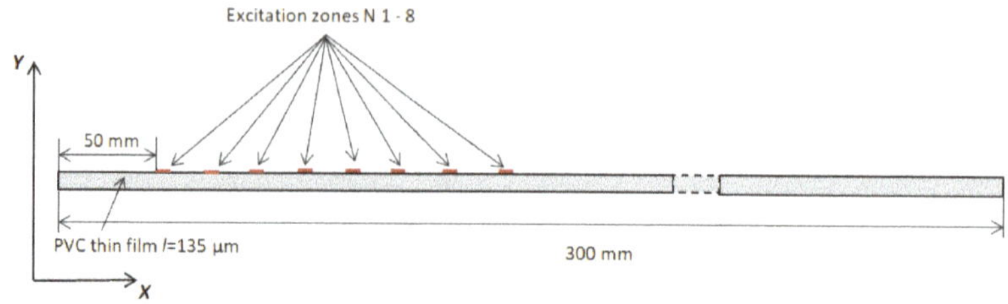

Figure 8. Simulation of excitation of guided waves by a linear phased array film.

For identification of the excited Lamb wave modes shown in Figure 9, we calculated propagation velocities of the simulated modes similar to the single element case and compared them with velocities determined by the Semi Analytical Finite Element method (Figure 2). From the numerical simulation, it follows that the A_0 mode propagates with a group velocity c_{gr} = 459 m/s; the A_0 phase velocity is c_{ph} = 244 m/s. The S_0 mode propagates with the phase velocity of c_{ph} = 1591 m/s (Figure 9). The ultrasonic pulses at three different distances from the excitation zone (x = 50 mm, y = 0 mm) are presented in Figure 10a–c. The first pulse in those figures is the S_0 mode, the second pulse corresponds to the A_0 mode. The ultrasound velocities of the ultrasonic pulses at two different distances from the excitation zone (Figure 10b,c) are calculated by Equations (5) and (6). The obtained S_0 mode phase velocity in this case is c_{ph} = 1551 m/s, which is very close to the velocity obtained by the Semi Analytical Finite Element method.

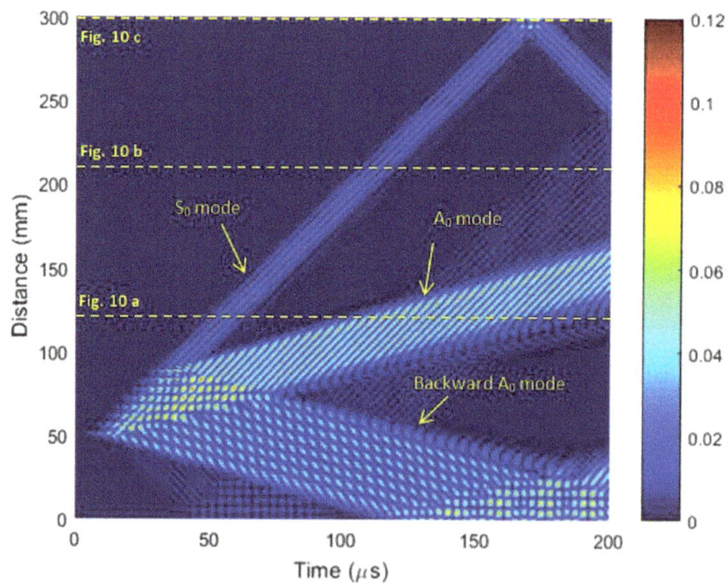

Figure 9. The B scan x component in-plane particle velocity in the thin PVC film.

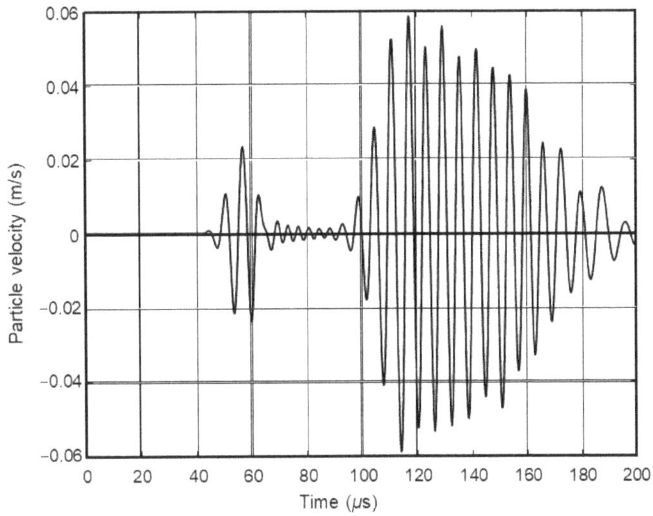

(a)

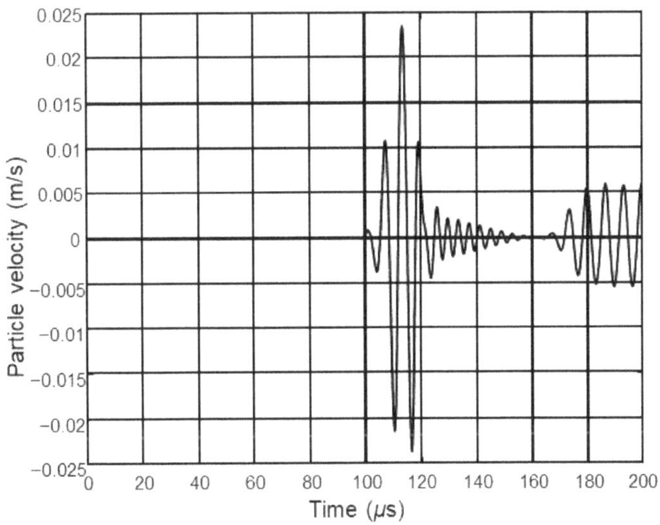

(b)

Figure 10. *Cont.*

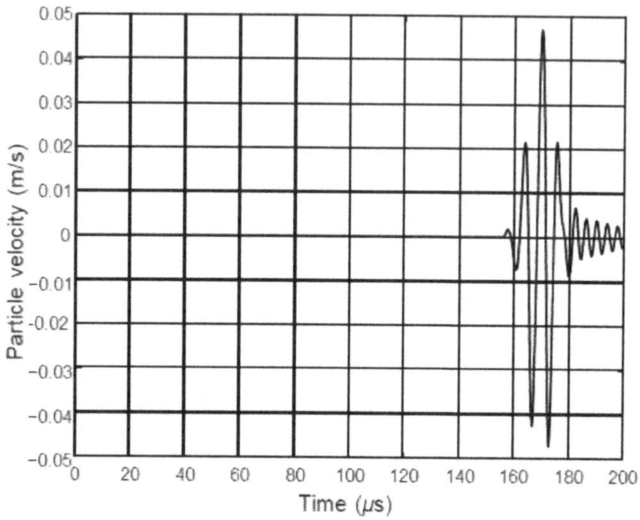

(c)

Figure 10. Ultrasonic pulses at two different distances from the excitation zone: (**a**) 120 mm, (**b**) 210 mm, (**c**) 300 mm.

The particle velocity amplitude at the excitation zone ($x = 50$ mm, $y = 0$ mm) is 0.05 m/s. The array excitation zone is presented in Figure 8. The planar linear array improves the excitation of S_0 mode eight times in comparison to when the excitation zone is 1 mm.

Although the amplitude of the A_0 mode is bigger than the S_0 mode at longer distances, both modes may be exploited for simultaneous measurements because they arrive at different time instants.

For the excitation of the A_0 mode at 150 kHz frequency we proposed to use the same phased array with spacings between array elements equal to $\lambda_{S_0}/2 = 5.3$ mm, however, with introduced delays between the excitation instants of the array elements necessary for the excitation of the A_0 mode:

$$\Delta\tau = \frac{\frac{\lambda_{S_0}}{2}}{c_{Aph}(f,l)}. \tag{8}$$

The delay time between elements in this case is $\Delta\tau = 22.8$ μs. The linear air-coupled array elements become excited successively with the time delay necessary for this mode to propagate the distance between the adjacent elements.

The B scan of the y component off-plane particle velocity of the array is presented in Figure 11. The A_0 mode is more strongly radiated along the y-axis direction. The start excitation point of the linear air-coupled array is shifted 50 mm from the origin of the coordinates $x = 0$ mm, $y = 0$ mm. Each element of the array excites two A_0 mode waves, propagating forward and backward (Figure 11).

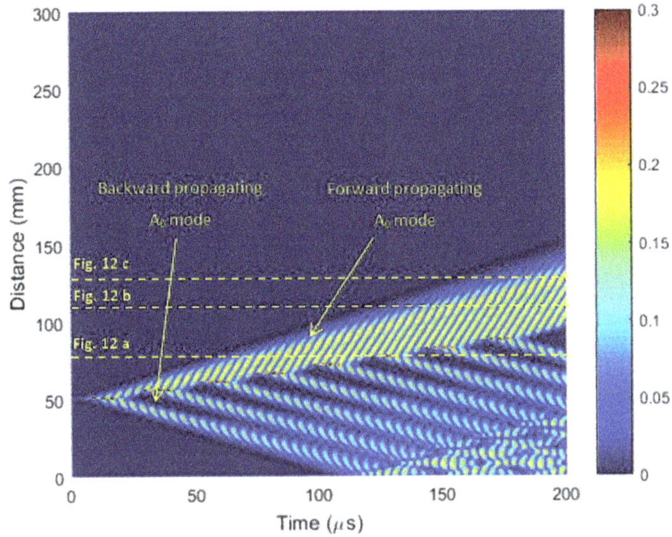

Figure 11. The B scan of the off-plane particle velocity (y component) in the thin PVC film.

The waveforms of the ultrasonic signals at different distances, 75 mm, 110 mm, and 125 mm are shown in Figure 12. The distances are indicated in Figure 11 by the horizontal dashed lines. The forward propagating A_0 mode is properly phased, which results in a single ultrasonic pulse at distances longer than the length of the phased array. Contrarily, the backward propagating wave is not phased and, in this case, the backward wave is propagating not a single ultrasonic pulse, but a series of ultrasonic pulses generated by each element of the array which are shown in the B scan by parallel beams.

It is necessary to point out that in the A_0 mode the off-plane particle velocity component prevails; therefore, it is presented in Figures 11 and 12. From the results presented it follows that there is only a strong A_0 mode excited, which was confirmed by calculating phase and group velocities from the simulated waveforms and comparing them with the corresponding velocities obtained by the Semi Analytic Finite Element method (Figure 2).

The phase and group velocities obtained from the numerical simulation results at the frequency 150 kHz are as follows: group velocity c_{gr} = 458 m/s and phase velocity c_{ph} = 248 m/s. For comparison, the phase and group velocities of the A_0 mode calculated by the Semi Analytic Finite Element method are c_{gr} = 448 m/s and c_{ph} = 232 m/s, which are very close to the results obtained from the numerical simulations. This means that by the proposed excitation method, only a strong A_0 mode is excited in a thin PVC film.

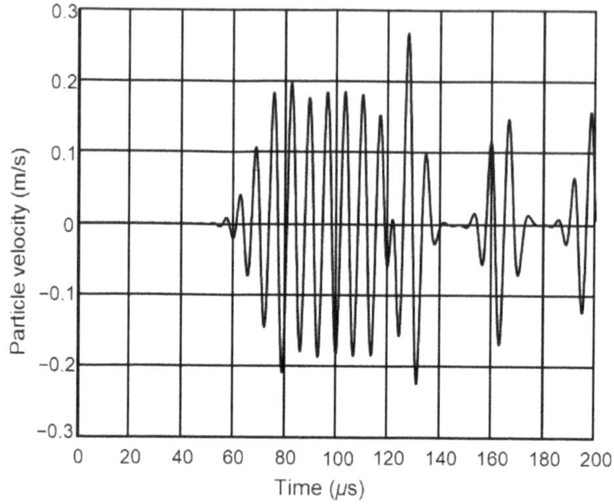

(**a**)

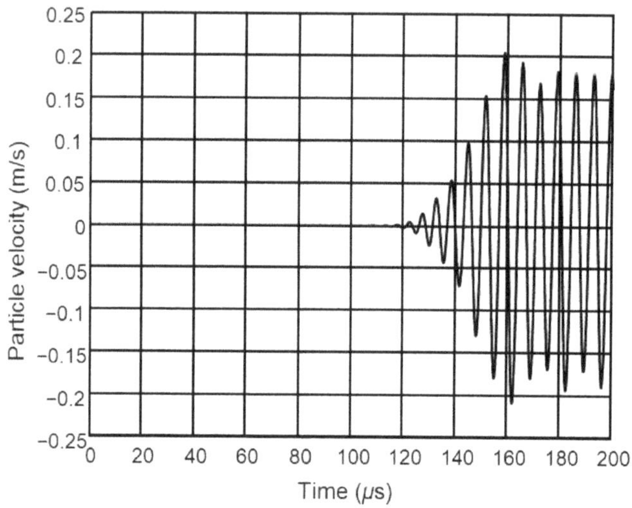

(**b**)

Figure 12. *Cont.*

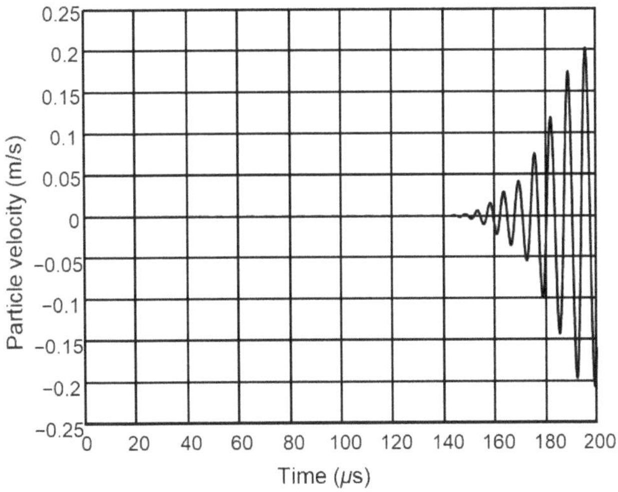

(c)

Figure 12. Ultrasonic pulses of the off-plane *y* component of the particle velocity at three different distances from the excitation zone: (**a**) 75 mm, (**b**) 110 mm, (**c**) 125 mm.

4. Conclusions and Discussion

There are some tasks in which applications of a few different guided waves enable us to obtain more information about structures under a test. For example, the measurement of phase velocities of symmetrical S_0 and antisymmetrical A_0 guided wave modes allowed us to determine the thickness and Young's modulus of thin PVC films [21]. The main problem is how to excite such modes in thin films especially using air-coupled methods.

The performed numerical modelling by 2D and 3D finite element methods showed that for such purpose, a contactless linear phased array might be successfully exploited. It was shown that the most serious problem is the excitation of the symmetrical S_0 mode, as in this mode the biggest challenge is the in-plane displacement component, which is almost impossible to excite by a single ultrasonic transducer. For a solution of this problem, we proposed using a linear phased array where the elements of which are excited by electric signals with properly selected delays. The delays of the excitation instants between adjacent elements are equal to the propagation time of the S_0 mode.

For the excitation of the A_0 mode we proposed using the same phased array, except the delays between adjacent elements of the array in this case were equal to the propagation time of the antisymmetrical A_0 mode. In our case, at the frequency 150 kHz this delay was $\Delta\tau = 22.8$ μs, e.g., 6.9 times longer than the delay for the excitation of the S_0 mode.

The obtained results clearly demonstrate feasibility to excite efficiently different guided wave modes with very different phase velocities using the proposed method based on an electric readjustment of the same phased array.

Author Contributions: R.J.K. supervised the research, wrote and edited the paper; J.S. carried out the simulations, analyzed the data, and wrote parts of the paper; E.Z. analyzed the data of simulations and edited the paper. All authors have read and agreed to the published version of the manuscript.

Funding: This research was funded by the European Social Fund under the No 09.3.3-LMT-K-712 "Development of Competences of Scientists, other Researchers and Students through Practical Research Activities" measure.

Institutional Review Board Statement: Not applicable.

Informed Consent Statement: Not applicable.

Data Availability Statement: Not applicable.

Conflicts of Interest: The authors declare no conflict of interest.

References

1. Fey, P.; Meiler, S.A.; Kreutzbruck, M. Non-Contact inline monitoring of thermoplastic CFRP tape quality using air-coupled ultrasound. In Proceedings of the 19th World Conference on Non-Destructive Testing, München, Germany, 13–17 June 2016; p. 129.
2. Rose, J.L. *Ultrasonic Waves in Solid Media*; Cambridge University Press: Cambridge, UK, 1999; pp. 3–32.
3. Wu, Y.; de Labachelerie, M.; Bastien, F. Investigations on excitation and detection methods for Lamb wave sensors. *Sens. Actuators A Phys.* **2002**, *100*, 214–222. [CrossRef]
4. Kim, D.; Cho, Y.; Lee, J. Assessment of wall-thinning in carbon steel pipe by using laser-generated guided wave. *Nucl. Eng. Technol.* **2010**, *42*, 546–551. [CrossRef]
5. Gao, H.; Ali, S.M.; Lopez, B. Efficient detection of delamination in multilayered structures using ultrasonic guided wave EMATs. *NDT E Int.* **2010**, *43*, 316–322. [CrossRef]
6. Gao, H.; Lopez, B. Development of Single-Channel and Phased Array EMATs for Austenitic Weld Inspection. *Mater. Eval.* **2010**, *68*, 821–827.
7. Rose, J.L.; Soley, L. Ultrasonic guided waves for the detection of anomalies in aircraft components. *Mater. Eval.* **2000**, *50*, 1080–1086.
8. Rose, J.L. Standing on the shoulders of giants: An example of guided wave inspection. *Mater. Eval.* **2002**, *60*, 53–59.
9. Böttger, W.; Schneider, H.; Weingarten, W. Prototype EMAT system for tube inspection with guided ultrasonic waves. *Nucl. Eng. Des.* **1987**, *102*, 369–376. [CrossRef]
10. Quaegebeur, N.; Masson, P.; Berry, A.; Ardin, C.; D'Anglade, P.-M. Ultrasonic non-destructive testing of cardboard tubes using air-coupled transducers. *NDT E Int.* **2018**, *93*, 18–23. [CrossRef]
11. Hirao, M.; Ogi, H. An SH-Wave EMAT Technique for Gas Pipeline Inspection. *NDT E Int.* **1999**, *32*, 127–132. [CrossRef]
12. Neuenschwander, J.; Furrer, R.; Roemmeler, A. Application of air-coupled ultrasonics for the characterization of polymer and polymer-matrix composite samples. *Polym. Test.* **2016**, *56*, 379–386. [CrossRef]
13. Ryden, N.; Lowe, M.J.S.; Cawley, P. Non-contact surface wave testing of pavements using a rolling microphone array. In Proceedings of the 7th International Symposium on Non-Destructive Testing in Civil Engineering, Nantes, France, 30 June–3 July 2009.
14. Worlton, D.C. Ultrasonic testing with Lamb waves. *Non Destr. Test.* **1957**, *15*, 218–222.
15. Kažys, R.; Stolpe, P. Ultrasonic non-destructive on-line estimation of the tensile strength of a running paper web. *NDT E Int.* **2001**, *34*, 259–267. [CrossRef]
16. Kazys, R.J.; Vilpisauskas, A.; Šeštoke, J. Application of Air-Coupled Ultrasonic Arrays for Excitation of a Slow Antisymmetric Lamb Wave. *Sensors* **2018**, *18*, 2636. [CrossRef]
17. Kazys, R.J.; Vilpisauskas, A. Air-Coupled Reception of a Slow Ultrasonic A_0 Mode Wave Propagating in Thin Plastic Film. *Sensors* **2020**, *20*, 516. [CrossRef]
18. Chimenti, D.E. Review of air-coupled ultrasonic materials characterization. *Ultrasonics* **2014**, *54*, 1804–1818. [CrossRef]
19. Mazzotti, M.; Marzani, A.; Bartoli, I. Dispersion analysis of leaky guided waves in fluid-loaded waveguides of generic shape. *Ultrasonics* **2014**, *54*, 408–418. [CrossRef] [PubMed]
20. Chimenti, D.E.; Martin, R.W. Nondestructive evaluation of composite laminates by leaky Lamb waves. *Ultrasonics* **1991**, *29*, 13–20. [CrossRef]
21. Kažys, R.J.; Tumšys, O. Simultaneous Measurement of Thickness and Elastic Properties of Thin Plastic Films by Means of Ultrasonic Guided Waves. *Sensors* **2021**, *21*, 6779. [CrossRef] [PubMed]
22. Bathe, K.J. The finite element method. In *Wiley Encyclopedia of Computer Science and Engineering*; Wah, B., Ed.; Wiley and Sons: Hoboken, NJ, USA, 2009; pp. 1253–1264.
23. Lee, J.F.; Lee, R.; Cangellaris, A. Time-domain finite-element methods. *IEEE Trans. Antennas Propag.* **1997**, *45*, 430–442.
24. Bathe, K.J. The inf-sup condition and its evaluation for mixed finite element methods. *Comput. Struct.* **2001**, *79*, 243–252. [CrossRef]
25. Bathe, K.J. *Finite Element Procedures*; Prentice Hall: Hoboken, NJ, USA, 1996.
26. Zienkiewicz, O.C.; Taylor, R.L. *The Finite Element Method*, 6th ed.; Butterworth-Heinemann: Oxford, UK, 2005; Volume 2.
27. Komatitsch, D.; Vilotte, J.P. The spectral element method: An efficient tool to simulate the seismic response of 2D and 3D geological structures. *Bull. Seismol. Soc. Am.* **1998**, *88*, 368–392.
28. Komatitsch, D.; Vilotte, J.P.; Vai, R.; Castillo-Covarrubias, J.M.; Sanchez-Sesma, F.J. The spectral element method for elastic wave equations-application to 2-D and 3-D seismic problems. *Int. J. Numer. Methods Eng.* **1999**, *45*, 1139–1164. [CrossRef]
29. Chapelle, D.; Bathe, K.J. *The Finite Element Analysis of Shells-Fundamentals*; Springer: Berlin, Germany, 2011.
30. ABAQUS. *ABAQUS Documentation*; Dassault Systemes: Providence, RI, USA, 2011.
31. Molnár, G.; Gravouil, A. 2D and 3D Abaqus implementation of a robust staggered phase-field solution for modeling brittle fracture. *Finite Elem. Anal. Des.* **2017**, *130*, 27–38. [CrossRef]

32. Barbero, E.J. *Finite Element Analysis of Composite Materials Using Abaqus*; International Standard Book Number-13: 978-1-4665-1663-2 (eBook—PDF); Taylor & Francis Group: Boca Raton, FL, USA, 2013.
33. Lowe, P.S.; Fateri, S.; Sanderson, R.; Boulgouris, N.V. Finite element modelling of the interaction of ultrasonic guided waves with coupled piezoelectric transducers. *Insight-Non-Destr. Test. Cond. Monit.* **2014**, *56*, 505–509. [CrossRef]
34. Barauskas, R.; Daniulaitis, V. Simulation of the ultrasonic wave propagation in solids. *Ultragarsas Ultrasound* **2000**, *37*, 34–39.
35. United States Plastics Corporation. Typical Physical Properties: Vintec®Clear PVC. Available online: http://www.usplastic.com/catalog/files/specsheets/Clear%20PVC%20-%20Vycom.pdf (accessed on 10 April 2021).

applied sciences

MDPI

Article

3D Modelling of the Scattering of the Fundamental Anti-Symmetric Lamb Mode (A_0) Propagating within a Point-Impacted Transverse-Isotropic Composite Plate

Dilbag Singh [1,2], Mourad Bentahar [1,*], Charfeddine Mechri [1,3] and Rachid El Guerjouma [1]

1 Laboratoire d'Acoustique de l'Université du Mans (LAUM), Le Mans Université, CNRS UMR 6613, Avenue Olivier Messiaen, 72085 Le Mans, France; kahlondilbag80@gmail.com (D.S.); cmechri@cttm-lemans.com (C.M.); rachid.elguerjouma@univ-lemans.fr (R.E.G.)
2 Department of Mathematics, Panjab University, Chandigarh 160014, India
3 Centre de Transfert de Technologie du Mans (CTTM), 20 Rue Thalès de Milet, 72000 Le Mans, France
* Correspondence: mourad.bentahar@univ-lemans.fr

Abstract: The present paper deals with an effort to model impact damage in 3D-FE simulation. In this work, we studied the scattering behavior of an incident A_0 guided wave mode propagating towards an impacted damaged zone created within a quasi-isotropic composite plate. Besides, barely visible impact damage of the desired energy was created and imaged using ultrasonic bulk waves in order to measure the size of the damage. The 3D-FE frequency domain model is then used to simulate the scattering of an incident guided wave at a frequency below an A_1 cut off with a wavelength comparable to the size of the damaged zone. The damage inside the plate is modeled as a conical-shaped geometry with decayed elastic stiffness properties. The model was first validated by comparing the directivity of the scattered fields for the A_0 Lamb mode predicted numerically with the experimental measurements. The modeling of the impact zone with conical-shape geometry showed that the scattering directivity of the displacement field depends significantly on the size (depth and width) of the conical damage created during the point-impact of the composite with potential applications allowing the determination of the geometric characteristics of the impacted areas.

Keywords: guided waves; CFRC composite plate; impact point damage; 3D-FE modeling

Citation: Singh, D.; Bentahar, M.; Mechri, C.; El Guerjouma, R. 3D Modelling of the Scattering of the Fundamental Anti-Symmetric Lamb Mode (A_0) Propagating within a Point-Impacted Transverse-Isotropic Composite Plate. *Appl. Sci.* **2021**, *11*, 7276. https://doi.org/10.3390/app11167276

Academic Editor: Habil. Michel Darmon

Received: 21 June 2021
Accepted: 3 August 2021
Published: 7 August 2021

1. Introduction

There is a rapid increase in the use of carbon-fiber-reinforced composite materials in the aerospace industry over the last decade [1,2]. The main advantage includes low weight, high static and fatigue strength, and the possibility to manufacture large integral shell structures. In present time, commercial aircraft contain such composites for primary and secondary wing and fuselage components. However, the main drawback of carbon fiber/epoxy is that these are inherently brittle and usually exhibit a linear elastic response up to failure with little or no plasticity. This makes composite structures very vulnerable to impact damage and must therefore satisfy certification procedures for high-velocity impact from runway debris or bird strike. Impact damage is considered to be a serious damage mechanism in composite structures, which limits performance and reliability for further use. The problem with impact damage is that it may not be visible from the surface of the material, and even so, small scratches at the surface may hide severe damages underneath the impacted surface. Indeed, the impact generates a shock wave whose mechanical energy propagates deep inside the material, thus causing different types of damage along its path. Generally, impact damage occurs during in-service applications or as a result of handling during manufacturing. Very common examples of impact events include tool drops during the manufacturing, preparation or storage of the structure, or during in-service use, e.g., birds' collisions for aeronautic components, hail impact, volcanic ash, runway debris etc. Meanwhile, it is also well known that mechanical properties of composite materials can be severely degraded by the initiation and propagation of structural damages (fiber breakage, delamination, matrix cracking, etc.), which may appear during the components' life cycles [1]. Since impacts can locally cause crack-like defects, they also may lead to local decays in material properties, i.e., in the vicinity of the impacted zone [1,3–5]. It is therefore necessary to reliably detect and evaluate impact damages, as well as their effects on residual resistance of the structure.

The use of non-destructive techniques to evaluate defects, such as the ones based on ultrasonic waves, is therefore suitable because the tested component remains available either for tests, as mentioned above, or for reuse in the whole structure. In particular, ultrasonic guided waves are very advantageous because they can finely interrogate the whole thickness of the material and be used for imaging internal hidden defects when appropriate post-processing is applied. Several authors have worked on locating the point of impact in composite plates [5–8]. Fromme and Rouge [9] investigated the directivity of the A_0 Lamb mode scattered at a crack-like notch defect in an isotropic plate. Moreau et al. [10,11] analytically and numerically discussed the scattering of guided waves partly through thickness and/or flat bottom cavities with irregular shapes in plates. Caminero et al. [12] studied the sensitivity of phased array techniques to detect internal damage caused by impacts at different energy levels on unidirectional and multidirectional laminates depending on the stacking sequence. The problem of impact damage evolution and penetration within thick-section composites was investigated by Gama and Gillespie [13] using explicit finite element (FE) analysis in 3D. Various studies on modeling guided waves in composite plates have been performed by Veidt and his coworkers [14]. Agreement between modelling and experimental data is, in general, achieved for the signals, which are less affected by the damage scattered waves. This is mainly due to the complexity of the involved phenomena during the scattering of ultrasonic guided waves in the presence of point impact damage in composites [15].

Analytical solutions corresponding to Lamb waves scattering in the presence of delamination do not exist for composite laminates due to their multilayer characteristics [16]. Within the framework of a plane strain assumption, the reflection of the fundamental symmetric (S_0) Lamb wave from a delamination in unidirectional and cross-ply laminates showed that S0 Lamb wave cannot be used to detect the delamination at locations with zero shear stress [17]. On the other hand, the use of a 2D method to study the reflection characteristics of S_0 and A_0 Lamb waves from a delamination showed that the A_0 Lamb wave is sensitive to delamination at all through-thickness locations [18]. However, it should be noticed that the wave scattering mechanism depends on the involved guided mode as well as on the depth and size of the damage zone. In the case of a delamination, various investigations have shown that A_0 and S_0 modes propagate individually in the two regions separated by delamination and then interact with each other after exiting the delamination area [19,20]. However, the scattering of guided waves in the case of an impact damage is different from the one corresponding to a delamination created along the full width of the composite laminates. Indeed, Cantwell et al. [21] found that in thin composite specimens, the damage is initiated in the bottom layers, whereas in thick specimens, the damage is initiated in the top layers. As a result of the stresses induced by the hemispherical impactor, fiber breakage is expected in the top plies. The impact damage in the bottom layers is mainly dominated by delamination due to bending stresses, which does not cover the whole thickness of the composite [22,23]. For a given impact energy, the delamination area is expected to be proportional to the bending stiffness mismatching coefficient between the matrix and reinforcements. Therefore, high mismatching coefficients are to be expected when the difference between young's moduli in the fiber and matrix directions is important [24]. Scatter amplitudes and scatter directivity distributions depend on the delamination size to wavelength ratio and the through-thickness location of the delamination damage. This effect is often studied by bonding masses to the surface of composite laminates in order to simulate delamination damage [16]. Different techniques and approaches have been used in order to model the propagation of ultrasonic guided waves within impacted composite plates. Zhang et al. proposed a modelling that combines low-velocity impact damage with the analysis of guided wave propagation, where delamination and matrix cracking are both directly meshed in the guided-wave model [25]. However, different types of damage could be created when composites are submitted to a low velocity impact. Indeed, impact damages in CFRP structures include delamination, shear cracks, transverse cracks, etc., which makes the precise modelling of all these micro-damages difficult [26–28]. The complexity of the multiple scattering phenomena, local resonances, etc., makes the use of a physical approach important and more realistic. Indeed, the effective medium approximation offers a good opportunity to simplify the modeling of complex media, where the inhomogeneous medium can be considered as homogeneous with effective properties. An application of the homogenization approach in the case of a strongly scattering medium can be found in the following references [29–32].

This work investigates the directivity pattern of the scattered field of incident A_0 guided wave by the impact damage and provide much more material and a more in-depth study in connection with new experimental results to our previous contribution [33]. Unlike most of the studies in the literature, the damage zone considered in this paper is not across the full width of the laminates. Therefore, the three-dimensional (3D) characteristics of the damage zone are presented and used in a 3D frequency domain finite element model [34] to simulate the scattering phenomenon of fundamental incident A_0 Lamb mode sent towards the damaged zone. The latter has been modeled as a right, circular, conical-shaped geometry with decayed material stiffness properties. Instead of only studying the backward and/or forward scattering [35], the present work investigates the scattering characteristics in different directions, where the frequency of the incident Lamb mode is chosen below A_1 cut-off. It is found that in composite plates, the directivity of scattered A_0 mode depends on the defect diameter to wavelength ratio, and on the through thickness of the conical damage as well. The numerically predicted scattered wave fields are compared with experiment measurements showing good agreement. Finally, the numerical finite-element-based-scattering characteristics is discussed for different geometrical parameters with a view towards modeling impact damage using conical-shape geometry with decay material properties.

2. Composite Plate and Material Characterization

The CFRC composite plate sample consists of 20 plain weave pattern laminas of 0.3 mm thickness each oriented in $[45/0/0/45/0/45/0/45/0/45]_S$ directions. The average thickness of the 50×50 cm^2 composite plate is approximately 6.2 mm. The homogenized elastic moduli are optimized by matching them with the experimental group velocities of A_0 and S_0 Lamb modes for a given set of frequencies and the through thickness ultrasonic bulk wave velocities. The schematic of the composite plate as well as the measurement directions of the excited Lamb waves are shown in Figure 1. The homogenized composite plate is transversely isotropic (quasi-isotropic) with Y-axis as axis of symmetry. The A_0 and S_0 Lamb wave velocities are measured around the emitter (shown in red dot) along different angular positions and in a straight line from the emitter (shown in green and blue dots respectively) in XZ-plane. Velocities were found to be unchanged for all directions in the considered frequency range, namely 10 kHz–150 kHz. Under the transversely isotropic condition, the inversion method was performed by fitting the analytic group velocity dispersion curve with the experimental group velocities corresponding to A_0, S_0 and SH_0 modes as well as bulk ultrasonic waves generated around the abovementioned frequency domain. The density ρ of the plate is first measured as 1.55 ± 0.05 g/cm^3. The elastic constant C_{22} (along Y-axis) is obtained by measuring the longitudinal wave velocity along Y-axis where $V_L = 2860 \pm 10$ m/s, which gives $C_{22} = V_L^2 * \rho = 12.6$ GPa. The measured shear wave velocity propagating along Y-axis and vibrating in Z-axis is found to be equal to $V_T = 1550 \pm 10$ m/s giving the elastic constant $C_{44} = V_T^2 * \rho = 3.72$ GPa. The group velocity related to SH_0 wave is found to be equal to 3300 ± 100 m/s. The value of the constant C_{13} is then adjusted in the dispersion curve to achieve the same velocity, where C_{13} is found to be equal to 17.8 GPa. On the other hand, the group velocity of S_0 Lamb mode at 100 kHz is 5700 ± 100 m/s and the one corresponding to A_0 mode at 40 kHz is 1500 ± 100 m/s. The speed of A_0 mode primary depends on the constant C_{44}, which is obtained using the abovementioned shear wave velocity V_T. The speed of S_0 mode (at 100 kHz) depends on the constants C_{11} and C_{12}, whose optimization is essential to the constant C_{55}, where $C_{55} = 1/2 * (C_{11}\text{-}C_{13})$ in the case of a transversely isotropic symmetry. The constants C_{11}, C_{12} (or C_{23}) are optimized using the S_0 wave velocity. It is observed that a linear increase of C_{11} (keeping other constants fixed) will result in a linear increase in the group velocity of S_0 mode while a linear increase of C_{12} (or C_{23}) will result in a linear decrease in the group velocity of S_0 mode (keeping other constants fixed). The change in C_{12} from 2 GPa to 10 GPa results in a linear decrease from 5800 to 5000 m/s in the group velocity of S_0. The values of C_{11} and C_{12} are adjusted to fit the measured velocity of 5700 m/s. The optimized values of the stiffness tensor are given in Table 1.

Table 1. Homogenized elastic properties of the carbon fiber reinforced composite plate.

ρ (g/cm^3)	C_{11} (GPa)	C_{22} (GPa)	C_{33} (GPa)	C_{12} (GPa)	C_{13} (GPa)	C_{23} (GPa)	C_{44} (GPa)	C_{55} (GPa)	C_{66} (GPa)
1.55 ± 0.05	52.4 ± 2%	12.6 ± 2%	52.4 ± 2%	3.1 ± 2%	17.8 ± 2%	3.1 ± 2%	3.7 ± 2%	17.3 ± 2%	3.7 ± 2%

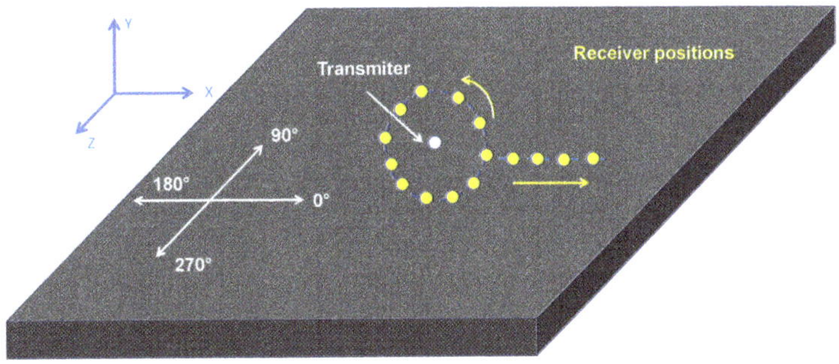

Figure 1. Schematic for measuring group velocities of A_0 and S_0 Lamb wave modes.

Frequency dispersion curves corresponding to phase velocities are computed using the semi-analytic finite element (SAFE) method for the composite plate using the optimized elastic constants given in Table 1 and shown in Figure 2. Finally, we note that the attenuation is determined by measuring the amplitude variation for a given propagation distance in the far field. Measurements have shown that for the same propagation distance, the amplitudes of the considered fundamental Lamb modes decrease by approximately ~12% regardless of the direction of propagation, which corresponds to ~0.13 Np/m.

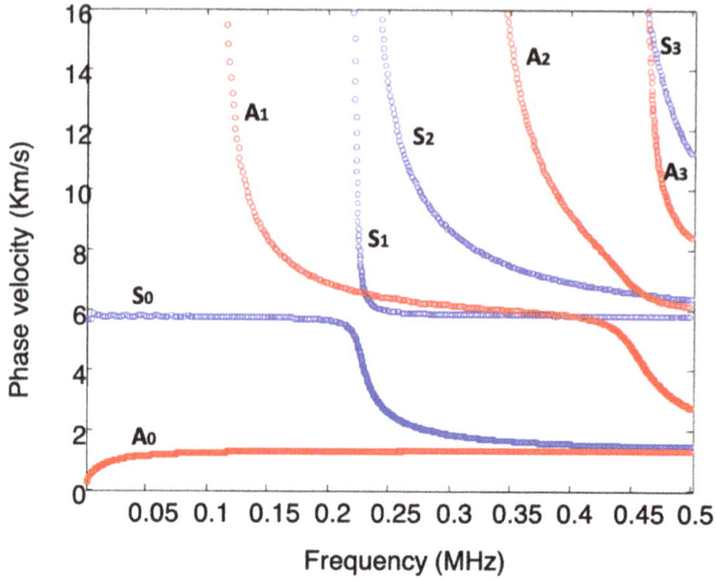

Figure 2. Optimized phase velocity dispersion curves corresponding to 6.2 mm-thick composite plate.

3. Impact Damage Setup and Ultrasonic C-Scan

A controlled laboratory experiment is performed to create point impact damage on the given sample plate as shown in Figure 3. The impact tests are carried out using an Imatek IM10 ITS drop tower with a capacity of 4000 Joules according to the ISO standard [ISO standard 6603-2, 2000]. The test is carried out using a falling mass in the form of a trolley with a mass of 8 kg fitted with a hemispherical impactor with a diameter of 20 mm. The composite plate is fixed inside a metallic part with an open area for impact. An anti-bouncing device is implemented to avoid a second shock, which could further damage the structure at a random energy. In order to obtain the desired energy, the drop height is changed. The specimen is placed in a compressed air clamping system on a circular support with an internal diameter of 40 mm and an external diameter of 60 mm in order to hold the specimen during the impact. The quantities measured, using sensors, are the impact force as well as the displacement of the falling mass. Then, the energies, initial (E_i), absorbed (E_{abs}) and restored (E_r), are post-processed thanks to the displacement–time and effort–displacement graphs. Preliminary impact tests were carried out on similar composite plates in order to create a non-visible damage. We have found that 40 J of impact energy is sufficient to damage the plate without penetrating through, where damage is not visible on the impact side and barely visible on the other side.

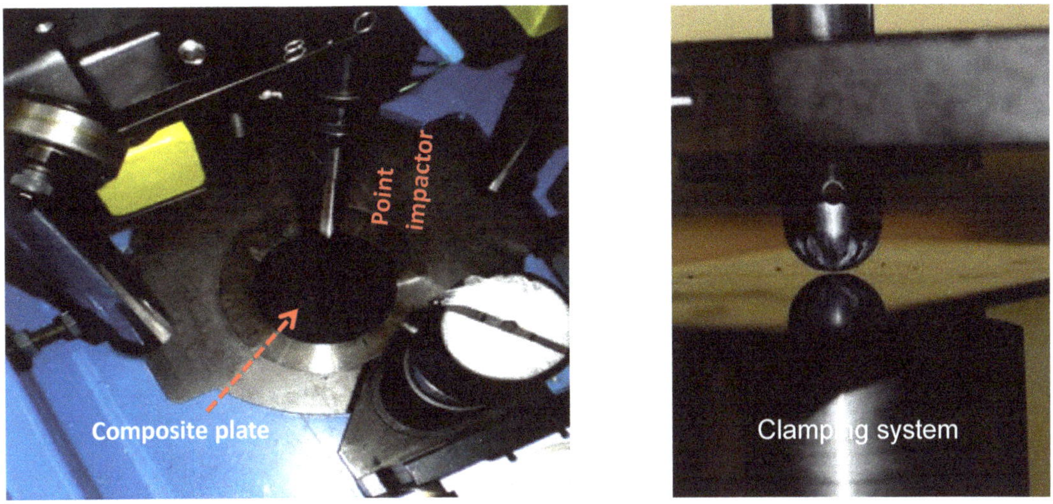

Figure 3. Experiment setup for controlled impact damage of energy equal to 40 Joule.

Immersion c-scan ultrasonic images are obtained using 5 MHz longitudinal pulses propagating through the composite plate. The ultrasonic c-scan image is obtained using a pulse-echo configuration, where gates 1 and 3 correspond to the front and back wall echoes as presented in Figure 4a. The c-scan image shown in Figure 4b is performed on the basis of the change in the back-wall amplitude of the ultrasonic wave, which drops in the presence of damage. Indeed, the defect area appearing in blue color indicates that the 40-joules point-impact test creates a damaged area of ~30 mm diameter around the impact point. In the literature, the conical-shape cavity has been found in different contributions during the impact tests performed on composites. Indeed, the overall damage pattern through the thickness in composite materials follows a conical shape as found by [35–37]. The size of the defect region as well as the involved damage mechanisms depend on the used impactors (hemispherical, conical, etc.) and on the composite characteristics as well. The use of hemispherical impactors was found to produce larger delamination areas compared to a conical impactor in laminates [23]. For instance, Lee et al. [38] showed that the different impactor shapes produced different damage mechanisms, which directly affect the energy absorption characteristics of the material. This result was also confirmed by Zhou et al. [39]. Using a metallographic microscope, Shyr et al. found that the structure of the fibers affects the delamination pattern and size [40]. Since the composite studied is transverse-isotropic, we can reasonably suppose, on the basis of the presented ultrasonic c-scan imaging as well, that the structure of the fibers in our case affect the delamination in an isotropic way. The 3D finite elements (3D-FE) model will then reasonably consider a damage area with a conical geometry and the same diameter, the value of which is equal to 30 mm according to the c-scan image.

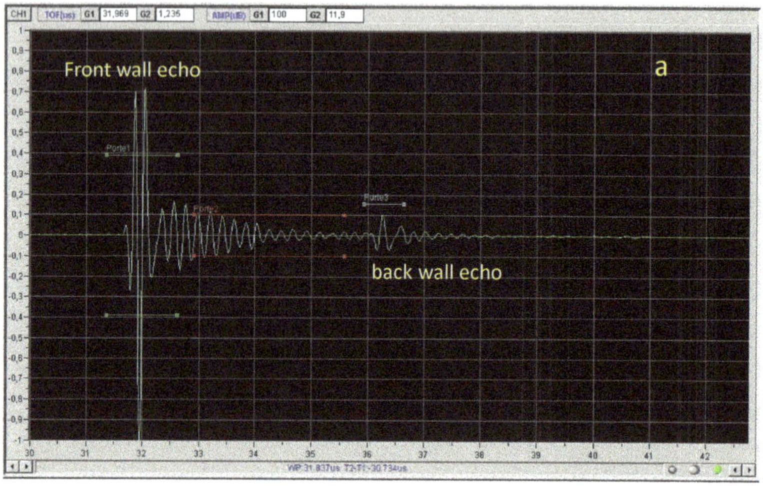

(**a**)

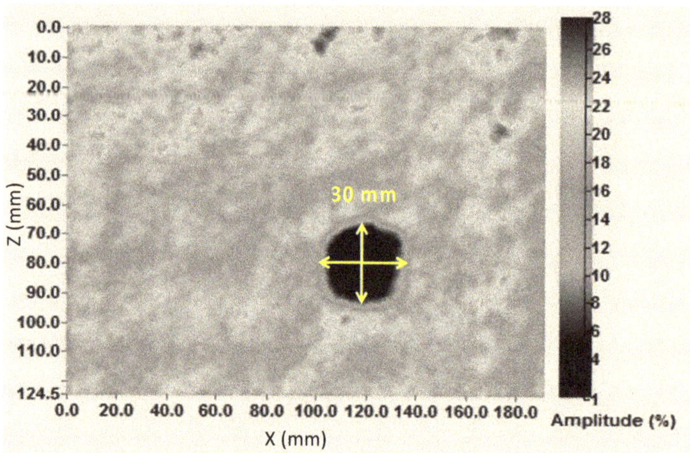

(**b**)

Figure 4. (**a**) A-Scan signals corresponding to the reflection of a longitudinal ultrasonic wave at the defect position of the impacted CFRC plate; (**b**) C- Scan image based on the A-scans performed at intact and damaged regions of the CFRC plate. The amplitude of the back-wall echo is strongly decreasing in the damaged area. The damage was found to be ~30 mm wide.

4. Impact Damage Shape and Modeling in 3D-FE Model

Due to point impact damage and based on the ultrasonic imaging, the composite plate is seriously micro-cracked inside the impact damaged zone. In order to model the damage zone in 3D-FE model, we can reasonably assume that a uniform point impact in the quasi-isotropic plate would produce a damage zone, the shape of which would be a right circular cone due to the impact conditions and the transversely isotropic nature of the plate. Damage in composites is a collection of various types of cracks of different characteristics (matrix voids, delamination, broken fibers, etc.), depending on the applied stress and the composite architecture. At the time when the strength can be clearly defined in metals (ex. unstable growth of a crack at the origin of the brittle fracture), failure in composite materials is still an active research topic, which often generates lively debates in the scientific community [41].

The development of a simple modeling method can be proposed when the wavelength of a Lamb wave is at least one order of magnitude greater than the microscopic defects created within the composite. In such a case, the homogenization procedure of the elastic properties can be expressed through a degradation of the elastic properties. In the literature [42], it has been observed that a decay of 80% in the elastic properties is quite representative of a severely cracked and delaminated zone, in terms of wave scattering. The choice of 80% decay in the material stiffness rather than other smaller values is motivated by the fact that the damaged zone in FE model is considered as a homogeneous medium with mass density equal to that of undamaged material (unchanged thickness), although it would likely be a heterogeneous region made of cracks, voids and delamination, which are known to be efficient scatterers of elastic waves. In the present FE model, we considered an 80% decrease in the material stiffness properties within the cracked zone, which is modeled as a right circular cone with height (h) and diameter (2 r). Finally, note that it was not possible through the ultrasonic imaging experiment to see how deep the damage occurred penetrating the plate through the thickness from the above surface. However, as the impact did not penetrate through the whole thickness, and for comparison with experimental data, the undamaged thickness below the conical damage is considered to be ~1 mm from the bottom surface of composite plate. Finally, we note that the effect of the damage depth on the scattering of the A_0 mode will be presented in the present contribution.

5. 3D FE Model and Schematic of the Problem

To simulate the dynamic response of the viscoelastic composite plate in 3D, a commercially available numerical analysis package Comsol Multiphysics [43] based on the Finite Element method is used. Preliminary tests on guided waves dispersion curves were performed using the abovementioned software and revealed that all simulations matched well with the theory and also agreed fairly well with experiments [44]. In the proposed simulations, the PDE mode in coefficient form is used to model the composite plate, and Neumann boundary conditions have been applied on all faces of the plate. At the interface between the conical damage zone and the composite plate, the continuity of stress and displacement is used to solve the numerical model. Besides, one external routine is developed for computing normal displacement fields at the midplane around the point impact zone in a circular region at 360 different angular positions. The equation of the dynamic equilibrium for orthotropic viscoelastic material, with nine complex moduli, is written and solved in the Fourier domain in the 3D FE model as: [13]

$$\sum_{j,\,k,l=1}^{3} C_{ikjl}\left[\frac{\partial^2 u_j}{\partial x_k \partial x_l}\right] + \rho\omega^2 u_i = 0, \quad i = 1,\ 2,\ 3, \tag{1}$$

where u_i is the Fourier transform of the components of the displacement vector $\boldsymbol{u}$, with $i = 1, 2, 3$, representing the direction of the coordinate axis as shown in Figure 5. Here C_{ikjl} is the component of the complex stiffness tensor, ρ is the material density and ω is the angular frequency. The above partial differential equations (PDE) have been written in the following form as: [13]

$$\nabla(c\nabla u) - au = 0 \tag{2}$$

where c is a 3×3 matrix. For an orthotropic material, the axis of symmetry coincides with the coordinate axis, and this matrix is composed of the following sub-matrices c_{kl}, $k, l = 1, 2, 3$ as shown is Equation (3): [13]

$$c_{11} = \begin{pmatrix} C_{11} & 0 & 0 \\ 0 & C_{66} & 0 \\ 0 & 0 & C_{55} \end{pmatrix}, c_{12} = \begin{pmatrix} 0 & C_{12} & 0 \\ C_{66} & 0 & 0 \\ 0 & 0 & 0 \end{pmatrix}, c_{13} = \begin{pmatrix} 0 & 0 & C_{13} \\ 0 & 0 & 0 \\ C_{55} & 0 & 0 \end{pmatrix},$$

$$c_{21} = \begin{pmatrix} 0 & C_{66} & 0 \\ C_{21} & 0 & 0 \\ 0 & 0 & 0 \end{pmatrix}, c_{22} = \begin{pmatrix} C_{66} & 0 & 0 \\ 0 & C_{22} & 0 \\ 0 & 0 & C_{44} \end{pmatrix}, c_{23} = \begin{pmatrix} 0 & 0 & 0 \\ 0 & 0 & C_{23} \\ 0 & C_{44} & 0 \end{pmatrix}, \quad (3)$$

$$c_{31} = \begin{pmatrix} 0 & 0 & C_{55} \\ 0 & 0 & 0 \\ C_{31} & 0 & 0 \end{pmatrix}, c_{32} = \begin{pmatrix} 0 & 0 & 0 \\ 0 & 0 & C_{44} \\ 0 & C_{32} & 0 \end{pmatrix}, c_{33} = \begin{pmatrix} C_{55} & 0 & 0 \\ 0 & C_{44} & 0 \\ 0 & 0 & C_{33} \end{pmatrix},$$

The expression of 'a' is written as:

$$a = \begin{pmatrix} -\rho\omega^2 & 0 & 0 \\ 0 & -\rho\omega^2 & 0 \\ 0 & 0 & -\rho\omega^2 \end{pmatrix} \quad (4)$$

where C_{ij} are the elastic moduli. As the 3D-FE model consists of modeling the propagation in the frequency domain, absorbing regions are used all around the plate to avoid undesired reflections. The absorbing regions (AR) in the 3D model are defined by a gradual increase of damping properties as explained in [45], where its formula is written as,

$$C_{ij}^{AR} = C'_{ij}\left(1 + i * \left(\frac{Ra}{La}\right)^3\right) + i * C''_{ij}, \ C_{ij} = C'_{ij} + iC''_{ij} \quad (5)$$

Here, 'Ra' and 'La' are the interval and length of absorbing region respectively. The size of the AR used in this FE model is 1.5 λ_{MAX}, where λ_{MAX} is the maximum wavelength for all modes existing in the plate in the studied frequency range. The full schematic of the 3D-FE model is shown in Figures 5 and 6. According to this configuration, the schematics of the laboratory experiment are shown in Figure 7. Since the frequency of the incident Lamb wave mode is chosen below the A_1 cut-off, λ_{MAX} is the wavelength of the S_0 mode at the frequency of interest.

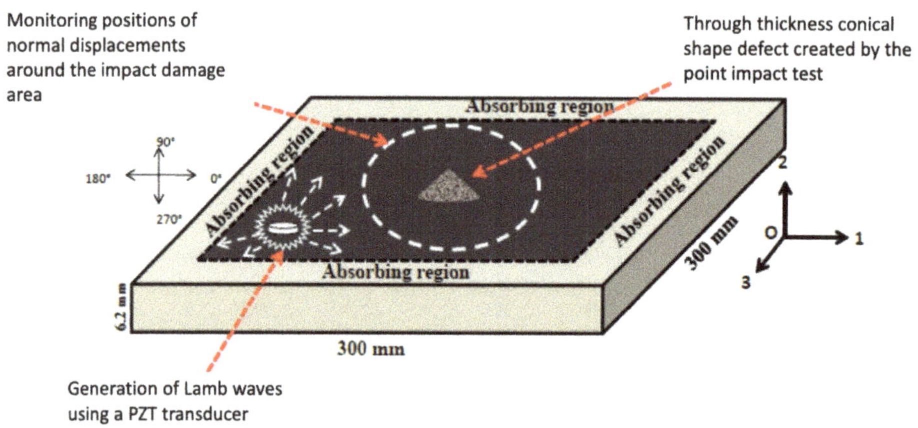

Figure 5. Schematic of the 3D-finite element model.

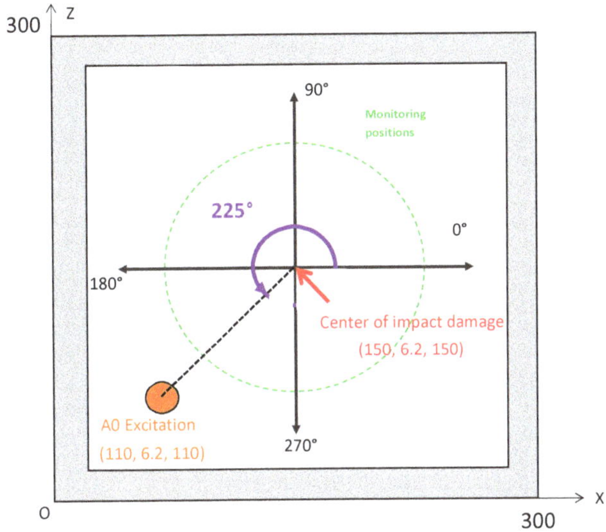

Figure 6. Schematic of the considered positions of excitation and reception in the model and experiments.

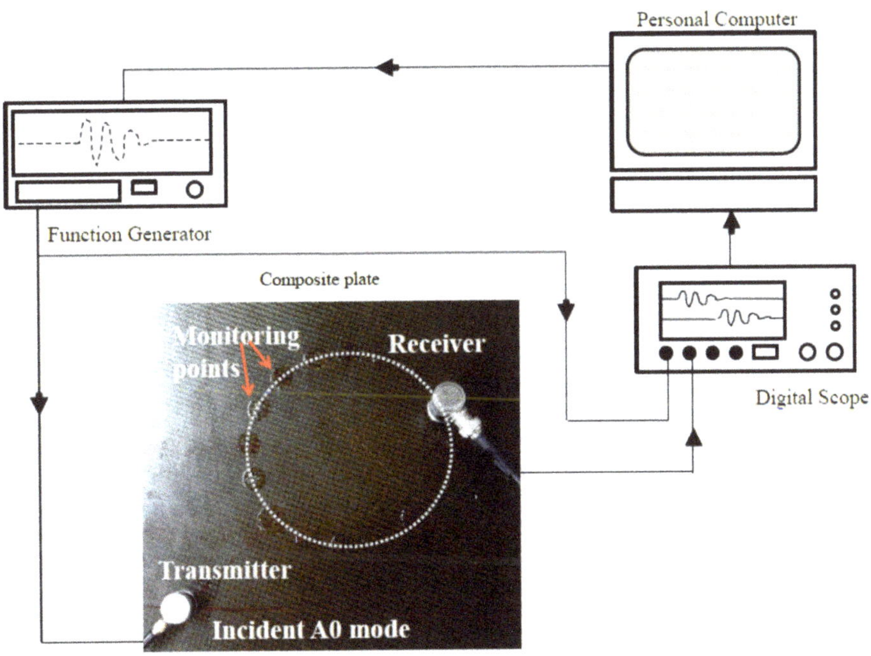

Figure 7. Schematic of the experimental setup for measuring the out-of-plane amplitude of the A_0 mode.

It is observed in the phase velocity dispersion curves in Figure 2 that the higher wave mode A_1 starts to appear right after 120 kHz. Accordingly, the excitation frequency of 100 kHz is chosen to avoid the appearance of higher-order Lamb modes. At 100 kHz, the wavelengths of A_0 and S_0 modes are equal to 13.9 mm and 57.7 mm, respectively. Therefore, the size of the AR is considered equal to 87 mm (≈1.5 × 57.7) all around the four sides of the 3D-FE model.

6. Interaction of A_0 Mode with Impact Damage and Comparison between Numerical Predictions and Experiment

To compute the scattered displacement field of the incident A_0 mode sent towards the impacted zone, two different 3D-FE models are developed. At first, a 3D-FE model, taken as a reference, consisting of an undamaged plate, is solved to compute the incident wave field corresponding to the A_0 mode. The second 3D FE model, which takes into account the conical defect representing impact damage, is solved to compute the total wave field of the incident A_0 mode. The incident field computed from the first model is then subtracted from the total wave field computed from the second model to obtain the complex scattered wave field of incident A_0 mode. In both undamaged and damaged cases, the dimensions of 3D-FE models are equal to 300 mm × 300 mm × 6.2 mm. The excitation of the pure A_0 mode is achieved by giving the unit norm$_a$l stress at a single frequency (in out of phase manner) using two circular regions, with diameters equal to 13 mm, placed on the opposite sides of the plate and centered at points (110, 0, 110) and (110, 6.2, 110) as shown in Figure 6. The reference model contains 724,365 tetrahedral mesh elements with 3,132,252 degrees of freedom that makes at least four elements per wavelength along the thickness of the plate and at least four elements along the *XZ* plane insuring good convergence and accuracy of the FE solution. For the validation of the FE model with the experiment, the model is first solved for a Gaussian profile sinusoidal signal with seven cycles centered at a frequency of 100 kHz by inserting its respective complex Fourier amplitudes as normal stresses on both sides of the plate for 63 different frequencies running between 76 kHz and 123 kHz. After solving the FE model using a linear solver with Lagrange quadratic elements and applying Neumann boundary conditions, the excitation and reception of signals are performed using the inverse Fourier transform of the normal components of the displacement at (110, 6.2, 110) and (210, 6.2, 210), respectively. These two points are separated by a distance of 142 mm. The reconstructed signals of excitation and reception predicted by the FE model are shown in red in Figure 8. Some angular signals around the center at (150, 6.2, 150) are also reconstructed and are shown in Figure 9. The latter figure shows that the amplitude at 225° is the highest and that the ones corresponding to the symmetrical positions at (180° and 270°) and (0° and 90°) have lower amplitudes and are equivalent. Since our study only deals with the scattering of amplitude, all 3D-FE models are solved for a single excitation frequency of 100 kHz to save computation time and cost. The amplitude of the incident A_0 mode, i.e., the normal component of the displacement vector 'v_{inci}', is monitored at the mid plane (at y = 3.1 mm), where 360 positions around a circle centered at (150, 3.1, 150) with a radius equal to 20 mm are considered. The center position corresponding to excitation and the center of the monitoring circle causes an incident angle of 225° in such an arrangement of the sensors as shown earlier in Figure 6. The numerically computed normalized incident field in the absolute value (shown in solid line) is shown in Figure 10. In the second FE model (damaged case), impact damage is introduced as a right circular cone with a diameter of 30 mm and a depth of 5.2 mm eep, with 80% decayed material stiffness. The top tip of the conical geometry is situated at (150, 6.2, 150) and at a depth reaching 5.2 mm through the thickness of the plate. In the presence of impact damage, the 3D-FE model contains 794,937 tetrahedral mesh elements in total with 3,419,892 degrees of freedom. The maximum mesh element size inside and around the impact zone is set to 1 mm while the remaining area is set to be at least 4 mm per wavelength. The model is solved for the same incident frequency of 100 kHz using the same linear solver with Lagrange quadratic elements and applying Neumann boundary conditions. The total field (i.e., the normal component of the displacement vector 'v_{tot}') from the 3D-FE model is monitored at the mid plane around the same circular path at 360 positions at a distance of 20 mm from the center of the damaged zone. The numerically calculated normalized total field in absolute value (shown in solid line) is shown in Figure 11. The amplitude of the scattered A_0 wave 'v_{scat}' around the impacted zone is calculated as the absolute difference of the complex magnitudes ($v_{scat} = |v_{tot} - v_{inc}|$) at each point between the reference and the damaged state for all 360 monitored points around the circle. The normalized absolute difference of the scattered A_0 mode is shown in Figure 12.

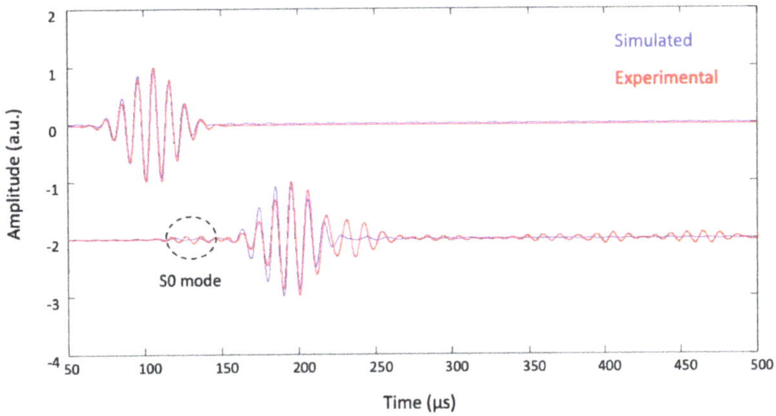

Figure 8. Comparison between reconstructed signal from the FE model and experimental measurements for two positions 142 mm apart.

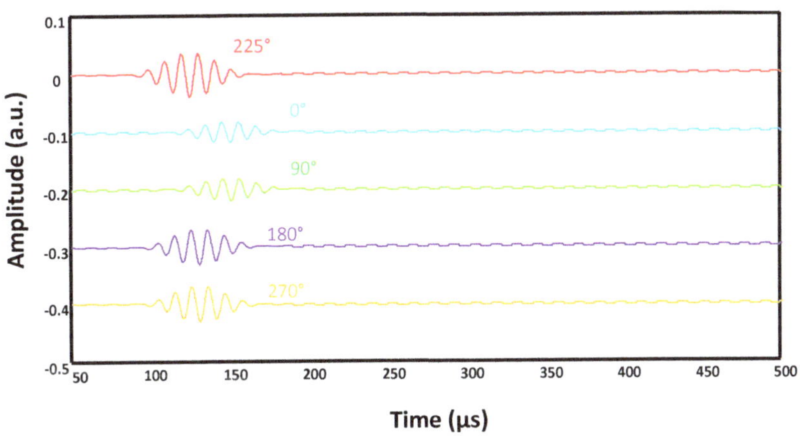

Figure 9. Reconstructed signal from the FE model at some angular positions 20 mm away from the center of impact damage.

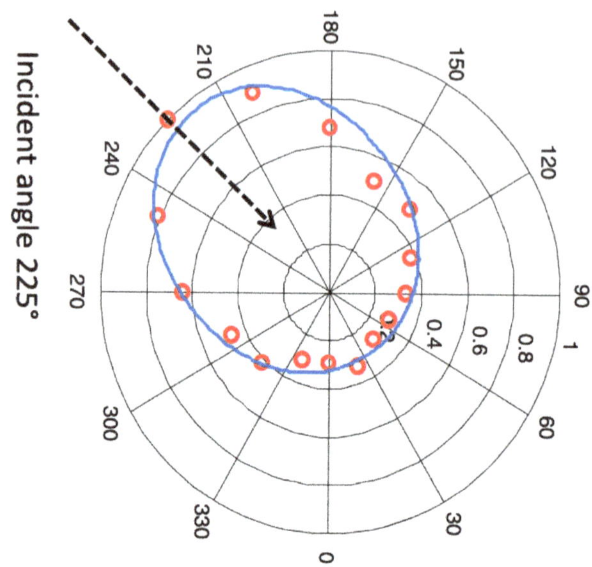

Figure 10. Experimental measurements and numerical predictions of the incident A_0 mode propagating in the intact CFRC plate.

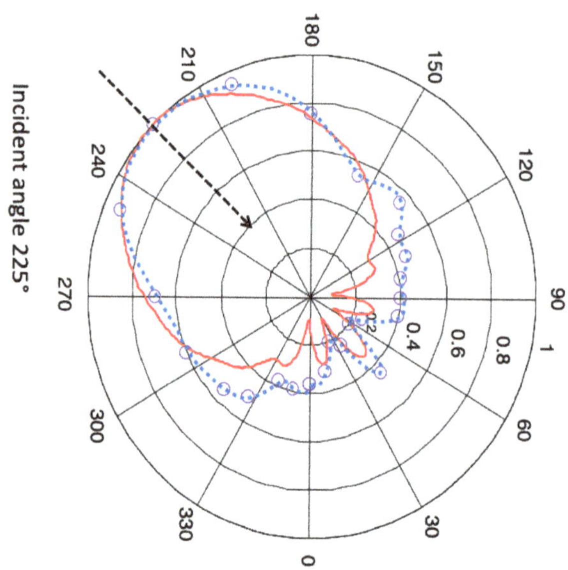

Figure 11. Experimental measurements (dashed) and numerical predictions (solid) of the total displacement field of A_0 mode propagating in the impacted CFRC plate.

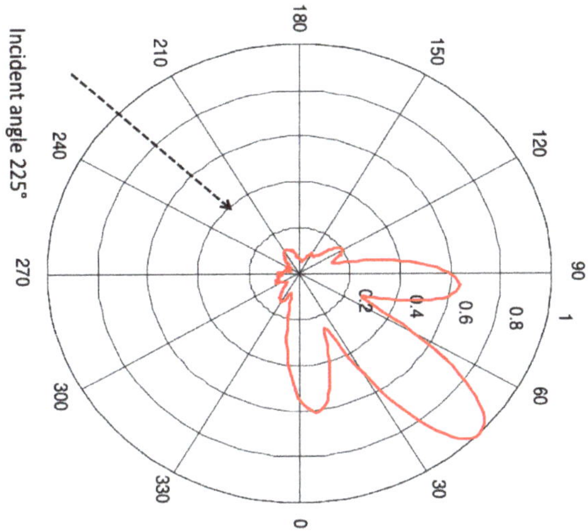

Figure 12. Directivity pattern of the scattered field of the A_0 mode deduced from total displacement fields on intact and damaged CFRC plates (See Figures 10 and 11).

The laboratory experimental setup (shown in Figure 7) for measuring the incident and total fields of the incident A_0 mode includes the same Gaussian profile sinusoidal signal with seven cycles centered at a frequency of 100 kHz (the same as that used in the FE model) where excitation and reception are performed via two identical transducers (Panametrics^{TM} V106). Acoustic field amplitudes are monitored at 32 different positions around the circle for intact and damaged plates. A Labview^{TM} code is programmed along with Matlab^{TM} to store and process the experimental measurements. The transmitter is attached to the composite plate with the help of a thermal resin. It is very difficult to generate a pure A_0 mode in the laboratory. Indeed, when a pulse is sent towards the transmitter to generate the A_0 mode, a small portion of the S_0 mode (shown inside the dotted circle in Figure 8) is also being excited. With S_0 being three times faster than A_0, one can easily choose the propagating distance to avoid any overlapping between the generated modes. Furthermore, the ratio of amplitudes between S_0 to A_0 is around 4%, which suggests neglecting the S_0 contribution, in accordance with the literature [25]. Therefore, the received mode will be almost a pure A_0 mode with negligible interference of the S_0 mode. A comparison between the reconstructed signals from the FE model and the experimental measurements (shown in red color) for same distance of 142 mm is shown in Figure 8. Indeed, by considering 12% attenuation into the 3D-FE model, the simulated amplitudes are found to be in good agreement with the experimental amplitudes for the A_0 mode. We also note a difference in simulated and experimental signals between ~225 µs and ~250 µs. The latter could be a consequence of the homogenization of the elastic properties corresponding to the different layers of the composite, which appears more clearly in the case of small wavelengths i.e., A_0 mode. However, these minor discrepancies have a very limited effect on the estimation of the amplitude of the A_0 mode as it can be observed in Figure 8. For the undamaged plate, the measured incident amplitudes (blue circles) are normalized and compared with the simulated normalized incident amplitudes (solid line) as shown in Figure 10. The latter shows that the scattered field remains symmetrical in terms of directivity and amplitude with respect to the direction 225°. For the damaged plate, the total displacement field is measured at the same 32 points to make the comparison possible. The normalized measured amplitude of the total field (blue circles) and simulated total field amplitudes (solid line) are shown in Figure 11. The directivity behavior of the measured total displacement fields is similar to simulations using the 3D-FE model. The A_0 scattered wavefield clearly shows the impact-induced damage area where the amplitude in the damaged area is considerably higher than in the intact area. Furthermore, the A_0 scattered wavefield is also direction dependent. The existing slight difference between simulated and measured total fields may be due to the considered geometrical and stiffness properties of the damaged zone in the 3D-FE model, which can be slightly different from the ones of the real impacted plate. However, one should note that the small discrepancy in the forward directions between simulated and experimental data has already been observed and was attributed to bonding considerations between the layers [14]. Nevertheless, the symmetry of the directivity along 45°–225° in measured total field amplitudes reinforces the hypothesis of a homogeneous distribution of the elastic properties as predicted numerically inside the conical damage. Furthermore, the scattered field is obtained by considering the absolute difference between the total and the incident wave fields. The numerically calculated normalized amplitude of scattered A_0 is shown in Figure 12. It is noted that the maximum scattered amplitude is towards the 45° direction i.e., in the direction of the incident A_0 mode. It is also found that there exist two positions near 30° and 60°, where the amplitude of the scattered A_0 mode is very small as compared to the 45° direction. The scattered field is symmetric along the 225°–45° direction, which corresponds to the direction of the wave vector of the generated guided wave.

7. Influence of the Damage Depth and Size on the Scattering Directivity Pattern

This section deals with the numerical study to observe the effect of the geometrical size of conical damage on the scattered wave fields in the case of an incident A_0 mode. For this purpose, two different studies have been performed. The first study considers the effect of the through-thickness variation (depth of the cone) on the scattered fields of the incident A_0 mode while fixing the diameter at 30 mm in order to conform with the impact test presented earlier. Four different models with different depths of damaged zones equal to 1/4, 1/2, 3/4 and 1 of the total thickness, i.e., 1.55, 3.1, 4.65 and 6.2 mm, are considered in the 3D-FE models, respectively. The schematic of the different depths is shown in Figure 13a. The four different models are solved for the same incident A_0 mode (100 kHz) using the same properties of the FE model as explained earlier in Section 5. The normalized scattered field corresponding to the A_0 mode is shown in Figure 13b in the case of the abovementioned depths. In all four cases of through-thickness depths, it is observed that there is a negligible amplitude scattered between 120° and 330° angles towards the anticlockwise direction. For the first three cases of through-thickness depth equal to 1/4, 1/2 and 3/4, the maximum scattered amplitude is towards the 45° direction and the extent of the scattered field gets narrow for increasing through-thickness depths of the conical damage. It is also noted that with the increase in through-thickness depth, side lobes start to appear in the scattered displacement field amplitudes and increases with the increase in depth. In the case of all through-thickness depth i.e., for depth equal to 6.2 mm, the side lobes have more scattered amplitude towards 80° and 10° directions. In this case, the maximum scattered amplitude is not towards the 45° direction but is oriented toward the side lobe directions. In all four cases, the directivity diagrams are found to be almost symmetrical along the 225°– 45° direction due to the isotropic nature of the plate in the XZ plane (and symmetrical geometry of the damage). The variation of amplitude of the scattered A_0 mode with through-thickness depths of the conical damaged zone in the direction of 45° is shown in Figure 14. The amplitude increases with the increase of depth until half the total through-thickness and then starts decreasing and decreases to the minimum value at all through-thickness depths. With this scattered field information for a fixed diameter of the damaged zone, one can gain an idea about the depth of the conical damage zone inside the composite plate. According to the literature, the micro-cracks due to impact damage (ex. matrix cracks) mainly contribute to the scattering of the wavefield in the damage region. The delamination reshapes the wavefield in the damaged and undamaged regions. In particular, it was found that the contribution of the delamination in the scattered wavefield is ~50% higher than the one corresponding to the matrix cracks [25].

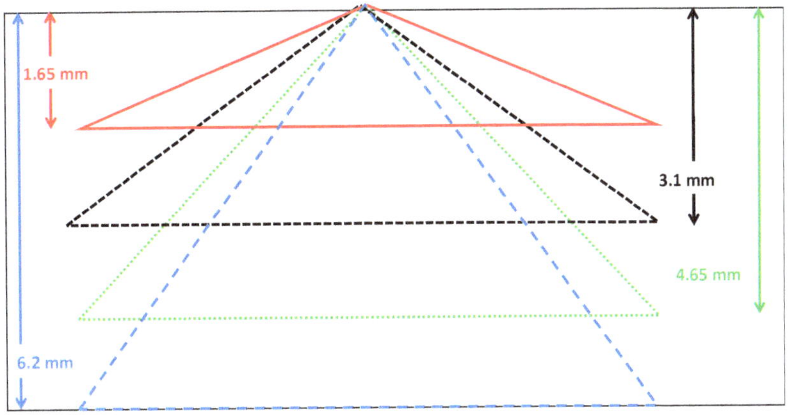

(a)

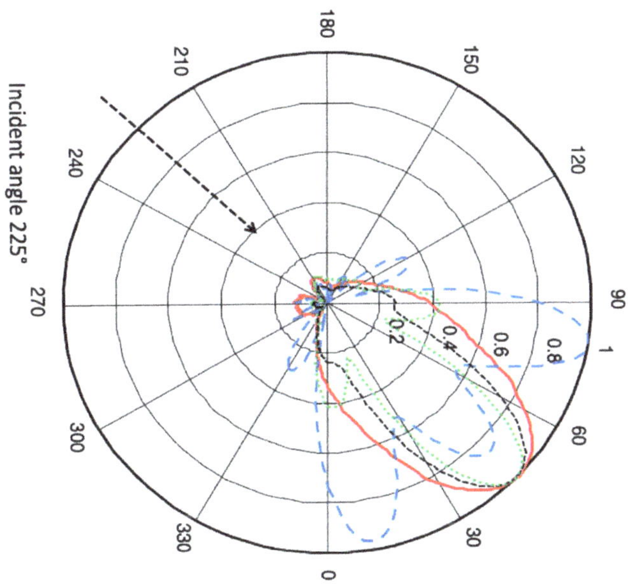

(b)

Figure 13. Effect of the conical depths: (**a**) Schematics of four different through-thickness depths (1.65 mm: red, 3.1 mm: black, 4.65 mm: green and 6.2 mm: blue), (**b**) normalized scattered amplitudes A_0 mode for different depths.

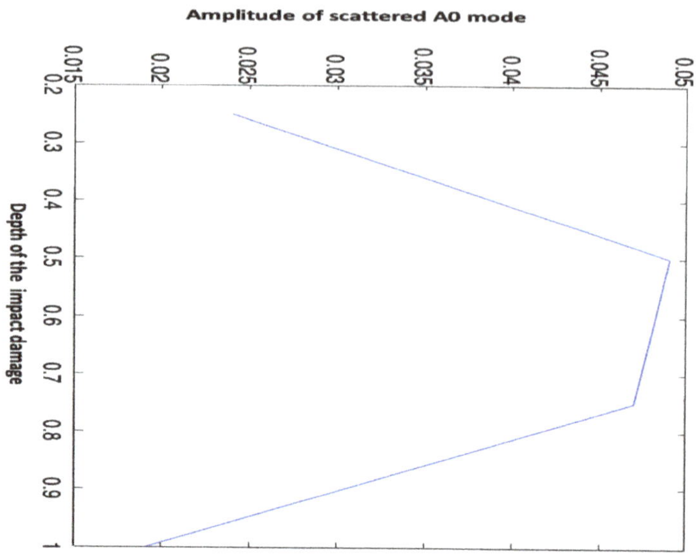

Figure 14. Effect of the through thickness of the conical defect on the scattered A_0 mode towards the 45° direction.

In the second case, the effect of the aspect ratio ($2\ r/\lambda$) on the scattered wave fields is studied, where r is the radius of the cone and λ is the wavelength of the A_0 mode. The study is performed by taking a through-thickness depth equal to 5.2 mm. Four different 3D-FE models with aspect ratios equal to 1/2, 1, 3/2 and 2, respectively, have been solved. The schematic for different aspect ratios is shown in Figure 15a. The numerically calculated normalized scattered field for different aspect ratios is shown in Figure 15b. In all four cases, the maximum scattered amplitude of A_0 mode is towards the 45° direction. It is observed that as the aspect ratio increases, the extent of scattered wave fields gets narrow in size. For an aspect ratio equal to 1/2 (shown in solid line), the angles of the minimum scattered amplitude are near ~120° and ~330°. For the case of an aspect ratio equal to 2, the angles of minimum scattered amplitudes are near ~20° and ~ 70°. Except for the aspect ratio of 1/2, the scattered field along the direction 120° towards 330° (in anticlockwise manner) is negligible. The size of the extent of the scattered field can give information about the size of the base of the conical defect. From the scattering directivity pattern, one can have an idea about the base size of the conical damage inside the composite plate, once the through-thickness depth is known. Finally, the variation of amplitude of the scattered A_0 mode with respect to aspect ratios in the direction of 45° is shown in Figure 16. It is observed that with the increase in aspect ratio, the amplitude of scattered field also increases. In real experiments, it is naturally expected that the increase of the impact energy will simultaneously increase the depth and width of the damage area. According to the literature, as the damage area increases, the delamination and matrix crack contributions in the scattered wavefield will then increase with the same proportions in most directions. The contribution of the delamination is expected to be more important and the wavefield possesses a larger amplitude on the transmission side than in the reflection side [16,25,46].

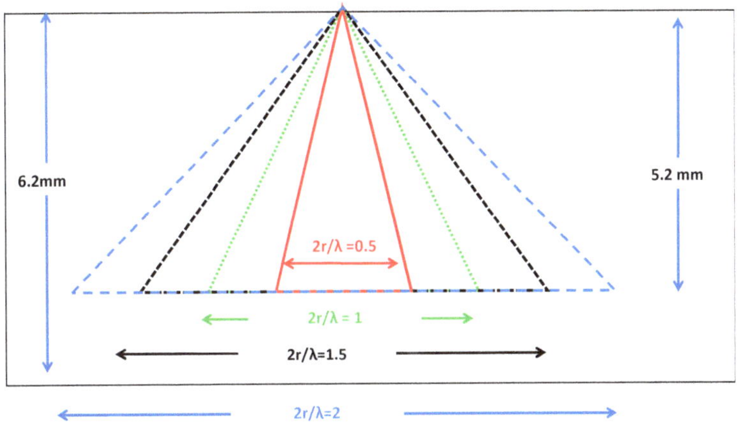

(a)

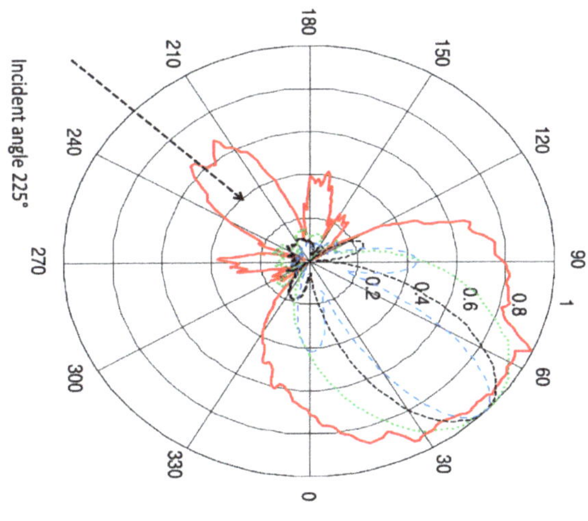

(b)

Figure 15. Effect of aspect ratios (2 r/λ): (**a**) Schematics for four different aspect ratios (0.5: red, 1: green, 1.5: black, 2: blue), (**b**) normalized scattered amplitude of A_0 mode.

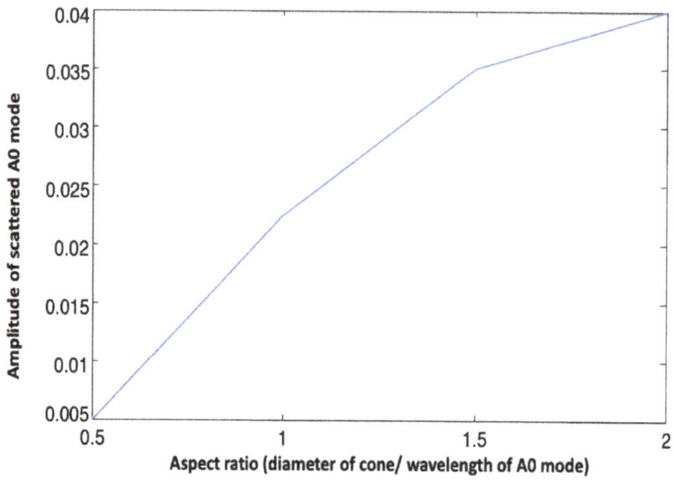

Figure 16. Variation of amplitude of scattered A_0 mode with respect to aspect ratio (2 r/λ) towards 45° direction.

8. Conclusions

In this paper, we have explored the possibility of modeling impact damages as a conical shape geometry with decayed stiffness properties in a homogeneous CFRC plate. This work is mainly motivated by the fact that scattering characteristics of Lamb waves in composites are more complicated than the scattering at defects in isotropic plates. The developed 3-dimensional finite element method allowed to determine the scattering pattern of an incident A_0 Lamb wave in intact and impacted CFRC plates, where the excitation frequency was set below the A_1 cut-off frequency. Besides, the numerically predicted scattered displacement fields were compared with the experimental measurements performed on CFRC plates with the same characteristics. A comparison between reconstructed signals from the FE model and experimental measurements allowed us to validate the model and to study the effect of different geometrical configurations (through-thickness depths and diameters of the base) of a conical damaged zone on the scattering diagrams. Results revealed that scattering directivity of the displacement field significantly depends on the depth and width of the conical damage created during the point-impact of the composite. Indeed, on the basis of the results, the scattering amplitude is maximum when the aspect ratio is beyond one and the depth of the impact damage is half the thickness of the composite. However, the detection of impact damage with a small aspect ratio and a relatively small depth (~0.3 or ~0.4) seems to be more difficult. The scattering physical insight at the impact-damaged area provided by the above-presented results can be used to optimize the monitoring of composite structures by improving the transducers' location in connection with the involved Lamb mode(s). This work also showed that the study of backward scattering is not always possible. Indeed, results revealed that backward scattering is smaller in amplitude than the forward scattering and its amplitude is only important in the case of a small aspect ratio. Finally, we note that the post-impact damage study performed herein can be extended to include uncertainties related to the structural and mechanical properties of the composite. Indeed, the uncertainties we have taken into account are essentially related to the velocities (position and time of flight). Under the transversely isotropic condition, the optimized elastic constants as well as their uncertainties are determined by fitting the analytic group velocity dispersion curve with the experimental velocities. At the wavelength scale, the considered elastic constants include the average properties of the constituents of the composite as well as the defects therein such as micro-porosities, etc. This can affect the results obtained when the wave propagates in certain directions, especially those where the defects are most present. However, we note that taking uncertainties into account in the modeling is beyond the scope of this work. Indeed, the main objective of this contribution is to study the interaction of the A_0 mode with a conical impact defect. Taking into account uncertainties is extremely expensive, computationally, in view of the large number of required analyses and requires the development of a specific study [46,47]. Indeed, the complex structure of composites and their manufacturing processes create different sources such as fiber waviness, ply wrinkling, machine tolerance, etc., which can be at the origin of uncertainty [48,49]. The statistical nature and distribution of imperfections in composites requires therefore the development of simulations, which are able to take into account the complexity of the material and perform calculations in reasonable times [50]. In that sense, Monte Carlo sampling methods are often used in different configurations aiming at reducing the calculation time while performing the uncertainty propagation analysis [47,51]. Other techniques based, for instance, on a metamodel approach [52,53] or a unified uncertain analysis approach [54] or else [55–58] can also be used. At present, there is still a need to develop an exact solution in the case of insufficient input data to determine the influence of uncertain elastic elements on the propagation of guided elastic waves, especially in the presence of damage in light of the work of Peng et al. based on the use of data-driven polynomial chaos expansion [59]. Indeed, a generalization and/or extension of our approach depends on such a development in order to perform reliable quantitative detection of damage created by impacts in isotropic and anisotropic composite structures with complex geometries.

Author Contributions: Formal analysis, M.B. and C.M.; Funding acquisition, M.B. and R.E.G.; Investigation, M.B. and C.M.; Methodology, D.S., M.B. and C.M.; Project administration, M.B. and R.E.G.; Software, D.S.; Supervision, M.B. and R.E.G.; Validation, D.S. and M.B.; Writing–original draft, D.S.; Writing–review & editing, M.B. All authors have read and agreed to the published version of the manuscript.

Funding: This work was supported by the project FUI Ecowingbox, France.

Institutional Review Board Statement: Not applicable.

Informed Consent Statement: Not applicable.

Acknowledgments: The authors thank Bertrand Lascoup and ESTACA Institute, Laval, France for the point impact experiments performed on the composite plates. This work was supported by the project FUI Ecowingbox, France.

Conflicts of Interest: The authors declare no conflict of interest.

References

1. Forintos, N.; Czigany, T. Multifunctional application of carbon fiber reinforced polymer composites: Electrical properties of the reinforcing carbon fibers—A short review. *Compos. Part B Eng.* **2019**, *162*, 331–343. [CrossRef]
2. Abbas, S.; Li, F.; Qiu, J. A Review on SHM Techniques and Current Challenges for Characteristic Investigation of Damage in Composite Material Components of Aviation Industry. *Mater. Perform. Charact.* **2018**, *7*, 224–258. [CrossRef]
3. Bar-Cohen, Y.; Mal, A.; Lih, S.S.; Chang, Z. Composite materials stiffness determination and defects characterization using enhanced leaky Lamb dispersion data acquisition method. *Proc. SPIE Nondestruct. Eval. Aging Aircr. Airpt. Aerosp. Hardw. III* **1999**, *3586*, 250–255.
4. Seale, M.D.; Madaras, E.I. Ultrasonic measurements of stiffness in thermal- mechanically fatigued IM7/5260 composites. *J. Mater. Eng. Perform.* **1999**, *8*, 429–436. [CrossRef]
5. Basri, R.; Chiu, W. Numerical analysis on the interaction of guided Lamb waves with a local elastic stiffness reduction in quasi-isotropic composite plate structures. *Compos. Struct.* **2004**, *66*, 87–99. [CrossRef]
6. Kundu, T.; Das, S.; Martin, S.A.; Jata, K.V. Locating point of impact in anisotropic fiber reinforced composite plates. *Ultrasonics* **2008**, *48*, 193–201. [CrossRef]
7. Hajzargerbashi, T.; Kundu, T.; Bland, S. An improved algorithm for detecting point of impact in anisotropic inhomogeneous plates. *Ultrasonics* **2011**, *51*, 317–324. [CrossRef]
8. Park, B.; Sohn, H.; Olson, S.E.; DeSimio, M.P.; Brown, K.S.; Derriso, M.M. Impact localization in complex structures using laser-based time reversal. *Struct. Health Monit.* **2012**, *11*, 577–588. [CrossRef]

9. Fromme, P.; Rougé, C. Directivity of guided ultrasonic wave scattering at notches and cracks. *J. Phys. Conf. Ser.* **2011**, *269*, 012018. [CrossRef]
10. Moreau, L.; Caleap, M.; Velichko, A.; Wilcox, P.D. Scattering of guided waves by flat-bottomed cavities with irregular shapes. *Wave Motion* **2012**, *49*, 375–387. [CrossRef]
11. Moreau, L.; Caleap, M.; Velichko, A.; Wilcox, P.D. Scattering of guided waves by through-thickness cavities with irregular shapes. *Wave Motion* **2011**, *48*, 586–602. [CrossRef]
12. Caminero, M.A.; García-Moreno, I.; Rodríguez, G.P.; Chacón, J.M. Internal damage evaluation of composite structures using phased array ultrasonic technique: Impact damage assessment in CFRP and 3D printed reinforced composites. *Compos. Part B Eng.* **2019**, *165*, 131–142. [CrossRef]
13. Gama, B.A.; Gillespie, J.W. Finite element modeling of impact, damage evolution and penetration of thick-section composites. *Int. J. Impact Eng.* **2011**, *38*, 181–197. [CrossRef]
14. Ng, C.-T.; Veidt, M.; Rose, L.; Wang, C.-H. Analytical and finite element prediction of Lamb wave scattering at delaminations in quasi-isotropic composite laminates. *J. Sound Vib.* **2012**, *331*, 4870–4883. [CrossRef]
15. de Luca, A.; Perfetto, D.; de Fenza, A.; Petrone, G.; Caputo, F. Guided wave SHM system for damage detection in complex composite structure. *Theor. Appl. Fract. Mech.* **2020**, *105*, 102408. [CrossRef]
16. Ng, C.-T.; Veidt, M. Scattering of the fundamental anti-symmetric Lamb wave at delaminations in composite laminates. *J. Acoust. Soc. Am.* **2011**, *129*, 1288. [CrossRef]
17. Guo, N.; Cawley, P. The interaction of Lamb waves with delamination in composite laminates. *J. Acoust. Soc. Am.* **1993**, *94*, 2240–2246. [CrossRef]
18. Hayashi, T.; Kawashima, K. Multiple reflections of Lamb waves at a delamination. *Ultrasonics* **2002**, *40*, 193–197. [CrossRef]
19. Cegla, F.B.; Rohde, A.; Veidt, M. Analytical prediction and experimental measurement for mode conversion and scattering of plate waves at non-symmetric circular blind holes in isotropic plate. *Wave Motion* **2008**, *45*, 162–177. [CrossRef]
20. Moreau, L.; Castaings, M. The use of an orthogonality relation for reducing the size of finite element models for 3D guided waves scatter- ing problems. *Ultrasonics* **2008**, *48*, 357–366. [CrossRef] [PubMed]
21. Cantwell, W.; Morton, J. Impact perforation of carbon fiber reinforced plastic. *Compos. Sci. Technol.* **1990**, *38*, 119. [CrossRef]
22. Mines, R.A.W.; Roach, A.M.; Jones, N. High velocity perforation behaviour or polymer composite laminates. *Int. J. Impact Eng.* **1999**, *22*, 561. [CrossRef]
23. Mitrevski, T.; Marshall, I.H.; Thomson, R. The influence of impactor shape on the damage to composite laminates. *Compos. Struct.* **2006**, *76*, 116–122. [CrossRef]
24. Liu, D. Impact induced delamination- A view of bending stiffness mismatching. *J. Compos. Mater.* **1988**, *22*, 674–792. [CrossRef]
25. Zhang, B.; Sun, X.C.; Eaton, M.J.; Marks, R.; Clarke, A.; Featherston, C.A.; Kawashita, L.F.; Hallett, S.R. An integrated numerical model for investigating guided waves in impact-damaged composite laminates. *Compos. Struct.* **2017**, *176*, 945–960. [CrossRef]
26. González, E.V.; Maimí, P.; Camanho, P.P.; Turon, A.; Mayugo, J.A. Simulation of drop- weight impact and compression after impact tests on composite laminates. *Compos. Struct* **2012**, *94*, 3364–3378. [CrossRef]
27. Feng, D.; Aymerich, F. Finite element modelling of damage induced by low-velocity impact on composite laminates. *Compos. Struct.* **2014**, *108*, 161–171. [CrossRef]
28. Sun, X.; Wisnom, M.; Hallett, S. Interaction of inter- and intralaminar damage in scaled quasi-static indentation tests: Part 2—Numerical simulation. *Compos. Struct.* **2016**, *136*, 727–742. [CrossRef]
29. Ammari, H.; Zhang, H. SIAM. *J. Math. Anal.* **2017**, *49*, 3252.
30. Leroy, V.; Strybulevych, A.; Page, J.H.; Scanlon, M.G. Influence of positional correlations on the propagation of waves in a complex medium with polydisperse resonant scatterers. *Phys. Rev. E* **2011**, *83*, 046605. [CrossRef]
31. Sharma, G.S.; Skvortsov, A.; MacGillivray, I.; Kessissoglou, N. Sound scattering by a bubble metasurface. *Phys. Rev. B* **2020**, *102*, 214308. [CrossRef]
32. Sharma, G.S.; Skvortsov, A.; MacGillivray, I.; Kessissoglou, N. On superscattering of sound waves by a lattice of disk-shaped cavities in a soft material. *Appl. Phys. Lett.* **2020**, *116*, 041602. [CrossRef]
33. Singh, D.; El Guerjouma, R.; Bentahar, M. Interaction of Fundamental Lamb Modes with a Point Impact Damaged Zone in Composite Plates. In Proceedings of the Acoustics 2012 Nantes Conference, Nantes, France, 23–27 April 2012.
34. Ke, W.; Castaings, M.; Bacon, C. 3D finite element simulations of an air-coupled ultrasonic NDT system. *NDT E Int.* **2009**, *42*, 524–533. [CrossRef]
35. Deng, P.; Saito, O.; Okabe, Y.; Soejima, H. Simplified modeling method of impact damage for numerical simulation of T Lamb wave propagation in quasi-isotropic composite structures. *Compos. Struct.* **2020**, *243*, 112150. [CrossRef]
36. Sjoblom, P.O.; Hartness, J.T.; Cordell, T.M. On low-velocity impact testing of composite materials. *Compos. Mater.* **1988**, *22*, 30–52. [CrossRef]
37. Tan, K.T.; Watanabe, N.; Iwahori, Y. X-ray radiography and micro-computed tomography examination of damage characteristics in stitched composites subjected to impact loading. *Compos. Part B Eng.* **2011**, *42*, 874–884. [CrossRef]
38. Lee, S.M.; Cheon, J.S.; Im, Y.T. Experimental and numerical study of the impact behavior of SMC plates. *Compos. Struct.* **1999**, *47*, 551–561. [CrossRef]
39. Zhou, G.; Lloyd, J.C.; McGuirk, J.J. Experimental evaluation of geometric factors affecting damage mechanisms in carbon/epoxy plates. *Compos. Part A Appl. S* **2001**, *32*, 71. [CrossRef]

40. Shyr, T.; Pan, Y. Impact resistance and damage characteristics of composite laminates. *Compos. Struct.* **2003**, *62*, 193. [CrossRef]
41. Talreja, R.; Phan, N. Assessment of damage tolerance approaches for composite aircraft with focus on barely visible impact damage. *Compos. Struct.* **2019**, *219*, 1–7. [CrossRef]
42. Castaings, M.; Singh, D.; Viot, P. Sizing of impact damages in composite materials using ultrasonic guided waves. *NDT E Int.* **2012**, *46*, 22–31. [CrossRef]
43. COMSOL, User's Guide. Version 3.5a by COMSOL AB 2008. Available online: http://www.comsol.com/ (accessed on 23 February 2012).
44. Leckey, C.A.; Wheeler, K.R.; Hafiychuk, V.N.; Hafiychuk, H.; Timuçin, D.A. Simulation of guided-wave ultrasound propagation in composite laminates: Benchmark comparisons of numerical codes and experiment. *Ultrasonics* **2018**, *84*, 187–200. [CrossRef] [PubMed]
45. Drozdz, M.; Moreau, L.; Castaings, M.; Lowe, M.; Cawley, P. Efficient Numerical Modelling of Absorbing Regions for Boundaries of Guided Waves Problems. In *AIP Conference Proceedings*; AIP Publishing: Melville, NY, USA, 2006; p. 126.
46. Murat, B.I.S.; Khalili, P.; Fromme, P. Scattering of guided waves at delaminations in composite plates. *J. Acoust. Soc. Am.* **2016**, *139*, 3044–3052. [CrossRef]
47. Richard, B.; Timothy, J.; Dodwell, R.T.; Haftka, K.N.H.; Kim, T.; Kynaston, S.; Scheichl, R. Uncertainty Quantification of Composite Structures with Defects using Multilevel Monte Carlo Simulations. In Proceedings of the 17th AIAA Non-Deterministic Approaches Conference, Kissimmee, FL, USA, 5–9 January 2015.
48. Melchers, R.E. *Structural Reliability Analysis and Prediction*; Wiley: Hoboken, NJ, USA, 1999.
49. Dodwell, T.; Butler, R.; Hunt, G. Out-of-plane ply wrinkling defects during consolidation over an external radius. *Compos. Sci. Technol.* **2014**, *105*, 151–159. [CrossRef]
50. Rhead, A.T.; Dodwell, T.J.; Butler, R. The effect of tow gaps on compression after impact strength of robotically laminated structures. *Comput. Mater. Contin.* **2013**, *35*, 1–16.
51. Li, Y.-W.; Elishakoff, I.; Starnes, J.H.; Bushnell, D. Effect of the thickness variation and initial imperfection on buckling of composite cylindrical shells: Asymptotic analysis and numerical results by BOSOR4 and PANDA2. *Int. J. Solids Struct.* **1997**, *34*, 3755–3767. [CrossRef]
52. Chiachio, M.; Chiachio, J.; Rus, G. Reliability in composites—A selective review and survey of current development. *Compos. Part B Eng.* **2012**, *43*, 902–913. [CrossRef]
53. Ghasemi, H.; Rafiee, R.; Zhuang, X.; Muthu, J.; Rabczuk, T. Uncertainties propagation in metamodel-based probabilistic optimization of CNT/polymer composite structure using stochastic multi-scale modeling. *Comput. Mater Sci.* **2014**, *85*, 295–305. [CrossRef]
54. Dey, S.; Mukhopadhyay, T.; Adhikari, S. Metamodel based high-fidelity stochastic analysis of composite laminates: A concise review with critical comparative assessment. *Compos. Struct.* **2017**, *171*, 227–250. [CrossRef]
55. Conceição, A.C.; Hoffbauer, L.N. Reliability-based, design optimization and uncertainty quantification for optimal conditions of composite structures with non-linear behavior. *Eng. Struct.* **2017**, *153*, 479–490. [CrossRef]
56. Dey, S.; Mukhopadhyay, T.; Sahu, S.; Li, G.; Rabitz, H.; Adhikari, S. Thermal uncertainty quantification in frequency responses of laminated composite plates. *Compos. Part B Eng.* **2015**, *80*, 186–197. [CrossRef]
57. Das Neves Carneiro, G.; Antonio, C.C. A RBRDO approach based on structural robustness and imposed reliability level. *Struct. Multidiscip. Optim.* **2017**, *57*, 2411–2429. [CrossRef]
58. Das Neves Carneiro, G.; Antonio, C.C. Robustness and reliability of composite struc- tures: Effects of different sources of uncertainty. *Int. J. Mech. Mater Des.* **2017**, *15*, 93–107. [CrossRef]
59. Peng, X.; Li, D.; Wu, H.; Liu, Z.; Li, J.; Jiang, S.; Tan, J. Uncertainty analysis of composite laminated plate with data-driven T polynomial chaos expansion method under insufficient input data of uncertain parameters. *Compos. Struct.* **2019**, *209*, 625–633. [CrossRef]

applied sciences

MDPI

Article

Multichannel Real-Time Electronics Platform for the Estimation of the Error in Impact Localization with Different Piezoelectric Sensor Densities †

Lorenzo Capineri *, Andrea Bulletti and Eugenio Marino Merlo

Department of Information Engineering, University of Florence, 50139 Florence, Italy; andrea.bulletti@unifi.it (A.B.); eugenio.marinomerlo@unifi.it (E.M.M.)
* Correspondence: lorenzo.capineri@unifi.it
† This paper is an extended version of our paper published in A. Bulletti, E. M. Merlo and L. Capineri, "Analysis of the accuracy in impact localization using piezoelectric sensors for Structural Health Monitoring with multichannel real-time electronics," 2020 IEEE 7th International Workshop on Metrology for AeroSpace (MetroAeroSpace), Pisa, Italy, 22–24 June 2020.

Abstract: The work presents a structural health monitoring (SHM) electronic system with real-time acquisition and processing for the determination of impact location in laminate. The novelty of this work is the quantitative evaluation of impact location errors using the Lamb wave guided mode S_0, captured and processed in real-time by up to eight piezoelectric sensors. The differential time of arrival is used to minimize an error function for the position estimation. The impact energy is correlated to the amplitudes of the antisymmetric (A_0) mode and the electronic design is described to avoid saturation for signal acquisition. The same electronic system is designed to acquire symmetric (S_0) low level signals by adequate gain, bandwidth, and signal-to-noise ratio. Such signals propagate into a 1.4 mm thick aluminum laminate at the group velocity of 5150 m/s with frequency components above 270 kHz, and can be discriminated from the A_0 mode to calculate accurately the differential arrival time. The results show that the localization error stabilizes at a value comparable with the wavelength of the S_0 mode by increasing the number of sensors up to six, and then remains constant at up to eight sensors. This suggests that a compromise can be found between sensor density and localization error.

Keywords: real-time electronics; structural health monitoring; Lamb wave; piezoelectric sensors; impact localization; ultrasonic guided waves

Citation: Capineri, L.; Bulletti, A.; Marino Merlo, E. Multichannel Real-Time Electronics Platform for the Estimation of the Error in Impact Localization with Different Piezoelectric Sensor Densities . *Appl. Sci.* **2021**, *11*, 4027. https://doi.org/10.3390/app11094027

Academic Editor: Habil. Michel Darmon

Received: 26 March 2021
Accepted: 27 April 2021
Published: 28 April 2021

1. Introduction

Structural health monitoring (SHM) has been the subject of different studies in the non-destructive testing (NDT) field aiming to identify the location of an impact point using piezoelectric sensors. Many plate-like geometries made in fiber-reinforced composite (CFRP) or aluminum—commonly used in aerospace structures—are suitable to support the guided propagation of ultrasound-guided waves over long distances; for plate-like structures a few mm thick we refer to extensional and flexural modes of Lamb waves. The possibility of identifying damage on a structure by piezoelectric sensor systems allows for the determination of its integrity, thus reducing downtime and maintenance costs. SHM is important because, by continuous monitoring, it permits real-time detection of damage in a structure, particularly of safety-critical components typical of the application of such compounds in aircrafts [1–5]. A real-time SHM system with a network of passive sensors was proposed to detect low- and high-impact events. For each event, the conversion of kinetic energy into propagating ultrasonic guided waves occurs and acoustic emission (AE) localization methods can be applied using piezoelectric transducers. Once a localization is performed, the SHM system can be switched to the active mode operation to investigate a restricted area around the estimated impact position. An array of sensors and real-time multichannel electronic processing are the enabling technologies for the application of SHM in aerospace, energy conversion, transportation, and automotive industries.

In the literature, different strategies for locating impact positions using Lamb waves were investigated with reference to the extraction of the differential time of arrival (DToA). Tobias in [6] used a triangulation technique; Ciampa and Meo in [7] evaluated the DToA with an algorithm based on continuous wavelet transform (CWT); Shukri Mohd et al. in [8] used a method based on wavelet transform analysis and modal location (WTML) with four sensors; Shenxin Yin et al. in [9] used eight sensors bonded in a Z-shaped arrangement. While CWT allows more sophisticated signal processing for the separation of the flexural and extensional dispersive modes, in a multisensory system with several nodes simpler solutions—like the first arrival signal detection with threshold method or short-time Fourier transform (STFT)—must also be considered. The latter requires less computational resources than CWT but does not permit the trade-off between the arrival time and spectral content.

For complex three-dimensional structures, analysis of the DToA, relative to propagation along multiple paths, is not straightforward and is thus difficult to determine with analytical models. In these cases, multiple sensors and artificial intelligence (AI) are often used; for example, the system based on neural networks presented by Worden et al. [10] used up to 17 sensors. Carrino et al. [11,12] proposed an innovative method based on nonlinear Lamb waves for locating disbonds in single-lap joints that uses PZT signals processed with a baseline-free algorithm and the localization of artificial defects placed inside the area, delimited by four PZT sensors.

Other strategies for locating impact positions, without knowledge of the characteristics of the material under test, are also reported in [13,14]. In a previous work, Kundu et al. [14] presented a study that considered different shapes of the wave front generated during an acoustic event and developed a methodology to localize the acoustic source in an anisotropic plate from those wave front shapes: an elliptical wave front shape-based technique was developed first, followed by the development of a parametric curve-based technique for non-elliptical wave front shapes. This methodology does not assume a straight-line wave propagation path and can predict the source location without any knowledge of the elastic properties of the material; however, this study reported only a theoretical approach without the development of a real-time system to validate the proposed technique.

In [15], the authors of this work presented a guided-wave ultrasound SHM system based on linear arrays of interdigital piezopolymer transducers bonded to a composite pressure vessel for spacecrafts. In that system, interdigital transducers were adopted to perform both damage assessment and impact detection/localization using a simple fixed-threshold technique. In a related research development [16], the authors presented a novel approach to estimate the DToA between the impact response signals collected by a triplet of sensors, overcoming the limitations of classical methods that rely on amplitude thresholds calibrated for a specific sensor type. Finally, in [17] is proposed a simple laboratory procedure based on a set-up with a pair of sensors that are symmetrically placed with respect to the impact point to estimate the uncertainty of the DToA and the propagation velocity.

In this work we have upgraded the real-time electronic system reported in [15] that allows for the accommodation of analog front-end electronic characteristics thanks to its modular design; different sensor types (piezoceramic, piezocomposites, and piezopolymers) need custom analog front-ends (AFE) to match their electrical impedance, sensitivity, and bandwidth. In this work, we designed and implemented 16 new receiver boards to collect information of the propagated extensional S_0 mode Lamb waves thanks to the improved bandwidth, adjustable gain, and signal-to-noise ratio. This new feature allowed us to overcome the limitation of using information for the localization carried by the higher level, but slower, A_0 mode. The detection of small amplitude (mV-10 mV) S_0 mode signal with threshold method simplifies the electronic design with benefits also to power consumption and rapid event detection. As observed in [18,19], A_0 mode propagates on longer paths with low attenuation but at lower propagation velocity and exhibits dispersion phenomena in the low-frequency range. Such characteristics complicate the interpretation of the impact-generated signals, especially considering large structures and multiple reflections from boundary, where in many cases S_0 is preferred. Moreover, the importance of sensor networks for aircraft SHM was reported in [20], considering the large dimensions and the trade-off between system performance, reliability, and cost.

The study presented in this paper differs from previously cited works because the aim was not to present the best method or technique for locating impact positions, but to report the analysis of errors on impact localization (see Section 2) using real-time electronics; this paper also evaluates the influence of the number of passive PZT sensors used (see Section 3) on the position error. The real-time electronics can also acquire multiple impacts on the same point for statistical analysis. We present the results of experimental tests with low-energy impact on a 1.4 mm thick aluminum plate using Physik Instrumente P-876.SP1 piezoceramic transducers, varying the number of channels from four to eight. The signals were then used to find the minimum of a cost function for the localization of the impact and, therefore, the evaluation of the positioning error relative to the size of the investigation area and the number of passive channels used. Finally, to help the reader in finding selected topics we include, a list of acronyms used in this paper.

2. Impact Detection and Localization

Considering a plate-like structure, a point impact on the surface will determine a mechanical response in the form of elastic waves of various natures propagating outward from the point of contact. Our work focused on detecting low-energy impacts (from 35 to 600 mJ) that did not result in permanent damage to the aluminum plate. Low-energy impacts are of interest for laboratory testing of system prototypes because they are non-destructive, while larger energy (>1 J) impacts are used for real-life SHM applications. The implication of sensor choice, regarding sensitivity and impact energy, is discussed in the next section.

In general, it is possible to detect the location of an impact event by measuring the DToAs of the elastic response wave front to a set of three or more sensors with knowledge of their position (x_i, y_i) on a planar structure. Using these data, the impact location can be numerically triangulated. Our system estimated the DToA of impact-generated elastic waves using a simple amplitude threshold thanks to the remarkable (better than 70 dB) signal-to-noise ratio (SNR) of the analog front-end.

The developed system performed impact detection by running in passive (listening) mode, i.e., by continuously sampling the transducer signals in a circular buffer, while waiting for a triggering event (a threshold-crossing) to occur. This threshold level (six times the root-mean-square noise level) had to be verified during the experimental tests to avoid spurious triggering, while maintaining a good sensitivity to low-energy impacts. As the data acquisition hardware detected an impact threshold-crossing event, it immediately froze the circular buffer and transferred the data to the acquisition system.

We defined a region of interest (ROI) of the aluminum plate equal to 250 mm × 240 mm (see Figure 1). This area was divided into a uniform grid of points spaced 1 mm apart with coordinates (x_p, y_p) for the subsequent analysis. For each of those points, the impact localization algorithm processed the received signals to extract the differential DToAs and calculated the value of the error function $E(x_p, y_p)$ [15–17]. Once the error function (1) was calculated for the whole grid, the impact point was assigned by the absolute minimum error value criterion. The error function $E(x_p, y_p)$ presents a minimum and has a monotonic behavior. In such cases, one can extrapolate a best-guess position of the impact by calculating the centroid of the points having a value within 1% of the absolute minimum.

The choice to use a grid of points spaced 1 mm apart was a good compromise between the accuracy of estimation of the impact point and the necessity to elaborate signals in real-time and immediately show the results on a PC display.

$$E(x_p, y_p) = \sum_{i=1}^{N_T-1} \sum_{j=i+1}^{N_T} \left| \frac{(t_i - t_j)v - \left(\sqrt{(\mathrm{x_i} - \mathrm{x_p})^2 + (\mathrm{y_i} - \mathrm{y_p})^2} - \sqrt{(\mathrm{x_j} - \mathrm{x_p})^2 + (\mathrm{y_j} - \mathrm{y_p})^2} \right)}{\sqrt{(\mathrm{x_i} - \mathrm{x_j})^2 + (\mathrm{y_i} - \mathrm{y_j})^2}} \right| \quad (1)$$

Here N_T is the total number of installed sensors whose positions are defined by the coordinates $\mathrm{P_i} = (\mathrm{x_i}, \mathrm{y_i})$ and $\mathrm{P_j} = (\mathrm{x_j}, \mathrm{y_j})$. The corresponding measured time of flights for each sensor are $\mathrm{t_i}$ and $\mathrm{t_j}$, respectively.

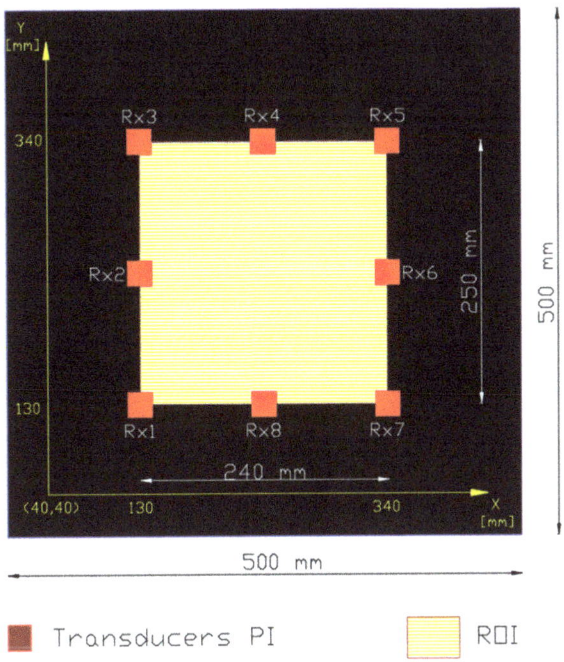

Figure 1. The 1.4 mm thick aluminum plate with 8-sensor arrangement (red squares). The picture also indicates the ROI (yellow area) and the area outside the ROI (black area) that corresponds to the aluminum laminate's dimensions.

3. Experimental Set-Up and Electronic Equipment

This section describes the electronic system designed for SHM application using multiple sensors in passive mode. The system can also be switched automatically to active mode for damage assessment using interdigital transducers as reported in [15].

3.1. Electronic Instrument Design for SHM Evaluation

The main architecture and design of the electronic instrument developed as part of this work has been presented in [21]. The instrument consists of a 16-channel device fully programmable in active or passive modes of operation. One of the main characteristics of this electronic system is the real-time acquisition and processing of ultrasonic signals received by 16 channels, which is an advantage in research of sensor layouts with more than 3 or 4 channels; the latter being quite a common experimental situation made available with use of a digital oscilloscope.

For the passive operation mode treated in this work, we designed 16 new receiver boards that easily replaced the previous ones thanks to the modular architecture of the system. A block-scheme of the new board is shown in Figure 2. The input attenuator is a fundamental block to adjust for different sensor sensitivities and impact energies. For example, impacts with 10 to 100 mJ of energy can be monitored with the 6.5 mm PVDF circular type (see [16,17]), producing voltage signals in the range of 100 mV to 1 V, while the highest sensitivity of piezoceramic sensors, such as the Physik Instrumente P-876.SP1, allows monitoring of signals up to 300 mVpp from the low-energy impact hammer or free-falling sphere typically used during non-destructive laboratory tests. The piezoceramic material of these types of sensors has a high electromechanical coupling factor and high sensitivity.

Figure 2. (**Left**) Piezoceramic sensor PI_876.SP1. Piezopolymer (PVDF) sensor protected by a white adhesive label and electrical connections on a PCB. (**Right**) Circular element of 6.45 mm diameter shown by the red circle on the CAD drawing.

The two different types of piezoelectric sensors used are shown in Figure 2, and their signal output amplitude was measured by direct connection to a digital oscilloscope; the comparison of the amplitudes is reported in Figure 3 for different impact energies from 0 to 45 mJ. In Figure 3, we can observe different energy values, $U = m \times g \times h$, corresponding to the same value of mass, m, and acceleration due to gravity, g, of the impacting steel sphere: these different values were obtained according to the above relationship by varying the falling height, h.

These amplitudes are relative to the peak amplitude of the large A_0 mode which is generated by the impact and has low-frequency components, typically below 50 kHz as predicted by Ross in [22]. The data reported in Figure 3 are important in making a decision as to the programmable threshold value depending on the expected impact energy. In this way, the real-time electronics capture the signals with a predefined minimum energy and, thanks to the programmed pre-trigger time, the signal trace also contains the information of signals preceding the large A_0 mode. This is a relevant feature of the electronic system because higher frequency components of the A_0 and S_0 propagate at higher velocities, as explained in Section 4.

The electronic analog front-end (AFE) (see Figure 4) is mainly composed of a programmable input attenuator (from 0 to −20 dB), a unity gain buffer based on an LF347 op amp, a non-inverting amplifier with voltage gain of 25 dB based on an LM6172 op amp, and a band pass filter (BPF). The BPF is designed to obtain an active (op amp LM6172) second-order Bessel HPF with the cut-off frequency (fc1) equal to 270 kHz at −3 dB, and a passive first-order LPF with the cut-off frequency (fc2) equal to 800 kHz at −3 dB. The last filtering block was inserted to limit the in-band noise; the noise measurements of the output of the analog chain with a standard 50-ohm input impedance for two different prototype boards are reported in Table 1.

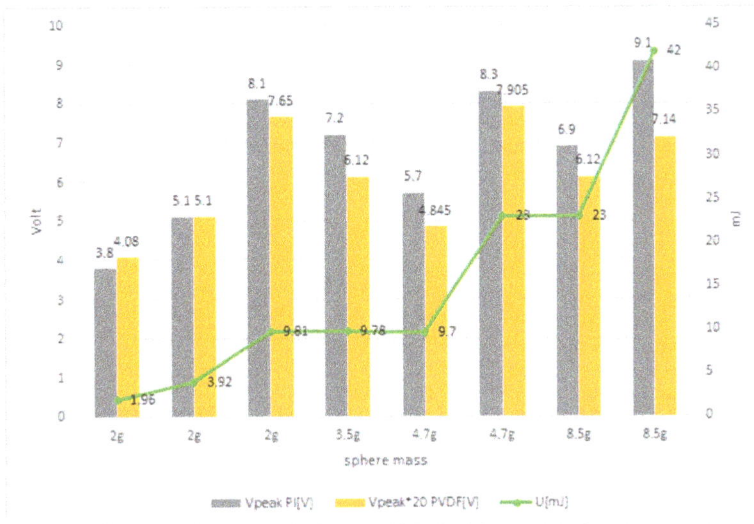

Figure 3. Peak voltage from impacts on the aluminum laminate generated by steel balls with mass ranging from 2 to 8.5 g in free-fall. The sensitivity of the two different sensors (see Figure 2) can be compared by the output voltages: the PVDF sensor amplitudes were multiplied by a constant factor, 20, to allow for comparison with the trend of the PZT amplitude. We can observe the good correlation between impact energy and output voltage.

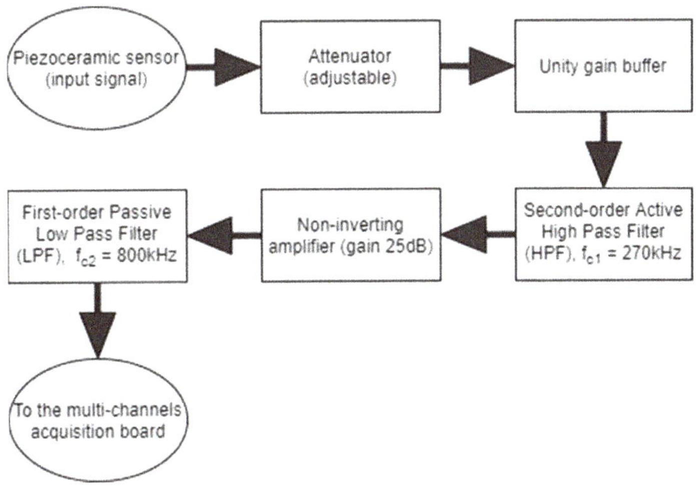

Figure 4. The block-scheme of the new design of the receiver board.

Table 1. V_{rms} output noise of the front-end channel with input impedance 50 Ω and attenuator selected at 0 dB.

	$V_{rms}out$ **Front-End #1**	$V_{rms}out$ **Front-End #2**
Zin = 50 Ω	1 m V	1 m V

These values are adequate considering the low-voltage analog-to-digital converter (ADC) has 12-bit resolution and an input dynamic of 3.3 V present on the electronic platform. Moreover, we observed that the programmable attenuator is necessary to limit the input signals in the AFE chain to avoid saturation and to protect the first amplifier from large signals exceeding the allowable common mode range at the operating frequency.

The main difference, with respect to the previous version [21], is the versatility of the receiver, which can be programmed to receive ultrasonic impact signals that in general include S_0 and A_0 modes. This makes it possible to receive higher frequencies, up to 270 kHz, with high signal-to-noise ratio (10 dB), to analyze S_0 Lamb modes that propagate at higher velocities, with respect to A_0 modes, and with lower amplitude.

An example of the output signal after the analog processing stage is shown in Figure 5. The fast S_0 mode and the high-frequency component of the A_0 mode are now present in the signal. The early S_0 mode signal is used later for the calculation of DToAs for the impact localization by triangulation algorithm.

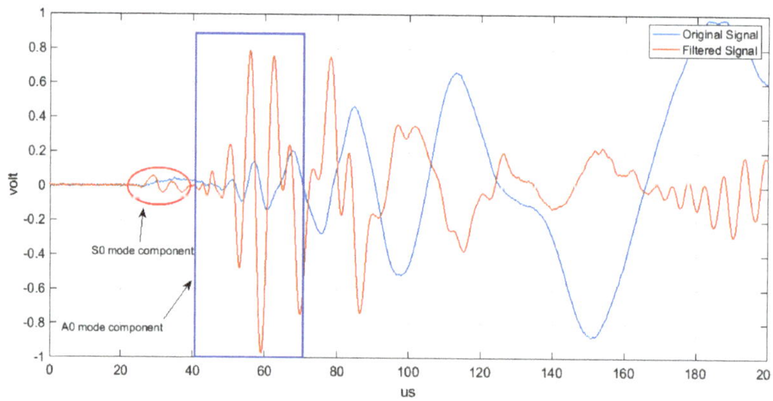

Figure 5. Comparison between original signal and filtered signal with high-pass second-order filter, with cut-off frequency 270 kHz followed by a 25-dB gain amplifier (see Figure 4). The signals were normalized and the distance between the impact point and the sensor was 10 cm. The two ultrasonic-guided modes A_0 (marked by blue box) and S_0 (marked by red oval) are well discriminated with high SNR.

3.2. Experimental Set-Up

In Figure 6 (right), the main blocks and connection of the SHM system are described. The main characteristics of the data acquisition system are listed below.

Programmable parameters:

- VGA gain from 54 to 90 dB to adjust for different sensor sensitivities
- High voltage, up to 100 V, for active mode operation
- Square wave burst (1–16 cycles), up to 500 kHz, for active mode operation

Data acquisition system characteristics:

- 12-bit resolution
- 16 channels at 20 MSps
- Storage: 1 MS per channel

The main components of the block diagram for the data acquisition system are the:

(1) sensor layout installed on the plate-like structure, coordinate reference system, and impact point markers (#1, #2, #3, #4, #5, etc.),
(2) signal-conditioning electronics,
(3) multichannel VGA with ADC evaluation module (Texas Instruments AFE5851EVM, Dallas, TX, USA), and

(4) Spartan-6 FPGA evaluation card (Xilinx SP605).

The experimental set-up adopted is reported in Figure 6 (left), where only 8 out of 16 channels were programmed for passive monitoring of impacts. Figure 6 (left) shows the low-energy hammer impactor in the center of the aluminum plate where the eight piezoelectric sensor arrangement was placed. The eight sensors were placed along the perimeter of the ROI (see Figure 1) and their arrangement is shown in the experimental results section below.

The high number of sensors were also able to reveal impacts outside the ROI, up to the full size of the aluminum plate. The four edges of the aluminum plate were covered by colored play dough (see the colored edges in Figure 6) to avoid spurious reflections of the ultrasonic signal due to the edges. This simple method turned out to be necessary to manage boundary reflection due to the free edges of the plate for estimation of impact position carried out outside of the ROI. The problem concerning multiple reflections from the plate edges can also be addressed by selecting a time according to the geometry of the plate-like structure [17]. More clever algorithms exploit, also, the information retrieved by multiple reflections from boundaries to minimize the number of sensors installed on the structure [23].

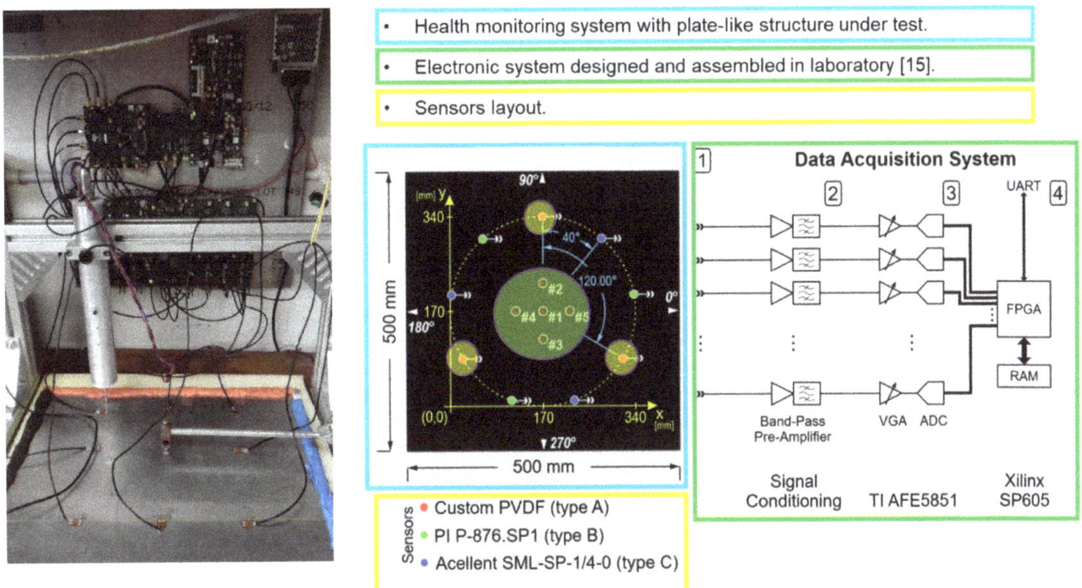

Figure 6. Experimental set-up with the eight piezoceramic transducers attached with a bi-adhesive tape to a 1.4 mm aluminum plate, shown in the bottom of the figure, and the real-time electronic acquisition system, shown in the top. In the center of the plate, the low-energy hammer impactor is visible. The figure also shows the colored play dough covering the edges of the aluminum plate; this method was adopted to attenuate the signals received from multipath. Reproduced from [24].

The next section reports the analysis of different errors obtained for different impact points (inside and outside the ROI) and for different numbers of channels used to find the minimum of function (1).

4. Experimental Results

The new custom real-time acquisition system was tested in a series of low-energy impact loading experiments so as to not damage the laminate. The aim of those experiments was to verify the actual capabilities of the SHM system (i.e., the localization accuracy) processing signals acquired from different numbers of sensors.

The fundamental symmetric (S_0) and anti-symmetric (A_0) group velocity dispersion curves for an aluminum plate with thickness of 1.4 mm are shown in Figure 7. These curves were obtained with the LAMB MATLAB toolbox GMM calculator. From Figure 7 we can evaluate the propagation velocity of the S_0 mode. This value was used to solve Equation (1). For our experiments we considered the propagation velocity v = 5150 m/s, corresponding to a frequency of fS0 = 650 kHz, which falls into a frequency range where the dispersive behavior of S_0 mode is negligible but still retains a small wavelength v/fS0 = 8 mm.

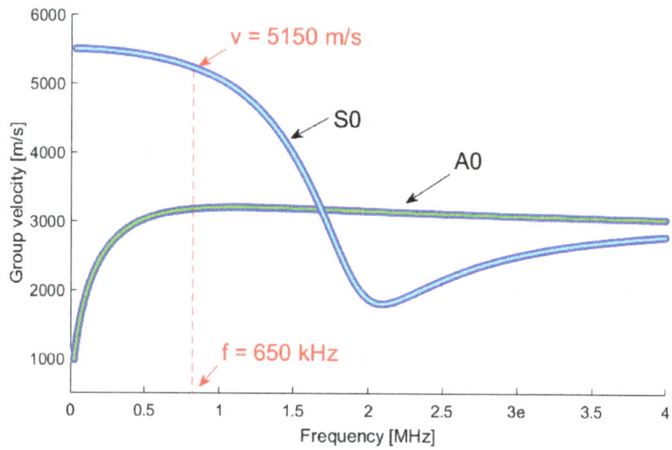

Figure 7. Viewgraph showing the calculated S_0 and A_0 group velocity dispersion curves for an aluminum plate with thickness of 1.4 mm.

We performed three impacts: impact P1 with position x_i = 190 mm, y_i = 150 mm, performed inside the ROI, delimited by sensors; impact P2 with position x_i = 193 mm, y_i = 32 mm, performed outside the ROI; and impact P3 with position x_i = 200 mm, y_i = 40 mm, performed outside the ROI.

In Figures 8–13 we report the results for impacts inside and outside the ROI.

In detail, Figure 8 shows the performance of the acquisition system detecting impact point P1 and processing the signal acquired by only four sensors: Rx1, Rx3, Rx5, and Rx7, whereas Figure 9 shows the result obtained by detecting impact point P1 and processing the signal acquired by all eight sensors (Rx1–Rx8). The estimated position of the impact point obtained by processing signals acquired by only four sensors (P1′-1 point) is also reported in Figure 8, whereas Figure 6 shows the estimated position of the impact point obtained by processing signals acquired by all sensors (P1′-2 point). To evaluate the trade-off between sensor density of an SHM system and error in impact localization, we processed the signal acquired by six sensors Rx1, Rx2, Rx3, Rx5, Rx6, and Rx7, and the estimated position of the impact point (P1′-3 point) is reported in Figure 10.

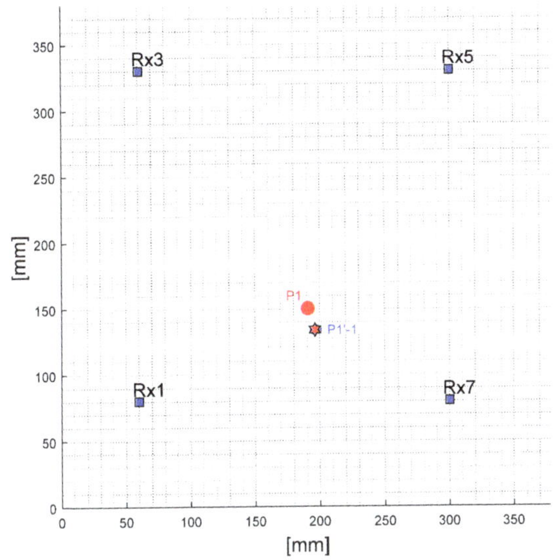

Figure 8. Position of impact P1 (red circle) inside the ROI with coordinates x_i = 190 mm, y_i = 150 mm. The detected impact position P1′-1 (red star) with coordinates x_p = 196 mm, y_p = 134 mm was obtained by processing the ultrasonic signals acquired by only four sensors (Rx1, Rx3, Rx5, and Rx7). Reproduced from [24].

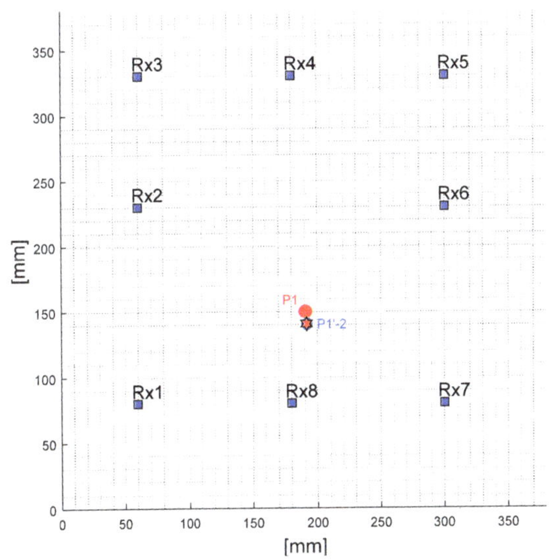

Figure 9. Position of impact P1 (red circle) inside the ROI with coordinates x_i = 190 mm, y_i = 150 mm. The detected impact position P1′-2 (red star) with coordinates x_p = 192 mm, y_p = 141 mm was obtained by processing the ultrasonic signals acquired by all sensors (Rx1–Rx8). Reproduced from [24].

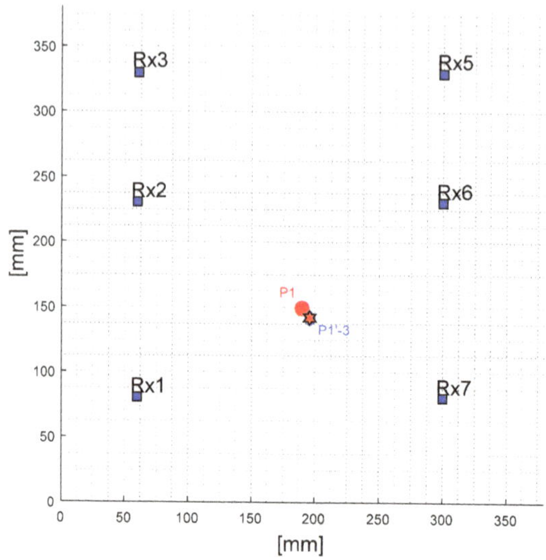

Figure 10. Position of impact P1 (red circle) inside the ROI with coordinates x_i = 190 mm, y_i = 150 mm. The detected impact position P1′-3 (red star) with coordinates x_p = 196 mm, y_p = 142 mm was obtained by processing the ultrasonic signals acquired by six sensors (Rx1, Rx2, Rx3, Rx5, Rx6, and Rx7).

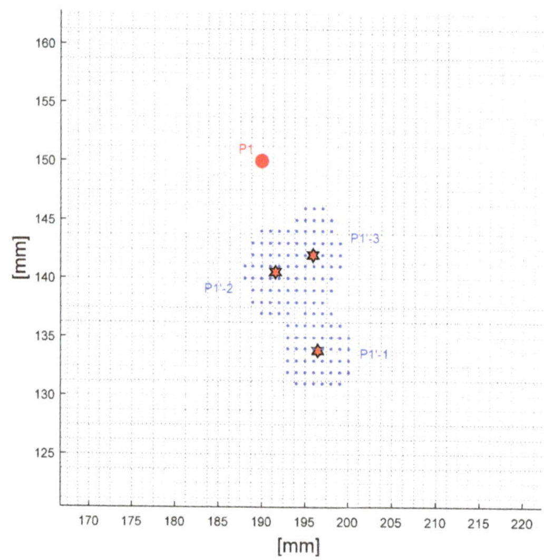

Figure 11. Zoom of results shown in Figures 8–10. The illustration reports the real impact point P1 (red circle), the detected impact points P1′-1, P1′-2, and P1′-3 (red stars) together with the points 1% above the minimum of the error function (blue dots). Reproduced from [24].

Results shown in Figures 8–10 are summarized in Figure 11.

In general, the installation of sensors is based on several constraints imposed by the target structure, while the impact event position is more unpredictable. It is, therefore, worthwhile analyzing impacts outside of the ROI. The full size of the area of the aluminum plate (500 mm × 500 mm) was divided into a uniform grid of points spaced 1 mm apart with coordinates (x_p, y_p). Figures 12 and 13 show the results obtained by processing the signals acquired from all sensors in the cases of impacts P2 and P3 outside of the ROI.

From Figures 8–11 we can observe that the accuracy of the localization of the impact coordinates is better using eight sensors, compared with using four sensors, but the analysis with six sensors could be a good compromise between sensor density and the error obtained in impact localization. With the eight sensor analysis the localization error was 2 mm for the x-axis and 9 mm for the y-axis, whereas when processing signals acquired by six sensors the error was 6 mm for the x-axis and 8 mm for the y-axis.

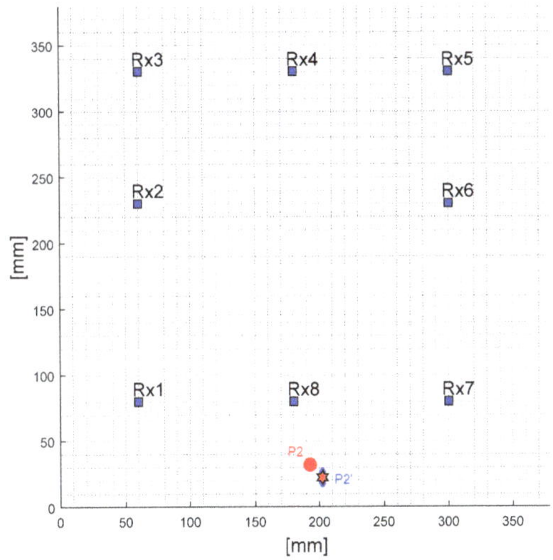

Figure 12. Position of impact P2 (red circle) outside the ROI. The detected impact position P2′ (red star) with coordinates x_p = 202 mm, y_p = 22 mm was obtained by processing signals acquired by all sensors. Reproduced from [24].

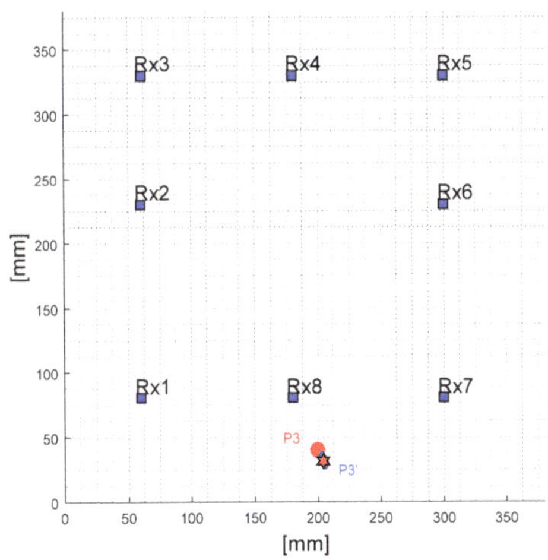

Figure 13. Position of impact P3 (red circle) outside the ROI. The detected impact position P3′ (red star) with coordinates x_p = 205 mm, y_p = 32 mm was obtained by processing signals acquired by all eight sensors.

Figures 12 and 13 demonstrate the ability of the electronic system to reveal, also, impacts outside the ROI with a good accuracy: in both cases, the error in the evaluation of x-axis and y-axis impact point coordinates was less than 10 mm.

Another observation concerns the placement of sensors. In several real applications, it is not possible to install the sensors on the vertices of regular geometric shapes (squares, circles, hexagons, etc.) and thus positions are decided according to these, and other, constraints. From Figures 8–13 we can observe the position of the eight sensors that are symmetrical along the x-axis (Rx1, Rx3, Rx4, Rx5, Rx7, and Rx8) but non-symmetric along the y-axis (Rx2 and Rx6): these results demonstrate that the calculated impact point is well-identified even in cases of non-symmetric positions of sensors.

The resulting estimates of the three impacts P1, P2, and P3 are summarized in Tables 2–4, respectively. The errors of the impact localization are reported in as a percentage with respect to the dimension of the monitored area, i.e., 500 mm.

Table 2. Estimated impact positions obtained for impact P1′ with coordinates x_i = 190 mm, y_i = 150 mm.

Predicted Impact	Estimated Position	Positioning Error	Error = Positioning Error/500 mm %
P1′-1	x = 196 mm y = 134 mm	6 mm (x-axis) 16 mm (y-axis)	1.2% 3.2%
P1′-2	x = 192 mm y = 141 mm	2 mm (x-axis) 9 mm (y-axis)	0.4% 1.8%
P1′-3	x = 196 mm y = 142 mm	6 mm (x-axis) 8 mm (y-axis)	1.2% 1.6%

Table 3. Estimated impact position obtained for impact P2′ with coordinates x_i = 193 mm, y_i = 32 mm.

Predicted Impact	Estimated Position	Positioning Error	Error = Positioning Error/500 mm %
P2′	x = 202 mm y = 22 mm	9 mm (x-axis) 10 mm (y-axis)	1.8% 2.0%

Table 4. Estimated impact position obtained for impact P3′ with coordinates x_i = 200 mm, y_i = 40 mm.

Predicted Impact	Estimated Position	Positioning Error	Error %
P3′	x = 205 mm y = 32 mm	5 mm (x-axis) 8 mm (y-axis)	1.0% 1.6%

5. Discussion and Final Remarks

This study presents an analysis of errors in impact location with different numbers of channels (from 4 to 8), connected to piezoelectric sensors on an aluminum plate and front-end electronics capable of detecting the early arrival signals of the S_0 mode. Impacts were carried out inside and outside of the defined ROI of the plate. To neglect the boundary reflections of the ultrasonic signal due to the free edges of the plate, an adhesive rubber was used with the aim of simulating a semi-infinite space outside the area delimited by the sensors. The DToA were obtained by the threshold method with real-time electronics. The results of the experiments explained well the compromise between sensor density of an SHM system and errors in impact location: the best accuracy was obtained with eight sensors, covering an area of 500 mm × 500 mm in the aluminum laminate, but a good compromise between sensor density and the error in impact localization was detected when the acquired signal was processed by six sensors—obtaining an error comparable with the assumed wavelength of the S_0 mode equal to 8 mm. We found that the error was below 3.2% and slightly better for the x-coordinate, which is probably due to a systematic error in sensor reference coordinates. The main benefit of the AFE designed for this application is the possibility to select the best signal processing chain to avoid input saturation, with large energy impacts generating 1 to 10 Vpp amplitudes and retaining high SNR, in excess of 72 dB, in the 270–800 kHz bandwidth. We also point out that a certain redundancy in the number of sensors is always useful when considering the possible failure of single channels. Finally, thanks to the programmable configuration of up to 16 channels, this criteria for the selection of optimal sensor density will be investigated also on CFRP plates with the same sensor layout to account for the variability of velocity in composite laminates in different directions.

Author Contributions: Conceptualization, E.M.M. and L.C.; methodology, A.B.; software, E.M.M.; validation, E.M.M. and A.B.; formal analysis, E.M.M.; investigation, A.B. and L.C.; resources, L.C.; data curation, A.B.; writing—original draft preparation, L.C.; writing—review and editing, L.C.; visualization, E.M.M.; supervision, L.C.; project administration, L.C. All authors have read and agreed to the published version of the manuscript.

Funding: This research received no external funding.

Institutional Review Board Statement: Not applicable.

Informed Consent Statement: Not applicable.

Acknowledgments: The authors wish to acknowledge the support of Marco Calzolai, senior technician at the Ultrasound and NDT laboratory of the University of Florence, in conducting this work.

Conflicts of Interest: The authors declare no conflict of interest.

Abbreviations

Acronym	
ADC	Analog-to-digital converter
AFE	Analog front-end
AI	Artificial intelligence
CFRP	Composite fiber reinforce polymer
CWT	Continuous wavelet transform
DToA	Differential time of arrival
IDT	Interdigital transducer
NDT	Non-destructive testing
PVDF	Polyvinylidene fluoride
PZT	Lead zirconate titanate
ROI	Region of interest
SHM	Structural health monitoring
SNR	Signal-to-noise ratio
STFT	Short-time Fourier transform

References

1. Giurgiutiu, V. *Structural Health Monitoring: With Piezoelectric Wafer Active Sensors*, 2nd ed.; Elsevier: Amsterdam, The Netherlands, 2014; 1024p, ISBN 9780124186910.
2. Rose, J.L. *Ultrasonic Guided Waves in Solid Media*; Cambridge University Press: Cambridge, UK, 2014.
3. Ostachowicz, W.; Kudela, P.; Krawczuk, M.; Zak, A. *Guided Waves in Structures for SHM: The Time-Domain Spectral Element Method*; John Wiley & Sons: New York, NY, USA, February 2012; 350p, ISBN 978-0-470-97983-9.
4. Purekar, A.S.; Pines, D.J. Damage detection in thin composite laminates using piezoelectric phased sensor arrays and guided lamb wave interrogation. *J. Intell. Mater. Syst. Struct.* **2010**, *21*, 995–1010. [CrossRef]
5. Qiu, L.; Deng, X.; Yuan, S.; Huang, Y.; Ren, Y. Impact monitoring for aircraft smart composite skins based on a lightweight sensor network and characteristic digital sequences. *Sensors* **2018**, *18*, 2218. [CrossRef] [PubMed]
6. Tobias, A. Acoustic-emission source location in two dimensions by an array of three sensors. *Non-Destr. Test.* **1976**, *9*, 9–12. [CrossRef]
7. Ciampa, F.; Meo, M. A new algorithm for acoustic emission localization and flexural group velocity determination in anisotropic structures. Compos. *Part A Appl. Sci. Manuf.* **2010**, *41*, 1777–1786. [CrossRef]

8. Mohd, S.; Holford, K.M.; Pullin, R. Continuous wavelet transform analysis and modal location analysis acoustic emission source location for nuclear piping crack growth monitoring. In Proceedings of the AIP Conference Proceedings, Punjab, India, 17–21 December 2014; pp. 61–68.
9. Yin, S.; Cui, Z.; Kundu, T. Acoustic source localization in anisotropic plates with "Z" shaped sensor clusters. *Ultrasonics* **2018**, *84*, 34–37. [CrossRef] [PubMed]
10. Worden, K.; Staszewski, W.J. Impact location and quantification on a composite panel using neural networks and a genetic algorithm. *Strain* **2000**, *36*, 61–68. [CrossRef]
11. Carrino, S.; Nicassio, F.; Scarselli, G. An innovative method based on nonlinear Lamb waves for locating disbonds in Single-Lap joints. In Proceedings of the 2019 IEEE 5th International Workshop on Metrology for AeroSpace (MetroAeroSpace), Torino, Italy, 19–21 June 2019; pp. 187–191.
12. Carrino, S.; Nicassio, F.; Scarselli, G. Non-linear Lamb Waves for Locating Defects in Single-Lap Joints. *Front. Built Environ. J.* **2020**, *6*. [CrossRef]
13. Kundu, T. Acoustic source localization. *Ultrasonics* **2014**, *54*, 25–38. [CrossRef] [PubMed]
14. Sen, N.; Kundu, T. A new wave front shape-based approach for acoustic source localization in an anisotropic plate without knowing its material properties. *Ultrasonics* **2018**, *87*, 20–32. [CrossRef] [PubMed]
15. Bulletti, A.; Giannelli, P.; Calzolai, M.; Capineri, L. An Integrated Acousto/Ultrasonic Structural Health Monitoring System for Composite Pressure Vessels. *IEEE Trans. Ultrason. Ferroelectr. Freq. Control* **2016**, *63*, 864–873. [CrossRef] [PubMed]
16. Merlo, E.M.; Bulletti, A.; Giannelli, P.; Calzolai, M.; Capineri, L. An Integrated Acousto/Ultrasonic Structural Health Monitoring System for Composite Pressure Vessels, A Novel Differential Time-of-Arrival Estimation Technique for Impact Localization on Carbon Fiber Laminate Sheets. *Sensors* **2017**, *17*, 2270. [CrossRef] [PubMed]
17. Merlo, E.M.; Bulletti, A.; Giannelli, P.; Calzolai, M.; Capineri, L. Analysis of Errors in the Estimation of Impact Positions in Plate-Like Structure through the Triangulation Formula by Piezoelectric Sensors Monitoring. *Sensors* **2018**, *18*, 3426. [CrossRef] [PubMed]
18. Lehmann, M.; Büter, A.; Frankenstein, B.; Schubert, F.; Brunner, B. Monitoring System for Delamination Detection-Qualification of Structural Health Monitoring (SHM) Systems. In Proceedings of the Conference on Damage in Composite Materials (CDCM), Stuttgart, Germany, 18–19 September 2006.
19. Ono, K. Review on Structural Health Evaluation with Acoustic Emission. *Appl. Sci.* **2018**, *8*, 958. [CrossRef]
20. Xinlin, Q.; Wenzhuo, L.; Yishou, W.; Hu, S. Piezoelectric Transducer-Based Structural Health Monitoring for Aircraft Applications. *Sensors* **2019**, *19*, 545. [CrossRef]
21. Capineri, L.; Bulletti, A.; Calzolai, M.; Giannelli, P.; Francesconi, D. Arrays of conformable ultrasonic Lamb wave transducers for structural health monitoring with real-time electronics. *Procedia Eng.* **2014**, *87*, 1266–1269. [CrossRef]
22. Ebrahimkhanlou, A.; Dubuc, B.; Salamone, S. A generalizable deep learning framework for localizing and characterizing acoustic emission sources in riveted metallic panels. *Mech. Syst. Signal Process.* **2019**, *130*, 248–272. [CrossRef]
23. Ross, R. Structural health monitoring and impact detection using neural networks for damage characterization. In Proceedings of the 47th AIAA/ASME/ASCE/AHS/ASC Structures, Structural Dynamics, and Materials Conference 14th AIAA/ASME/AHS Adaptive Structures Conference 7th, Newport, RI, USA, 1–4 May 2006; p. 2117.
24. Bulletti, A.; Merlo, E.M.; Capineri, L. Analysis of the accuracy in impact localization using piezoelectric sensors for Structural Health Monitoring with multichannel real-time electronics. In Proceedings of the 2020 IEEE 7th International Workshop on Metrology for AeroSpace (MetroAeroSpace), Pisa, Italy, 22–24 June 2020.

applied sciences

Article

Image Human Thorax Using Ultrasound Traveltime Tomography with Supervised Descent Method

Tong Zhang, Rui Guo, Haolin Zhang, Hongyu Zhou, Yeyu Cao, Maokun Li *, Fan Yang and Shenheng Xu

Beijing National Research Center for Information Science and Technology (BNRist), Institute for Precision Medicine, Department of Electronic Engineering, Tsinghua University, Beijing 100084, China; t-zhang20@mails.tsinghua.edu.cn (T.Z.); guor93@tsinghua.edu.cn (R.G.); zhanghl13@tsinghua.org.cn (H.Z.); zhouhy21@mails.tsinghua.edu.cn (H.Z.); yy-cao19@mails.tsinghua.edu.cn (Y.C.); fan_yang@tsinghua.edu.cn (F.Y.); shxu@tsinghua.edu.cn (S.X.)
* Correspondence: maokunli@tsinghua.edu.cn

Abstract: The change of acoustic velocity in the human thorax reflects the functional status of the respiratory system. Imaging the thorax's acoustic velocity distribution can be used to monitor the respiratory system. In this paper, the feasibility of imaging the human thorax using ultrasound traveltime tomography with a supervised descent method (SDM) is studied. The forward modeling is computed using the shortest path ray tracing (SPR) method. The training model is composed of homogeneous acoustic velocity background and a high-velocity rectangular block moving in the domain of interest (DoI). The average descent direction is learned from the training set. Numerical experiments are conducted to verify the method's feasibility. Normal thorax model experiment proves that SDM traveltime tomography can efficiently reconstruct thorax acoustic velocity distribution. Numerical experiments based on synthetic thorax model of pleural effusion and pneumothorax show that SDM traveltime tomography has good generalization ability and can detect the change of acoustic velocity in human thorax. This method might be helpful for the diagnosis and evaluation of respiratory diseases.

Keywords: thorax imaging; ultrasound traveltime tomography; supervised descent method

Citation: Zhang, T.; Guo, R.; Zhang, H.; Zhou, H.; Cao, Y.; Li, M.; Yang, F.; Xu, S. Imaging Human Thorax Using Ultrasound Traveltime Tomography with Supervised Descent Method. *Appl. Sci.* **2022**, *12*, 6763. https://doi.org/10.3390/app12136763

Academic Editor: Nikolaos Dikaios

Received: 30 April 2022
Accepted: 29 June 2022
Published: 4 July 2022

1. Introduction

Ultrasound imaging is widely used in industrial and biomedical imaging. Industrial imaging is important in nondestructive evaluation (NDE) [1]. It can be used to evaluate the health of material [2–5]. The information of liquid flow can also be obtained by ultrasound imaging [6–8]. In the medical field, ultrasound imaging is widely used for its low-cost, portability, radiationless, and real-time compared with other imaging methods, such as computed tomography (CT) and magnetic resonance imaging (MRI). Ultrasound is applied to imaging heart [9–11], liver [12,13], brain [14–16] and many other human parts [17]. In most cases, ultrasound imaging is realized by transmitting ultrasound at megahertz into the human body and analyzing the reflected signal. However, this workflow is hard to implement in thorax imaging, because complex reflection and refraction occur at the lung-pleura interface and among the millions of alveoli inside the lungs, resulting in both chaotic and low-energy reflected signal [18]. In addition, the high frequency will cause severe attenuation, resulting in the reflected signal's low signal-to-noise ratio (SNR). Current thorax imaging methods include X-ray [19], CT [20], MRI [21], and electrical impedance tomography (EIT) [22]. Microwave thorax imaging is still in the research stage [23]. These methods all have limitations. The radiation from CT and X-rays is harmful to human health. The equipment of MRI is not portable and its imaging quality is not that good because of the air in the lung. The resolution of EIT and microwave imaging is relatively low.

Though the reflected signal can hardly be used for thorax imaging, it contains information about the thorax state. B-lines are suggested to originate as ring-down artifacts from irregularities at the lung–pleura interface, which contains information inside the thorax and correlates with a variety of respiratory diseases [24]. However, the results obtained from B-lines remain qualitative, subjective, and suboptimal [25]. In addition, information in the frequency domain can be used to diagnose pulmonary fibrosis quantitatively [25]. Ultrasound can also be used to monitor the states of the respiratory system, see [26–28]. In addition to reflected signal, transmission signal can be applied to thorax examination. Transmission signal is the signal that propagates through the human thorax. Acoustic wave can permeate human thorax as long as the frequency is low enough. According to Rueter et al. [29], the propagation of acoustic wave in human thorax shows three distinct bands according to frequency. Sound of frequency around 1 kHz can propagate in human thorax at velocity of 30–50 m/s. At this band, the thorax model is established and verified by simulations and experiments, see [30,31]. The feasibility of diagnosing pulmonary diseases using sound at this band is also researched, see [32,33]. Sound of frequency between 1 and 10 kHz is hard to permeate thorax. Ultrasound with frequency between 10 kHz and 750 kHz can permeate human thorax at velocity of about 1500 m/s. An ultrasound at this band permits monitoring of the human respiratory system. The transmission ultrasound signal can be analyzed directly for diagnosing lung diseases, see [34,35]. In addition, the transmission ultrasound signal can be used to conduct traveltime tomography. This method is successfully implemented on breast imaging, see [36,37]. So far, to the best knowledge of the authors, traveltime tomography for thorax has not been well studied.

Ultrasound traveltime tomography reconstructs the acoustic velocity distribution from the measured time of the signal traveling from transmitter to receiver. This method is widely used in geophysics. It is robust at low SNR situations [38,39] and can mitigate the challenge aroused by the complexity of ultrasound propagation in complicated media [40,41]. The acoustic velocity is sensitive to the contents of gas, liquid, and their distribution, because their acoustic velocity differs a lot. Thus, it is possible to measure the amount of gas and fluid in the thorax by reconstructing the acoustic velocity distribution, which is crucial in the diagnosis of many diseases [42,43].

Machine learning techniques develop quickly in recent years and it has been widely used in inverse problems. One of the reasons for its popularity is that, through the training process, lots of work can be completed offline, which makes the online predicting process simple and efficient. Supervised descent method is one of these techniques, which is firstly used in face alignment [44]. The average descent direction of the cost function is learned from the training process and used to update the model in the online predicting process. The time-consuming matrix inversion in traditional gradient-based method is no longer needed. Previous studies show that SDM performs well in imaging, such as 2D magnetotelluric imaging [45], microwave thorax imaging [46], EIT thorax imaging [47], and corrosion imaging [48].

In this paper, we study the feasibility of imaging human thorax using ultrasound traveltime tomography with SDM. Numerical experiments based on normal thorax model and synthetic patient thorax model are conducted. The remaining of the paper is organized as follows. In Section 2, the overview of the forward modeling is described. In Section 3, details of traveltime tomography with SDM are discussed. In Section 4, our method is compared with the traditional gradient-based inversion method and its ability to detect the change of acoustic velocity distribution is tested.

2. Forward Modeling for Computing Traveltime

Forward modeling computes traveltime between the transducer T and the receiver R with the given acoustic velocity distribution. The traveltime can be computed as

$$t = \int_T^R \frac{1}{v(\boldsymbol{x})} dl \tag{1}$$

where v is the velocity, $\boldsymbol{x}$ is the position, and dl is the differential length [49]. This non-linear function can be solved by the shortest path ray tracing (SPR) method based on Fermat's principle. Fermat's principle says that the ultrasonic wave travels along the path that takes the shortest time. SPR converts computing traveltime into finding single-source shortest path in a graph, which can be solved by the Dijkstra algorithm. According to [50], SPR divides the DoI into rectangular elements. Acoustic velocity is defined on the element, which means the velocity in an element is a constant. Nodes are defined on the boundary of elements. Two nodes are connected only when they belong to the same element and are not located on the same element boundary. Figure 1 is an example. The area is divided into $2 \times 2 = 4$ elements, see the dashed squares. Three nodes lie on one edge of the elements. The solid lines are the connections between nodes.

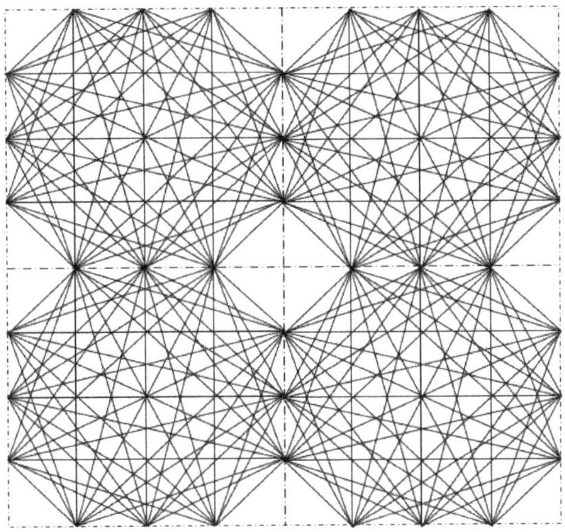

Figure 1. Example of element and possible ray path. Dashed lines: element boundaries. Solid lines: node connections.

The weight of the connection between two nodes is defined as the quotient of their distance and the velocity of the element they belong to. With nodes, connections and weight of connections, an undirected graph is established. Assume t and r is the closest nodes to T and R. Naturally, the shortest path between T and R is the path with the lowest weight between t and r, and the traveltime is the path's weight. The forward modeling can be formulated as:

$$\boldsymbol{d} = F(\boldsymbol{m}) \tag{2}$$

where $\boldsymbol{d}$ is the traveltime data, $\boldsymbol{m}$ is the model including distributions of acoustic velocity and positions of the transducers. With the acoustic velocity distribution, the graph can be constructed. Then, the transducers are assigned to the nodes that are closest to them and the traveltime between them can be calculated by $F(\cdot)$, which is the operator of forward modeling. $F(\cdot)$ calculates traveltime for each transmitter–receiver pair using Dijkstra algorithm.

3. Traveltime Tomography with SDM

3.1. Formulations of Inverse Problem with SDM

Solving the inverse problem is an optimization process to find the best $\boldsymbol{m}$ that minimizes the loss function. The loss function of this problem is defined as

$$L(\boldsymbol{m}) = \|\boldsymbol{d}_{obs} - F(\boldsymbol{m})\|^2 \tag{3}$$

where $\boldsymbol{d}_{obs}$ is the real traveltime data. The Gauss–Newton method can be used to solve this optimization problem [51]. The update of the model can be calculated as

$$\Delta \boldsymbol{m} = \left(\boldsymbol{J}^T \boldsymbol{J}\right)^{-1} \boldsymbol{J}^T (\boldsymbol{d}_{obs} - F(\boldsymbol{m})) = \boldsymbol{K}(\boldsymbol{d}_{obs} - F(\boldsymbol{m})) = \boldsymbol{K}\Delta \boldsymbol{d} \tag{4}$$

where $\boldsymbol{J}$ is the Jacobian matrix of $F(\boldsymbol{m})$ and $\boldsymbol{K} = \left(\boldsymbol{J}^T \boldsymbol{J}\right)^{-1} \boldsymbol{J}^T$ is a mapping from $\Delta \boldsymbol{d}$ to $\Delta \boldsymbol{m}$. The element of Jacobian matrix J_{ik} tells us the change in the ith data point (traveltime) due to a unit change in the kth model parameter (acoustic velocity). In this problem, J_{ik} is non-zero only when the ith path goes through the kth element, and it has an analytical form as

$$J_{ik} = -\frac{s_{ik}}{v_k^2} \tag{5}$$

where s_{ik} is the length of the ith path within the kth element. v_k^2 is the acoustic velocity of the kth element. The update direction $\boldsymbol{K}$ contains the local information of $\boldsymbol{F}$ near $\boldsymbol{m}$. The calculation of K is the key step in Gauss–Newton method, which is also the most time-consuming step, because the calculation of $\boldsymbol{J}$ and matrix inversion take a lot of time. SDM introduces an alternative approach to compute $\boldsymbol{K}$ in a shorter time.

In SDM, $\boldsymbol{K}$ is calculated offline in advance. From Equation (4), the optimization of Equation (3) can be converted to minimize the following function:

$$S = \|\Delta \boldsymbol{m} - \boldsymbol{K}\Delta \boldsymbol{d}_{obs}\|^2 \tag{6}$$

The size of $\Delta \boldsymbol{m}$ and $\boldsymbol{m}$ is $n_{ele} \times 1$, where n_{ele} is the number of elements in the grid, which is also the number of unknowns to be solved. The size of $\boldsymbol{d}_{obs}$ is $n_d \times 1$, where n_d is the number of traveltime data we use. The size of $\boldsymbol{K}$ is $n_{ele} \times n_d$.

We assume that the real model $\boldsymbol{m}$ is in the space expanded by N reference models $\boldsymbol{m}^i_{prior} (i = 1, 2, \cdots, N)$. The corresponding traveltime is $\boldsymbol{d}^i_{prior}$. Then, an average update direction $\tilde{\boldsymbol{K}}$ is calculated by minimizing

$$S^{Tr}(\tilde{\boldsymbol{K}}) = \sum_{i=1}^{N} \left\| \Delta \boldsymbol{m}^i_{prior} - \tilde{\boldsymbol{K}}\Delta \boldsymbol{d}^i_{prior} \right\|^2 \tag{7}$$

where

$$\Delta \boldsymbol{m}^i_{prior} = \boldsymbol{m}^i_{prior} - \boldsymbol{m}_0 \tag{8}$$

$$\Delta \boldsymbol{d}^i_{prior} = \boldsymbol{d}^i_{prior} - \boldsymbol{d}_0 \tag{9}$$

and $\boldsymbol{m}_0$ is an initial guess of the real model $\boldsymbol{m}$ and $\boldsymbol{d}_0 = F(\boldsymbol{m}_0)$. The process of computing $\tilde{\boldsymbol{K}}$ offline is called training.

3.2. Offline Training Process

In the training process, the initial model $\boldsymbol{m}_0$ is chosen as homogeneous background. The training set $\boldsymbol{M}_{prior}$ contains N training models. Equation (7) can be written in matrix form as

$$S^{Tr}(\tilde{\boldsymbol{K}}) = \left\| \Delta \boldsymbol{M} - \Delta \boldsymbol{D}\tilde{\boldsymbol{K}}^T \right\|^2 \tag{10}$$

where

$$\boldsymbol{M}_{prior} = \begin{bmatrix} {\boldsymbol{m}^1_{prior}}^T \\ {\boldsymbol{m}^2_{prior}}^T \\ \vdots \\ {\boldsymbol{m}^N_{prior}}^T \end{bmatrix}, \boldsymbol{M}_0 = \begin{bmatrix} {\boldsymbol{m}_0}^T \\ {\boldsymbol{m}_0}^T \\ \vdots \\ {\boldsymbol{m}_0}^T \end{bmatrix}, \Delta \boldsymbol{M} = \boldsymbol{M}_{prior} - \boldsymbol{M}_0 \tag{11}$$

and

$$D_{prior} = \begin{bmatrix} {d^1_{prior}}^T \\ {d^2_{prior}}^T \\ \vdots \\ {d^N_{prior}}^T \end{bmatrix}, D_0 = \begin{bmatrix} {d_0}^T \\ {d_0}^T \\ \vdots \\ {d_0}^T \end{bmatrix}, \Delta D = D_{prior} - D_0 \tag{12}$$

Here, the size of ΔM is $N \times n_{ele}$ and the size of ΔD is $N \times n_d$. $\tilde{K}$ is a $n_{ele} \times n_d$ matrix.

Because forward modeling of traveltime tomography is non-linear, it takes several iterations to minimize S^{Tr} and learn $\tilde{K}$. The prediction also computes in a iterative way [52]. Thus, in the kth training iteration we have

$$S^{Tr}_k(\tilde{K}) = \left\| \Delta M_k - \Delta D_k \tilde{K}^T_k \right\|^2 \tag{13}$$

where

$$\Delta M_k = M_{prior} - M_k \tag{14}$$

$$\Delta D_k = D_{prior} - D_k = D_{prior} - F(M_k) \tag{15}$$

To minimize Equation (13), pseudo-inverse can be applied as

$$\tilde{K}^T_k = \left(\Delta D^T_k \Delta D_k \right)^{-1} (\Delta D^T_k \Delta M_k) \tag{16}$$

With the update direction $\tilde{K}$, the model in the $(k+1)$th iteration is

$$M_{k+1} = M_k + \Delta D_k \tilde{K}^T_k \tag{17}$$

In practice, the difference between D and D_{prior} decreases rapidly. To stablize the training process, a regularization procedure is introduced [46]. Equation (16) becomes

$$\tilde{K}^T_k = \left(\Delta D^T_k \Delta D_k + \alpha I \right)^{-1} (\Delta D^T_k \Delta M_k) \tag{18}$$

where α is proportional to the largest singular value of ΔD_k and I is the identity matrix.

Normalized model misfit is checked every iteration for validating training, which defined as

$$Rms_M(M_k) = \frac{||\Delta M_k||}{||M_{prior}||} \tag{19}$$

3.3. Online Predicting Process

The online predicting process is to minimize the following function after offline training:

$$S^{Pr}(\Delta m) = ||\Delta m - \tilde{K} \Delta d_{obs}||^2 \tag{20}$$

With the model update direction $\tilde{K}$ obtained from offline training process, we can directly minimize Equation (20) as follows:

$$m_{k+1} = m_k + \tilde{K}(d_{obs} - F(m_k)) \tag{21}$$

However, traveltime tomography is a ill-posed problem. The result obtained from Equation (21) might have artifacts. Here, Tikhonov regularization is applied to mitigate the ill-posedness [53]. New objective function can be written as

$$S^{Pr}_R(m_{k+1}) = P(m_{k+1}) + \alpha_v R_v(m_{k+1}) + \alpha_h R_h(m_{k+1}) \tag{22}$$

where

$$P(m_{k+1}) = ||m_{k+1} - m_k - \tilde{K}_k(d_{obs} - F(m_k))||^2 \tag{23}$$

$$R_v(\boldsymbol{m}_{k+1}) = \frac{\|\boldsymbol{d_{obs}} - F(\boldsymbol{m}_k)\|}{\|\boldsymbol{d_{obs}}\|} \int_\Omega \left(|\nabla_v \boldsymbol{m}_{k+1}|^2 + \delta_k^2 \right) d\boldsymbol{r} \quad (24)$$

$$R_h(\boldsymbol{m}_{k+1}) = \frac{\|\boldsymbol{d_{obs}} - F(\boldsymbol{m}_k)\|}{\|\boldsymbol{d_{obs}}\|} \int_\Omega \left(|\nabla_h \boldsymbol{m}_{k+1}|^2 + \delta_k^2 \right) d\boldsymbol{r} \quad (25)$$

$$\delta_k^2 = \frac{\|\boldsymbol{d_{obs}} - F(\boldsymbol{m}_k)\|^2}{\|\boldsymbol{d_{obs}}\|^2 \hat{\Delta}} \quad (26)$$

R_v and R_h are regularization terms along vertical and horizontal direction and they are proportional to the data misfit. $\hat{\Delta}$ is a positive real parameter related to the mesh size [54]. At the begining of the inversion, the data misfit is large and the regularization terms are strong, which prevents the appearing of extreme values and helps to converge to a better model. As the iteration goes on, the data misfit decreases and the model is closed to the local optimal model, regularization terms becomes weak and the model converges faster.

Equation (22) can be minimized directly as follow:

$$\frac{\partial S_R^{Pr}(\boldsymbol{m}_{k+1})}{\partial \boldsymbol{m}_{k+1}} = 0 \quad (27)$$

We have

$$\boldsymbol{m}_{k+1} = \boldsymbol{A}_{k+1}^{-1} \boldsymbol{b}_{k+1} \quad (28)$$

where

$$\boldsymbol{A}_{k+1} - \boldsymbol{I} + \frac{\|\boldsymbol{d_{obs}} - F(\boldsymbol{m}_k)\|}{\|\boldsymbol{d_{obs}}\|} \left(\alpha_v \nabla_v^T \nabla_v + \alpha_h \nabla_h^T \nabla_h \right) \quad (29)$$

$$\boldsymbol{b}_{k+1} = \boldsymbol{m}_k + \tilde{\boldsymbol{K}}_k (\boldsymbol{d_{obs}} - F(\boldsymbol{m}_k)) \quad (30)$$

Equation (28) is used to update the model every iteration during the inversion.

4. Numerical Experiments

In this section, details of the test domain and thorax model, settings of the training set and various numerical experiments will be discussed. In numerical experiments, our method is tested on a normal thorax model and synthetic thorax model of patients with respiratory diseases. It is also compared with the traditional gradient-based method on normal thorax model. All the numerical experiments are carried out with MATLAB R2020a on the same computer with Intel i5-6500 CPU, 32GB RAM.

4.1. Description of the Test Domain and Forward Modeling

The size of the test domain is 28.5 cm in length and 19 cm in width, and we divide it into 57 × 38 elements with length of 5 mm. The thorax model is shown in Figure 2, containing lung, heart, rib, spine, etc. [55]. The acoustic velocity of each tissue can be checked in [56]. The velocity of primary tissues in the thorax model is shown in Table 1. The blue background area is set to be water. The 22 black points mark the positions of the transducer. The transducer is labeled counterclockwise and the top-middle transducer in Figure 2 is the first one. Each transducer serves as the transmitter in turn and others serve as receivers. Totally 22 × 21 = 462 sets of traveltime data are obtained.

Table 1. The acoustic velocity of primary tissues in the thorax model.

Tissue	Acoustic Velocity (m/s)
Heart	1561
Lung (deflated)	1500
Bone (cancellous)	2117.5
Bone Marrow (yellow)	1371.9

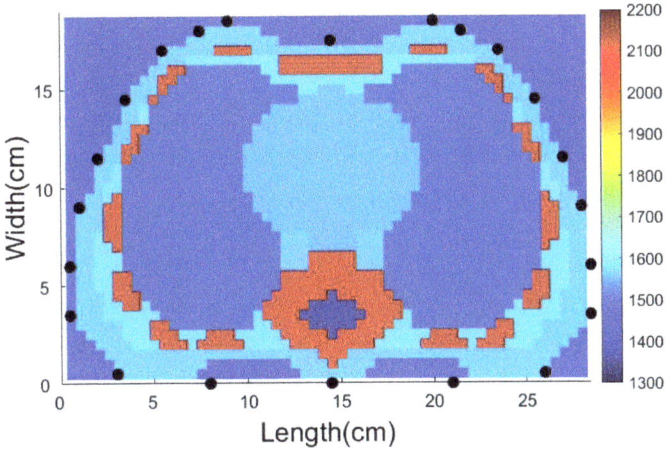

Figure 2. The thorax model in [55], containing lung, heart, rib, spine, etc. It is 28.5 cm in length and 19 cm in width. The blue background is set to be water. The black points around the thorax are where the transducers placed. The transducer is labeled counterclockwise and the top-middle transducer is the No. 1 transducer.

The number of nodes n_{nodes} on each edge of the element during the offline training process and online predicting process is chosen to be 3. n_{nodes} will influence the accuracy of the forward modeling and the time it takes. Increasing the number of nodes will improve the accuracy but at the same time increase the computation time. We calculate the traveltime data with $n_{nodes} = 1, 2, \ldots, 7$. The computation time for the forward modeling is shown in Figure 3a. We choose the traveltime obtained with $n_{nodes} = 7$ as the standard and calculate the relative error. Here, we plot the relative error of the traveltime data obtained from transmitter No. 6 in Figure 3b.

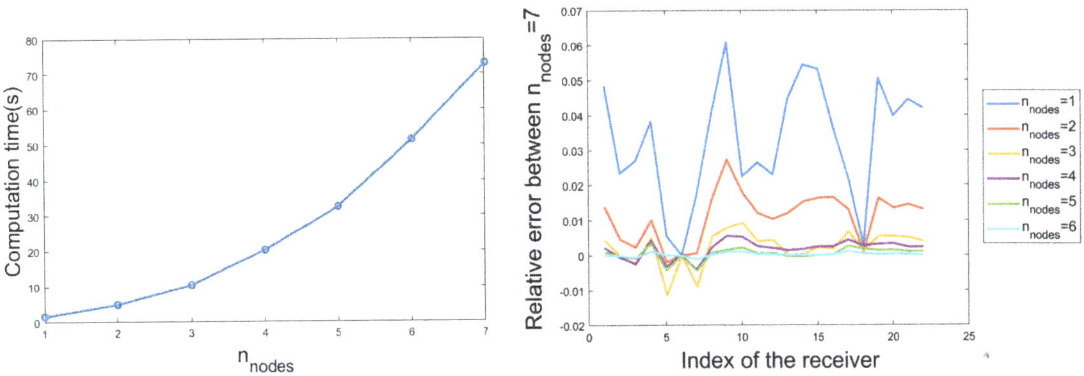

(a) (b)

Figure 3. (**a**) Computation time of the forward modeling versus n_{nodes}. (**b**) The relative error of the traveltime data from transmitter No.6. The standard is the traveltime data obtained with $n_{nodes} = 7$.

When $n_{nodes} = 3$, the relative error of each data point is almost smaller than 0.01, which is small enough. If we keep increasing n_{nodes}, the decrease in the relative error is small, while the increase in the computation time is large. The computation time of $n_{nodes} = 4$ is almost twice as long as $n_{nodes} = 3$. In consideration of both efficiency and accuracy, we choose $n_{nodes} = 3$. In addition, in order to avoid inverse crime, the traveltime data used for inversion in the following numerical experiments computed by SPR method is obtained from the grid with $n_{nodes} = 5$, because increasing n_{nodes} is equivalent to using a higher-resolution grid for forward modeling.

The distribution of the traveltime data obtained from normal thorax is shown in Figure 4. Maximum traveltime data are less than 180 μs. The median is 101.58 μs and its position is marked by the red line. To further avoid inverse crime, zero-mean white Gaussian noise with $\sigma = 2$ μs is added to the data.

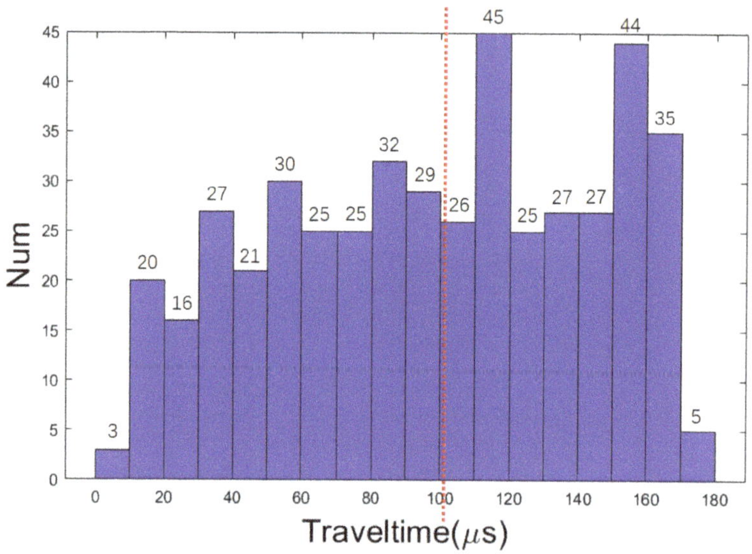

Figure 4. The distribution of 462 traveltime data. Maximum traveltime data is less than 180 μs. The red line marks the position of the median, which is 101.58 μs.

Then, we compare the traveltime data obtained from SPR method with the data obtained from k-Wave simulation [57]. We check tissues' density and attenuation coefficient in [56]. The thorax model is upsampled to 114 × 76 and put into a 128 × 128 domain with grid size of 2.5 mm. The timestep is set to be 0.25 μs and the simulation lasts for 500 μs. Gaussian waveform input signal is assigned to the transmitter and the rest transducers serve as receivers. We apply the Akaike information criteria (AIC) method to extract traveltime from the received waveform [58]. The transmitted waveform from transmitter No. 9 and the received waveform from receiver No.15 are shown in Figure 5. The position of the marker is calculated using AIC method. The distance between the two positions indicates the traveltime of the signal, which is 155.75 μs. The computed traveltime using the SPR method with $n_{nodes} = 3$ is 155.99 μs, which agrees well with the k-Wave simulation. The absolute value of the difference between the traveltime data computed by the SPR method and the k-Wave simulation is shown in Figure 6. Most are smaller than 2.5 μs, which verifies that the SPR method is accurate and can be used to compute traveltime in thorax imaging.

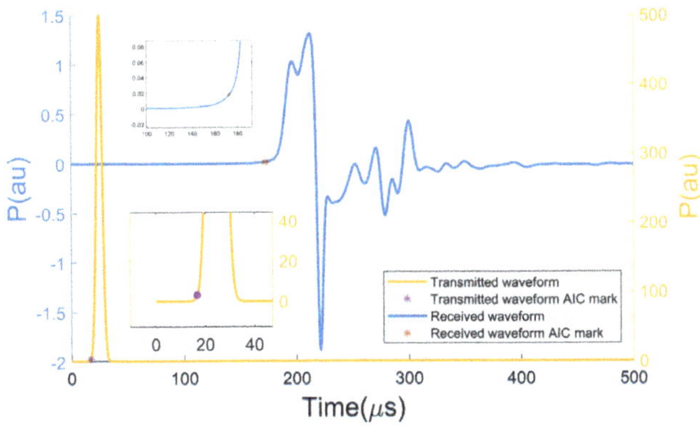

Figure 5. The transmitted waveform and the received waveform from transmitter No. 9 and receiver No. 15. The blue line is the received waveform corresponding to the left blue scale. The yellow line is the transmitted waveform corresponding to the right yellow scale. The position of the marker is calculated using AIC method, which marks the waveform is received/transmitted. The difference of the two positions is the traveltime of the signal, which is 155.75 µs.

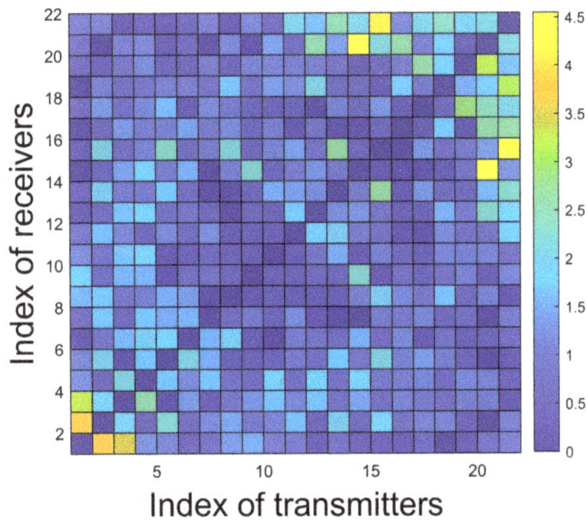

Figure 6. The absolute value of the traveltime data difference between SPR method with $n_{nodes} = 3$ and k-Wave simulation. Colorbar: 0–4.5 µs.

4.2. Details of the Training Set and the Training Process

When using the traditional gradient-based method to solve the traveltime tomography problem, the key step is computing the Jacobian matrix and the inverse of the Hessian matrix. However, only elements traversed by paths can be sensed in the traditional method. Usually, only few elements are traversed, which means the Hessian is generally sparse and hard to invert. In a traditional method, computing the inverse of Hessian needs the help of regularization and is time-consuming. To solve these problems, information of all the elements is added to the training set and the average descent direction is learned offline, which makes the online predicting fast and stable.

The training set is composed of a homogeneous low-velocity background (1500 m/s) and a high-velocity rectangular block (2200 m/s). The block has two sizes (4×8 and 8×4). The block moves one element at a time and covers the whole DoI (38×57) to make sure that all the elements in the DoI are sensed. When moving to the edge of the DoI, the part of the block out of the region is cut. Totally $(38+7)\times(57+3)+(38+3)\times(57+7)=5324$ training models are generated. One of the training models is shown in Figure 7. Thus, the training process learns the sensitivity of traveltime with respect to velocity, which is inherently embedded with the prior knowledge on spatial correlations of velocity.

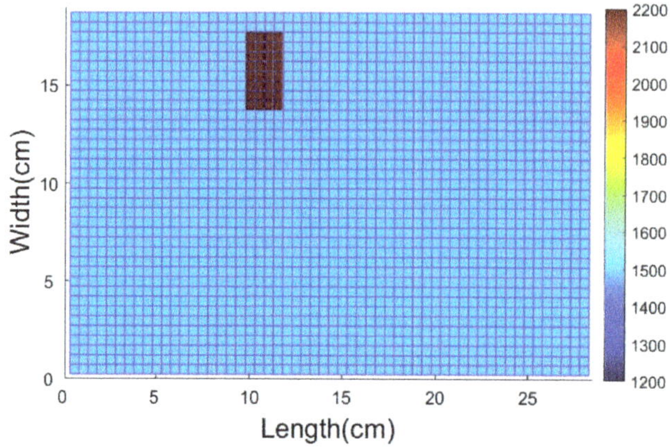

Figure 7. Example of the training model. Red Block: high-velocity block (2200 m/s). Blue region: low-velocity (1500 m/s) background.

During the training process, the initial model is set to be a homogeneous background with velocity of 1500 m/s. The maximum training step is set to be 5 for convenience. The predicting process will also last for 5 iterations and use the same initial value. The normalized model misfit is shown in Figure 8.

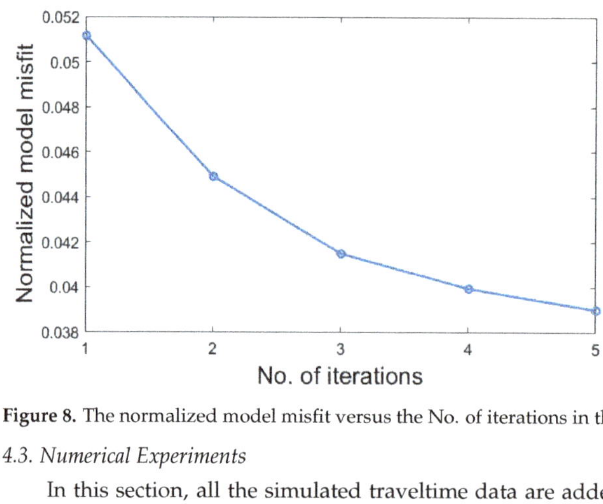

Figure 8. The normalized model misfit versus the No. of iterations in the training stage.

4.3. Numerical Experiments

In this section, all the simulated traveltime data are added with zero-mean white Gaussian noise with $\sigma = 2$ µs. The initial value of the inversion is a homogeneous background with acoustic velocity of 1500 m/s. The maximum iteration number is 5.

4.3.1. Comparison with Traditional Gradient-Based Method

The performance of SDM inversion is compared with traditional Gauss–Newton inversion. The ground truth is the thorax model shown in Figure 2. The results of the two methods are shown in Figure 9. We conduct inversion based on SPR method traveltime data and k-wave simulation traveltime data.

For inversion based on SPR method data, see Figure 9a,b. SDM finishes the inversion in 49.9 s. Although the Gauss–Newton method takes 102.9 s. SDM takes only half of the time the traditional method takes. In addition, from the results we find that the acoustic velocity distribution computed by SDM has fewer artifacts around the thorax and more precise tissue boundaries for ribs. The edges of lungs, ribs, and spine are clear. Heart and lungs are hard to distinguish because their acoustic velocity is close. However, both methods cannot correctly reconstruct the acoustic velocity of marrow because few paths traverse this region. According to Fermat's Principle, the wave will propagate along the path that takes the shortest time. The acoustic velocity of the marrow is significantly smaller than the spine around it, which means path would like to propagate in the spine rather than the marrow and result in low sensitivity of the marrow region. This example verifies the feasibility of imaging human thorax using ultrasound traveltime tomography with SDM. It also shows the strong generalization ability of SDM because little prior information on the shape of the thorax is added to the training set. The training model contains no structure information of thorax but only a rectangular high-velocity block and a homogeneous background. For inversion based on k-Wave simulation traveltime data, see Figure 9c,d, the results are similar to the inversion based on SPR method data, which further verifies the modeling and inversion algorithm. The traveltime data used in the following numerical experiment is all computed by SRP method with $n_{nodes} = 5$.

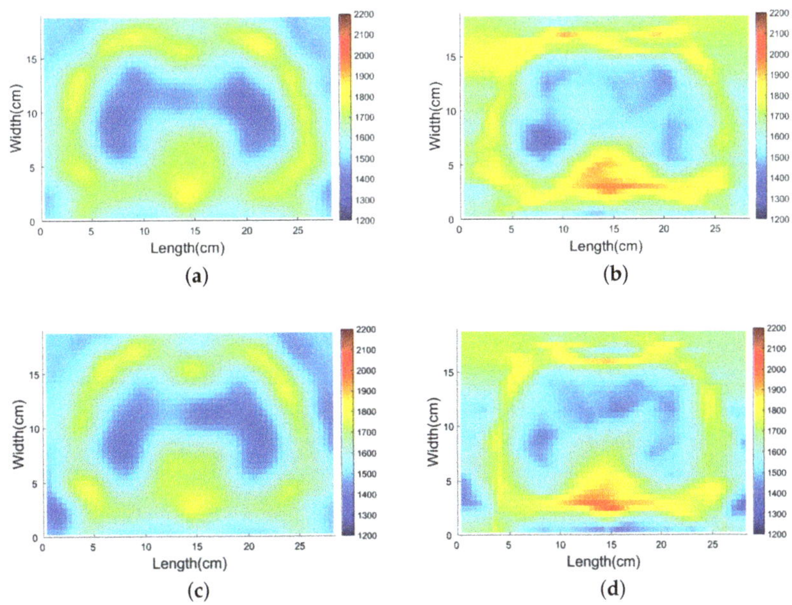

Figure 9. The inversion results of SDM inversion and Gauss–Newton inversion based on different traveltime data. (**a**) SDM inversion based on SPR method traveltime data, (**b**) Gauss–Newton inversion based on SPR method traveltime data, (**c**) SDM inversion based on k-Wave simulation traveltime data, (**d**) Gauss–Newton inversion based on k-Wave simulation traveltime data. Colorbar: 1200–2200 m/s. As a result (**a**), SDM takes 49.9 s to finish the inversion. As a result (**b**), the Gauss–Newton method takes 102.9 s.

A better initial model would help the inversion. The new initial model is obtained by smoothing the true model with a Gauss filter, see Figure 10a. The result of SDM inversion is shown in Figure 10b. Comparing the new result with the result shown in Figure 9a, we found that the area of the heart, marrow, and bones is better reconstructed. In the future, the initial model can be derived from CT or other imaging modalities.

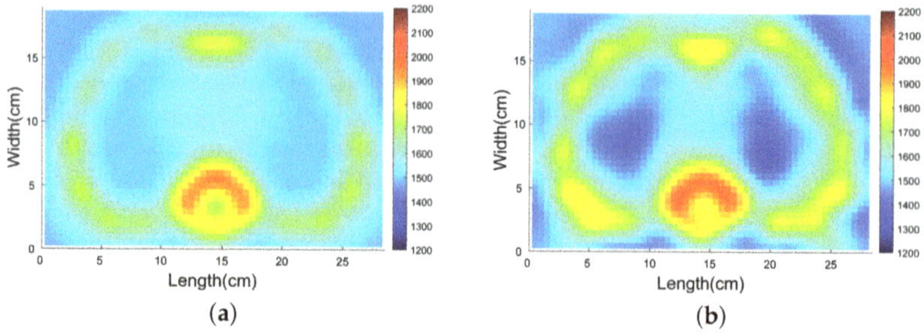

Figure 10. The new initial model and the resulf of SDM inversion. (**a**) The new initial model. (**b**) The result of SDM inversion. Colorbar: 1200–2200 m/s.

The inversion results cannot be improved by adding more transducers nor using a higher-resolution grid. We increase the number of transducers and the new array with 40 transducers is shown in Figure 11a. We also prepare a new grid whose size is 57 × 86. We conduct reconstruction on the same normal thorax model with previous array and new grid, new array and previous grid, and new array and new grid. New training sets are constructed for each task using the previous method. The inversion results are shown in Figure 11b–d. Compared with the inversion results shown in Figure 9b, no significant improvement can be seen. The reason the results have not improved might be that this traveltime tomography problem is too ill-posed. The current method cannot obtain the model out of the local minima. In addition, more transducers and higher-resolution grid will increase the time of inversion. The computation time of these three results is 132.9 s, 160.1 s, and 407.9 s, which are much longer than the previous 49.9 s.

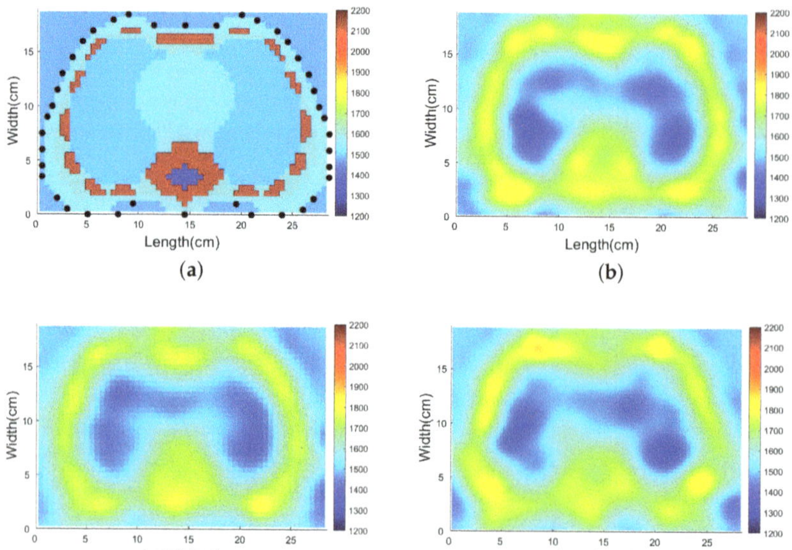

Figure 11. The new transucer array and the SDM inversion results with different array and different grid. (**a**) The new transducer array with 40 transducers, (**b**) inversion result with previous array and new grid, (**c**) inversion result with new array and previous grid, (**d**) inversion result with new array and new grid. Colorbar: 1200–2200 m/s.

4.3.2. SDM Traveltime Inversion of Thorax on Different States

Some respiratory system diseases might change the amount of gas and fluid in the thorax, which will cause changes in acoustic velocity because acoustic velocity in air is significantly smaller than in water. In this section, SDM traveltime tomography will be used to test whether these changes can be detected. Here, we take pleural effusion and pneumothorax as examples. Patients with pleural effusion tend to have more fluid in the thorax [59]. Although patients with pneumothorax will have more gas in the thorax [60].

Firstly, a pleural effusion patient's thorax acoustic velocity distribution is reconstructed, see the first column of Figure 12. We assume that the acoustic velocity of lungs when breathing in air will decrease to 1200 m/s (lower than average tissue acoustic velocity of 1540 m/s and higher than air acoustic velocity of 340 m/s). We suppose that the velocity distribution of a pleural effusion patient when breathing in air is like Figure 12a. The lower part of the left lung is the pleural effusion area. We suppose that the velocity of this area is 1540 m/s, same as the average tissue acoustic velocity. We want to reconstruct this high-velocity area near the low-velocity area through SDM traveltime tomography. The reconstruction result is shown in Figure 12c. The reconstruction result shows that both lungs' boundaries are clear and the left lung is smaller than the right one because the effusion area's velocity is higher than the normal lung area. For comparison, the result of the Gauss–Newton method is shown in Figure 12e, which has more artifacts. This example verifies that the change of acoustic velocity resulting from pleural effusion can be detected by SDM traveltime inversion.

Then, thorax acoustic velocity distribution of a pneumothorax patient is reconstructed, see the second column of Figure 12. We suppose that the velocity distribution of a pneumothorax patient when breathing out air is like Figure 12b. We suppose that the pneumothorax area is also located in the lower-left part of the thorax. The acoustic velocity of the pneumothorax area is assumed to be 1200 m/s, the same as the normal thorax region when breathing in air. In this example, a low-velocity region among high-velocity region will be reconstructed. The reconstruction result is shown in Figure 12d. The result shows that the boundaries of both lungs are clear and the pneumothorax area is reconstructed. The lower part of the left lung is darker than the lower part of the right lung. For comparison, the result of the Gauss–Newton method is shown in Figure 12f, where the boundary between normal area and pneumothorax area is not clear. This example verifies that the change of acoustic velocity resulting from pneumothorax can be detected by SDM traveltime inversion. In addition, the generalization ability of the method can also be demonstrated in these experiments. The learned descent direction can be applied to thorax under different conditions.

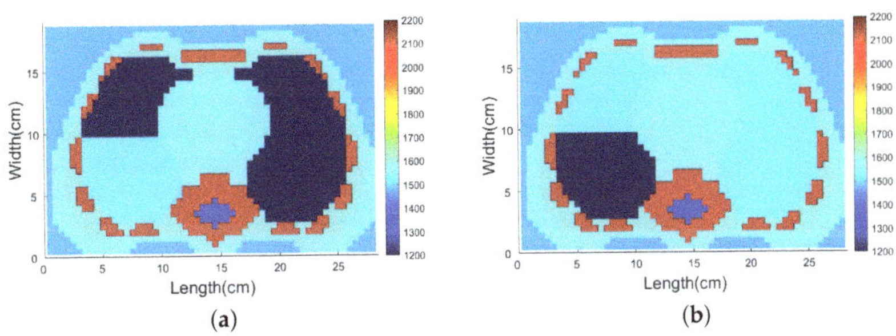

Figure 12. *Cont.*

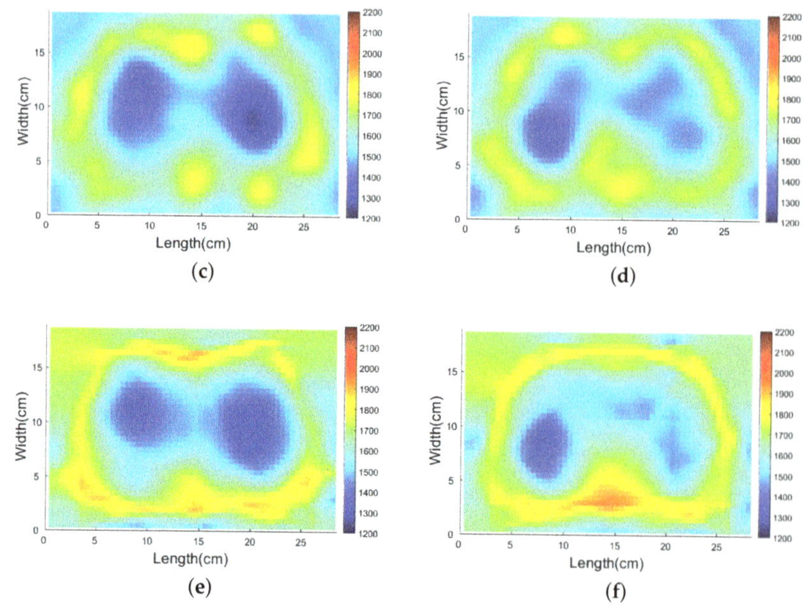

Figure 12. Thorax acoustic velocity distribution of a pleural effusion patient breathing in air and a pneumothorax patient breathing out air and the corresponding reconstruction results. (**a**) Thorax acoustic velocity distribution of a pleural effusion patient when breathing in air. (**b**) Thorax acoustic velocity distribution of a pneumothorax patient when breathing out air. (**c**) SDM inversion result of the thorax model (**a**). (**d**) SDM inversion result of the thorax model (**b**). (**e**) Gauss–Newton inversion result of the thorax model (**a**). (**f**) Gauss–Newton inversion result of the thorax model (**b**). Colorbar: 1200–2200 m/s.

5. Conclusions

In this paper, we verify the feasibility of imaging the human thorax using ultrasound traveltime tomography with SDM and detecting the change of acoustic velocity distribution resulting from respiratory diseases. A training set containing training models with homogeneous background and a high-velocity block covering the whole DoI is introduced. The average descent direction is learned from the training set and all elements in the DoI are sensed. Thanks to the offline training process, the online predicting process is fast and stable. In numerical experiments, SDM traveltime inversion is compared with the traditional gradient-based method on a normal thorax model. A slightly better result is obtained in half the time as the Gauss–Newton method, which means SDM traveltime inversion can be used to image human thorax efficiently. Then, we test its ability to detect the change in acoustic velocity distribution resulting from respiratory diseases. Thorax velocity distribution of pleural effusion and pneumothorax is assumed and inversed by our method. The results show that the synthetic change in acoustic velocity distribution can be detected. SDM also shows good generalization ability. No prior information about thorax structure is added to the training set and the learned descent direction can be applied to inverse thorax at different states. Computing the acoustic velocity distribution can help to monitor human respiratory system, as the acoustic velocity is closely related to tissues' physical properties. In conclusion, ultrasound traveltime tomography with SDM can be used to image human thorax and detect the change in thorax acoustic velocity distribution efficiently. It might be helpful for the diagnosis and quantitative evaluation of respiratory diseases and the localization of the lesion area.

Author Contributions: Project administration, T.Z., M.L., F.Y. and S.X.; Resources, T.Z.; software, T.Z., R.G., H.Z. (Hongyu Zhou), H.Z. (Haolin Zhang) and Y.C.; Supervision, M.L.; Writing—original draft preparation, T.Z. All authors have read and agreed to the published version of the manuscript.

Funding: This research was funded by National Natural Science Foundation of China (61971263), the National Key R&D Program of China (2018YFC0603604), and Institute for Precision Medicine, Tsinghua University, Beijing, China, and Biren Tech, Beijing, China.

Conflicts of Interest: The authors declare no conflict of interest.

Abbreviations

The following abbreviations are used in this manuscript:

SDM	Supervised Descent Method
SPR	Shortest Path Ray
DoI	Domain of Interest
NDE	Nondestructive Evaluation
CT	Computed Tomography
MRI	Magnetic Resonance Imaging
SNR	Signal to Noise Ratio
EIT	Electrical Impedance Tomography

References

1. Chiao, R.Y.; Thomas, L.J. Analytic evaluation of sampled aperture ultrasonic imaging techniques for NDE. *IEEE Trans. Ultrason. Ferroelectr. Freq. Control* **1994**, *41*, 484–493. [CrossRef]
2. Sun, H.; Zhu, J. Nondestructive evaluation of steel-concrete composite structure using high-frequency ultrasonic guided wave. *Ultrasonics* **2020**, *103*, 106096. [CrossRef]
3. Liu, H.; Xia, H.; Zhuang, M.; Long, Z.; Liu, C.; Cui, J.; Xu, B.; Hu, Q.; Liu, Q.H. Reverse time migration of acoustic waves for imaging based defects detection for concrete and CFST structures. *Mech. Syst. Signal Process.* **2019**, *117*, 210–220. [CrossRef]
4. Gardner, P.; Fuentes, R.; Dervilis, N.; Mineo, C.; Pierce, S.; Cross, E.; Worden, K. Machine learning at the interface of structural health monitoring and non-destructive evaluation. *Philos. Trans. R. Soc. A* **2020**, *378*, 20190581. [CrossRef] [PubMed]
5. Ricciardi, L.; Perissinotto, A.; Dabala, M. Mechanical monitoring of fracture healing using ultrasound imaging. *Clin. Orthop. Relat. Res.* **1993**, *293*, 71–76. [CrossRef]
6. Dong, F.; Jiang, Z.; Qiao, X.; Xu, L. Application of electrical resistance tomography to two-phase pipe flow parameters measurement. *Flow Meas. Instrum.* **2003**, *14*, 183–192. [CrossRef]
7. Tan, C.; Li, X.; Liu, H.; Dong, F. An ultrasonic transmission/reflection tomography system for industrial multiphase flow imaging. *IEEE Trans. Ind. Electron.* **2019**, *66*, 9539–9548. [CrossRef]
8. Yang, M.; Schlaberg, H.I.; Hoyle, B.S.; Beck, M.S.; Lenn, C. Real-time ultrasound process tomography for two-phase flow imaging using a reduced number of transducers. *IEEE Trans. Ultrason. Ferroelectr. Freq. Control* **1999**, *46*, 492–501. [CrossRef]
9. Gonçalves, L.F.; Espinoza, J.; Romero, R.; Kusanovic, J.P.; Swope, B.; Nien, J.K.; Erez, O.; Soto, E.; Treadwell, M.C. Four-dimensional ultrasonography of the fetal heart using a novel Tomographic Ultrasound Imaging display. *J. Perinat. Med.* **2006**, *34*, 39–55. [CrossRef]
10. Brekke, S.; Tegnander, E.; Torp, H.; Eik-Nes, S. Tissue Doppler gated (TDOG) dynamic three-dimensional ultrasound imaging of the fetal heart. *Ultrasound Obstet. Gynecol. Off. J. Int. Soc. Ultrasound Obstet. Gynecol.* **2004**, *24*, 192–198. [CrossRef]
11. Papadacci, C.; Finel, V.; Villemain, O.; Tanter, M.; Pernot, M. 4D ultrafast ultrasound imaging of naturally occurring shear waves in the human heart. *IEEE Trans. Med. Imaging* **2020**, *39*, 4436–4444. [CrossRef] [PubMed]
12. Xue, L.Y.; Jiang, Z.Y.; Fu, T.T.; Wang, Q.M.; Zhu, Y.L.; Dai, M.; Wang, W.P.; Yu, J.H.; Ding, H. Transfer learning radiomics based on multimodal ultrasound imaging for staging liver fibrosis. *Eur. Radiol.* **2020**, *30*, 2973–2983. [CrossRef] [PubMed]
13. Berzigotti, A.; Castera, L. Update on ultrasound imaging of liver fibrosis. *J. Hepatol.* **2013**, *59*, 180–182. [CrossRef] [PubMed]
14. Imbault, M.; Chauvet, D.; Gennisson, J.L.; Capelle, L.; Tanter, M. Intraoperative functional ultrasound imaging of human brain activity. *Sci. Rep.* **2017**, *7*, 7304. [CrossRef]
15. Demene, C.; Baranger, J.; Bernal, M.; Delanoe, C.; Auvin, S.; Biran, V.; Alison, M.; Mairesse, J.; Harribaud, E.; Pernot, M.; et al. Functional ultrasound imaging of brain activity in human newborns. *Sci. Transl. Med.* **2017**, *9*, eaah6756. [CrossRef]
16. Macé, E.; Montaldo, G.; Cohen, I.; Baulac, M.; Fink, M.; Tanter, M. Functional ultrasound imaging of the brain. *Nat. Methods* **2011**, *8*, 662–664. [CrossRef]
17. Szabo, T.L. *Diagnostic Ultrasound Imaging: Inside Out*; Academic Press: Cambridge, MA, USA, 2004.
18. Zhou, B.; Yang, X.; Zhang, X.; Curran, W.J.; Liu, T. Ultrasound elastography for lung disease assessment. *IEEE Trans. Ultrason. Ferroelectr. Freq. Control.* **2020**, *67*, 2249–2257. [CrossRef]

19. Wang, X.; Peng, Y.; Lu, L.; Lu, Z.; Bagheri, M.; Summers, R.M. Chestx-ray8: Hospital-scale chest X-ray database and benchmarks on weakly-supervised classification and localization of common thorax diseases. In Proceedings of the IEEE Conference on Computer Vision and Pattern Recognition, Honolulu, HI, USA, 21–26 July 2017; pp. 2097–2106.
20. Nehmeh, S.; Erdi, Y.; Pan, T.; Pevsner, A.; Rosenzweig, K.; Yorke, E.; Mageras, G.; Schoder, H.; Vernon, P.; Squire, O.; et al. Four-dimensional (4D) PET/CT imaging of the thorax: 4D PET/CT. *Med. Phys.* **2004**, *31*, 3179–3186. [CrossRef]
21. Ates, O.F.; Taydas, O.; Dheir, H. Thorax magnetic resonance imaging findings in patients with coronavirus disease (COVID-19). *Acad. Radiol.* **2020**, *27*, 1373–1378. [CrossRef]
22. Zhang, K.; Li, M.; Yang, F.; Xu, S.; Abubakar, A. Three-dimensional electrical impedance tomography with multiplicative regularization. *IEEE Trans. Biomed. Eng.* **2019**, *66*, 2470–2480. [CrossRef]
23. Zhang, H.; Li, M.; Yang, F.; Xu, S.; Zhou, H.; Yang, Y.; Chen, L. A low-profile compact dual-band l-shape monopole antenna for microwave thorax monitoring. *IEEE Antennas Wirel. Propag. Lett.* **2020**, *19*, 448–452. [CrossRef]
24. Martelius, L.; Heldt, H.; Lauerma, K. B-lines on pediatric lung sonography: Comparison with computed tomography. *J. Ultrasound Med.* **2016**, *35*, 153–157. [CrossRef] [PubMed]
25. Mento, F.; Soldati, G.; Prediletto, R.; Demi, M.; Demi, L. Quantitative lung ultrasound spectroscopy applied to the diagnosis of pulmonary fibrosis: The first clinical study. *IEEE Trans. Ultrason. Ferroelectr. Freq. Control* **2020**, *67*, 2265–2273. [CrossRef] [PubMed]
26. Dargent, A.; Chatelain, E.; Kreitmann, L.; Quenot, J.P.; Cour, M.; Argaud, L.; COVID-LUS Study Group. Lung ultrasound score to monitor COVID-19 pneumonia progression in patients with ARDS. *PLoS ONE* **2020**, *15*, e0236312. [CrossRef]
27. Wang, G.; Ji, X.; Xu, Y.; Xiang, X. Lung ultrasound: A promising tool to monitor ventilator-associated pneumonia in critically ill patients. *Crit. Care* **2016**, *20*, 1–10. [CrossRef]
28. Buonsenso, D.; Brancato, F.; Valentini, P.; Curatola, A.; Supino, M.; Musolino, A.M. The use of lung ultrasound to monitor the antibiotic response of community-acquired pneumonia in children: A preliminary hypothesis. *J. Ultrasound Med.* **2020**, *39*, 817–826. [CrossRef]
29. Rueter, D.; Hauber, H.P.; Droeman, D.; Zabel, P.; Uhlig, S. Low-frequency ultrasound permeates the human thorax and lung: A novel approach to non-invasive monitoring. *Ultraschall Der Med.-Eur. J. Ultrasound* **2010**, *31*, 53–62. [CrossRef]
30. Dai, Z.; Peng, Y.; Mansy, H.A.; Sandler, R.H.; Royston, T.J. Comparison of poroviscoelastic models for sound and vibration in the lungs. *J. Vib. Acoust.* **2014**, *136*, 050905. [CrossRef]
31. Peng, Y.; Dai, Z.; Mansy, H.A.; Henry, B.M.; Sandler, R.H.; Balk, R.A.; Royston, T.J. Sound transmission in porcine thorax through airway insonification. *Med. Biol. Eng. Comput.* **2016**, *54*, 675–689. [CrossRef]
32. Mansy, H.A.; Balk, R.A.; Warren, W.H.; Royston, T.J.; Dai, Z.; Peng, Y.; Sandler, R.H. Pneumothorax effects on pulmonary acoustic transmission. *J. Appl. Physiol.* **2015**, *119*, 250–257. [CrossRef]
33. Palnitkar, H.; Henry, B.M.; Dai, Z.; Peng, Y.; Mansy, H.A.; Sandler, R.H.; Balk, R.A.; Royston, T.J. Sound transmission in human thorax through airway insonification: An experimental and computational study with diagnostic applications. *Med. Biol. Eng. Comput.* **2020**, *58*, 2239–2258. [CrossRef] [PubMed]
34. Morenz, K.; Biller, H.; Wolfram, F.; Leonhadt, S.; Rüter, D.; Glaab, T.; Uhlig, S.; Hohlfeld, J.M. Detection of air trapping in chronic obstructive pulmonary disease by low frequency ultrasound. *BMC Pulm. Med.* **2012**, *12*, 8. [CrossRef] [PubMed]
35. Pohlmann, A.; Sehati, S.; Young, D. Effect of changes in lung volume on acoustic transmission through the human respiratory system. *Physiol. Meas.* **2001**, *22*, 233. [CrossRef] [PubMed]
36. Li, C.; Duric, N.; Littrup, P.; Huang, L. In vivo breast sound-speed imaging with ultrasound tomography. *Ultrasound Med. Biol.* **2009**, *35*, 1615–1628. [CrossRef] [PubMed]
37. Duric, N.; Littrup, P.; Poulo, L.; Babkin, A.; Pevzner, R.; Holsapple, E.; Rama, O.; Glide, C. Detection of breast cancer with ultrasound tomography: First results with the Computed Ultrasound Risk Evaluation (CURE) prototype. *Med. Phys.* **2007**, *34*, 773–785. [CrossRef]
38. Zhang, H.; Thurber, C.; Rowe, C. Automatic P-wave arrival detection and picking with multiscale wavelet analysis for single-component recordings. *Bull. Seismol. Soc. Am.* **2003**, *93*, 1904–1912. [CrossRef]
39. Li, C.; Huang, L.; Duric, N.; Zhang, H.; Rowe, C. An improved automatic time-of-flight picker for medical ultrasound tomography. *Ultrasonics* **2009**, *49*, 61–72. [CrossRef]
40. Fatemi, A.; Måsøy, S.E.; Rodriguez-Molares, A. Row–Column-Based Coherence Imaging Using a 2-D Array Transducer: A Row-Based Implementation. *IEEE Trans. Ultrason. Ferroelectr. Freq. Control* **2020**, *67*, 2303–2311. [CrossRef]
41. Wodicka, G.R.; Stevens, K.N.; Golub, H.L.; Cravalho, E.G.; Shannon, D.C. A model of acoustic transmission in the respiratory system. *IEEE Trans. Biomed. Eng.* **1989**, *36*, 925–934. [CrossRef]
42. Picano, E.; Pellikka, P.A. Ultrasound of extravascular lung water: A new standard for pulmonary congestion. *Eur. Heart J.* **2016**, *37*, 2097–2104. [CrossRef]
43. Porcel, J.M.; Light, R.W. Pleural effusions due to pulmonary embolism. *Curr. Opin. Pulm. Med.* **2008**, *14*, 337–342. [CrossRef] [PubMed]
44. Xiong, X.; De la Torre, F. Supervised descent method and its applications to face alignment. In Proceedings of the IEEE Conference on Computer Vision and Pattern Recognition, Portland, OR, USA, 23–28 June 2013; pp. 532–539.
45. Guo, R.; Li, M.; Yang, F.; Xu, S.; Abubakar, A. Application of supervised descent method for 2D magnetotelluric data inversion. *Geophysics* **2020**, *85*, WA53–WA65. [CrossRef]

46. Zhang, H.; Li, M.; Yang, F.; Xu, S.; Yin, Y.; Zhou, H.; Yang, Y.; Zeng, S.; Shao, J. A feasibility study of 2-d microwave thorax imaging based on the supervised descent method. *Electronics* **2021**, *10*, 352. [CrossRef]
47. Zhang, K.; Guo, R.; Li, M.; Yang, F.; Xu, S.; Abubakar, A. Supervised descent learning for thoracic electrical impedance tomography. *IEEE Trans. Biomed. Eng.* **2020**, *68*, 1360–1369. [CrossRef]
48. Lin, M.; Liu, Y. Guided Wave Tomography Based on Supervised Descent Method for Quantitative Corrosion Imaging. *IEEE Trans. Ultrason. Ferroelectr. Freq. Control* **2021**, *68*, 3624–3636. [CrossRef]
49. Rawlinson, N.; Sambridge, M. Seismic traveltime tomography of the crust and lithosphere. *Adv. Geophys.* **2003**, *46*, 81–199.
50. Moser, T. Shortest path calculation of seismic rays. *Geophysics* **1991**, *56*, 59–67. [CrossRef]
51. Zhang, H.; Zhang, T.; Zhou, H.; Li, M.; Yang, F.; Xu, S.; Cao, Y. A Preliminary Experiment Based on One-step Measurement-trained Supervised Descent Method for Microwave Thorax Imaging. In Proceedings of the 2021 Photonics & Electromagnetics Research Symposium (PIERS), Hangzhou, China, 21–25 November 2021; pp. 1921–1930.
52. Guo, R.; Li, M.; Yang, F.; Xu, S.; Abubakar, A. First arrival traveltime tomography using supervised descent learning technique. *Inverse Probl.* **2019**, *35*, 105008. [CrossRef]
53. Tikhonov, A.N.; Arsenin, V.Y. *Solutions of Ill-Posed Problems*; VH Winston & Sons: New York, NY, USA, 1977.
54. Song, X.; Li, M.; Yang, F.; Xu, S.; Abubakar, A. Feasibility study of acoustic imaging for human thorax using an acoustic contrast source inversion algorithm. *J. Acoust. Soc. Am.* **2018**, *144*, 2782–2792. [CrossRef]
55. Song, X.; Li, M.; Yang, F.; Xu, S.; Abubakar, A. Study on joint inversion algorithm of acoustic and electromagnetic data in biomedical imaging. *IEEE J. Multiscale Multiphysics Comput. Tech.* **2019**, *4*, 2–11. [CrossRef]
56. Hasgall, P.; Di Gennaro, F.; Baumgartner, C.; Neufeld, E.; Lloyd, B.; Gosselin, M.; Payne, D.; Klingenböck, A.; Kuster, N. *IT'IS Database for Thermal and Electromagnetic Parameters of Biological Tissues*; version 4.1; Elsevier: Boston, FL, USA, 2022.
57. Treeby, B.E.; Budisky, J.; Wise, E.S.; Jaros, J.; Cox, B. Rapid calculation of acoustic fields from arbitrary continuous-wave sources. *J. Acoust. Soc. Am.* **2018**, *143*, 529–537. [CrossRef] [PubMed]
58. Maeda, N. A method for reading and checking phase times in autoprocessing system of seismic wave data. *Zisin* **1985**, *38*, 365–379. [CrossRef]
59. Kuhlman, J.E.; Singha, N.K. Complex disease of the pleural space: Radiographic and CT evaluation. *Radiographics* **1997**, *17*, 63–79. [CrossRef] [PubMed]
60. Currie, G.P.; Alluri, R.; Christie, G.L.; Legge, J.S. Pneumothorax: An update. *Postgrad. Med. J.* **2007**, *83*, 461–465. [CrossRef] [PubMed]

applied sciences

MDPI

Article

Edge and Notch Detection in a Plate Using Time Reversal Process of Leaky Lamb Waves

Jean-Christophe Vallée [1,2,*], Marie-Aude Ploix [2], François Baqué [1], Matthieu Cavaro [1] and Jean-François Chaix [2]

[1] CEA, DES, IRESNE, DTN, Center of Cadarache, F-13108 Saint-Paul-Lez-Durance, France; francois.baque@cea.fr (F.B.); matthieu.cavaro@cea.fr (M.C.)
[2] Aix Marseille Univ, CNRS, Centrale Marseille, LMA UMR 7031, F-13453 Marseille, France; marie-aude.ploix@univ-amu.fr (M.-A.P.); jean-francois.chaix@univ-amu.fr (J.-F.C.)
* Correspondence: jean-christophe.vallee@cea.fr

Abstract: Leaky Lamb waves are proven effective to carry out nondestructive testing especially on parallel and immersed plates. To detect and localize defects in such a set, this work associates for the first time the topological energy method and leaky Lamb waves. This methodology is applied in a single immersed plate to validate its application. Firstly, Lamb mode A1 is generated in the plate, and the reflected waves on the defect are measured. A first case is examined where the edge is considered as a defect to be localized. Then, measurements are taken on a plate where a notch is machined. The measurements are time reversed and reinjected in a finite-element simulation. The results are then correlated with the direct problem of the topological energy method that is also simulated. In both cases, the defects are precisely localized on the energy images. This work is the preliminary step to an application of the topological energy method to a set of two parallel and immersed plates where the research defect is located in the second plate.

Keywords: ultrasound; immersed guided waves; topological energy method; nondestructive testing

Citation: Vallée, J.-C.; Ploix, M.-A.; Baqué, F.; Cavaro, M.; Chaix, J.-F. Edge and Notch Detection in a Plate Using Time Reversal Process of Leaky Lamb Waves. *Appl. Sci.* **2022**, *12*, 228. https://doi.org/10.3390/app12010228

Academic Editor: Nico P. Avdelidis

Received: 23 November 2021
Accepted: 23 December 2021
Published: 27 December 2021

1. Introduction

Ultrasonic methods are proven to be particularly relevant within the inspection and monitoring of sodium-cooled fast reactors (SFR) due to opacity and the oxidizing property of liquid sodium that prevents optical inspection and the immersion of conventional ultrasonic transducers. Thus, it has been shown previously that nondestructive testing from the outside of the main vessel allows generating and propagating guided waves in the internal structures, similar to a layered structure (parallel steel plates) immersed in liquid [1,2]. The context implies a main restriction: the only available access to position the transducers is the outside face of the first plate.

These immersed guided waves, called leaky Lamb waves, have been widely studied in the literature [3–6] and applied to damage detection in plate-like structures [7,8]. In 1917, Sir Horace Lamb investigated the theory of vibration of thin plates and rods [4]. Later, Merkulov and Viktorov studied the vibration of immersed plates in depth [5,6]. The Lamb waves are called "leaky" because of the re-emission of bulk waves in the surrounding fluid. Attenuation due to the leakage is mostly preponderant versus the intrinsic attenuation due to the material [6]. The latter is then not taken into account in this work. This re-emission in the fluid propagates towards the second plate and generates Lamb waves in it.

In order to perform nondestructive testing, the knowledge of the behavior of the Lamb waves is essential. Multiple Lamb modes may coexist and propagate simultaneously at different group and phase velocities. Velocities and attenuations are frequency dependent so that a broadband signal presents a dispersive behavior. The dispersion leads to a spread of the wave packets during the propagation [9]. Nevertheless, by selecting a mode and a frequency range where the group velocity is quite constant, the dispersion can be limited.

The selection of the mode can be done by imposing an incidence to the generated acoustic beam through a wedge or a delay law if a linear array transducer is used [10].

Several imaging methods involving Lamb waves are proven effective. For example, a classical B-scan may provide an image of a defect assuming the knowledge of the propagating mode and its velocity in a single plate or in the second plate of the layered structure [2]. Other linear methods using delay-and-sum can also be applied, such as the Synthetic Aperture Focusing Technique (SAFT) [11–13], the Total Focusing Method (TFM) [14,15] and their adaptation to Lamb waves, Lamb-SAFT [16] to take into account the dispersive behavior of the Lamb waves, and the sparse-TFM [17] to carry out short range inspection. However, those techniques are focused on the time travel of a single mode and do not use information of the mode conversion that may occur at the reflection on a defect, such as a crack [18,19]. Later on, time-reversal techniques were introduced [20–24] to optimize the resolution of a reconstructed image of an inspected medium. For example, the DORT (Decomposition of the Time-Reversal Operator) method [21] is a selective detection of scatterers using the eigenvectors of the time-reversal operator, extracted from the full-matrix capture of a linear array of transducers. Each significant eigenvector is associated with a scatterer in the inspected medium, and the associated signal can be retropropagated independently in a numerical medium to locate the scatterer. This operation is made possible because of the completeness of the fundamental Lamb modes as proven by Kirrmann [3]. Meanwhile, the topological energy method [22,24–28] requires the numerical solution of only two problems: the direct problem where the experimental source is generated in the healthy corresponding structure and the adjoint problem where the source is the time-reversed difference between the ultrasonic field measured on the transducers of the inspected medium and the reference medium. This time-reversed source can be interpreted as the time-reversed signature of the defects and could provide data of very good quality.

This method has already been applied on Lamb waves in the literature. For example, Rodriguez et al. applied it to the inspection of a free anisotropic plate by a monomodal inspection. The transducers are located on the edge of the plate, and several defects are detected [29]. Further works of Sun et al. showed the efficiency of the topological imaging method towards other methods, notably the TFM [27]. Thus, this method is applicable and efficient in cases of free single plates. But no work was found concerning the leaky Lamb waves in a set of immersed plates.

The topological energy method was preferred in this work because the adjoint field and the time reversal phenomenon allow maintaining a good sensibility and a great localization. Moreover, it involves fewer simulations that can become very time consuming. The topological energy method is also more versatile regarding the complex geometry of two parallel and immersed plates, in particular considering the multiple reflections between the two plates. This work is the first step towards this final issue, and the purpose here is to prove the applicability and efficiency of this method to detect and localize a defect in a single immersed plate using experimental data, where the transducers are set in the fluid above one face of the plate only. The use of leaky Lamb waves associated with the topological energy method constitutes an innovative work in view of the existing literature.

Firstly, the generation and propagation of leaky Lamb waves in parallel and immersed plates are pointed out. Dispersion equations that apply to the Lamb wave propagation are solved numerically and discussed. Then, the topological energy imaging method is applied in order to detect and to localize two kinds of defects: the edge of the plate (representative of a through-crack) and a notch close to the edge. The mode A1 is experimentally generated, and a phased array system acquires the ultrasonic signature of the defect. Afterwards, the two propagation problems are solved numerically. Results eventually show the reconstructed images based on the experimental datasets in two cases.

2. Theoretical Background of the Leaky Lamb Wave Propagation and Generation

Lamb waves were first discovered by Sir Horace Lamb in 1917 [4]. Lamb waves are comparable to resonances resulting from the superposition of longitudinal and shear waves in thin plates. In free plates, Lamb waves can travel long distances that allow them to be a useful tool for long-range inspection. The Lamb modes can be separated into two categories: symmetrical (S) and antisymmetric (A). This denomination is based on the symmetrical nature of the displacement profile parallel to the surface of the plate. An infinite number of harmonics of the symmetrical and antisymmetric modes exist [6]. When a fluid surrounds the plate, a leaky attenuation appears as explained by Merkulov [5]. The energy of the Lamb waves is converted into compressional waves in the liquid as shown in Figure 1. Each Lamb wave mode is associated with an angle of re-emission (and by reciprocity to an angle of incidence, as explained later).

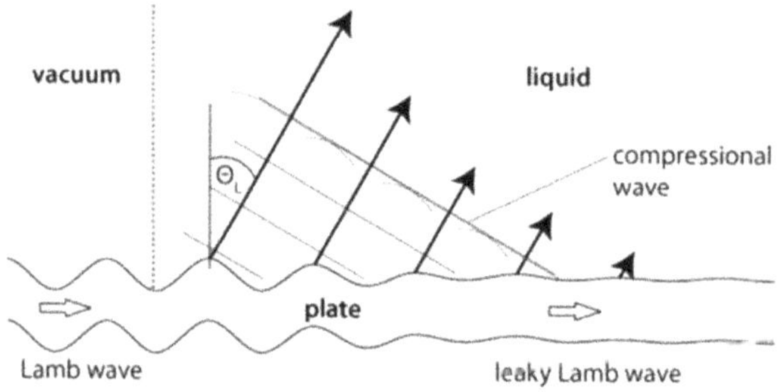

Figure 1. Generation of a compressional wave in the liquid by a leaky Lamb wave in the plate [30].

The dispersion equations that govern respectively symmetrical and antisymmetric leaky Lamb wave modes are recalled [6]:

$$\frac{\tan\left(\frac{qe}{2}\right)}{\tan\left(\frac{pe}{2}\right)}+\frac{4k^2pq}{(q^2-k^2)^2}-i\frac{\rho_0 k_T^4 p \tan\left(\frac{qe}{2}\right)}{\rho\sqrt{k_0^2-k^2}(q^2-k^2)^2}=0 \quad (1)$$

$$\frac{\tan\left(\frac{pe}{2}\right)}{\tan\left(\frac{qe}{2}\right)}+\frac{4k^2pq}{(q^2-k^2)^2}+i\frac{\rho_0 k_T^4 p}{\rho\sqrt{k_0^2-k^2}(q^2-k^2)^2\tan\left(\frac{qe}{2}\right)}=0 \quad (2)$$

where i is the imaginary unit. e is the thickness of the plate. $p^2 = k_L^2 - k^2$, $q^2 = k_T^2 - k^2$, $k_L = \frac{2\pi f}{c_L}, k_T = \frac{2\pi f}{c_T}, k_0 = \frac{2\pi f}{c_0}$, k (the wave number) and f (the frequency) are the solutions. k_L and k_T are respectively the longitudinal and transversal wave numbers in the plate. c_L and c_T are respectively the speed of the longitudinal and transversal wave in the plate. c_0 is the speed of waves in the fluid. ρ is the density of the plate, and ρ_0 is the density of the fluid.

One can notice that the real part of the equations corresponds to the dispersion equations of a free plate [4]. The imaginary part corresponds to the fluid–structure interaction and is proportional to the ratio ρ_0/ρ. The wave number solution of these equations is a complex number $k = k' + ik''$. Its real part k' represents the propagative properties of the wave, and its imaginary part k'' represents the leaky attenuation by re-emission in the fluid.

The duets (k, f) solutions of the dispersion equations are plotted in Figure 2a regarding the real part of k and in Figure 2b regarding the imaginary part of k.

In this work, the plate is 7.8 mm thick and made of stainless steel. The density of the plate is $\rho = 7950$ kg·m^{-3}, and the celerities of the longitudinal and transversal waves are respectively $c_L = 5738$ m·s^{-1} and $c_T = 3143$ m·s^{-1}. The density of the water is $\rho_f = 1000$ kg·m^{-3}, and the velocity of the compressional wave in the water is $c_0 = 1490$ m·s^{-1}. On the one hand, a normalization on the frequency is performed because the dispersion equations are invariant regarding the product frequency × thickness. On the other hand, the wave number and the attenuation are inversely proportional to the thickness, and a normalization is also performed.

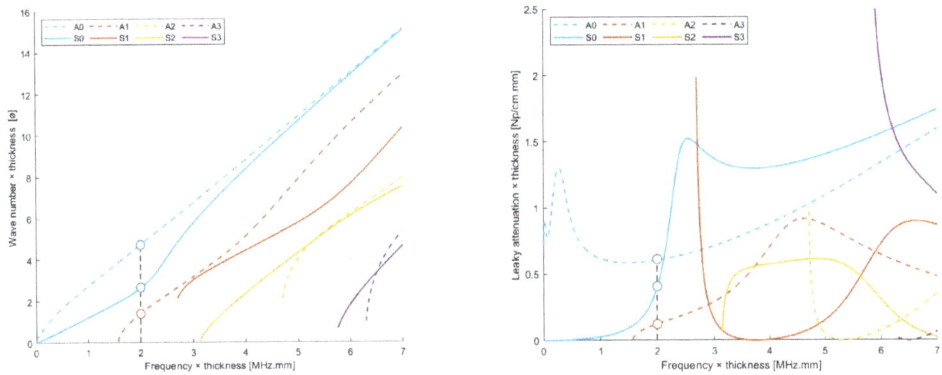

(**a**) (**b**)

Figure 2. (**a**) Wave number of leaky Lamb waves normalized by the plate thickness; (**b**) Attenuation of leaky Lamb waves, as a function of the product frequency × thickness.

At a given frequency × thickness, several modes may coexist and propagate at the same time. For example, in Figure 2a, at 2 MHz·mm, the modes A0, S0, and A1 may propagate together. One can note in Figure 2b that the modes A0 and S0 present a high attenuation, whereas the mode A1 presents a lower leakage by re-emission.

The phase velocity of the modes is evaluated using the real part of the wave numbers $c_p = \frac{2\pi f}{k'}$. The group velocity c_g is derived from the phase velocity: $c_g(\omega) = c_p(\omega)^2 \left[c_p(\omega) - \omega \frac{dc_p(\omega)}{d\omega} \right]^{-1}$, where $\omega = 2\pi f$. Both phase and group velocities are plotted respectively in Figure 3a,b. A real signal is never purely monochromatic. Its spectrum is defined by a band. That is why group velocities are of interest. In the example given by Figure 3b around 2 MHz·mm, the modes have different group velocities. On the bandwidth, the group velocity of the mode A0 is rather constant, whereas the group velocity of the mode S0 varies by up to 100%. This highlights the dispersive property of the Lamb wave, described in detail in [9].

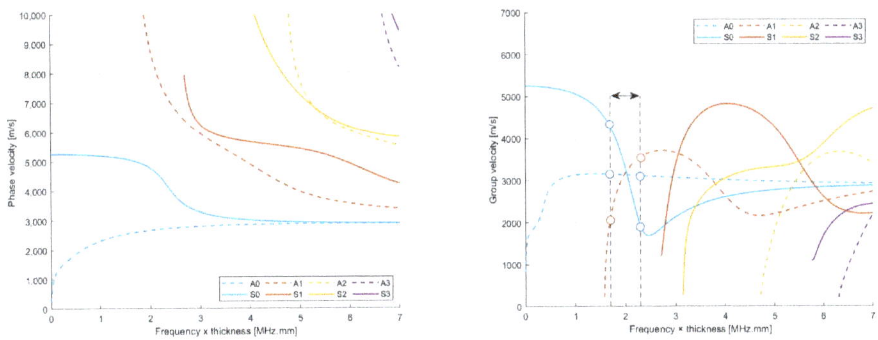

(**a**) (**b**)

Figure 3. (**a**) Phase velocity of leaky Lamb waves; (**b**) Group velocity of leaky Lamb waves, as a function of the product frequency × thickness. Arrow indicates the bandwidth of a real 2 MHz·mm signal and highlights the dispersion phenomenon in the propagation of the Lamb waves.

By reciprocity of the re-emission shown in Figure 1, Lamb waves can be generated in an immersed plate by a steered incident beam. The acoustic wave propagates in the fluid to the plate, where a reflected wave and a transmitted wave appear [31]. One can find the angles θ that maximize the transmitted wave in the plate [32]. Those angles tie in with the angles that can be found using Snell's law, and the phase velocity of each mode is

$$\frac{c_0}{\sin(\theta_i)} = \frac{c_p}{\sin(\theta_{Lamb})} \tag{3}$$

where θ_i is the beam incidence angle in the fluid, $\theta_{Lamb} = \frac{\pi}{2}$ in the plate. c_0 is the wave velocity in the fluid. c_p is the phase velocity of the aimed mode at a given product frequency × thickness. The beam angles in the case of our experiments are plotted in Figure 4. For example, at 2 MHz·mm, the A1 mode can be generated by an incident wave with an angle of 10°.

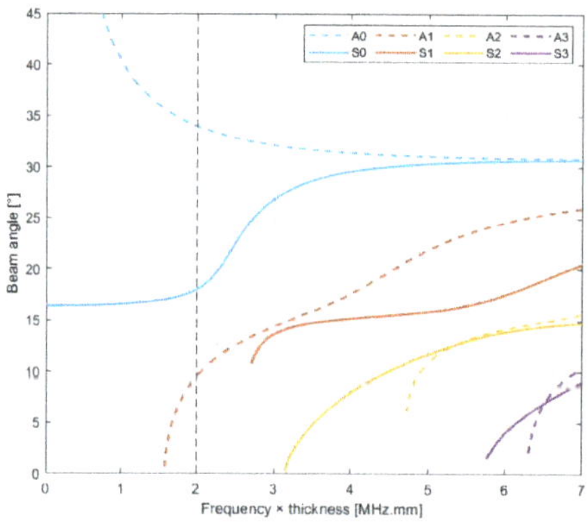

Figure 4. Beam incidence angle needed to generate Lamb modes in an immersed plate.

3. Imaging Method and Configuration

3.1. Topological Energy Imaging Computation

The topological energy imaging method comes from the optimization for inverse problems [33] and was first introduced by Dominguez [22]. This method is defined as a minimization problem of the mathematical distance between an unknown inspected medium and a simulated reference medium. One can assume that the reference undamaged medium has the same material properties as the damaged unknown one. In our case, the simulation is achievable and workable using a finite-element model with Comsol Multiphysics®.

The principle of the method is to minimize a cost function defined as the distance between the two media so that the reference medium, by inserting defects at the right locations, converges to the unknown one. The mathematical proofs are not presented in this paper and can be found in [22,33]. Nevertheless, a summary of the steps needed in the application of the topological energy method to obtain a map of the unknown medium is proposed (Figure 5 illustrates the steps in the application of the topological imaging method in a general case of bulk waves in a homogeneous material):

1. Measurement in the experimental unknown medium of the acoustic field $p_{exp}\left(\vec{r},t\right)$, $t \in [0,T]$ on the surface Γ_{mes}.
2. Numerical resolution in the reference medium of the direct problem by emitting the same initial signal. Thus, measurement of the acoustic field $p_{ref}\left(\vec{r},t\right)$, $t \in [0,T]$ in the whole medium, and especially on the surface Γ_{mes}.
3. Numerical resolution in the reference medium of the adjoint problem. The source term of the adjoint problem is defined as the time reversal of the difference between

$p_{exp}\left(\vec{r},\, t\right)$ and $p_{ref}\left(\vec{r}, t\right)$. The acoustic field $p_{adj}\left(\vec{r},\, T-t\right)$, $t \in [0, T]$ is then measured in the whole medium and on the surface Γ_{mes}.

4. Evaluation of the topological gradient defined by the limit conditions of the problem and the material properties, and p_{ref} and p_{adj} that are respectively the solutions of the direct and the adjoint problem.

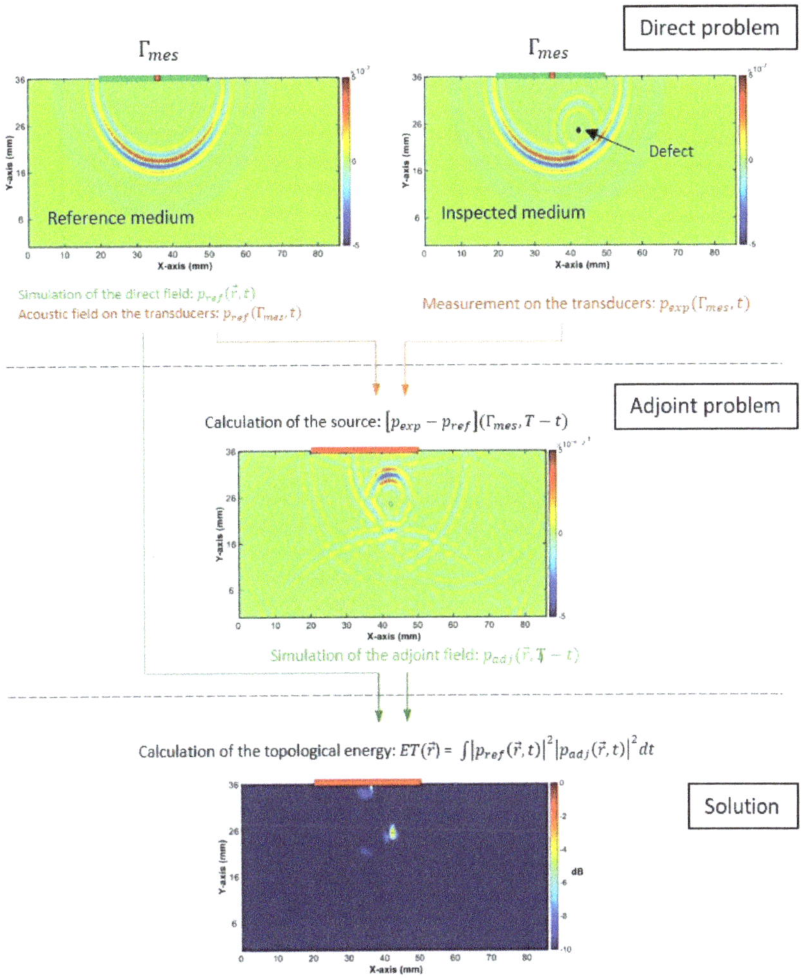

Figure 5. Diagram of the general principle of the topological imaging method in an isotropic medium [34]. Reproduced with permission from E. Lubeigt.

As said previously, the topological imaging method consists of an optimization; the optimization is iterative and can become very time consuming. A Fast Topological Imaging Method (FTIM) is used in [22,29] to overcome this problem and gives reliable results. Only the first iteration of the algorithm is computed. The result of this method is an image of the medium and is given for each point $\vec{r}$ by:

$$ET\left(\vec{r}\right) = \int \left|p_{ref}\left(\vec{r}, t\right)\right|^2 \left|p_{adj}\left(\vec{r},\, t\right)\right|^2 dt \tag{4}$$

In this equation, the square of the acoustic fields is used to enhance the high value of the topological energy on the final image.

This Fast Topological Imaging Method is used here.

One can interpret the topological energy as a temporal correlation between two simulated wave fields (direct and adjoint). The direct field corresponds to the propagation of the waves in the undamaged medium. The adjoint problem corresponds to the field induced by the backward propagation of the acoustic signature of the difference between the reference medium and the experimental medium (with defects). The topological energy between those fields takes maximum value at the location of the defects, where the two fields coincide.

3.2. Experimental Setup

The considered plate is made of stainless steel and is immersed in water (schematic configuration and picture of the experimental setup are represented in Figure 6). Its thickness and length are defined by $e = 7.8$ mm and $L = 70$ cm. A linear phased array transducer with 16 elements that defines the surface Γ_{mes} parallel to the plate, is used as a transmitter–receiver sensor. The central frequency is $f = 320$ kHz. The pitch of the linear array is $p = 3$ mm.

(a)

(b)

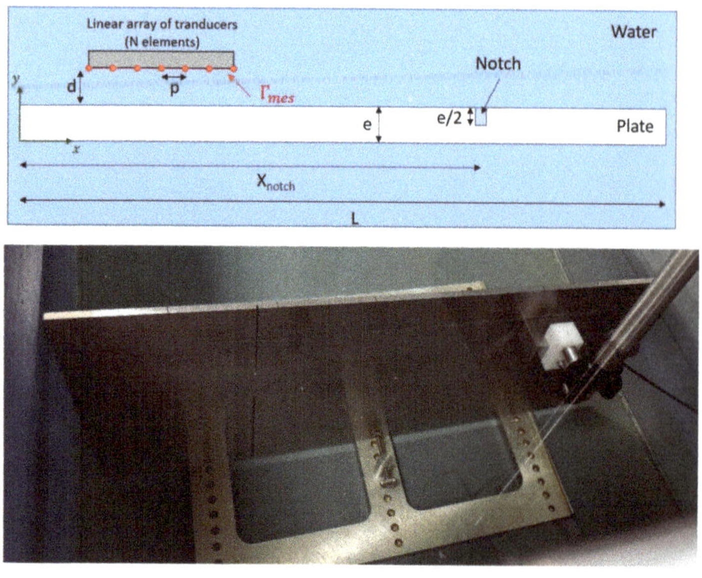

Figure 6. (**a**) Geometry of the experimental setup; (**b**) Experimental setup.

A delay law is applied to the linear array that imposes the beam angle at the emission, and thus the Lamb wave mode is generated in the plate. The linear array is positioned very close to the plate ($d = 0.5$ mm) to lower successive reflections between the linear array and the plate that would lead to the generation of multiple Lamb waves in the experience. The topological energy is calculated inside the plate that implies the use of the horizontal $u\left(\vec{r},t\right)$ or vertical $v\left(\vec{r},t\right)$ displacements in the calculation of the latter.

In an initial stage, a plate without any defect is studied. The objective is the detection of the edge of the plate at $L = 70$ cm. Then, the application of the method on a second plate containing a notch at $X_{notch} = 52$ cm is carried out.

The measurements are performed with a Sonaxis® ultrasonic linear array composed of sixteen 1.8 mm wide elements and a Lecoeur® OPEN system. The experimental acquisition and the simulations are done for a beam angle $\theta_i = 12.6°$ that corresponds to the mode A1.

The emitted signal is a ten-period sinus signal filtered by a Hanning window in order to lower the dispersion phenomenon. The mode A1 has been chosen because it presents a low attenuation along the propagation so that it is easily detectable. The linear array transducer is set at 40 cm of the left edge of the plate.

3.3. Finite-Element Simulation

In this section, the finite-element model is exposed—necessary for the implementation of the topological energy method—and developed with Comsol Multiphysics® and the postprocessing techniques in order to analyze the experimental signals.

As illustrated in Figure 7, the plate is immersed in the water. Perfectly Matched Layers (PMLs) are set to simulate an infinite domain, especially at the right end of the plate to simulate the flawless infinite plate and around the domain to avoid reflections on the calculation boundaries. In the experimentation, the tank edges are far enough apart to avoid reflected waves in the studied time interval.

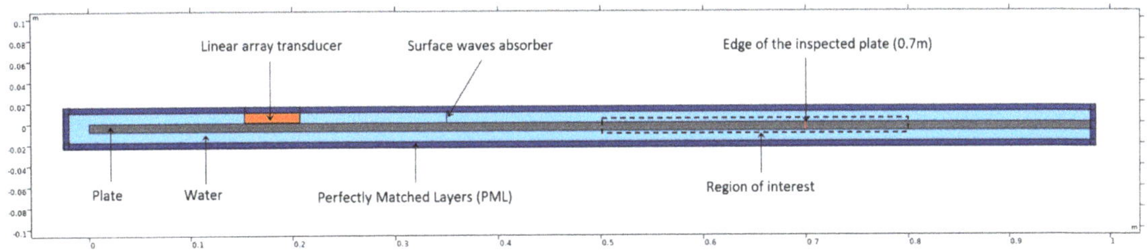

Figure 7. Sketch of the simulation under COMSOL Multiphysics®.

Segments that represent each element of the transducer define the linear phased array. This simple definition leads to a surface wave in the water in the numerical model that is not observed in the experimental propagation. That is why a thin absorbent layer in the upper part of the water is introduced to cancel this numerical surface wave.

The mesh is described by triangular elements. The maximum size of the elements is given by $\lambda_{min}/8$ in the plate and the water, where λ_{min} represents the smallest wavelength that can exist in the considered bandwidth.

4. Results and Discussion

Two kinds of defects are studied and discussed: the edge of the plate that approximates the behavior of a vertical through-crack and a machined mid-thickness notch. They allow evaluating the methodology in two typical calibrating situations in NDT when searching cracks.

4.1. Edge Plate Detection

Leaky Lamb waves reflect on the edge of the plate and propagate back toward the transducer. The experimental temporal signals are extracted from each element of the linear array, and each line in Figure 8a is the envelope of the signal acquired on an element. The signals are then processed with 2D Fast Fourier Transform (2D-FFT) to transform the space–time domain (x, t) into a wavenumber-frequency (k, f) representation as shown in Figure 8b. This transformation allows measuring the real part of the wavenumber, the leaky attenuation, and the phase velocity of the Lamb wave. Superimposing the theoretical dispersion curves allows verifying the nature of the Lamb wave mode involved in the propagation. The emitted wave packet propagates along the x-axis and is reflected by the right edge of the plate. The wavenumber in Figure 8b is negative because of the propagation of the reflected wave along the decreasing x-axis.

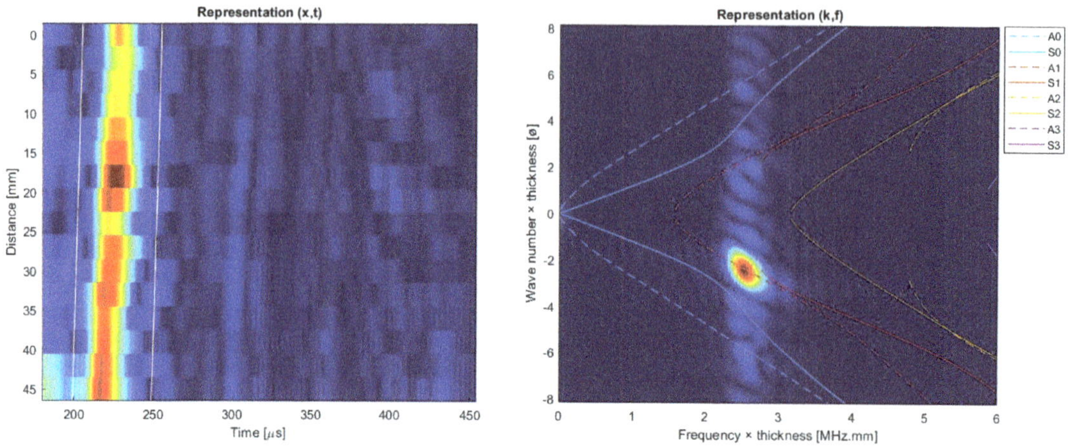

(**a**) (**b**)

Figure 8. (**a**) Envelope of the experimental signals recorded by each element of the linear array—(x,t) representation; (**b**) 2D-FFT of the signals marked by the white lines—(k,f) representation.

One can verify that the reflected mode corresponds to the initial propagative A1 mode: no conversion has occurred.

The selected wave packet for following the time reversal process is chosen as the more energetic one in the negative wavenumbers that corresponds to a reflective wave. Considering the mode A1 at this product frequency × thickness, the horizontal displacement has more amplitude than the vertical displacements. Thus, the topological energy is calculated using $u\left(\vec{r},t\right)$ at each point of the plate.

In this particular case, the computation of the source in the adjoint problem is different as described previously in Section 3. One can assume that in the reference medium in the direct problem, the plate is infinite, and there are no reflected waves ($u_{ref} = 0$). Thus, the calculation of the source of the adjoint problem is straightforward: the wave packet delimited by the white lines in Figure 8a corresponds to the signal that will be time reversed and reinjected in the adjoint problem. The topological energy is then computed using the horizontal displacements $u\left(\vec{r},t\right)$ in the zone of interest (represented in Figure 6), and the result is plotted in Figure 9. One can note that the distribution of the maximum of topological energy in the thickness of the plate follows the displacement profile of the considered mode.

The maximum of the topological energy occurs at $x = 70.2$ cm. The results are in good agreement with the real location of the edge ($x = 70$ cm). Indeed, the difference with the real location of the edge is under the uncertainty based on the regular grid used in the postprocessing ($\Delta x_{grid} = 0.61$ cm). At this location, the fields in the direct and the adjoint problems cross and lead to the highest value of the topological energy (Figure 10b). On the contrary, the correlation between the fields before and after the edge is much lower (examples on Figure 10a,c).

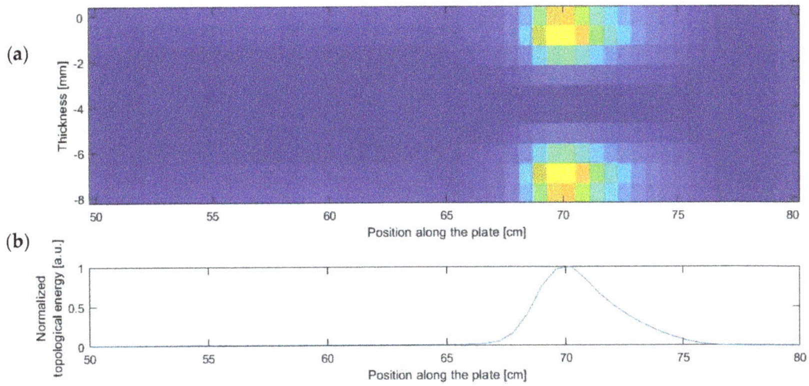

Figure 9. (**a**) Topological image in the zone of interest of the simulated plate; (**b**) Normalized sum of the topological energy at each position along the plate.

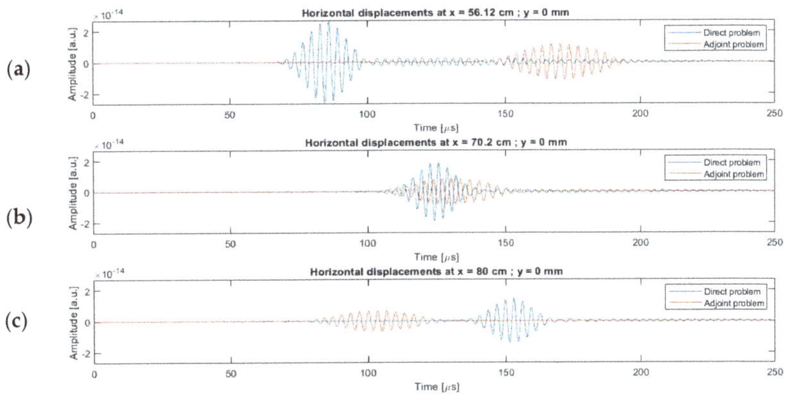

Figure 10. Horizontal displacements of the direct and adjoint problems at (**a**) x = 56.12 cm; y = 0 cm; (**b**) x = 70.2 cm; y = 0 cm; and (**c**) x = 80 cm; y = 0 cm.

One can notice the decrease in the amplitude of the direct problem along the propagation and the increase in the amplitude of the adjoint problem due to the leaky attenuation. To improve the contrast of the topological image, an idea would be to compensate the attenuation knowing the attenuation coefficient of the mode involved. Indeed, such a compensation can only work in a purely monomodal propagation. The compensation cannot be done if at least two modes propagate.

4.2. Notch Detection

The second studied case is the detection of a notch machined with depth e/2 (see Figure 6) in a plate that has the same physical properties and thickness as previously. The notch is located at x = 52 cm. The emitted signal is the same as previously. One can observe two wave packets on the spatial time representation on the signals plotted in Figure 11a. The selected wave packet for the time reversal is chosen again as the most energetic one. It corresponds to the reflection on the notch.

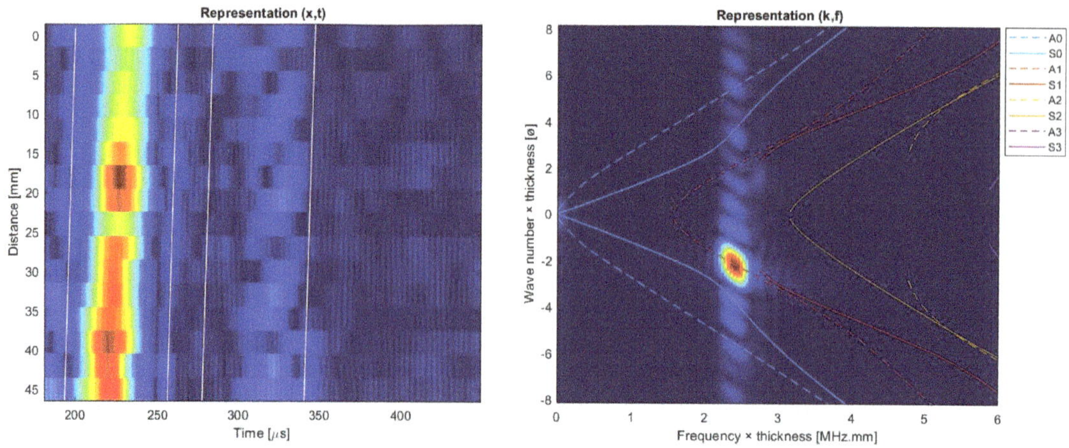

(**a**) (**b**)

Figure 11. (**a**) Envelope of the signals recorded by each element of the linear array—(x,t) representation; (**b**) 2D-FFT of the first wave packet—(k,f) representation.

The topological energy is computed using the horizontal displacements $u(\vec{r},t)$ in the region of interest, and the results are plotted in Figure 12. The maximum of the topological energy occurs at x = 52.86 cm. The error of the location is higher than the uncertainty due to the regular grid (Δx_{grid} = 0.61 cm), but the difference remains lower than the wavelength of the mode A1 at 320 kHz, λ = 2.13 cm. So, the result is considered acceptable.

(**a**)
(**b**)

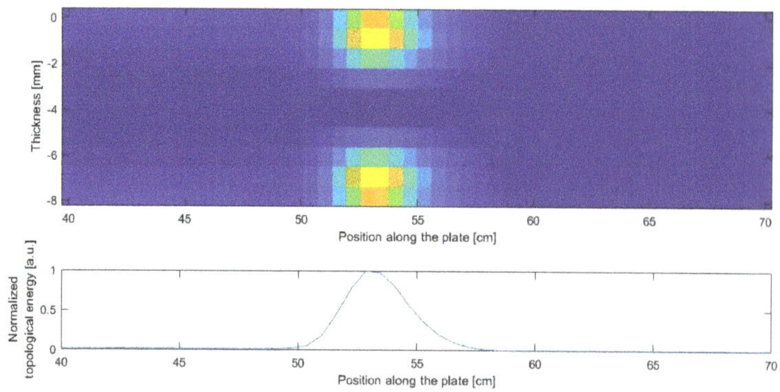

Figure 12. (**a**) Topological image in the region of interest of the simulated plate. (**b**) Normalized sum of the topological energy at each position along the plate.

The second wave packet shown in Figure 11a can also be time reversed and studied. In this case, the maximum of the topological energy plotted in Figure 13 occurs at x = 70.71 cm and corresponds to the edge of the plate. A local maximum is also present at x = 50.51 cm and is linked to the notch.

(a)

(b)

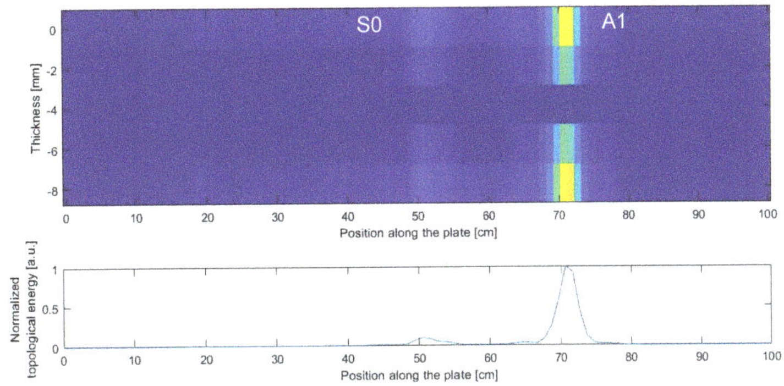

Figure 13. (**a**) Topological image in the region of interest of the simulated plate; (**b**) Normalized sum of the topological energy at each position along the plate.

One can thus assume that the second wave packet contains information about both defects (notch and edge). Further investigations have shown that two modes coexist: the S0 and A1 modes. The A1 mode comes from the initial A1 mode that reflects on the edge. The S0 mode comes from the mode conversion that takes place at the reflection at the notch.

The results obtained with the A1 mode on the edge and on the notch are reliable. The A1 mode has a low attenuation along the propagation at the studied frequency × thickness fe = 2.496 MHz·mm (Figure 2). Another mode, S0 or A0, would have leaked in the surrounded fluid before any reflected signal returned on the transducers. Lowly attenuated modes are preferred for a long-range inspection in a single plate. Nevertheless, considering the final aim of the work, that being the research of a defect in the second plate in a set of two parallel and immersed plates, the highly attenuated modes may be considered in order to maximize the energy incident to the second plate. This assumption will be studied in a future work.

The results presented in this paper are in good agreement with the results of Rodriguez et al., which is the most similar work found in the literature in a free isotropic plate [25] and a free anisotropic plate [29]. The locations of the defects are found to be good and accurate. Nevertheless, it has to be pointed out that those studies are not properly comparable. The emitter's positioning differs, and since the plates in Rodriguez's studies are free, the Lamb waves do not leak into the surrounding medium.

5. Conclusions and Perspectives

The theory of leaky Lamb waves was discussed in the first part. Dispersive and multimodal behaviors were highlighted by solving the dispersion equations and plotting the duets solution (k, f). The selective generation of a Lamb wave mode was explained. Then, the main steps of the topological energy method were outlined. Two problems had to be solved: the direct one and the adjoint one. The time reversal process in the adjoint problem allowed us to overcome the complexity induced by the dispersion and to sustain a good sensibility. To validate the process in a simple case, we performed experimental measurements on a plate without any defects. The detection and the localization of the edge were reached. In this configuration, we could assume that a through-crack has the same effect as the edge and could be detected. Then, measurements were performed on a plate with a machined half-thickness notch. By retropropagating separately the two wave packets acquired, both the notch and the edge were detected and localized. The mode conversion that occurs at the reflection on the notch provides more information than expected. That could lead to the detection and localization of cracks in the plate.

This work shows the applicability of the topological energy method on a single immersed plate. It constitutes the first step of the topological energy method implementation in a set of parallel and immersed plates with the goal of detecting and localizing a defect in the second and then the third plates. The transducers were positioned in the fluid above the upper face of the first plate, and the multiple reflections between the two plates had to be studied. The topological energy method appears to be a reliable method to filter those reflections in the calculation of the source term of the adjoint problem in order to retropropagate only the diffracted signal by the defect in the second plate.

Author Contributions: Methodology, J.-C.V., J.-F.C. and M.-A.P.; software, J.-C.V.; writing—original draft preparation, J.-C.V.; writing—review and editing, all authors; supervision, J.-F.C. and M.-A.P. project administration, F.B. and M.C. All authors have read and agreed to the published version of the manuscript.

Funding: This research received no external funding.

Institutional Review Board Statement: Not applicable.

Informed Consent Statement: Not applicable.

Acknowledgments: This work was developed within the framework of the MISTRAL joint research laboratory between Aix-Marseille University, CNRS, Centrale Marseille and CEA.

Conflicts of Interest: The authors declare no conflict of interest.

References

1. Kauffmann, P.; Ploix, M.-A.; Chaix, J.-F.; Potel, C.; Gueudre, C.; Corneloup, G.; Baque, F. Multi-Modal Leaky Lamb Waves in Two Parallel and Immersed Plates: Theoretical Considerations, Simulations, and Measurements. *J. Acoust. Soc. Am.* **2019**, *145*, 1018–1030. [CrossRef] [PubMed]
2. Kauffmann, P. Étude de la propagation, réémission et transmission des ondes de Lamb en vue du Contrôle Non Destructif de structures de réacteurs nucléaires immergées en sodium liquide. Ph.D. Thesis, Aix-Marseille Université, Marseille, France, 2018.
3. Kirrmann, P. On the Completeness of Lamb Modes. *J. Elast.* **1994**, *37*, 39–69. [CrossRef]
4. Lamb, H. On Waves in an Elastic Plate. *Proc. R. Soc. Lond. A Math. Phys. Eng. Sci.* **1917**, *93*, 114–128. [CrossRef]
5. Merkulov, L.G. Damping of Normal Modes in a Plate Immersed in a Liquid. *Sov. Phys. Acoust.* **1964**, *10*, 169–173.
6. Viktorov, I.A. *Rayleigh and Lamb Waves: Physical Theory and Applications*; Softcover Reprint of the Original 1st ed. 1967 edition; Springer: Berlin/Heidelberg, Germany, 2013; ISBN 978-1-4899-5683-5.
7. Cawley, P.; Alleyne, D. The Use of Lamb Waves for the Long Range Inspection of Large Structures. *Ultrasonics* **1996**, *34*, 287–290. [CrossRef]
8. Carboni, M.; Gianneo, A.; Giglio, M. A Lamb Waves Based Statistical Approach to Structural Health Monitoring of Carbon Fibre Reinforced Polymer Composites. *Ultrasonics* **2015**, *60*, 51–64. [CrossRef]
9. Alleyne, D.; Cawley, P. Optimization of Lamb Wave Inspection Techniques. *NDT E Int.* **1992**, *25*, 11–22. [CrossRef]
10. Wilcox, P.D.; Lowe, M.J.S.; Cawley, P. Mode and Transducer Selection for Long Range Lamb Wave Inspection. *J. Intell. Mater. Syst. Struct.* **2001**, *12*, 553–565. [CrossRef]
11. Jensen, J.A.; Nikolov, S.I.; Gammelmark, K.L.; Pedersen, M.H. Synthetic Aperture Ultrasound Imaging. *Ultrasonics* **2006**, *44*, e5–e15. [CrossRef]
12. Schickert, M.; Krause, M.; Müller, W. Ultrasonic Imaging of Concrete Elements Using Reconstruction by Synthetic Aperture Focusing Technique. *J. Mater. Civ. Eng.* **2003**, *15*, 235–246. [CrossRef]
13. Schmitz, V.; MÜller, W. Synthetic Aperture Focusing Technique for Industrial Applications. In *Acoustical Imaging*; Arnold, W., Hirsekorn, S., Eds.; Acoustical Imaging; Springer: Dordrecht, The Netherlands, 2004; Volume 27, pp. 57–68, ISBN 978-90-481-6652-7.
14. Muller, A.; Robertson-Welsh, B.; Gaydecki, P.; Gresil, M.; Soutis, C. Structural Health Monitoring Using Lamb Wave Reflections and Total Focusing Method for Image Reconstruction. *Appl. Compos. Mater.* **2017**, *24*, 553–573. [CrossRef]
15. Holmes, C.; Drinkwater, B.W.; Wilcox, P.D. Post-Processing of the Full Matrix of Ultrasonic Transmit–Receive Array Data for Non-Destructive Evaluation. *NDT E Int.* **2005**, *38*, 701–711. [CrossRef]
16. Sicard, R.; Goyette, J.; Zellouf, D. A Numerical Dispersion Compensation Technique for Time Recompression of Lamb Wave Signals. *Ultrasonics* **2002**, *40*, 727–732. [CrossRef]
17. Zhang, H.; Liu, Y.; Fan, G.; Zhang, H.; Zhu, W.; Zhu, Q. Sparse-TFM Imaging of Lamb Waves for the Near-Distance Defects in Plate-Like Structures. *Metals* **2019**, *9*, 503. [CrossRef]
18. Alkassar, Y. Simulation of Lamb Wave Modes Conversions in a Thin Plate for Damage Detection. *Procedia Eng.* **2017**, *173*, 948–955. [CrossRef]
19. Castaings, M.; Le Clezio, E.; Hosten, B. Modal Decomposition Method for Modeling the Interaction of Lamb Waves with Cracks. *J. Acoust. Soc. Am.* **2002**, *112*, 2567–2582. [CrossRef]
20. Chakroun, N.; Fink, M.A.; Wu, F. Time Reversal Processing in Ultrasonic Nondestructive Testing. *IEEE Trans. Ultrason. Ferroelect. Freq. Contr.* **1995**, *42*, 1087–1098. [CrossRef]

21. Prada, C.; Wu, F.; Fink, M. The Iterative Time Reversal Mirror: A Solution to Self-focusing in the Pulse Echo Mode. *J. Acoust. Soc. Am.* **1991**, *90*, 1119–1129. [CrossRef]
22. Dominguez, N.; Gibiat, V.; Esquerre, Y. Time Domain Topological Gradient and Time Reversal Analogy: An Inverse Method for Ultrasonic Target Detection. *Wave Motion* **2005**, *42*, 31–52. [CrossRef]
23. Minonzio, J.-G.; Prada, C.; Chambers, D.; Clorennec, D.; Fink, M. Characterization of Subwavelength Elastic Cylinders with the Decomposition of the Time-Reversal Operator: Theory and Experiment. *J. Acoust. Soc. Am.* **2005**, *117*, 789–798. [CrossRef]
24. Rodriguez, S.; Sahuguet, P.; Gibiat, V.; Jacob, X. Fast Topological Imaging. *Ultrasonics* **2012**, *52*, 1010–1018. [CrossRef]
25. Rodriguez, S.; Deschamps, M.; Castaings, M.; Ducasse, E. Guided Wave Topological Imaging of Isotropic Plates. *Ultrasonics* **2014**, *54*, 1880–1890. [CrossRef]
26. Dominguez, N.; Gibiat, V. Non-Destructive Imaging Using the Time Domain Topological Energy Method. *Ultrasonics* **2010**, *50*, 367–372. [CrossRef]
27. Sun, L.-J.; Zhu, W.-F.; Shao, W.; Zheng, S.-B.; Fan, G.-P.; Zhang, H.; Li, Z.-W. Using Lamb Wave TDTE Method to Realize Ultrasonic Array Super-Resolution Imaging of Multiple Asymmetric Defects. *Sens. Actuators A Phys.* **2021**, *332*, 113102. [CrossRef]
28. Zhu, W.-F.; Shao, W.; Peng, L.-L.; Fan, G.-P.; Chen, X.-J.; Zheng, S.-B.; Zhang, H.-Y. Time-Domain Topological Energy Imaging Method of Concrete Cavity Defect by Lamb Wave. *Shock Vib.* **2019**, *2019*, 6294603. [CrossRef]
29. Rodriguez, S.; Castaings, M.; Deschamps, M.; DUCASSE, E. Topological Imaging of Defects in Anisotropic Plates. In *EWSHM—7th European Workshop on Structural Health Monitoring*; Le Cam, V., Mevel, L., Schoefs, F., Eds.; IFFSTTAR, Inria, Université de Nantes: Nantes, France, 2014.
30. Schmitt, M.; Schmidt, K.; Olfert, S.; Rautenberg, J.; Lindner, G.; Henning, B.; Reindl, L.M. Detection of Coatings within Liquid-Filled Tubes and Containers by Mode Conversion of Leaky Lamb Waves. *J. Sens. Sens. Syst.* **2013**, *2*, 73–84. [CrossRef]
31. Bertoni, H.L.; Tamir, T. Unified Theory of Rayleigh-Angle Phenomena for Acoustic Beams at Liquid-Solid Interfaces. *Appl. Phys.* **1973**, *2*, 157–172. [CrossRef]
32. Ngoc, T.D.K.; Mayer, W.G. A General Description of Ultrasonic Nonspecular Reflection and Transmission Effects for Layered Media. *IEEE Trans. Sonics Ultrason.* **1980**, *27*, 229–235. [CrossRef]
33. Garreau, S.; Guillaume, P.; Masmoudi, M. The Topological Asymptotic for PDE Systems: The Elasticity Case. *SIAM J. Control Optim.* **2001**, *39*, 1756–1778. [CrossRef]
34. Lubeigt, E. Imagerie Topologique de Domaines Élastiques Bornés: Application Au Contrôle Non Destructif Des Soudures. Ph.D. Thesis, Aix-Marseille Université, Marseille, France, 2017.

applied sciences

MDPI

Article

Towards Explainable Augmented Intelligence (AI) for Crack Characterization

Larissa Fradkin [1,*], Sevda Uskuplu Altinbasak [1,†] and Michel Darmon [2]

1 Sound Mathematics Ltd., 11 Mulberry Close, Cambridge CB4 2AS, UK; sevda.uskuplu@gmail.com
2 Université Paris-Saclay, CEA, List, F-91120 Palaiseau, France; Michel.DARMON@cea.fr
* Correspondence: l.fradkin@soundmathematics.com
† Current Address: ARM, 110 Fulbourn Road, Cambridge CB1 9NJ, UK.

Abstract: Crack characterization is one of the central tasks of NDT&E (the Non-destructive Testing and Evaluation) of industrial components and structures. These days data necessary for carrying out this task are often collected using ultrasonic phased arrays. Many ultrasonic phased array inspections are automated but interpretation of the data they produce is not. This paper offers an approach to designing an explainable AI (Augmented Intelligence) to meet this challenge. It describes a C code called AutoNDE, which comprises a signal-processing module based on a modified total focusing method that creates a sequence of two-dimensional images of an evaluated specimen; an image-processing module, which filters and enhances these images; and an explainable AI module—a decision tree, which selects images of possible cracks, groups those of them that appear to represent the same crack and produces for each group a possible inspection report for perusal by a human inspector. AutoNDE has been trained on 16 datasets collected in a laboratory by imaging steel specimens with large smooth planar notches, both embedded and surface-breaking. It has been tested on two other similar datasets. The paper presents results of this training and testing and describes in detail an approach to dealing with the main source of error in ultrasonic data—undulations in the specimens' surfaces.

Keywords: Non-destructive Testing/Evaluation (NDT/NDE); ultrasonic imaging and inversion; ultrasonic characterization; explainable Augmented Intelligence

Citation: Fradkin, L.; Uskuplu Altinbasak, S.; Darmon, M. Towards Explainable Augmented Intelligence (AI) for Crack Characterization. *Appl. Sci.* **2021**, *11*, 10867. https://doi.org/10.3390/app112210867

Academic Editor: Stefano Invernizzi

Received: 23 October 2021
Accepted: 12 November 2021
Published: 17 November 2021
Corrected: 20 January 2022

1. Introduction

The aim of this paper is to address a challenge of developing an explainable AI for semi-automatic crack characterization, with a view to its ultimate deployment in ultrasonic units for NDT&E (the Non-destructive Testing and Evaluation) of industrial components and structures. Since the most advanced units are phased arrays of ultrasonic transducers all the experimental data used to train and test the AI discussed below have been collected using linear arrays of this nature. Moreover, the experiments have been designed to emulate cracks and inspection surfaces typically encountered in walls of nuclear reactors. It is particularly important to minimize human involvement in interpretation of NDT data in nuclear industry: With the new nuclear build already under way, NDT practitioners anticipate a severe shortage of suitably qualified and experienced personnel. Also there is pressure in industry for both speeding up the inspections and increasing their reliability. Interestingly, even though ultrasonic inspections have been conducted for decades, a study conducted by TWI (The Welding Institute) a few years ago has demonstrated that although their reliability is high it is not as high as many believe or wish it to be [1]. The most surprising outcome of the study was the fact that human inspectors experienced the greatest difficulty when characterizing large planar cracks. A less surprising finding was that the most difficult cracks to identify were those normal to inspection surface—the responses of their tips are known to be weak. A desire to respond to this study has been another rationale for the work reported here.

In order to carry out crack characterization NDT inspectors rely mostly on TOFD (Time of Flight Diffraction) configurations, in which the most prominent features are the diffraction spots surrounding crack tips. By contrast the approaches pursued by those who work towards automating crack characterization often rely on specular reflections. There have been attempts to develop general but time-consuming model-based data processing algorithms, see e.g., [2,3] as well as pure signal processing approaches, such as CS (Compressed Sensing) algorithms [4]. The approach meeting a practical need best seems to be TFM (Total Focusing Method) based on FMC (Full Matrix Capture) [5,6]. Briefly, every element of the Full Matrix is an A-scan (a sequence of ultrasonic pulses) received by an array transducer after this or another array transducer fires a single pulse. TFM uses this matrix to create images that lend themselves to a relatively easy interpretation by both human and artificial intelligence. However, TFM images are often contaminated by noise and various strategies have been offered to modify the TFM algorithm to eliminate false indications [7,8] and reduce noise [8–10], enabling real-time imaging with portable NDT devices [8,11]. Researchers also began to explore application of machine learning to NDT [12–15]. However, at present, standard machine learning approaches have limited value: Firstly, most researchers have no access to big data such approaches require and even a few laboratory datasets used below have required a considerable effort and expense to collect. Secondly, standard approaches often lead to results that are unexplainable, and a highly regulated branch of industry, such as NDT of nuclear reactors is unlikely to adopt results of this nature. In this paper we present an alternative: a code that combines a signal processing algorithm based on a simple modification of the TFM with the well-known image processing algorithms as well as a decision tree. The latter is an AI module, which mimics thought processes followed by human inspectors in writing standard inspection reports. The code has been designed to deal with the scatter from large planar cracks, whether specular reflections from crack surfaces or echoes from crack tips.

We demonstrate the efficacy of the approach using laboratory data. To collect such data engineers manufacture test blocks to contain flaws with known characteristics and use the the NDT procedure they want to investigate to establish whether it can generate reasonable estimates of these characteristics [1]. The paper is organized as follows: in Section 2 we describe the relevant experiments; in Section 3 we present our composite signal/image processing/AI algorithm for crack characterization and in Section 4 we present results of its training on 16 datasets and testing on two. Since it is known that in many industrial situations the main source of error is undulations in component surfaces, one of the test blocks has been deliberately chosen to have a qualitatively different surface to the test block used for the AI training. In the last section we discuss our findings and present recommendations.

2. The Experimental Set-Up

This paper builds on the original feasibility study reported in [16], with the experimental set-up presented in Figure 1. The RF (radio-frequency) data used there were collected by DPS (Doosan Power Systems) engineers with a demonstrator multiplexed to an 128 element IMASONIC linear transducer array with the pitch $D_e = 0.8$ mm, the central pulse frequency $f = 5$ MHz and sampling frequency $f_s = 50$ MHz. The specimen probed was a steel block, 30 mm thick, 200 mm wide and 350 mm long, with four surface-breaking notches and four further notches buried underneath the notched surface. Four notches out of eight were non-tilted and four, tilted at 110° to the surface. The longitudinal speed in steel varies with composition. In the steel used in this experiment it was $c_l = 5.89$ km/s.

The experiments have been performed in immersion, with the water temperature of 22 °C, so that the speed in water was $c_w = 1.48$ km/s. The water path standoff distance was about 13 mm. A typical input pulse (a pulse transmitted by a transducer) is presented in Figure 2a, and a typical A-scan (a train of pulses received at a transducer), in Figure 2b. The full matrix of A-scans, $[A_{kln}, k, l, = 1, 2, \ldots, K, n = 1, \ldots, N]$ has been collected, where the first index denotes the transmitter, second—the receiver and third—the time sample.

Both transmitters and receivers are numbered from the left. Let us introduce $t_{k,x,l}$, the time of travel from transmitter k to receiver l through one of the evenly spread nodes **x** = (x, z) and signal $A_{kl}(t) \approx A_{kln}|_{n=\lfloor t/\Delta t \rfloor}$, where Δt is the time increment defined by the sampling frequency; $\lfloor t/\Delta t \rfloor = \text{floor}\ (t/\Delta t)$. In order to reduce the processing burden [17], instead of A-scans AutoNDE uses their Hilbert transform,

$$h(A_{kl})(t) = \frac{1}{\pi} p.v. \int \frac{A_{kl}(\tau)}{t-\tau} d\tau. \quad (1)$$

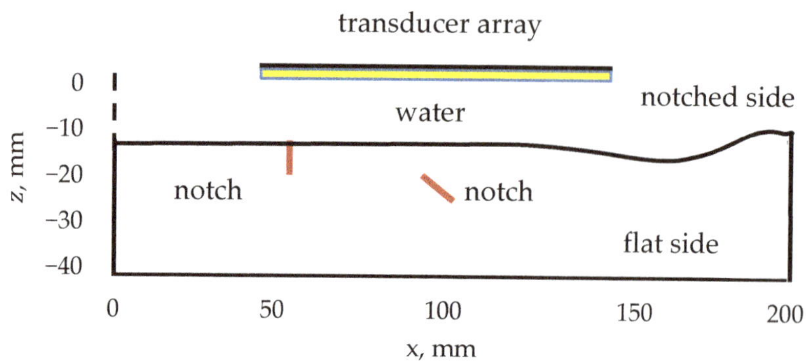

Figure 1. A schematic of the DPS experiment.

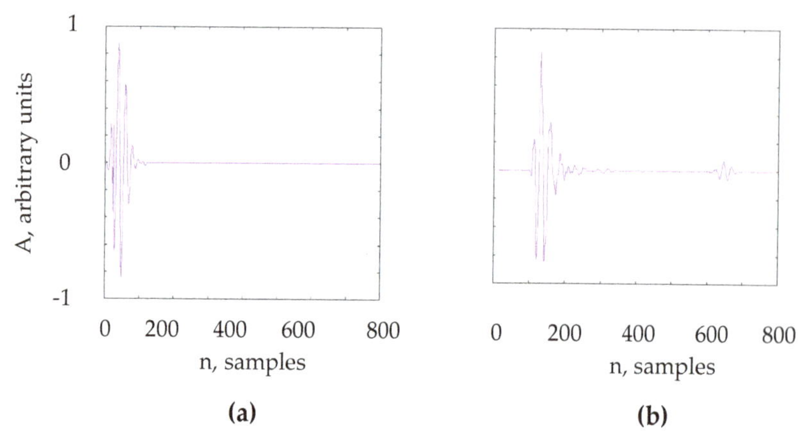

Figure 2. (**a**) A typical input pulse A_n; and (**b**) a typical A-scan A_n.

Note that the A-scan in Figure 2b is $A_{82,82,n}$. Taking into account that the offset was n = 743 the first pulse is the echo of the pulse transmitted by element 82 arriving back to this element from the point x_1 on the upper surface of the specimen and the second pulse is the echo arriving back from point x_2 on the backwall. The distance d between x_1 and x_2 can be calculated using the standard formula $d = 0.5\ n\ c_l/f_s$, where n is the number of time samples between arrivals of two pulses.

3. The AutoNDE Code for Semi-Automatic Crack Characterization

The original version of the code described in [16] was written in LabView and contained only a rudimentary AI module. In this paper we present a more advanced version written in C. Its flowchart is presented in Figure 3.

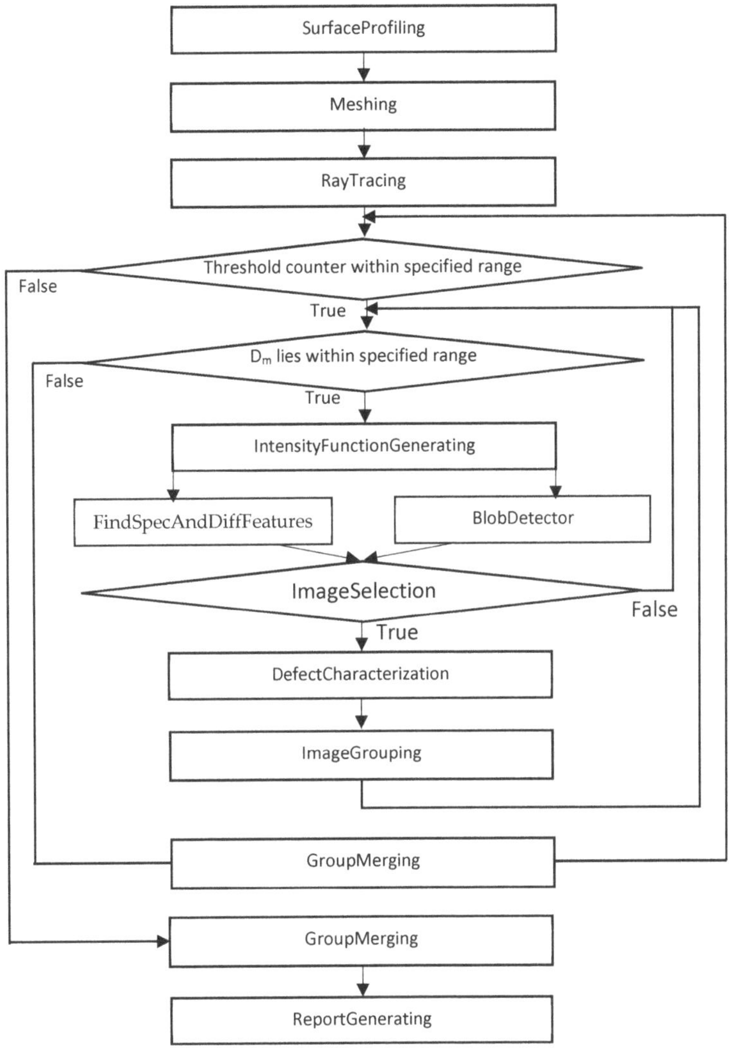

Figure 3. A flowchart of AutoNDE.

Let us describe the submodules presented there in more detail:

3.1. Signal Processing

The submodules of the Signal Processing module are used to create 2D images of the tested specimen:

1. *SurfaceProfiling* effects profiling by (1) locating for each array element the surface point directly underneath and (2) interpolating the acquired surface points using polynomial regression. The first step is performed by convolving $h(A)(N\Delta t - t)$, the Hilbert transform of the time inverse of the input pulse with the corresponding pulse scattered by the surface. Only A-scans received by the same transducers that

transmitted them are utilized. Hence the maximum number of surface points collected during the first step is K. The regression model used in interpolation is

$$\mathbf{z} = \mathbf{X}\,\mathbf{b} + \mathbf{e}, \tag{2}$$

where the response vector **z**, the parameter vector **b**, the design matrix **X** and the error vector **e** are given, respectively, by

$$\mathbf{z} = \begin{bmatrix} z_1 \\ z_2 \\ \cdots \\ z_J \end{bmatrix}, \mathbf{X} = \begin{bmatrix} 1\ x_1\ x_1^2\ \ldots\ x_p^p \\ 1\ x_2\ x_2^2\ \ldots\ x_2^p \\ \cdots\cdots\cdots\cdots \\ 1\ x_J\ x_J^2\ \ldots\ x_J^p \end{bmatrix}, \mathbf{b} = \begin{bmatrix} b_1 \\ b_2 \\ \cdots \\ b_p \end{bmatrix}, \mathbf{e} = \begin{bmatrix} e_1 \\ e_2 \\ \cdots \\ e_J \end{bmatrix} \tag{3}$$

- Originally, the polynomial degree to produce good results with the DPS data has been found by trial and error to be $p = 8$.
- In the latest version of AutoNDE the degree p is selected automatically. There is a number of approaches recommended for this purpose in the literature on machine learning. We have found that the most common of those, the bias-variance trade-off leads to ill-conditioned the Vandermonde matrix $X^T X$ and overfitting of the DPS data. Since for all DPS datasets *SurfaceProfiling* acquires surface points whose location error is random it is reasonable to assume that their underlying error distribution is normal. Therefore, we attempted and found satisfactory a method that involves the Wald test [18] based on the t-statistic of the leading coefficient.
- In order to apply this method we first estimate p_{max} and q_{max}, where p_{max} is the highest polynomial degree that can be reliably estimated from the available data and q_{max} is the maximum number of digits of accuracy on top of what would be lost to the numerical method due to loss of precision from arithmetic methods [19]. A well-known rule-of-thumb suggests that $p_{max} = J/5$ and the training of AutoNDE on the DPS data suggests that for realistic random surface undulations used in this experiment $q_{max} = 6$.
- The suggested method utilizes the algorithm presented in Figure 4. Note that since all X_{j2} values are non-zero and distinct for every $p = p'$ all Vandermonde matrices $X^T X$ are invertible [20]. Note too that the t-statistic is normally applied to assess significance of regression parameters, while here $t_{p'}$ is used to test the null hypothesis that the leading coefficient $b_{p'} = 0$. It follows that the algorithm selects the polynomial of the highest significant degree. The threshold $t_p = 1.96$ assures that if the error in location of surface points has a normal distribution, the null hypothesis that the leading coefficient is zero can be rejected at the 95% significance level.

The spline method has been tried too but was found to be too sensitive to the choice of smoothing parameters and thus not amenable to automation

2. *Meshing* of the specimen is performed by specifying a regular grid of evenly spaced rows and columns, covering the portion of the specimen, which lies underneath the probe. The meshing module also specifies the region of interest. If the measurements are taken only when the crack is located more or less underneath the array center the region of interest is reduced to the central region underneath the probe. Any reduction of the region of interest speeds up the crack characterization process.
3. *RayTracing* starts by issuing a fan of rays from each array element. The central angle of each fan is $-90°$ to the x-axis, the optimal vertex angle has been found to be $60°$ and the optimal difference between the angles of neighboring rays, $0.057°$. These values effect a trade-off between the code accuracy and speed. For each ray the *RayTracing* submodule locates the point where it hits the upper surface, finds the refracted ray issuing from this point (in the current version no shadowing is accounted for) and calculates the time it takes the ray to reach each row in the region of interest. In the

present version of AutoNDE mode conversion is allowed as well as one reflection from the backwall.

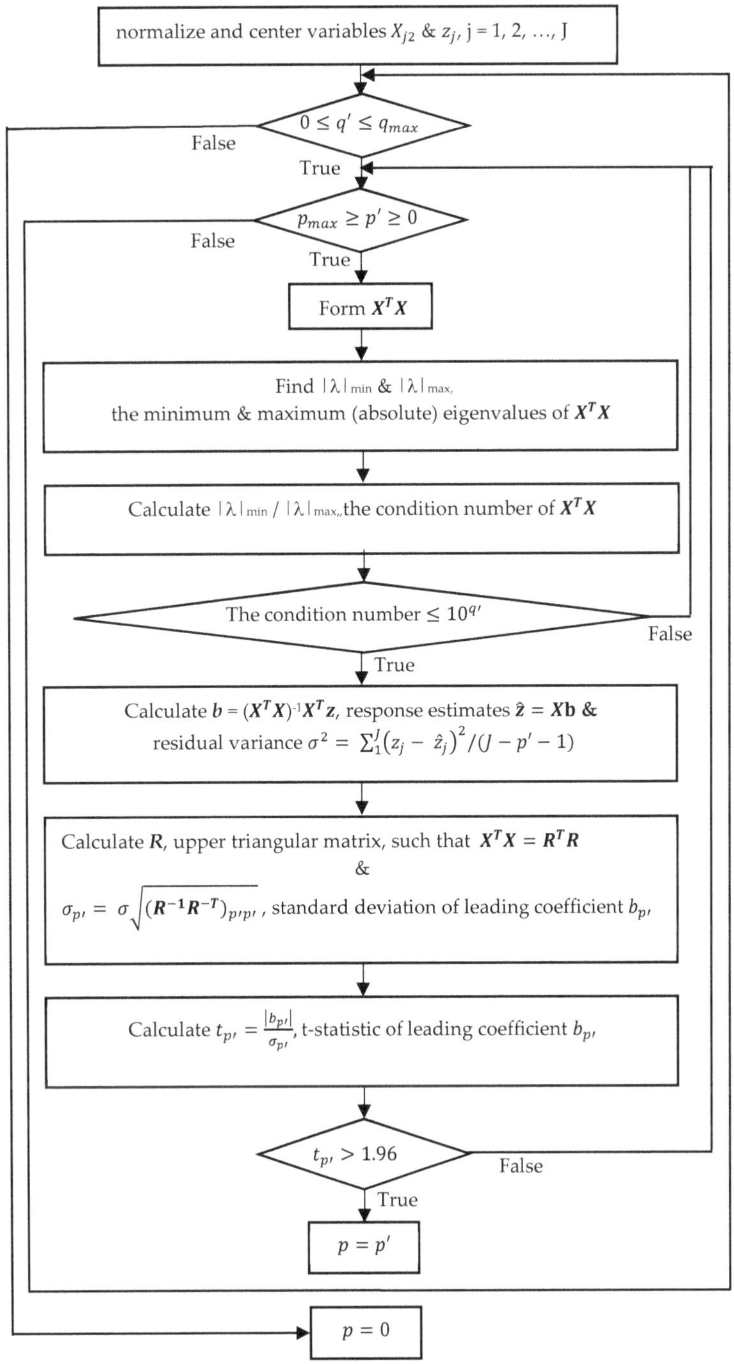

Figure 4. An algorithm for automatic identification of the interpolating polynomial.

4. *IntensityFunctionGenerating* utilizes the matrix A of A-scans to generate the intensity function

$$I(x) = |\sum_{k,l} h(A_{kl})(t_{k,x,l})| \quad (4)$$

where time $t_{k,x,l}$ is the moment of time the corresponding pulse is at its peak. In the standard TFM (Total Focusing Method) the summation in (4) is carried out over the whole probe. In addition to TFM we use an MTFM (a Modified TFM), a signal processing approach developed by trial and error to produce not just one image for one position of the probe as in TFM but a series of images m: Inside each such image, each vertical segment x is scanned with a "partial probe" $[k + D_m, k + D_m + L]$, $k < K - D_m - L$ when $D_m > 0$ (a blue colored portion of the transducer array in Figure 5 or $[k + D_m - L, k + D_m]$, $k > L - D_m$ when $D_m < 0$, where the transducer element $k = \lfloor x/D_e \rfloor$ or $k = \lceil x/D_e \rceil$, respectively. Here $\lceil x/D_e \rceil = ceil(x/D_e)$. This allows us to use the same amount of information to image each vertical segment of the specimen, except for the segments close to the array ends. However, as a rule, the end portions of the array lie outside the region of interest. The approach often filters out the "blinding" surface reflections and enhances images of diffraction spots. Both the TFM and MTFM images are produced using the normalized version of intensity function, $I_1(x) = 256\ I(\mathbf{x})/\max_x I(x)$. Each image is stored in the standard way, using 256 different intensities, the highest indicated by red color and the lowest, by blue.

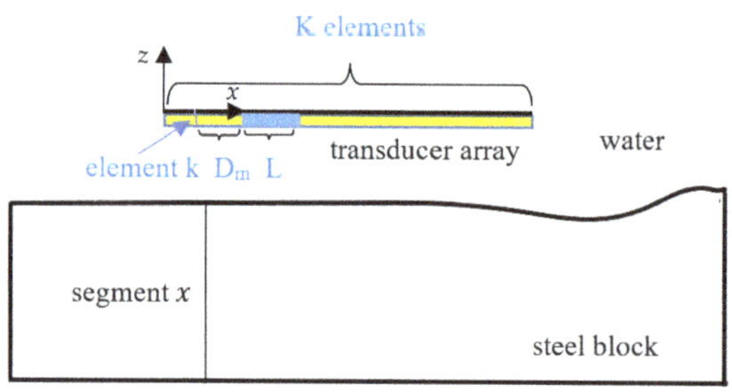

Figure 5. A schematic of MTFM.

3.2. Image Processing

The image processing module is used to select those MTFM images, which lend themselves to easy interpretation. The basis for selection is *a priori* knowledge that the crack to be characterized is large and plane. Therefore the crack image is expected to contain a straight segment, which is a specular reflection from the crack, or else two diffraction spots surrounding the crack tips. Sometimes only one crack tip can be picked up. AutoNDE differentiates the possible diffraction spots from the possible specular features by size, allowing for some overlap.

The *ImageProcessing* module of AutoNDE uses a variety of intensity thresholds. As mentioned above, the maximum intensity is 256. Thresholding is a standard tool in image processing, which is used to filter out noise. During the AutoNDE training, in most cases 125 has been found to produce the best results. However, some significant weak features could only be picked up at lower thresholds, while some noise could be filtered out only at thresholds that are higher. For this reason, AutoNDE normally utilizes three thresholds, 65, 125 and 185.

AutoNDE analyse the resulting images using OpenCV (Open Source Computer Vision Library) functions [21]. Two submodules are involved, *FindSpecAndDiffFeatures* and *BlobDetector*.

1. *FindSpecAndDiffFeatures* relies on the OpenCV *FindContour* function to select two types of features, large (longer than 7 mm) and small (between 1 mm and 7 mm long).

If one of the features is 7 mm or slightly smaller and there are several other small features smaller than 3 mm in extent the small features are neglected and the larger one is treated as a specular reflection.

2. The *BlobDetector* relies on the OpenCV *DetectBlob* function to filter blobs by size between 80 and 160 pixels. The *BlobDetector* is particularly useful when dealing with surface-breaking cracks, because in these situations the probe often picks up only one crack tip. When only one blob is picked up the final crack characterization can be made only by a human inspector. In cases like this the AutoNDE flags the situation by putting the question mark after every defect characteristic and estimate of the report quality (the definition of quality is given below). All the feature and blob parameters mentioned above have been chosen by trial and error to maximize the number of true positives and minimize the number of false positives selected by the code.

3.3. Explainable AI

The AI module of AutoNDE is a decision tree, which selects images that appear to contain defects, characterizes these defects and then groups similar images. Note that by their nature, decision trees produce explainable results: all the reasoning can be traced. The decision tree comprises the following submodules:

1. *ImageSelection* submodule selects images containing one or two blobs (bright spots), two small contour selected features, one large feature or maybe, one blob and one small feature. If a blob and a contour selected feature are detected at the same location it is the feature parameters that are used to characterize the potential diffraction spot. If one of the contour selected features is slightly bigger than 7 mm it is still treated as a possible diffraction spot.
2. *DefectCharacterization* carries out calculations of the extent (notch length in the imaged plane), depth (the smallest of distances between notch tips and specimen surfaces) and orientation (the angle the notch makes in the imaged plane with the mean specimen surface) of the detected planar defect. The calculations are based on parameters of the bounding boxes, which the *FindContour* OpenCV function draws around the objects or else on parameters of blobs detected by the *DetectBlob* function. Planar cracks are expected to produce two types of images, specular reflections and TOFD (Time of Diffraction) images, which contain two diffraction spots surrounding notch tips. When the image contains one large feature (interpreted as a specular notch image), the extent is calculated as the longest box side; the depth, as the shortest distance between box vertices and specimen surfaces; and orientation as orientation of box's longest side. For TOFD, the extent is calculated as the largest distance between vertices of their bounding boxes; the depth, as the shortest distance between vertices of these boxes and specimen surfaces; and orientation, as orientation of the line connecting the gravity centers of the boxes. If only two small features are identified the code draws a straight yellow line connecting their gravity centers.
3. *ImageGrouping* checks whether each selected image appears to be similar to the pivot image in the group $g = 1, 2, \ldots, G$, that is contains a notch with a similar extent E and orientation O at a similar location C (so that the coordinates of the gravity centers of the notches are similar). The pivot is the image with the smallest D_m in the group. If the image is similar to the pivot it is added to the group; otherwise, it is used as a pivot for the next group. The crack parameters are referred to below as $v = E$, O, C, respectively. For each group a preliminary report is compiled, describing the weighted averages $\overline{v}_g$.

$$\overline{v}_g = \frac{\sum_{m=1}^{M_g} \hat{w}_{g,m} v_{g,m}}{\hat{M}_g}, \quad \hat{w}_{g,m} = w_{e,g,m} \cdot w_{o,g,m} \cdot w_{l,g,m} \cdot w_{add}, \tag{5}$$

$$w_{v,g,m} = w(\Delta v_{g,m}, t_v).$$

Above $\hat{M}_g = \sum_{m=1}^{M_g} \hat{w}_{g,m}$ is a modified number of images in group g, with M_g—the number of images in group; $\Delta v_{g,m} = |v_{g,m} - v_{g,0}|$ is the deviation of parameter $v_{g,m}$ in the group g and image m from the corresponding parameter $v_{g,0}$ in the pivot image (in case of the gravity center location, this deviation is the distance between centers); t_v is the acceptable threshold for this deviation; and the weighting function, which smoothes transition over this threshold is

$$w(u,\ U) = \begin{cases} 1, \text{ if } u < U \\ e^{-\left(\frac{u}{U}-1\right)^2}, \text{ otherwise,} \end{cases} \tag{6}$$

see Figure 6. The following thresholds have been established by trial and error:

$$t_e = t_l = 2.1 \text{ mm},\ t_o = 21^{\circ}. \tag{7}$$

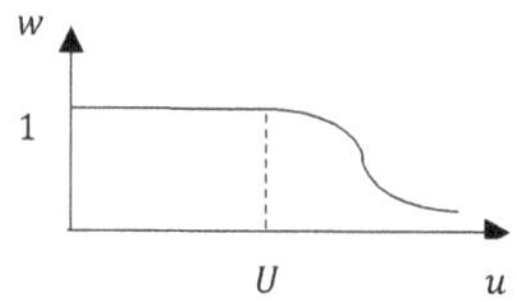

Figure 6. Modification weights.

The weight w_{add} is used to taper off the probability of almost horizontal cracks situated very close to the top surface or backwall. The quality of the resulting group report is assessed by using the subjective probability (rounded up to the nearest multiple of 10),

$$Q_g = \max\left\{10\%, \min\left[90\%,\ 100\%\frac{\hat{M}_g}{\hat{M}}\right]\right\}, \tag{8}$$

where $\hat{M} = \sum_{g=1}^{G} \hat{M}_g$ is the sum of modified numbers of images in all G groups identified. Thus, one of the advantages of MTFM is the fact that various images it produces allow AutoNDE to assess the quality of crack characterization.

4. *GroupMerging* employs similar principles to *ImageGrouping*, working with group averages instead of individual crack characteristics. Group merging is performed first for each intensity threshold: the first of all groups on the list is chosen as a pivot, the next group on the list is merged with it if the extents of their defects differ by no more than 2.1 mm; the distance between the gravity centers of these defects is no more than 2.5 mm; and their orientations differ by no more 21°. The remaining groups form a new list and the merging process is repeated. For a given intensity threshold, only groups detected by the same method (*FindContour* or *BlobDetector*) can be merged. No such restriction is used when merging groups identified using different intensity thresholds. Otherwise, this last merging is performed using the same principles as above but with deviations in extents and distances allowed to reach 3 mm.
5. *ReportGenerating* reports the group(s) with the maximum probability. If more than one group with the maximum probability is reported the final choice has to be made by the human inspector on scrutinizing the TFM image.

4. Training AutoNDE

The AutoNDE was trained using sixteen datasets produced by DPS (Doosan Power Systems) and then tested on one dataset collected by AMEC and one, by CEA (The French Commission for Atomic and Alternative Energies).

4.1. Training AutoNDE on DPS Data

It has been established by trial and error that the best images of the specimen used in DPS experiment (see Figure 1) could be obtained by specifying its thickness as 29.5 mm and distance from the probe to the specimen, as 12.5 mm. In other words, it has been established that the 0.5 mm difference in these parameters has a significant effect on the image quality. The number of array elements and length of A-scan have been already specified above as K = 128 and N = 800. The optimal length of the "partial probe" has been established by trial and error to be 25 elements, covering the aperture of 20 mm. This aperture is large with respect to the typical length of the longitudinal wave: Given the longitudinal speed within the steel specimen of 5.89 km/s and the central pulse frequency of 5 MHz, this typical length is 1 mm. Similarly, it has been found that enough information could be collected with 25 images, D_m varying between 0 to 24 array elements. No interpretable images were produced for larger values of D_m. Finally, whatever the dataset, quality results have been obtained for the same region of interest. This has been chosen as the central region, roughly 20% of the area underneath the probe, symmetrical with respect to the probe center. The resulting estimates of notch characteristics are compared to their known experimental values in Table 1. The experimental values were established using standard approaches used in experiments of this nature, see, e.g., [1].

Table 1. The DPS data set: Estimated (est) and experimental (exp) crack characteristics. The crack position is specified as distance between the notch and left edge of the specimen. The "hc" stands for Human Choice and indicates that the estimates are chosen by a human and not the AI.

Inspection Surface/ Approximate Notch Depth/ Notch Distance from Edge	Report Quality/Comments	Defect Parameters		
		Extent, in mm Est/Exp	Orientation, in deg Est/Exp	Depth, in mm Est/Exp
Flatside/Buried/24 mm	60%	9/10	105/110	8/5
Flatside/Buried/62 mm	40%	6/5	105/110	5/5
Flatside/Buried/113 mm	40%	12/10	90/90	5/5
Flatside/Buried/149 mm	20%hc	5/5	85/90	6/5
Flatside/Breaking/25 mm	60%	6/5	80/90	0/0
Flatside/Breaking/64 mm	90%	11/10	90/90	0/0
Flatside/Breaking/113 mm	60%	7/5	100/110	0/0
Flatside/Breaking/150 mm	90%	12/10	110/110	1/0
Notchside/Buried/24 mm	60%	11/10	110/110	5/5
Notchside/Buried/62 mm	70%	6/5	110/110	5/5
Notchside/Buried/113 mm	30%	12/10	90/90	4/5
Notchside/Buried/149 mm	30%	5/5	80/90	3/5
Notchside/Breaking/25 mm	40%	5/5	100/90	1/0
Notchside/Breaking/70 mm	30%hc	8/10	90/90	2/0
Notchside/Breaking/113 mm	30%	5/5	115/110	1/0
Notchside/Breaking/155 mm	50%	8/10	115/110	2/0

Table 1 shows that depths of the notches in DPS data could be estimated with the error of up to 2 mm (in one instance, 3 mm), and orientations—with the error of 5–10°. We note here that assuming the inspection surface plane would produce similar estimates for most of the notches but the second entry would be 7 mm in extent, located at 1 mm depth, oriented at 75° and the fourth entry would be 7 mm in extent, located at 0 mm depth, oriented at 125°. It follows that results are more reliable when small surface undulations are taken into account.

Typical MTFM images are presented in Figure 7: both Figure 7a,b contain three bright spots, with the two brightest ones joined by a thin yellow line. However, while the top spot is bright in both images, in Figure 7a the brightest lower spot is found at a distance of 4 mm from the backwall, while in Figure 7b the brightest spot lies on the backwall. We know that both bright spots in Figure 7a represent diffraction spots surrounding tips of the planar notch, while the lowest bright spot in Figure 7b is spurious, probably due to a defect in the backwall: The noise is similar to the signal and inside any given image the code cannot always distinguish between the two. However, in this case most MTFM images allow it to make the correct choice. This leads to a reasonable entry for the correponding notch in Table 1. The accompanying AutoNDE inspection report is presented in Figure 8.

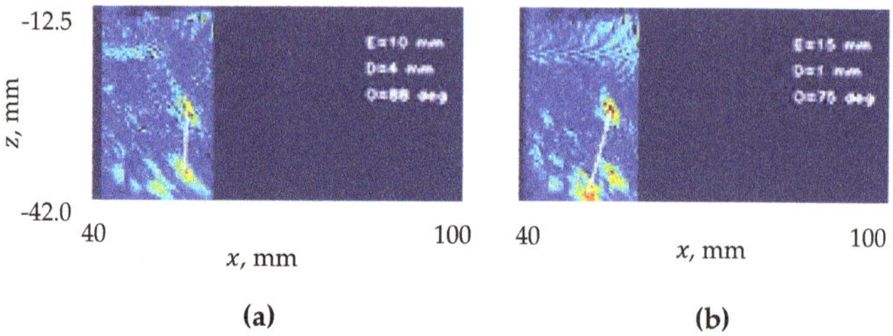

Figure 7. DPS data set: the buried notch located 113 mm from the left edge, imaged from the flat side. MTFM images have been obtained with intensity threshold = 185 and Dm = (**a**) 24; (**b**) 21. Key: E—notch extent, D—notch depth and O—notch orientation. The portion of the image to the left of the region of interest is cut off.

POSSIBLE INSPECTION REPORT

Folder: FSBuried113mm

TFM image

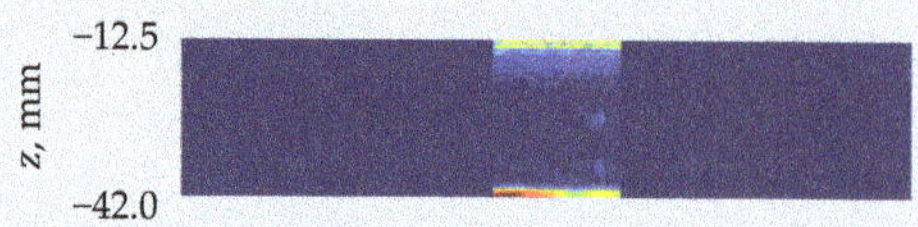

Upper surface (interpolated using the polynomial degree 9)

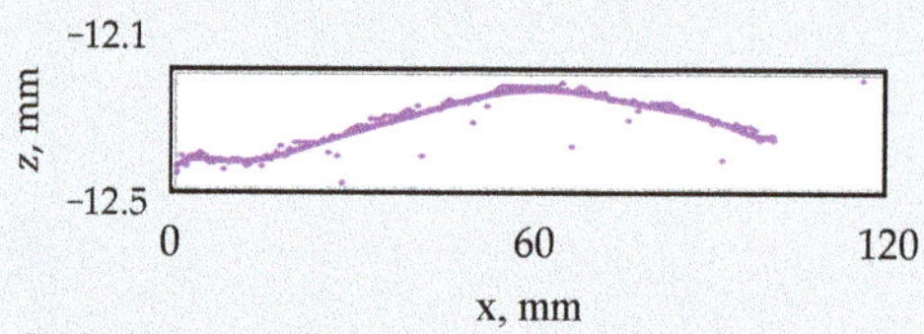

GROUP-1:

Number of images in group = 11
DM(s) in groups: 14 17 24 22 23 24 -20
A possible planar defect is detected.
Defect depth = 5 mm
Defect extent = 12 mm
Defect orientation = 90 deg
Report Quality = 40% (the group displays the shooting star effect)

Figure 8. Inspection report for buried notch located 113 mm from the left edge. Key (here and below): dots–surface points acquired by SurfaceProfiling; line interpo lating polynomial.

Note that unlike MTFM images in Figure 7, the TFM image in Figure 8 contains bright reflections from both top surface and backwall and the portion of the image to the left of the region of interest is not cut off. Unlike with MTFM images the diffraction spots surrounding notch tips are very faint. The presence of the TFM image in the report allows a human inspector to make an immediate assessment of the validity of the AI conclusions. Note too that the upper surface points presented in the second figure of this report have been obtained using the *Profiling* submodule and solid line is the interpolating polynomial. Finally, the order of DMs listed in the report indicates that the first four interpretable images have been obtained with the intensity threshold of 65, the next four—with the intensity threshold of 125, and the last two—with the intensity threshold of 185.

As mentioned above, when an AutoNDE report lists several possibilities, it is for a human inspector to select the most probable. Let us illustrate this by the report for the surface-breaking notch situated 113 mm from the left edge of the specimen and inspected from the notched side. The corresponding AutoNDE report can be seen in Figure 9.

POSSIBLE INSPECTION REPORT

****

Folder: NSBreaking113mm

TFM image

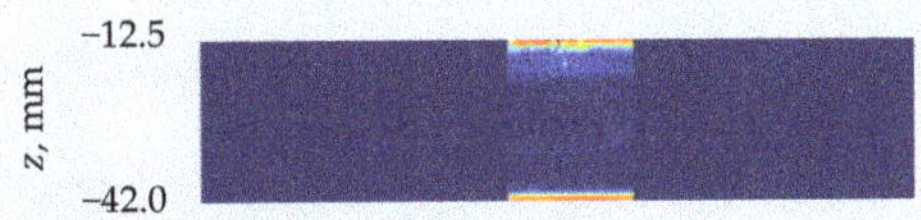

Upper surface (interpolated using the polynomial degree 9)

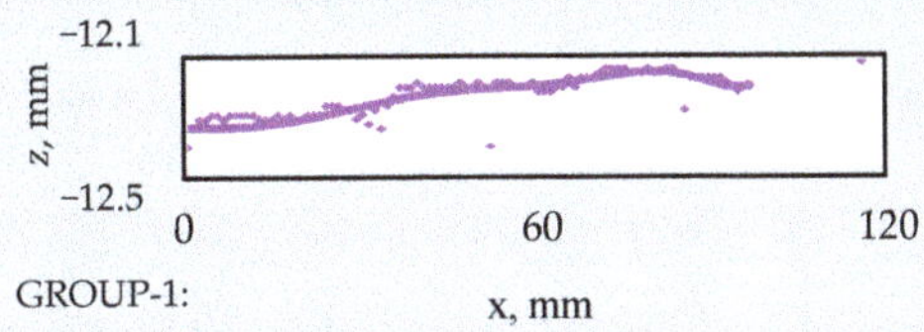

GROUP-1:

Number of images in group = 3
DM(s) in groups: -12 -13 -11
A possible planar defect is detected.
Defect depth = 3 mm
Defect extent = 19 mm
Defect orientation = 30 deg
Report Quality = 30%

GROUP-2:

Number of images in group = 10
DM(s) in groups: 3 4 6 10 22 15 17 18 20 21
A possible planar defect is detected.
Defect depth = 1 mm
Defect extent = 5 mm
Defect orientation = 115 deg
Report Quality = 30% (the group displays the shooting star effect)

Figure 9. Inspection report for the surface-breaking notch located 113 mm from the left edge.

The presence of Group 1 is due to the fact that some MTFM images pick up two spurious spots, see Figure 10a. Group 2 contains slightly skewed specular images, see Figure 10b. We emphasize here that all TFM images obtained with DPS data contain either clear diffraction spots as above or else clear specular images, see Figure 11. While the present version of AutoNDE has not been trained to mask the images of upper and lower surfaces this will be done in future. It would then become possible to characterize these images without employing MTFM. Thus, the main advantage of MTFM is the fact that unlike TFM it allows to produce many images instead of one, allowing to estimate the quality of a notch image by how often it is reproduced.

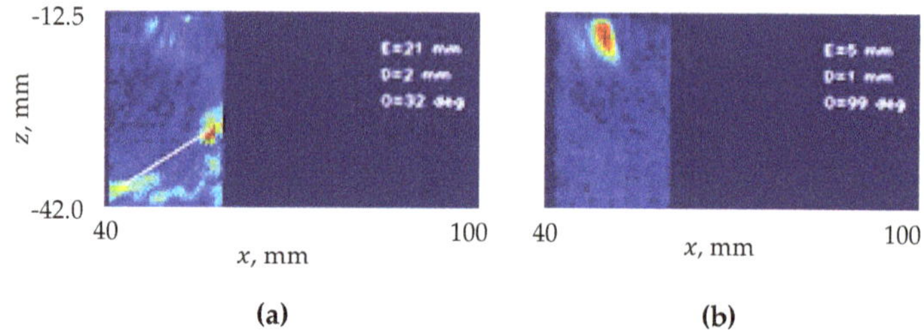

Figure 10. The DPS data set: the surface-breaking notch located 113 mm from the left edge, imaged from the notched side. MTFM images have been obtained with (**a**) intensity threshold = 125, Dm = −11, and (**b**) intensity threshold = 185, Dm = 15. Key as in Figure 7.

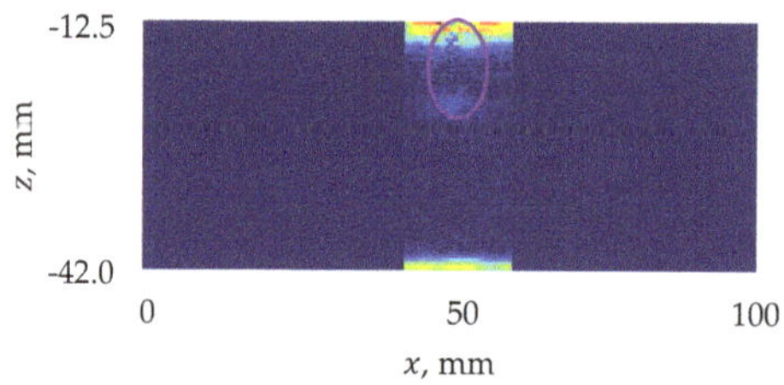

Figure 11. The DPS data set: the TFM image of the surface-breaking notch (circled) located 70 mm from the left edge, imaged from the notched side. Key as in Figure 7.

Typical run times involved in creating Table 1 under the Ubuntu 64-bit operating system on the VMware workstation 16.x with an i7-1165G7 @ 2.80 GHz and 16 GB of Ram are presented in Table 2.

Table 2. Maximum run times (in seconds) involved in creating Table 1.

Reading Data	Profiling	Raytracing	Creating an Image	I/O Handling	Total
1.5	2	2	1.5	4	12

4.2. Testing AutoNDE on AMEC Data

AutoNDE has been tested on a data set collected by AMEC engineers using a 64 element phased transducer array with the pitch D_e = 0.63 mm and sampling frequency f_s = 25 MHz, placed in direct contact with a 55.5 mm deep steel specimen. The geometry of the experiment, input pulse and typical A-scans are similar to the ones in the DPS experiment and are not reproduced.

The standard TFM image of the AMEC specimen is presented in Figure 12a. It contains two diffraction spots, but they are too faint to be identified by the current version of AutoNDE. The code picks the diffraction spots up only when we cut off minimum 20% of the specimen thickness from the bottom of the image, see Figure 12b. The latter displays the following estimates of the crack characteristics: extent—4 mm, depth—16 mm, orientation —100°. The necessity to reduce the region of interest appears to be due to the defect in the backwall, which produces a response that is too bright. After the region of interest is cropped as described, AutoNDE produces the report reproduced in Figure

13. The parameters of the manufactured notch are as follows: extent—5 mm, depth—16 mm, orientation—101°. We can see that the AutoNDE estimates are of the same quality as estimates obtained with the DPS data.

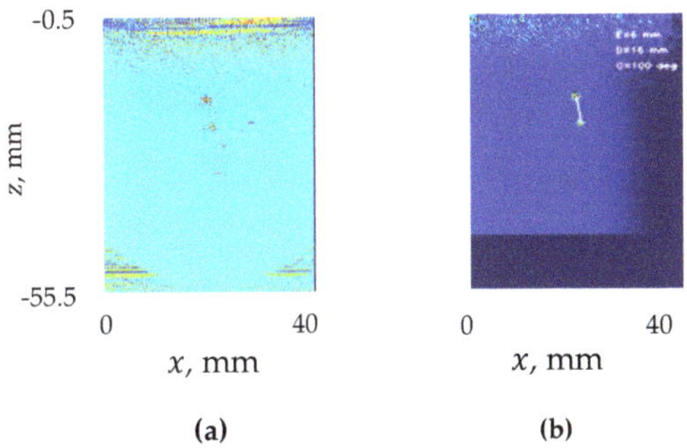

Figure 12. The AMEC data set. (**a**) The standard TFM image of the full specimen; (**b**) an MTFM image, with 20% thickness cut off the bottom, intensity threshold = 125, Dm = 3.

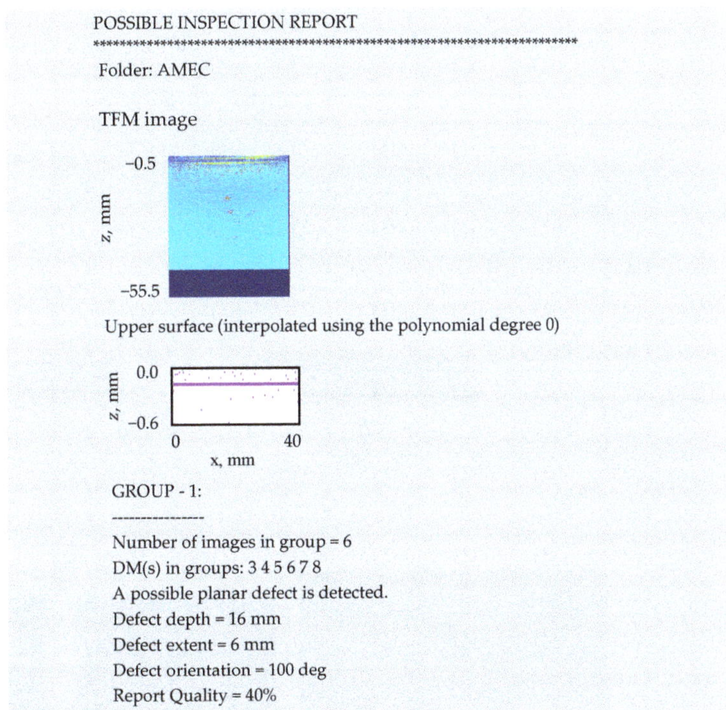

POSSIBLE INSPECTION REPORT

****

Folder: AMEC

TFM image

Upper surface (interpolated using the polynomial degree 0)

GROUP - 1:

Number of images in group = 6
DM(s) in groups: 3 4 5 6 7 8
A possible planar defect is detected.
Defect depth = 16 mm
Defect extent = 6 mm
Defect orientation = 100 deg
Report Quality = 40%

Figure 13. The inspection report for the AMEC dataset.

4.3. Testing AutoNDE on CEA Data

AutoNDE has been also tested on a data set collected by CEA engineers using a 64 element phased transducer array, with the pitch D_e = 0.6 mm and sampling frequency f_s = 50 MHz, imaging in immersion a 42 mm deep steel specimen. The geometry of the experiment is presented in Figure 14, the input and typical A-scans are similar to the ones in the DPS experiment and are not reproduced.

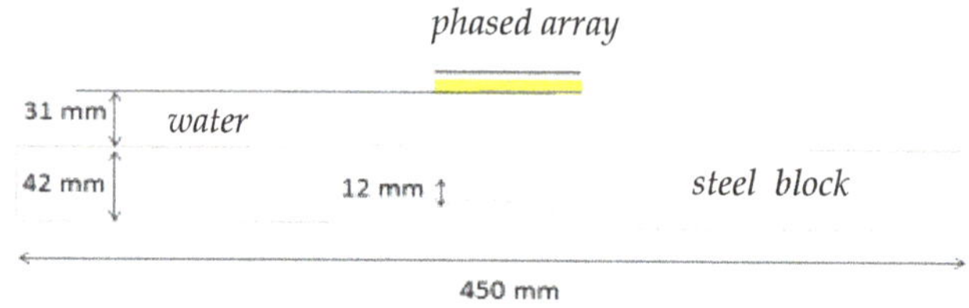

Figure 14. A schematic of the CEA experiment.

The set-up is more challenging than the one used in the DPS experiments: both the top surface and blackwall are more indulated and there is a backwall breaking flaw, creating additional response, see Figures 15b and 16a. The numerical experiments conducted with CIVA [22] (a commercial package for analyzing and simulating NDT [23–28], in particular, NDT of components with irregular surfaces [7,8,28,29]) showed that the notch fabricated for the purposes of this experiment was best imaged using the half-skip LTT mode, with the L (longitudinal) transmitted pulse converting at the backwall to the T (transverse) pulse and then reflecting from the notch, so that the received pulse is also T, see Figure 16a.

Note that the CEA experiment was designed to investigate the effect of highly undulated surfaces, with the distribution of undulations different to normal. Both types of surfaces, those reproduced in the CEA experiment and those reproduced in the DPS experiment are realistic, but our analysis confirmed that they have to be modeled differently. The CIVA code was provided with precise descriptions of both the inspection surface and backwall obtained with a flexible probe. AutoNDE relied instead on a rather crude *Profiling* submodule described above. Moreover, the offset of 2095 samples in A-scans obtained with the 64 element transducer array was too high, eliminating reflections from the higher portions of the inspection surface. For this reason, the quality of the *Profiling* output was very low. A trial and error approach was used to establish that the best results could be obtained when the upper surface was assumed to be plane and the backwall was represented by a parabola, cf. Figure 15a,b with the surfaces in the AutoNDE report presented in Figure 17.

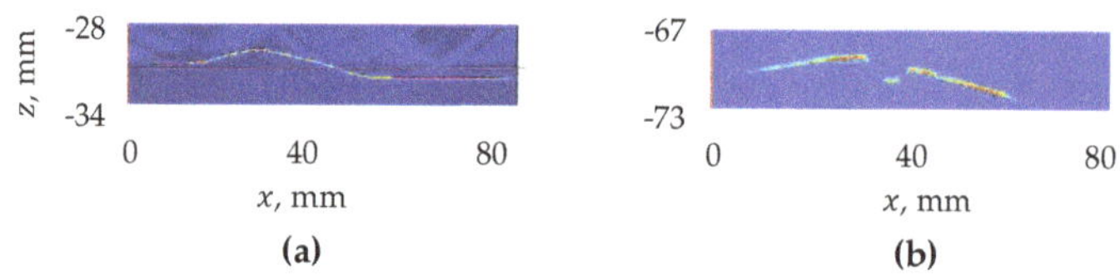

Figure 15. CEA experimental specimen. (**a**) The upper surface and (**b**) the backwall.

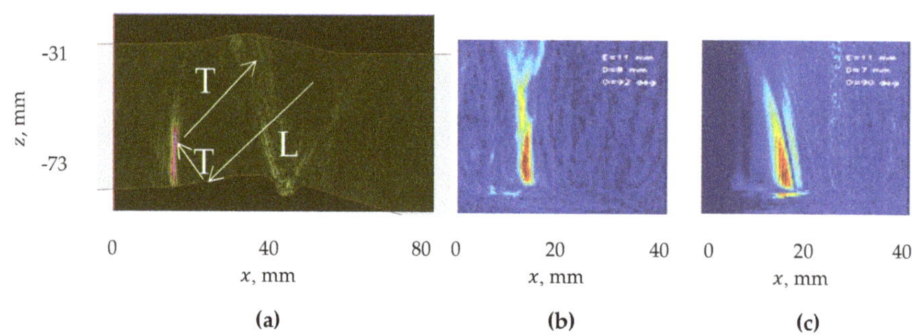

Figure 16. CEA data set. The backwall notch imaged with a 64 element transducer array, using the half-skip LTT mode as processed by (**a**) CIVA and AutoNDE, (**b**) employing the standard TFM algo-rithm and (**c**) MTFM (intensity threshold = 185 and Dm = −8).

This report was obtained using the partial probe of 33 transducerelements, that is of 19.8 mm aperture. All other parameter values were the same as described in the previous sections. Inspecting the TFM image confirms that the second group provides a more reliable characterization of the 12 mm surface-breaking notch normal to the backwall. The depth is overestimated due to distortions introduced by crude of the surfaces.

POSSIBLE INSPECTION REPORT

****

Folder: CEA64

TFM image

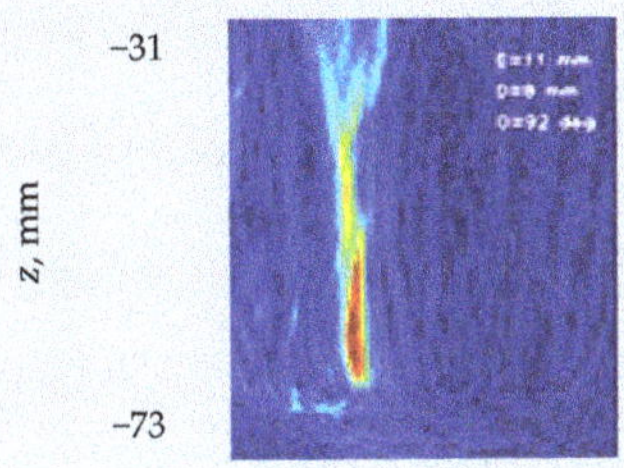

Upper surface (interpolated using the polynomial degree 0)

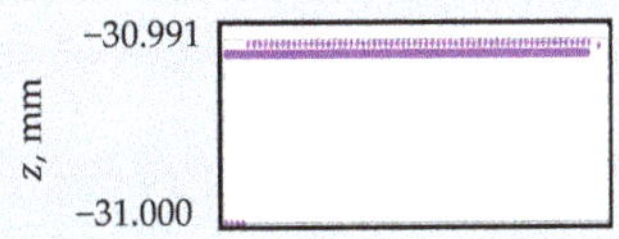

Backwall (interpolated using the polynomial degree 2)

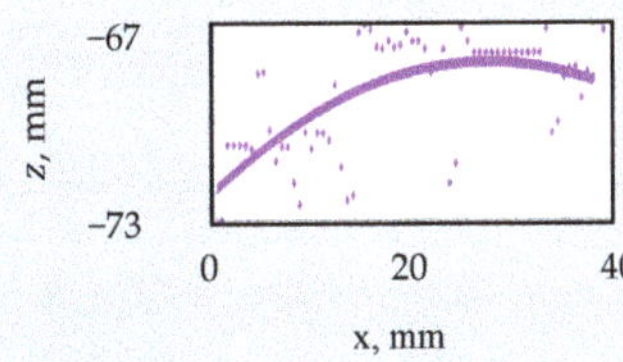

GROUP-1:

Number of images in group = 8
DM(s) in groups: 16 17 18 19 23 24 25 26
A possible planar defect is detected.
Defect depth = 1 mm
Defect extent = 8 mm
Defect orientation = 110 deg
Report Quality = 20%

GROUP-2:

Number of images in group = 7
DM(s) in groups: -31 -4 -5 -6 -7 -8 -9
A possible planar defect is detected.
Defect depth = 7 mm
Defect extent = 12 mm
Defect orientation = 95 deg
Report Quality = 20%

Figure 17. The inspection report for the CEA dataset.

5. Conclusions

A novel code containing a decision tree, that is, an explainable AI has been designed and developed for characterizing single large planar cracks. For the component surfaces whose undulation errors can be described using a normal distribution, we developed a method for automatic estimation of the degree of the interpolating polynomial. The code has been trained on 16 experimental data sets and tested on two. The inspection surface and backwall used in training had realistic small undulations whose distribuation could be considered normal. One test dataset was collected using a specimen with plane surfaces and another, a specimen with surfaces whose undulations were smooth and large and could not be described using a normal distribution.

It has been demonstrated that every type of material and inspection configuration requires preliminary investigation to establish not only how to model the surfaces but also most appropriate values of such hyperparameters as the component thickness, distance to the probe and portion of the image to be analyzed. Numerous other parameters described in this paper have been optimized manually. Remarkably, they perform well on all datasets described in the paper. It is important to realize that in some configurations only one crack tip can be picked up and in others no crack localization is possible.

Once suitable parameters and limitations are established the code can be used to generate possible inspection reports. These contain an assessment of their own quality based on the subjective probability of the report being correct. The probabiltiy is calculated by analysing a variety of images (rather than one) produced by a particular modification of the TFM offered in this paper. It is expected that the human inspectors would still have to examine the AutoNDE reports, particularly the TFM images they contain, to ascertain whether they agree with the preliminary conclusions made by the AI module.

Despite the initial success reported here, just like any other artificial intelligence system, the code can be guaranteed to analyze well only the type of data used for its training, so that, say, the random undulations of the component surface follow the same probability distribution as in the training data set. Also, so far AutoNDE has been trained to process only the regions of interest, which contain one crack or else several cracks parallel to the inspection surface. It is clear that many more data sets are required for testing AutoNDE before it is accepted by the NDT community as a practical tool. To widen the AutoNDE applicability we have plans to automate the choice of the hyperparameters described above too. It is also clear that other methodologies have to be developed for modeling surfaces with undulations that do not obey a normal distribution. Notwithstanding these challenges, AutoNDE shows a great promise, demonstrating feasibility of an explainable AI, suitable for applications in industrial NDE, increasing its accuracy and efficiency.

Author Contributions: Conceptualization, methodology, validation, writing, supervision, project administration, funding acquisition, L.F.; software, S.U.A. and L.F.; data interpretation, review and editing, M.D. All authors have read and agreed to the published version of the manuscript.

Funding: The research was supported by the Innovate UK grants nos. 131671, 132928 and 2553 as well as the EURIKA/Innovate UK grant no 106151.

Data Availability Statement: Restrictions apply to the availability of these data. Data was obtained from Doosan Power Systems, AMEC and CEA are available from L.F. with the permission of these organisations.

Acknowledgments: The authors are grateful to the Innovate UK monitoring officers Tom Harris and Chris Marshall for their critical feedback and constant support and to the colleagues from the Ultrasonics and NDT group, Bristol University for providing the AMEC data and Sébastien Robert for the CEA data acquisition. They are also grateful to Nicolas Pignet, who undertook translation of the original LabView code into C when a second year ENSEIRB-MATMECA engineering student seconded under the Internship Agreement to Sound Mathematics Ltd., funded by a European Union Erasmus grant. Finally the authors would like to thank anonymous reviewers for many helpful suggestions on improving the presentation of the paper.

Conflicts of Interest: The authors declare no conflict of interest.

References

1. Schneider, C.; Bird, C. Reliability of Manually Applied Phased Array Inspection. In Proceedings of the 4th European-American Workshop on Reliability NDE, Berlin, Germany, 24–26 June 2009.
2. Hunter, A.J.; Drinkwater, B.W.; Wilcox, P.D. Least-squares estimation of imaging parameters for an ultrasonic array using known geometric image features. *IEEE Trans. Ultrason. Ferroelectr. Freq. Control.* **2011**, *58*, 414–426. [CrossRef]
3. Marengo, E.A.; Gruber, F.K.; Simonetti, F. Time-Reversal MUSIC Imaging of Extended Targets. *IEEE Trans. Image Process.* **2007**, *16*, 1967–1984. [CrossRef] [PubMed]
4. Champagnat, F.; Goussard, Y.; Idier, J. Unsupervised deconvolution of sparse spike trains using stochastic approximation. *IEEE Trans. Signal Process.* **1996**, *44*, 2988–2998. [CrossRef]
5. Holmes, C.; Drinkwater, B.W.; Wilcox, P. Advanced post-processing for scanned ultrasonic arrays: Application to defect detection and classification in non-destructive evaluation. *Ultrasonics* **2008**, *48*, 636–642. [CrossRef] [PubMed]
6. Hunter, A.J.; Drinkwater, B.W.; Wilcox, P. The wavenumber algorithm for full-matrix imaging using an ultrasonic array. *IEEE Trans. Ultrason. Ferroelectr. Freq. Control.* **2008**, *55*, 2450–2462. [CrossRef] [PubMed]
7. Iakovleva, E.; Chatillon, S.; Bredif, P.; Mahaut, S. Multi-mode TFM imaging with artifacts filtering using CIVA UT forwards models. *AIP Conf. Proc.* **2014**, *1581*, 72–79. [CrossRef]
8. Le Jeune, L.; Robert, S.; Villaverde, E.L.; Prada, C. Plane Wave Imaging for ultrasonic non-destructive testing: Generalization to multimodal imaging. *Ultrasonics* **2016**, *64*, 128–138. [CrossRef]
9. Bannouf, S.; Robert, S.; Casula, O.; Prada, C. Noise filtering in the total focusing method by decomposition of the time reversal operator and the virtual array approach. *AIP Conf. Proc.* **2013**, *1511*, 857–864. [CrossRef]
10. Villaverde, E.L.; Robert, S.; Prada, C. Ultrasonic imaging of defects in coarse-grained steels with the decomposition of the time reversal operator. *J. Acoust. Soc. Am.* **2016**, *140*, 541–550. [CrossRef]
11. Robert, S.; Casula, O.; Roy, O.; Neau, G. Real-time nondestructive testing of composite aeronautical structures with a self-adaptive ultrasonic technique. *Meas. Sci. Technol.* **2013**, *24*, 074011. [CrossRef]
12. Pyle, R.J.; Rhodri, L.T.; Bevan, R.R.; Hughes, R.K.; Rachev, A.A.S.A.; Wilcox, P.D. Deep Learning for Ultrasonic Crack Characterization in NDE. *IEEE Trans. Ultrason. Ferroelectr. Freq. Control* **2021**, *68*, 1854–1865. [CrossRef] [PubMed]
13. Bai, L.; Le Bourdais, F.; Miorelli, R.; Calmon, P.; Velichko, A.; Drinkwater, B.W. Ultrasonic Defect Characterization Using the Scattering Matrix: A Performance Comparison Study of Bayesian Inversion and Machine Learning Schemas. *IEEE Trans. Ultrason. Ferroelectr. Freq. Control.* **2021**, *68*, 3143–3155. [CrossRef] [PubMed]
14. Shipway, N.; Huthwaite, P.; Lowe, M.; Barden, T. Using ResNets to perform automated defect detection for Fluorescent Penetrant Inspection. *NDT E Int.* **2021**, *119*, 102400. [CrossRef]
15. Nageswaran, C. The Snooker Algorithm for Ultrasonic Imaging of Fatigue Cracks in order touse Parameter-Spaces to Aid Machine Learning. *Preprints* **2021**, 2021070269. [CrossRef]
16. Fradkin, L.; Zernov, V.; Elston, G.; Taneja, R.; Bell, I.; Lines, D.; Wharrie, J.; Fitzgerald, P.J. Towards semi-automated crack characterisation. *J. Phys. Conf. Ser.* **2013**, *457*, 012008. [CrossRef]
17. Badeau, N.; Painchaud-April, G.; Le Duff, A. Use of the Total Focusing Method with the Envelope Feature. Available online: https://www.olympus-ims.com/en/resources/white-papers/use-of-the-total-focusing-method-with-the-envelope-feature/ (accessed on 1 November 2021).
18. Fahrmeir, L.; Kneib, T.; Lang, S.; Marx, B. *Regression: Models, Methods and Applications*; Springer: Berlin, Germany, 2013.
19. Cheney, W.; Kincaid, D. *Numerical Mathematics and Computing*, 7th ed.; BROOKS/COLE Sengale Learning: Boston, MA, USA, 2013.
20. Macon, B.; Spitzbart, X. Inverses of Vandermonde Matrices. *Am. Math. Mon.* **1958**, *65*, 95–100. [CrossRef]
21. Home—OpenCV. Available online: https://opencv.org (accessed on 1 November 2021).
22. Available online: http://www.extende.com/ (accessed on 1 November 2021).
23. Mahaut, S.; Chatillon, S.; Darmon, M.; Leymarie, N.; Raillon, R.; Calmon, P. An Overview of Ultrasonic Beam Propagation and Flaw Scattering Models in the Civa Software. *AIP Conf. Proc.* **2010**, *1211*, 2133–2140.
24. Toullelan, G.; Raillon, R.; Chatillon, S. Results of the 2013 UT modeling benchmark obtained with models implemented in CIVA. *AIP Conf. Proc.* **2014**, *1581*, 2093–2100. [CrossRef]
25. Raillon-Picot, R.; Toullelan, G.; Darmon, M.; Calmon, P.; Lonné, S. Validation of CIVA Ultrasonic Simulation in Canonical Configurations. In Proceedings of the 18th World Conference on Non-destructive Testing (WCNDT), Durban, South Africa, 16–20 April 2012.
26. Raillon, R.; Bey, S.; Dubois, A.; Mahaut, S.; Darmon, M. Results of the 2010 Ut Modeling Benchmark Obtained with Civa: Responses of Backwall And Surface Breaking Notches. *AIP Conf. Proc.* **2011**, *1335*, 1777–1784. [CrossRef]
27. Raillon, R.; Bey, S.; Dubois, A.; Mahaut, S.; Darmon, M. Results of the 2009 Ut Modeling Benchmark Obtained with Civa: Responses of Notches, Side-drilled Holes and Flat-bottom Holes of Various Sizes. *AIP Conf. Proc.* **2010**, *1211*, 2157–2164. [CrossRef]
28. Ferrand, A.; Darmon, M.; Chatillon, S.; Deschamps, M. Modeling of ray paths of head waves on irregular interfaces in TOFD inspection for NDE. *Ultrasonics* **2014**, *54*, 1851–1860. [CrossRef] [PubMed]
29. Darmon, M.; Dorval, V.; Baqué, F. Acoustic Scattering Models from Rough Surfaces: A Brief Review and Recent Advances. *Appl. Sci.* **2020**, *10*, 8305. [CrossRef]

applied sciences

MDPI

Correction

Correction: Fradkin et al. Towards Explainable Augmented Intelligence (AI) for Crack Characterization. *Appl. Sci.* 2021, *11*, 10867

Larissa Fradkin [1,*], Sevda Uskuplu Altinbasak [1,†] and Michel Darmon [2]

[1] Sound Mathematics Ltd., 11 Mulberry Close, Cambridge CB4 2AS, UK; sevda.uskuplu@gmail.com
[2] Université Paris-Saclay, CEA, List, F-91120 Palaiseau, France; Michel.DARMON@cea.fr
* Correspondence: l.fradkin@soundmathematics.com
† Current Address: ARM, 110 Fulbourn Road, Cambridge CB1 9NJ, UK.

Citation: Fradkin, L.; Uskuplu Altinbasak, S.; Darmon, M. Correction: Fradkin et al. Towards Explainable Augmented Intelligence (AI) for Crack Characterization. *Appl. Sci.* 2021, *11*, 10867. *Appl. Sci.* **2022**, *12*, 1043. https://doi.org/10.3390/app12031043

Received: 27 December 2021
Accepted: 4 January 2022
Published: 20 January 2022

Error in Figure

The authors wish to make the following corrections to their paper [1]:

In the original article, there was a mistake in Figure 5 as published. The "*x*" was not in the correct place. The corrected Figure 5 appears below.

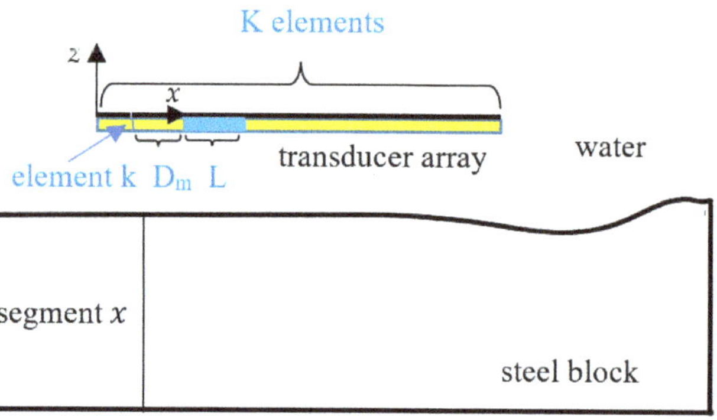

The authors also wish to replace all figures with higher clarity versions and make sure that Figure 6 does not look unnaturally enlarged.

Text Correction

Corrections have been made as follows:

In Section 3.1, the authors wish to change the phrase "an approach" on line 7 of paragraph 4 (page 5) to "a method"; change the word "it" on the first line of paragraph 5 (page 5) to "this method"; change the word "Wald test" on the first line of paragraph 6 (page 5) to "method"; add the sentence "The spline method has been tried too but was found to be too sensitive to the choice of smoothing parameters and thus not amenable to automation." after paragraph 6 (page 5); replace the word "Finally" with "Both" on line 14 of paragraph 9 (page 7).

In Section 3.2, the authors wish to replace "The code" on line 5 of paragraph 1 (page 7) to a more detailed word "AutoNDE"; modify the phrase "we normally consider" in paragraph 2 (page 7) to "AutoNDE normally ultilizes"; modify the typo "We analyse" to "AutoNDE analyze" in paragraph 3 (page 7).

In Section 3.3, the authors wish to change the word "us" on the last line of point 3 (page 9) to "AutoNDE"; replace the word "defect" on line 2 of point 4 (page 9) to "crack".

In Section 4.1 (page 10), the authors wish to add the phrase "by trial and error" after the word "established" on line 7 of paragraph 1; add the phrase "whatever the dataset" after the word "Finally" on line 12 of paragraph 1; replace the word "These" with "The experimental values" on line 16 of paragraph 1.

In Section 4.2 (page 13), the authors wish to replace the word "technicians" on the first line of paragraph 1 with "engineers"; add the word "minimum" before "20%" on line 3 of paragraph 2.

In Section 4.3, the authors wish to replace the sentence "(The French Alternative Energies and Atomic Energy Commission)" on the first line of paragraph 1 (page 14) to "engineers"; replace all words "signal" in paragraph 2 (page 15) with "pulse".

In Section 5 (page 17), the authors wish to move up the last sentence in paragraph 1 to line 2 as the second sentence. The sentence is "For the component surfaces whose undulation errors can be described using a normal distribution, we developed a method for automatic estimation of the degree of the interpolating polynomial".

The authors apologize for any inconvenience caused and state that the scientific conclusions are unaffected.

Reference

1. Fradkin, L.; Uskuplu Altinbasak, S.; Darmon, M. Towards Explainable Augmented Intelligence (AI) for Crack Characterization. *Appl. Sci.* **2021**, *11*, 10867. [CrossRef]

applied sciences

MDPI

Article

Quantitative Inspection of Complex-Shaped Parts Based on Ice-Coupled Ultrasonic Full Waveform Inversion Technology

Wenjin Xu, Maodan Yuan *, Weiming Xuan, Xuanrong Ji and Yan Chen

State Key Laboratory of Precision Electronic Manufacturing Technology and Equipment, Guangdong University of Technology, Guangzhou 510006, China; w.j.xu@mail2.gdut.edu.cn (W.X.); xwm772446573@163.com (W.X.); xr.ji@gdut.edu.cn (X.J.); yanchen@gdut.edu.cn (Y.C.)

* Correspondence: mdyuan@gdut.edu.cn; Tel.: +86-151-1262-5974

Featured Application: Nondestructive testing and structural health monitoring for complex-shaped parts with a high resolution.

Abstract: Ultrasonic methods have been extensively developed in nondestructive testing for various materials and components. However, accurately extracting quantitative information about defects still remains challenging, especially for complex structures. Although the immersion technique is commonly used for complex-shaped parts, the large mismatch of acoustic impedance between water and metal prevents effective ultrasonic transmission and leads to a low signal-to-noise ratio(SNR). In this paper, a quantitative imaging method is proposed for complex-shaped parts based on an ice-coupled full waveform inversion (FWI) method. Numerical experiments were carried out to quantitatively inspect the various defects in a turbine blade. Firstly, the k-space pseudospectral method was applied to simulate ice-coupled ultrasonic testing for the turbine blade. The recorded full wavefields were then applied for a frequency-domain FWI based on the Limited-memory Broyden–Fletcher–Goldfarb–Shanno (L-BFGS) method. With a carefully selected iterative number and frequency, a successive-frequency FWI can well detect half wavelength defects. Extended studies on an open notch with different orientations and multiple adjacent defects proved its capability to detect different types of defects. Finally, an uncertainty analysis was conducted with inaccurate initial velocity models with a relative error of ±2%, demonstrating its robustness even with a certain inaccuracy. This study demonstrates that the proposed method has a high potential to inspect complex-shaped structures with an excellent resolution.

Keywords: ice-coupled ultrasonic testing; full waveform inversion; complex structure; defect detection

Citation: Xu, W.; Yuan, M.; Xuan, W.; Ji, X.; Chen, Y. Quantitative Inspection of Complex-Shaped Parts Based on Ice-Coupled Ultrasonic Full Waveform Inversion Technology. *Appl. Sci.* **2021**, *11*, 4433. https://doi.org/10.3390/app11104433

Academic Editor: Habil. Michel Darmon

Received: 5 April 2021
Accepted: 10 May 2021
Published: 13 May 2021

1. Introduction

Rapid development in the fields of computer technology, manufacturing techniques and materials science makes the geometry of key components more complex. For example, instead of an assembly of a disk and removable blades, the integratedbladed disk (BLISK) technology has been widely applied in the aeroengine design due to the merits of a light weight, high aerodynamic efficiency, excellent fatigue resistance and high efficiency of fuel consumption [1]. Recently, such monocoque components have been conveniently designed and fabricated by the emerging additive manufacturing (AM), also known as 3D printing technology. Although AM has an excellent ability to manufacture complex-shaped parts with an optimal structure design, the quality of manufactured products may be influenced by buckling, residual stress and internal defects during the cooling process [2]. Moreover, there is a higher tendency to bring in defects in complex-shaped parts due to a much greater surface area. Therefore, to guarantee the safety of these key components, special attention should be paid to the quantitative nondestructive testing (NDT) of complex-shaped parts. X-ray computed tomography (XCT) is widely used to inspect such parts in an industrial application. Although XCT possesses a high sensitivity to volumetric defects such as gas pores and inclusions, it is insensitive to planar defects such as cracks and delamination [3]. In addition, a high energy X-ray source is required for thick metals, resulting in radiation hazards and expensive equipment.

On the other hand, ultrasound plays an irreplaceable role in NDT with the advantages of a large penetration depth, large area coverage and high sensitivity to both internal and surface defects. Both a single transducer with a scanning system and a transducer array have been widely applied to develop methods to visualize ultrasonic results with various imaging algorithms. Synthetic imaging methods including a common source method (CSM) [4,5], a synthetic aperture focusing technique (SAFT) and a total focusing method (TFM) [6,7] have been well developed to produce high resolution images for the objects below the measurement based on the wave reflections. However, when the surface of the tested object is uneven or irregular, the deployment of contact transducers will be difficult and poor coupling will lead to inaccurate or false detection. Currently, there are three common schemes to tackle this issue. The first method is to integrate the transducer with a wedge or "shoe" to match the surface to allow direct contact. This method is indeed simple but a wedge is only suitable for one specific surface. When the surface profile is varying or too complex, this method is difficult or even impossible to apply. The second method is to use flexible transducers [8], which show a greater compatibility than wedges. However, a few potential defects may go undetected in the dead zone because of the near field. The third method is to place the tested object into a water-filled chamber and the water acts as an acoustic couplant [9]. As a whole water-filled regular structure, objects with highly curved surfaces can be inspected. However, there is a large difference of acoustic impedance between the water and metal materials that hinders ultrasonic energy from effectively penetrating into the tested metal object. Thus, the performance is limited for metal parts with complex shapes.

Recently, to bridge the gap between the immersion method and the contact method, Simonetti et al. proposed a cryo-ultrasonic method to inspect complex-shaped parts, which were encased in ice with a cylinder container [10]. As the longitudinal wave speed of ice is about 2.5 times that of water, the acoustic impedance mismatch with the metal material can be greatly reduced so that the transmission energy will significantly increase. Their preliminary results demonstrated its superiority compared with the immersion method. Furthermore, they presented experimental schemes to avoid the formation of crack and bubble nucleation when water is freezing [11]. To reconstruct the internal complex structure of such an ice-encased part, traditional focusing imaging methods are no longer suitable. Focusing imaging can precisely localize the defects but it is difficult to obtain the information of the shape and appearance. Moreover, these methods ignore multiple reflection signals and are unable to obtain the information of lower and side surfaces. Simonetti et al. applied a diffraction mitigation technique to carry out ultrasonic imaging/tomography [10,11]. However, the performance of diffraction tomography is limited due to its assumption of the first-order diffraction. It can achieve a better resolution than ray tomography but it is only suitable for low-contrast defects. Therefore, to quantitatively evaluate real defects with arbitrary forms in complex-shaped structures, a full wavefield containing all of the possible linear and higher-order diffraction effects should be considered in the tomography.

To solve the inversion problem with a recorded full wavefield, full waveform inversion (FWI) tomography was firstly derived in geophysics for seismic wave imaging [12–15]. The FWI is performed via an iterative minimization of the misfit between an experimentally measured wavefield and a numerically calculated wavefield generated with a forward model. Based on a certain optimization algorithm, the estimated model of the tested object is iteratively updated. The iteration process will end when the convergence criterion is satisfied; then, the final model is the approximation of the tested object. FWI automatically considers higher-order scattering effects; thus, it has been widely studied in medical applications for breast cancer screening and diagnosis [16], bone quantitative imaging [17] and delamination detection in civil infrastructures [18,19]. Recently, FWI has also been applied in guided wave technology to quantify corrosion defects in plate-like structures [20,21]. In this article, a frequency-domain FWI is introduced as an imaging algorithm for the quantitative defect detection in a complex structure combined with ice-coupled ultrasonic testing.

The rest of this paper is structured as follows. In Section 2, the fundamental theory of FWI including forward modelling and the successive-frequency FWI procedure will be briefly introduced. The numerical experiment setup and data calibration of the ice-coupled ultrasonic testing for turbine blades are then illustrated in Section 3. The reconstructed results for various defects are shown and discussed in Section 4. Finally, concluding remarks are drawn in Section 5.

2. Theory of Full Waveform Inversion

A detailed description of the FWI algorithm can be found in the literature [13,15]; thus, only the key techniques and steps are briefly introduced here. Generally, FWI imaging can be regarded as a composition of three segments: forward modeling to generate calculated data by solving a wave equation, the computation of the gradient and Hessian matrix of the misfit function and updating the model with a suitable optimization method.

2.1. Forward Modelling for Ultrasonic Wave Propagation

Reliable forward modelling for wave propagation is a primary prerequisite for ultrasonic FWI tomography. Ultrasonic wave propagation in solids is typically described by an elastic wave equation [22]. To boost the whole efficiency of FWI, the scalar acoustic wave equation is usually used to approximate the ultrasonic wave propagation [17]. The formalism of a 2D acoustic wave equation in a frequency-domain is expressed as:

$$(\nabla^2 + \omega^2 m^2)\psi(r,\omega) = -s(r,\omega) \tag{1}$$

where $\nabla^2 = \frac{\partial^2}{\partial x^2} + \frac{\partial^2}{\partial z^2} = \partial_x^2 + \partial_z^2$ is the Laplace operator, ω is the angular frequency, $m = 1/c$ is the slowness, c is the wave speed, $r = (x,z)$ is the spatial position in the Cartesian coordinate system and $\psi(r,\omega)$ is the displacement wavefield. $s(r,\omega) = s(\omega)\delta(r - r_s)$ is the excitation signal where $s(\omega)$ refers to the Fourier transform of the source, $\delta(r - r_s)$ denotes the Dirac function and r_s is the excitation position.

The finite difference technique is the most common method to solve partial differential equations [23,24] and is applied in forward modelling here. In order to simplify the denotation, Equation (1) can be expressed by the matrix equation $\mathbf{F}\boldsymbol{\psi} = \mathbf{s}$, where $\mathbf{F} = \nabla^2 + \omega^2 m^2$ is a $l \times l$ complex-valued impedance matrix, which is related to the frequency, material properties, discrete format and boundary conditions. $\boldsymbol{\psi}$ and $\mathbf{s}$ are $l \times 1$ column vectors at the nodal points of the discrete grid.

To reduce the computation cost, the LU factorization is utilized to solve the wavefield $\boldsymbol{\psi}$:

$$\mathbf{LU}\boldsymbol{\psi} = \mathbf{s} \tag{2}$$

where $\mathbf{L}$ and $\mathbf{U}$ denote the lower and upper triangular matrices from the LU factorization. As the solution is reusable for different sources to solve the forward problem, multiple source problems can be handled efficiently [25]. To alleviate the computation burden and avoid reflection interference, a perfectly matched layer (PML) boundary condition is applied around the computational domain to absorb outgoing waves in a similar configuration as done in [26].

2.2. Full Waveform Inversion

The aim of inversion is to find a model parameter $\mathbf{m} = [m_1, m_2, \ldots, m_{n_x n_z}]^t$ in the model space to reduce the residual between the calculated and measured data. A least-square norm is chosen to define this misfit as the object function:

$$\min_m E(\mathbf{m}) = \frac{1}{2} \parallel P_r(\boldsymbol{\psi}) - \mathbf{d} \parallel^2 = \frac{1}{2}(\Delta\mathbf{d}^t) * \Delta\mathbf{d} \tag{3}$$

where P_r is the sampling operator to extract the wavefield at the receivers from the forward model. $\boldsymbol{\psi} = \{\psi_i\}$, $i = (1, 2, \ldots, n_r)$ and $\mathbf{d} = \{d_i\}$, $i = (1, 2, \ldots, n_r)$ are the calculated and measured data vectors, respectively. $\Delta\mathbf{d} = \{P_r(\psi_i) - d_i\}$, $i = (1, 2, \ldots, n_r)$ is the residual vector. The superscript t is the vector or matrix transposed and $*$ is the complex conjugate to convert the object function to real numbers.

Various optimization algorithms can be applied to minimize the object function $E(\mathbf{m})$. First and foremost, the gradient of $E(\mathbf{m})$ should be defined with respect to the model parameters such as density, wave speed and attenuation. It offers the direction to iteratively update the vector space to minimize the misfit [27]. Based on Taylor's theorem, we have:

$$E(\mathbf{m} + \delta\mathbf{m}) \approx E(\mathbf{m}) + \mathbf{g} \cdot \delta\mathbf{m} + \frac{1}{2}\delta\mathbf{m}^t \cdot \mathbf{H} \cdot \delta\mathbf{m} = P(\delta\mathbf{m}) \tag{4}$$

where $\mathbf{g} = \frac{\partial E(\mathbf{m})}{\partial \mathbf{m}} = \Re e(\mathbf{J}^t \Delta\mathbf{d}^*)$ is the gradient with a size of $l \times 1$. $\mathbf{J} = [\frac{\partial \psi_p(m_q)}{\partial m_q}], p = (1, 2, \ldots, n_r); q = (1, 2, \ldots, n_x n_z)$ is the $n_r \times l$ order Fréchet derivative matrix and $\Re e$ denotes the real part of the complex value. $\mathbf{H} = \frac{\partial^2 E(\mathbf{m})}{\partial \mathbf{m}^2}$ is the Hessian matrix. To minimize the object function $E(\mathbf{m})$, let $\frac{dP(\delta\mathbf{m})}{d(\delta\mathbf{m})} = \mathbf{g} + \mathbf{H}\delta\mathbf{m}$ be zero then $\delta\mathbf{m} = -\mathbf{H}^{-1} \cdot \mathbf{g}$. Therefore, the model can be updated as:

$$\mathbf{m}_{k+1} = \mathbf{m}_k - \varepsilon\mathbf{H}^{-1}\mathbf{g} \tag{5}$$

where ε is the step length. When only the first-order derivative is considered, the corresponding method is called the steepest descent method. It is a straightforward way, but its convergence rate is so slow that it limits the application for large-scale problems [17]. To boost the convergence, the second-order derivative should be taken into consideration. However, the computation amount and storage space increase sharply because of the calculation of the inverse Hessian matrix $\mathbf{H}^{-1}$. To handle this problem, quasi-Newton methods have been developed to approximate the Hessian matrix. The limited-memory Broyden–Fletcher–Goldfarb–Shanno (L-BFGS) method is one of the most popular methods. This method calculates the approximation of the inverse Hessian matrix by a limited number (n_{lim}) of the gradient-difference and model-difference vectors associated with the n_{lim} latest iteration (n_{lim} is generally defined between 3 and 20) [28].

At every iteration $\hat{n}_{lim} = \min\{n_{lilm}, k-1\}$, correction pairs $\{\boldsymbol{\alpha}_i, \boldsymbol{\beta}_i\}, i = k - \hat{n}_{lim}, k - \hat{n}_{lim} + 1, \ldots, k-1$ are stored, where $\boldsymbol{\alpha}_i = \mathbf{m}_{i+1} - \mathbf{m}_i$, $\boldsymbol{\beta}_i = \mathbf{g}_{i+1} - \mathbf{g}_i$. Let $\gamma_i = \frac{1}{\boldsymbol{\beta}_i^t \boldsymbol{\alpha}_i}$, $\boldsymbol{\eta}_i = (\mathbf{I} - \gamma_i \boldsymbol{\beta}_i \boldsymbol{\alpha}_i^t)$, $\mathbf{B} = \mathbf{H}^{-1}$, if $k + 1 <= l$, then

$$\begin{aligned} \mathbf{B}_k &= \boldsymbol{\eta}_{k-1}{}^t \mathbf{B}_{k-1} \boldsymbol{\eta}_{k-1} + \gamma_{k-1} \boldsymbol{\alpha}_{k-1} \boldsymbol{\alpha}_{k-1}{}^t \\ &= \boldsymbol{\eta}_{k-1}^t \boldsymbol{\eta}_{k-2}^t \cdots \boldsymbol{\eta}_0^t \mathbf{B}_k^0 \boldsymbol{\eta}_0 \cdots \boldsymbol{\eta}_{k-2} \boldsymbol{\eta}_{k-1} + \boldsymbol{\eta}_{k-1}^t \boldsymbol{\eta}_{k-2}^t \cdots \boldsymbol{\eta}_1^t \gamma_0 \boldsymbol{\alpha}_0 \boldsymbol{\alpha}_0^t \boldsymbol{\eta}_1 \cdots \boldsymbol{\eta}_{k-2} \boldsymbol{\eta}_{k-1} + \ldots + \boldsymbol{\eta}_{k-1}^t \gamma_{k-2} \boldsymbol{\alpha}_{k-2} \boldsymbol{\alpha}_{k-2}^t \boldsymbol{\eta}_{k-1} + \gamma_{k-1} \boldsymbol{\alpha}_{k-1} \boldsymbol{\alpha}_{k-1}^t \end{aligned} \tag{6}$$

If $k + 1 > l$, the iterative expression is given by:

$$\begin{aligned} \mathbf{B}_k &= \boldsymbol{\eta}_{k-1}{}^t \mathbf{B}_{k-1} \boldsymbol{\eta}_{k-1} + \gamma_{k-1} \boldsymbol{\alpha}_{k-1} \boldsymbol{\alpha}_{k-1}{}^t \\ &= \boldsymbol{\eta}_{k-1}^t \boldsymbol{\eta}_{k-2}^t \cdots \boldsymbol{\eta}_{k-l}^t \mathbf{B}_k^0 \boldsymbol{\eta}_{k-l} \cdots \boldsymbol{\eta}_{k-2} \boldsymbol{\eta}_{k-1} + \boldsymbol{\eta}_{k-1}^t \boldsymbol{\eta}_{k-2}^t \cdots \boldsymbol{\eta}_{k-l+1}^t \gamma_{k-l} \boldsymbol{\alpha}_{k-l} \boldsymbol{\alpha}_{k-l}^t \boldsymbol{\eta}_{k-l+1} \cdots \boldsymbol{\eta}_{k-2} \boldsymbol{\eta}_{k-1} \cdots + \boldsymbol{\eta}_{k-1}^t \gamma_{k-2} \boldsymbol{\alpha}_{k-2} \boldsymbol{\alpha}_{k-2}^t \boldsymbol{\eta}_{k-1} + \gamma_{k-1} \boldsymbol{\alpha}_{k-1} \boldsymbol{\alpha}_{k-1}^t \end{aligned} \tag{7}$$

In practical cases, the initial guess of $\mathbf{B}_k^0$ can be computed by $\mathbf{B}_k^0 = \frac{\boldsymbol{\beta}_k{}^t \boldsymbol{\alpha}_k}{||\boldsymbol{\beta}_k||_2^2}\mathbf{I}$, where $\mathbf{I}$ is the identity matrix [28]. With the approximation of the inverse Hessian matrix, it is easy to update the model with Equation (5). When proper convergence criteria or the maximum iteration numbers are set, the final model can be obtained as the inversion results.

The flow chart of the FWI algorithm is illustrated in Figure 1. When only one frequency is applied to the update model, no external loop is needed to perform and this method is called monochromatic inversion. In actual ultrasonic testing, the ultrasonic signal contains not just a single frequency. When several frequencies are used simultaneously during the inversion, the robustness will be improved but it comes at the cost of an increased computational time. In this case, they can be classified as a successive-frequency inversion and a simultaneous multi-frequency inversion. In this research, the successive-frequency inversion is applied, as shown in Figure 1.

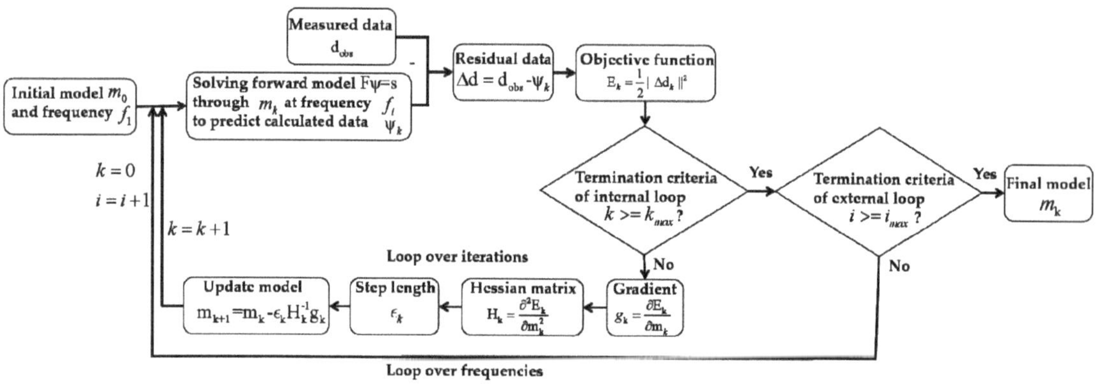

Figure 1. Workflow of the ultrasonic full waveform inversion algorithm.

3. Numerical Experiments on a Turbine Blade

To verify the feasibility and performance of the proposed ice-coupled ultrasonic FWI tomography to evaluate the defects in complex-shaped parts, a series of numerical experiments were conducted. The turbine blade is a typical safety-critical component in the aerospace industry with a complex geometry, as shown in Figure 2a. In conventional ultrasonic testing, phased array ultrasonic transducers can only be placed to directly contact smooth surfaces at the upper part. It makes the defects at the lower part difficult to inspect, such as the red circles in Figure 2a. To inspect the defects at different positions, the transducer should be moved to different positions, shown as the red line. This complex object was applied in this study. To save computational costs and storage space, only 2D tomography was carried out for a typical cross section of the turbine blade in this study, as shown in Figure 2b. To facilitate the fast acquisition of a full ultrasonic wavefield for FWI tomography, an acoustic model rather than an elastic model was adopted here to simulate the process of ice-coupled ultrasonic testing because the acoustic model is commonly employed in ultrasonic NDT research [21,29]. All of the wave phenomena including multiple reflections from the ice-steel interface and the wave-defect interaction can be revealed in such a model. It is necessary to note that in a real case shear waves could exist in a solid due to mode conversion at the interface and defects. Numerical experiments were carried out to optimize the key parameters such as the frequency selection and iteration number to quantitatively detect different types and numbers of defects in a turbine blade.

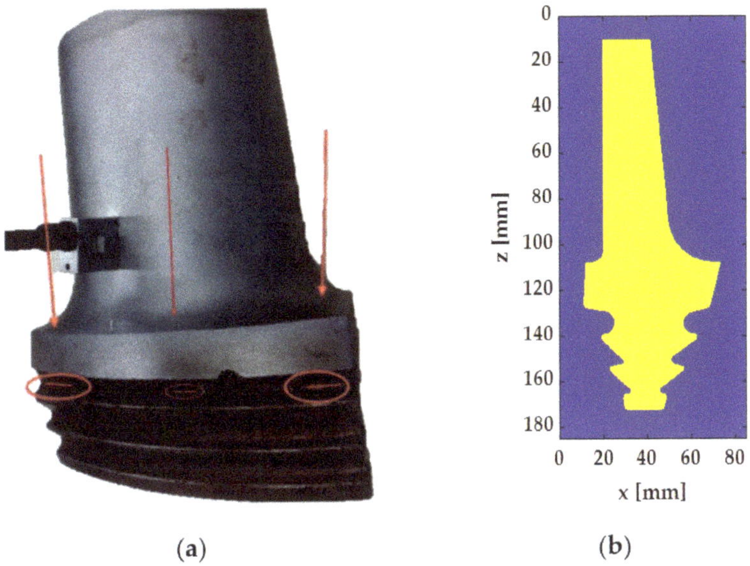

Figure 2. (**a**) Photo of a turbine blade and (**b**) typical cross section.

3.1. *Simulation of Ice-Coupled Ultrasonic Testing in a Turbine Blade*

In this study, an open-source code k-wave toolbox was applied to conduct the numerical simulation of ice-coupled ultrasonic testing to record the full wavefield [30]. This toolbox solves the acoustic wave equation based on a k-space pseudospectral method. Compared with the conventional finite difference method, the memory and computational cost were reduced dramatically.

The turbine blade was encased by ice as shown in Figure 2b. The material properties of the turbine blade and ice are shown in Table 1 [11]. The size of the whole model was 185 mm × 85 mm. A series of ultrasonic transducers were deployed evenly around the turbine blade. Considering that the length of the specimen in the horizontal direction was approximately half that of the vertical direction, 6 transducers were arranged at the top and bottom sides and 12 transducers at the left and right sides, shown in Figure 3. Such an arrangement of transducers can reduce the transducer number and assure the imaging quality. At four corners, the spacing of two adjacent transducers was set as $d_1 = d_2 = 5$ mm. The PML boundary was set around the calculation domain to absorb outgoing waves and avoid reflection. To avoid uncertain factors further from the absorbing layer, all transducers were shifted inward from all sides by 1 mm. It was an effective simplification to simulate the wave propagation in a large domain because the reflections from the boundaries beyond the area of interest could be excluded by applying appropriate time gates to the recorded signals. The spacing of the transducers was $d_x = 14.6$ mm in a horizontal direction and $d_z = 15.73$ mm in a vertical direction.

Table 1. Acoustic properties of the turbine blade and ice [11].

	Mass Density ρ (kg/m^3)	Wave Speed c (m/s)
Steel	7800	5875
Ice (at −5 °C)	917	4000

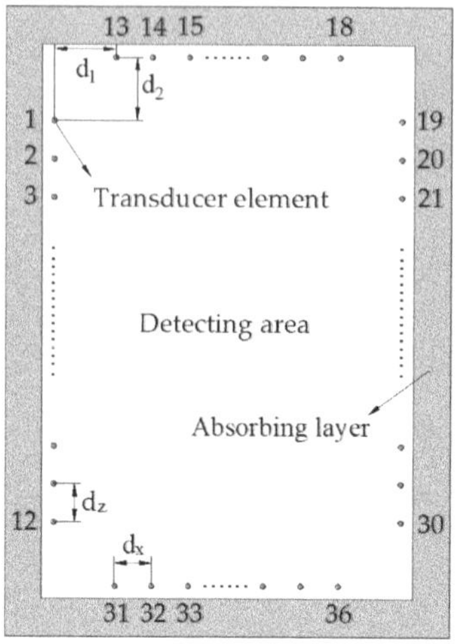

Figure 3. Configuration of the ultrasonic transducer array.

The ultrasonic wave was excited with the following signal:

$$R(t) = [1 - 2\pi^2 f_c^2 (t - t_0)^2] e^{-\pi^2 f_c^2 (t-t_0)^2} \tag{8}$$

where f_c was the center frequency and t_0 referred to the time delay. As the theoretical resolution of the FWI is a half wavelength [20,21], an ultrasonic wave with a center frequency of 1 MHz was applied to inspect a defect as small as 3 mm in the turbine blade. When $f_c = 1\text{MHz}$ and $t_0 = 1.5/f_c$, this excitation waveform and the corresponding amplitude spectrum are depicted in Figure 4. It could be seen that the dominant energy (above the value of a -6 dB drop from the maximum amplitude) was within the effective frequency band of 0.48 MHz–1.6 MHz. A quadrilateral mesh was applied to discretize the spatial domain and the mesh size was carefully chosen to satisfy the Courant–Friedrichs–Lewy (CFL) condition. The mesh size was chosen as $L_{element} = 0.2$ mm in both directions for an accurate acoustic wave calculation up to a frequency of 2 MHz (about 10 elements per minimum wavelength) [31]. To ensure numerical stability, the time step should be chosen carefully according to the smallest element and the wave speed. Here, time step was set as $\Delta t = 0.5 L_{element}/c_{max}$, where c_{max} was the wave speed in the steel.

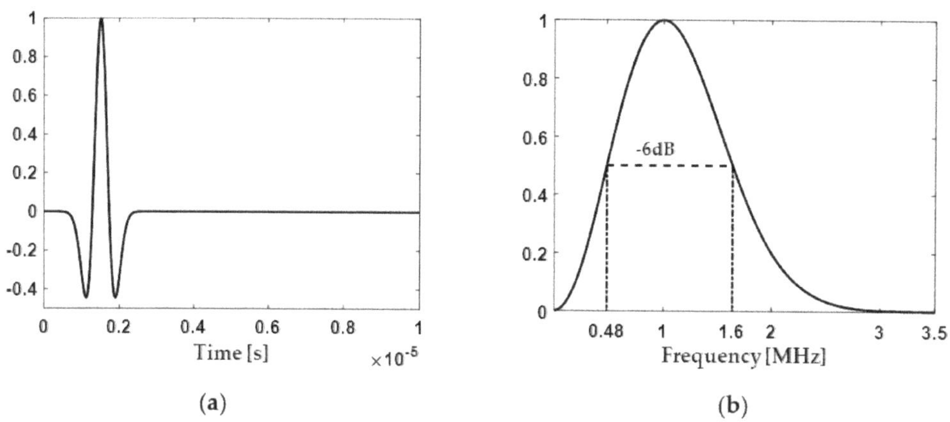

Figure 4. Excitation signal ($f_c = 1$ MHz and $t_0 = 1.5/f_c$). (**a**) Waveform in the time domain and (**b**) the corresponding normalized amplitude spectrum.

The resolution and sensitivity of tomography strongly depend upon 'ray density' [32]. For this reason, a full matrix capture (FMC) technique was adopted to perform the ultrasonic data acquisition, which collected complete datasets including all transmitter-receiver combinations. Therefore, each transducer was excited successively and all transducers recorded the ultrasonic signals simultaneously. Finally, the resulting wavefield consisted of $N_{rec} \times N_{src} = 1296$ waveforms. In a k-wave simulation, a series of $N_{src} = 36$ simulations should be performed independently to obtain this full wavefield. These calculations could be practically and conveniently distributed via the different processors of a cluster. Thus, a parallelization scheme was adopted to accelerate the simulation. The sampling frequency was set to be approximately 60 MHz to satisfy the Nyquist theorem. The total time in the simulation was set as 110 μs to guarantee that the receiver at one corner could receive the reflected signal from the opposite corner.

3.2. Data Calibration

As the inversion procedure was performed in the frequency-domain, the recorded full wavefield should be transformed into the frequency-domain by utilizing a fast Fourier transform (FFT). It should be noted that all full waveform datasets were transformed directly and there was no need to identify and truncate direct waves collected by the receivers.

To obtain an accurate inversion result, the frequency components of the observed or recorded data must match those of the calculated data from the forward model. There was some inevitable deviation of the amplitude and phase spectrum between the recorded data and the calculated data. This deviation may bring a few errors into the inversion procedure. Therefore, a prior data calibration should be performed to match the recorded data and calculated data with an intact specimen. The correction factors for amplitude and phase were defined as [21]:

$$Q_{amp} = \frac{abs(fft(\psi_0))}{abs(fft(\mathbf{d}_{obs,0}))} \tag{9}$$

and:

$$Q_{pha} = \arg(fft(\psi_0)) - \arg(fft(\mathbf{d}_{obs,0})) \tag{10}$$

where $abs()$ represented the modulus of a complex value, arg() denoted the phase of a complex value and $fft()$ was the fast Fourier transform. ψ_0 and $\mathbf{d}_{obs,0}$ were the calculated and observed data for the intact specimen. With these two factors, the recorded amplitude and phase were firstly calibrated for the following inversion process.

4. Results and Discussion

4.1. Single Internal Defect

At first, a single internal defect was embedded in a turbine blade to perform data processing and a parametric study of ultrasonic FWI tomography. For simplicity, a circular hole with a diameter of 3 mm was located at (40,120). In the first simulation, an ultrasonic wave was excited with the first transducer located at (1,6). The corresponding wave propagations at different time slots are shown in Figure 5. From the simulated wavefield, the ultrasonic wave reflected at the ice-steel interface and diffracted with the defect. At $t = 6$ μs, the incident wave began to reach the edge of the blade. The reflection from the ice-steel interface can be seen in Figure 5b, indicated by a red arrow. The transmitted wave and several interface waves propagated inside the blade. When $t = 21$ μs, the direct wave arrived at the receiving transducers as shown in Figure 5c. At $t = 26$ μs, the propagating waves interacted with the defect, generating diffracted waves, which can be seen in Figure 5d. Finally, the ultrasonic wavefield was recorded by all of the 36 transducers. Without a loss of generality, the waveform recorded by the 19th receiver placed at (84,6) and its normalized amplitude spectrum are shown in Figure 6. The first wave package was the direct wave from the 1st transducer to the 19th transducer. In addition, there were two more waves due to the reflection at the boundary and the diffraction from the internal defects. The path of the diffraction wave from the defect is depicted in Figure 5d with blue dashed lines. The corresponding time-of-flight was about 26 μs. Thus, the second wave package was the diffracted wave from the defect, followed by multiple waves reflected from the lower boundary of the blade. All of the reflections from the ice-steel interface and the diffraction signal from the defect can be seen in the A-scan signal in Figure 6a. Due to the complex geometry of the turbine blade, reflection signals from the boundaries were complicated and the energy was very high, making the diffracted wave from the internal defect difficult to identify. Moreover, compared with the excitation, the amplitude spectrum was distorted to a certain degree due to the existence of wave scattering phenomena, as shown in Figure 6b. Nevertheless, the recorded signal still contained information at frequencies around 1 MHz with a large bandwidth.

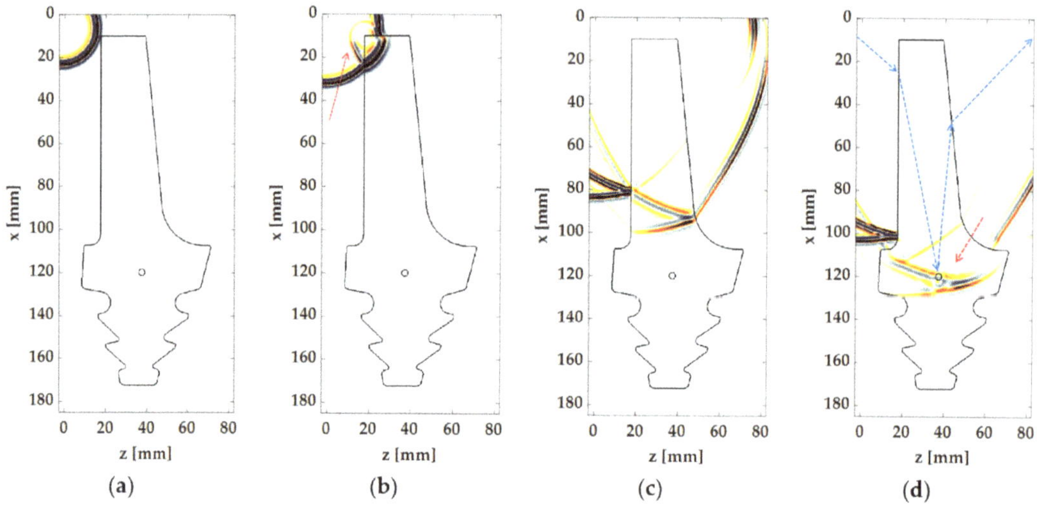

Figure 5. Ultrasonic wave propagation excited with the 1st transducer at different time slots: (**a**) t = 6 μs; (**b**) = 8 μs; (**c**) = 21 μs; (**d**) = 26 μs.

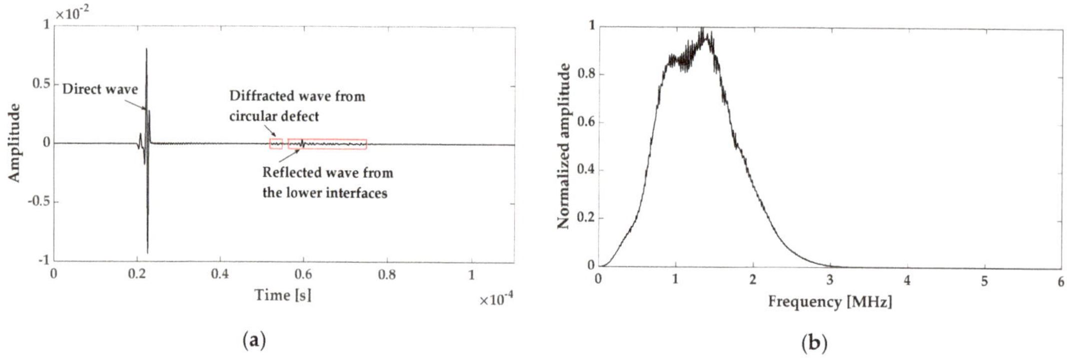

Figure 6. The recorded signal of the 19th transducer excited with the 1st transducer. (**a**) The A-scan waveform and (**b**) the corresponding amplitude spectrum.

The number of iterations is one of the key parameters in the inversion process. With the increase of iterations, the accuracy of the inversion result will be improved but the computing time will increase linearly. Therefore, it is important to determine the number of iterations to balance the accuracy and the computation cost. The variation of the misfit function, as described by Equation (3), as a function of iterations is given in Figure 7.

Obviously, the misfit monotonically decreases as the iterative number increases. After 20 iterations, the misfit was small enough (lower than 1%), indicating that the accuracy of the inversion result was high enough. The FWI tomographs for the different iterations are also shown in Figure 8. When the iteration was set to as few as $k = 5$, the quality of the FWI image was poor and the defect was difficult to observe. After 10 iterations, the defect became clearly visible but the detailed information of the reconstructed defect changed. When the number of iterations was larger than 20, the inversion results became stable. Hence, the number of iterations was set as 20 for the following studies. Although the inversion result converged after 20 iterations, there was still a remarkable noise in the final result.

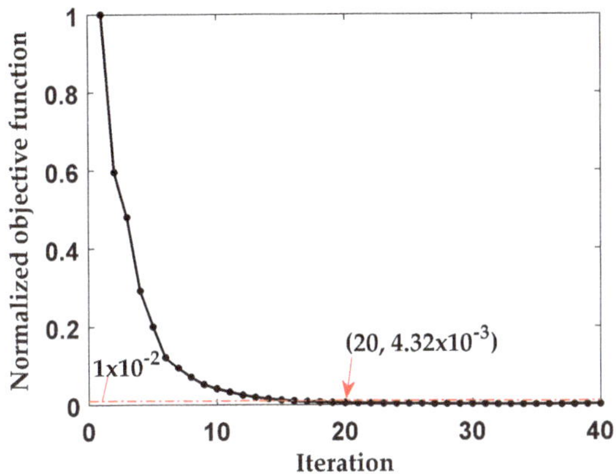

Figure 7. The evolution of the objective function with the number of iterations.

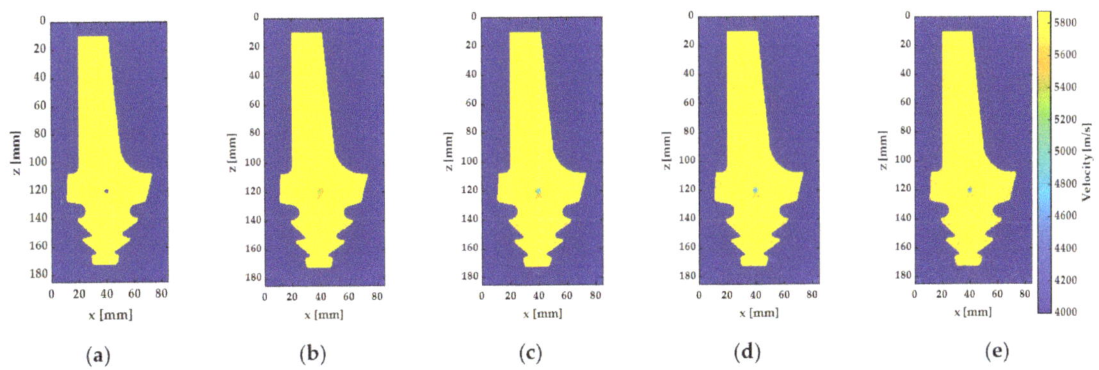

Figure 8. (**a**) True model with a 3 mm circular hole and tomographs with different iterations: (**b**) $k = 5$; (**c**) $k = 10$; (**d**) $k = 20$; (**e**) $k = 40$.

To improve the overall accuracy of the inversion, information of different frequencies should be taken into consideration. A lower frequency has a better tolerance for a relatively large-scale structure while a higher frequency should be considered for a small substructure [15]. Therefore, successive inversions with a moving frequency from low to high were applied to mitigate the nonlinearity of the inversion problem, resulting in a more robust result than a monochromatic inversion. In this study, a frequency group of seven discrete frequencies from 400 kHz to 1600 kHz were adopted because the excited and received signals showed an effective bandwidth of 480 kHz−1600 kHz, as shown in Figure 6b. Figure 9 shows the reconstructed tomographs when the inversion finished at frequencies of 0.4 MHz, 0.6 MHz, 0.8 MHz, 1 MHz, 1.2 MHz, 1.4 MHz and 1.6 MHz, respectively.

Compared with the monochromatic inversion of 1 MHz, the overall noise was reduced based on the successive inversion of low frequencies. It could be seen that with the increase of the frequency of successive inversions, the accuracy of the 3 mm diameter circular hole was improved. The cross sections through the defect along the horizontal and vertical direction are presented in Figure 10. The FWI results showed a good agreement with the profile of the true defect in both directions. To quantitatively evaluate the reconstruction accuracy, the relative error of the full width at half maximum (FWHM) along the two directions along with their root mean square (RMS) error were considered. Therefore, the three indicators were defined as $E_V = \frac{FWHM_V}{D_{def}}$, $E_H = \frac{FWHM_H}{D_{def}}$ and $E_{RMS} = \sqrt{\frac{E_V^2+E_H^2}{2}}$, where D_{def} denoted the diameter of the circular defect. Figure 11 shows the variations of the relative errors of the FWHM and their RMS error as a function of the successive-frequency inversions for circular holes with different diameters. Generally, the errors decreased as the successive-frequency increased and almost converged after the frequency of 1 MHz. It meant that successive-frequency inversion could indeed improve the accuracy of FWI tomography. Especially for the 3 mm diameter defect, the relative errors sharply decreased from 28.3% to 8.5%. When the defects were larger than 3 mm, the relative errors decreased to smaller than 2%. The relatively large error for the 3 mm diameter defect was due to the fact that 3 mm is close to half of the interrogating wavelength at 1 MHz, which is the theoretical resolution of the FWI [22]. However, the final error was smaller than 8%, indicating that a successive-frequency FWI could obtain an accurate tomograph for internal defects and the resolution approached a half wavelength.

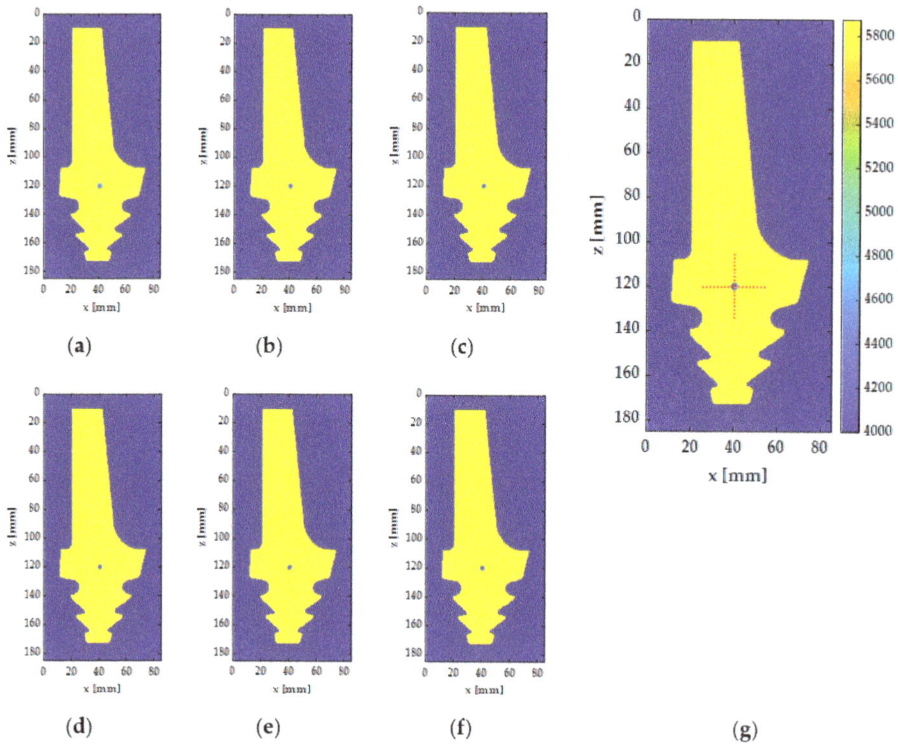

Figure 9. Tomographs of the 3 mm circular hole based on successive-frequency inversions at: (**a**) 0.4 MHz; (**b**) 0.6 MHz; (**c**) 0.8 MHz; (**d**) 1 MHz; (**e**) 1.2 MHz; (**f**) 1.4 MHz; (**g**) 1.6 MHz.

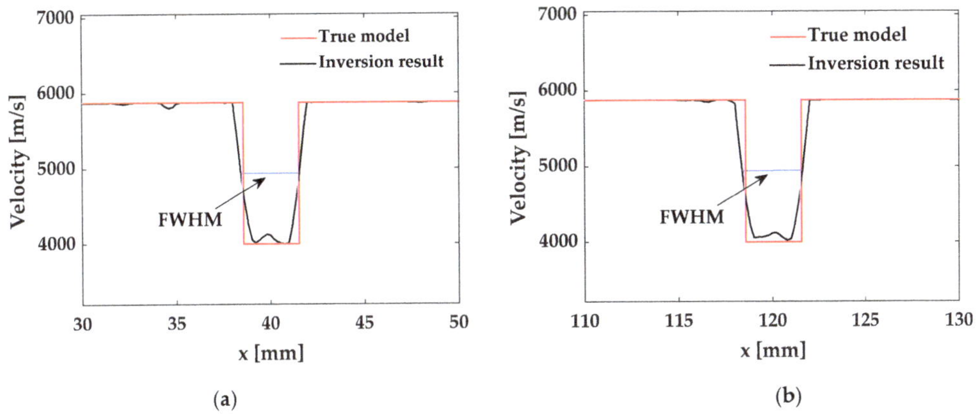

Figure 10. Cross sections of the reconstruction of the 3 mm diameter circular hole along the (**a**) horizontal and (**b**) vertical direction.

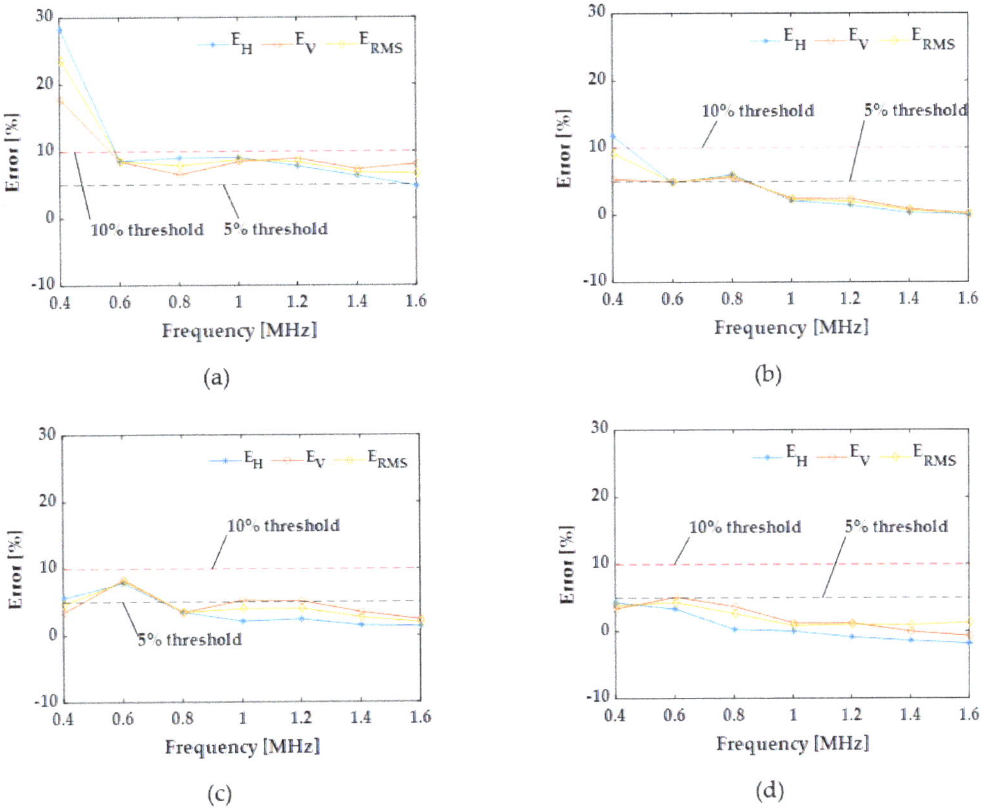

Figure 11. The variation of relative errors of the FWHM as a function of successive-frequency inversions for circular holes: (**a**) 3 mm; (**b**) 4 mm; (**c**) 5 mm; (**d**) 6 mm.

4.2. Open Notch Crack

In addition to the internal defects, a turbine blade also tends to suffer from surface cracks. Therefore, investigations were also conducted to evaluate the surface cracks based on the proposed method. Open notches with different orientations were introduced to mimic the surface crack. The size of the notch was set as a constant at 5 mm long and 2 mm wide. Only the influence of the orientation was studied and three directions ($\theta = 0^\circ, 45^\circ, 90^\circ$) were investigated here. To investigate the capability of FWI to detect open notch defects with different angles, three numerical experiments were conducted. The transducer configuration and inversion scheme remained the same as those in the previous cases. The FWI tomographs of the three notches are shown in Figure 12a,c,e. Generally, the successive-frequency FWI was able to reconstruct the surface crack with a high resolution. Similarly, to quantitatively evaluate the accuracy, relative errors defined as the profiles along the width and length were represented as E_S and E_L. The trends of the FWHM errors with a successive-frequency are presented in Figure 12b,d,f. E_L decreased significantly when the frequency was smaller than 1 MHz and converged to around 10%. However, different from the result of the circular defect in Section 4.1, E_S was generally much smaller than E_L and it increased with the successive-frequency. The reason might lie in the fact that ultrasound is much more capable of reconstructing the plane defects. Moreover, when the orientation of the crack was 45°, the final error was slightly higher than the other two cases. It implied that the alignment of the transducer network probably introduced some inaccuracy into the FWI results.

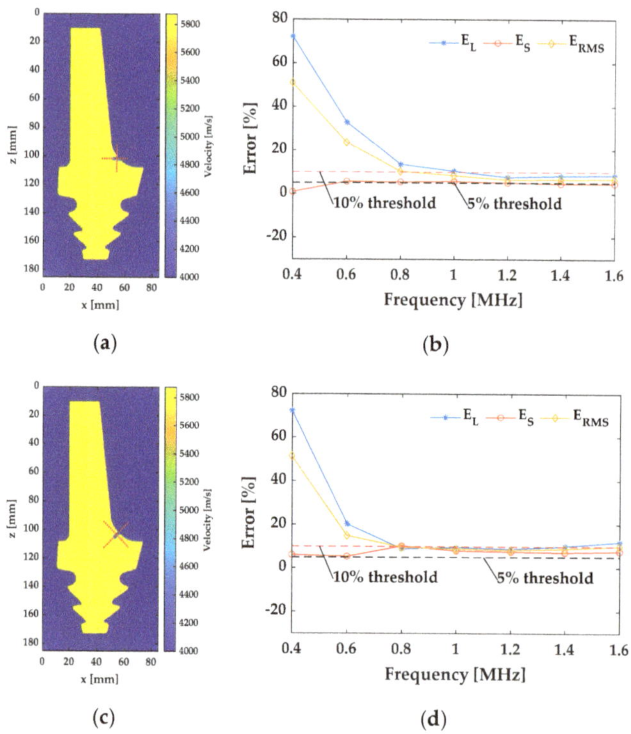

Figure 12. *Cont.*

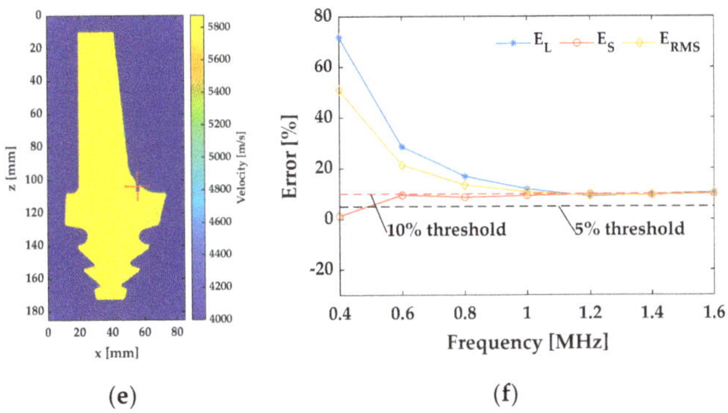

Figure 12. Inversion results for open notch cracks. Tomograph and error variation for (**a**,**b**) the horizontal notch, (**c**,**d**) the inclined notch at 45° and (**e**,**f**) the vertical notch.

4.3. Multiple Defects

In practical engineering applications, a component probably contains multiple defects simultaneously. It should be pointed out that it is a difficult problem for an accurate inspection because multiple scattering will occur if the distance of the two defects is closed. Hence, three circular defects with different sizes were considered, as shown in Figure 13a. A central defect (marked as #1 defect) was located at (40,120) with a diameter of 3 mm. The second one with a diameter of 4 mm (marked as #2 defect) was horizontally shifted by 6 mm from #1 defect and the third one with a diameter of 5 mm (marked as #3 defect) was vertically shifted by 6 mm from #1 defect.

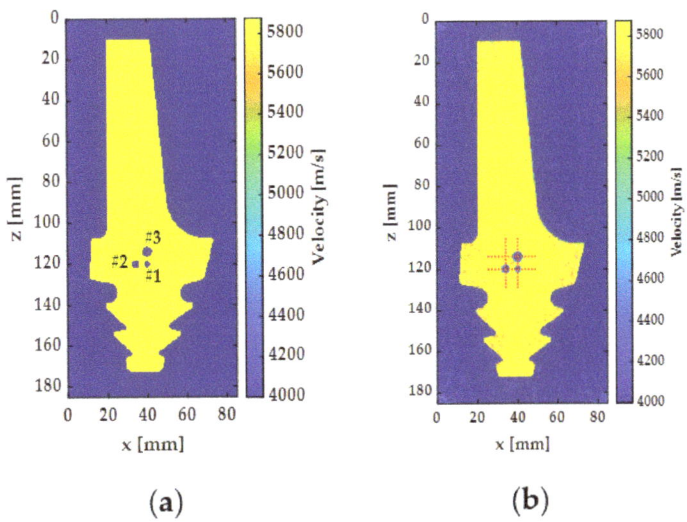

Figure 13. (**a**) True model and (**b**) FWI tomograph of three defects.

Similarly, a successive-frequency FWI was applied and the reconstructed defects are shown as Figure 13b. From the tomograph, these three circular defects were clearly discernable despite complex wave scattering due to the presence of the three adjacent defects. Furthermore, both vertical and horizontal profiles were compared with the true model as shown in Figure 14. The good agreement proved that the proposed method was capable of accurately detecting multiple defects in complex parts.

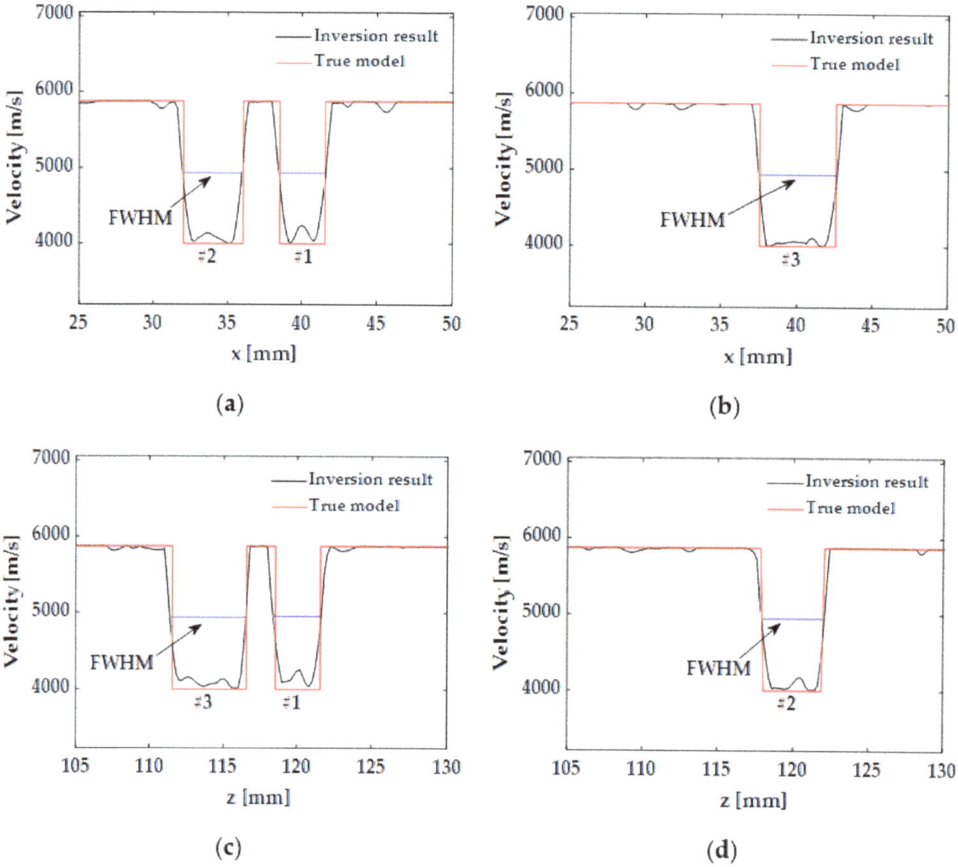

Figure 14. Cross sections (**a**) through #1 and #2 defects and (**b**) through #3 defect along the horizontal direction. Cross sections (**c**) through #1 and #3 defects and (**d**) through #2 defect along the vertical direction.

In addition, it is much more common that different types of defects might exist in the structure at the same time. In order to extend this study to more practical applications, an additional case study was also carried out to evaluate the structure with coexisting circular holes and cracks. Here, the circular holes were set as the same as Figure 13a and an incline notch of 45° was considered. The inversion results are presented in Figure 15. The FWI tomograph clearly showed the notch crack and the disperse holes. Moreover, due to the fact that the crack was relatively distant from the dispersed circular holes, the cross section profiles were very close to the results with only holes or cracks. The good agreement demonstrated that the proposed method was able to inspect different types of defects at the same time with a satisfactory accuracy.

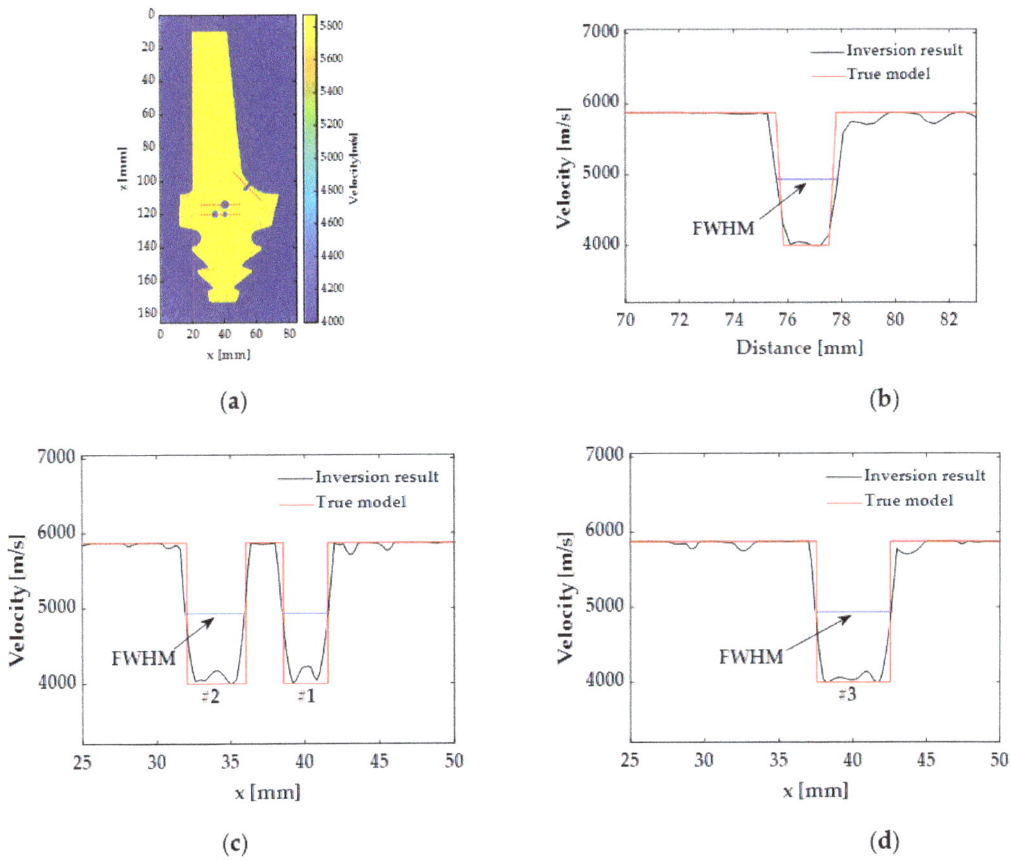

Figure 15. (**a**) FWI tomograph of three circular holes and a notch defect. Cross section (**b**) through the incline notch along the width direction, (**c**) through #1 and #2 defects along the horizontal direction and (**d**) through #3 defect along the horizontal direction. The positions are depicted in (**a**) with red dashed lines.

4.4. Uncertainty Analysis

Although a successive-frequency FWI shows a high accuracy in quantitative defect evaluation in a complex structure, an accurate or reliable initial model is very important for FWI. In previous studies, the exact wave speed was applied in the initial model for the inversion process. However, accurate material properties are not always available and might vary with the practical environments. Therefore, an uncertainty analysis should be conducted to assess the robustness of the proposed method. Here, two initial models with a −2% and +2% error for the velocity of the turbine blade were considered. Without a loss of generality, an internal defect with a diameter of 4 mm, slightly larger than the half wavelength, was introduced for FWI.

The tomograph results are shown in Figure 16a,c. In both cases, this defect could be accurately reconstructed. Compared with Figure 9 in Section 4.1, the reconstruction results showed more background noise. From the comparison of the vertical and horizontal profiles, as shown in Figure 16b,d, the RMS error of the FWHM was 5.42% for the −2% error in the initial model and 8.23% for the +2% error in the initial model. Compared with the relative error, which was smaller than 2% in Section 4.1, the inaccuracy of the initial model increased the relative error for the inversion. Vice versa, accurate parameters for the initial model could improve the final result. Nevertheless, the overall error was smaller than 10% and it was acceptable for the practical application in NDT even with a slight variation in the initial model.

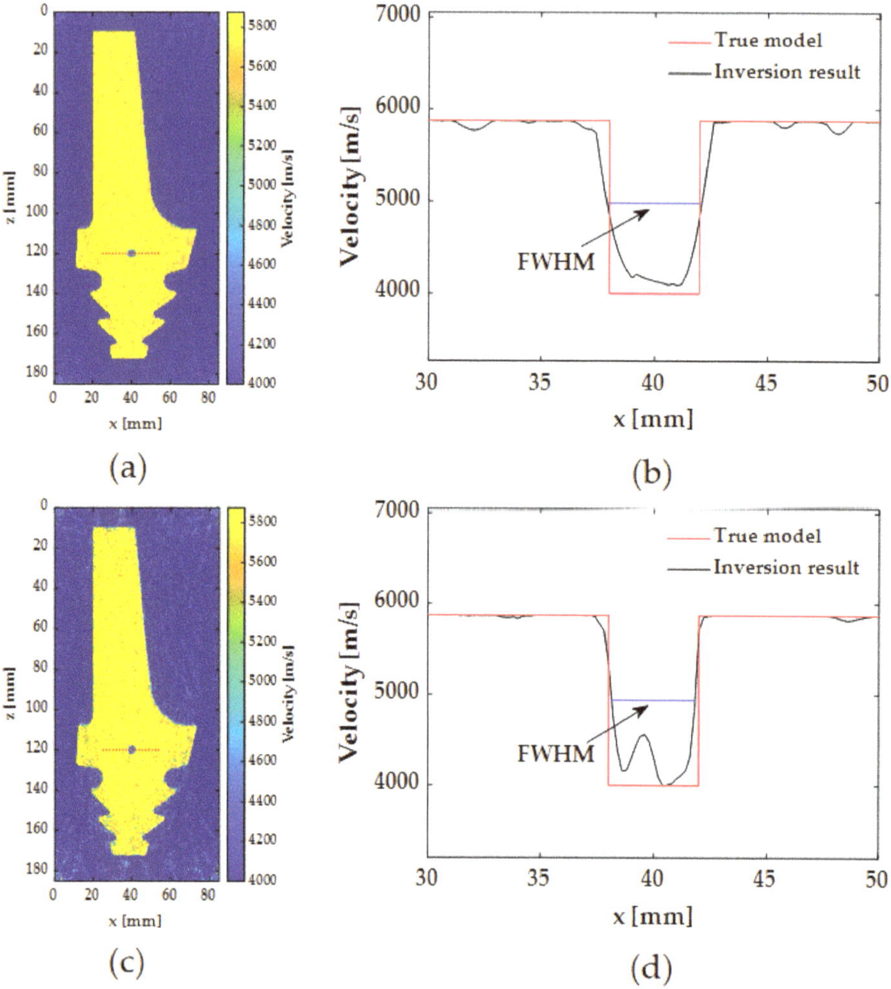

Figure 16. Inversion results with an inaccurate initial model. (**a**) Tomograph and (**b**) horizontal profile for an initial model with a −2% error. (**c**) Tomograph and (**d**) horizontal profile for an initial model with a +2% error.

5. Conclusions

In this paper, an ice-coupled ultrasonic tomography based on successive-frequency full waveform inversion technology was proposed to quantitively inspect the defects in complex-shaped parts. An ice-coupled ultrasound was able to treat the complex-shaped part as a regular structure and sharply reduced the mismatch of acoustic impedance compared with the immersion method. A series of numerical experiments were carried out to demonstrate the feasibility of the proposed method to quantitatively inspect the defects in a turbine blade. Firstly, the k-space pseudospectral method was applied to simulate ice-coupled ultrasonic testing for the turbine blade. A full matrix capture was then performed to record the ultrasonic signals with an evenly deployed transducer array. After data calibration, the recorded full wavefield was applied for a frequency-domain full waveform inversion based on the L-BFGS method. A single circular defect was applied to validate the proposed method and determine the iterative number and frequency. Compared with the monochromatic inversion, successive-frequency FWI could obtain more accurate results and the resolution approached a half wavelength. When the defects were larger than a half wavelength, the relative error was smaller than 2%. Extended studies on open notch cracks with different orientations and multiple adjacent defects proved that the proposed method was able to quantitatively detect various types of defects. Finally, inaccurate initial velocity models with a relative error of $\pm 2\%$ were also investigated. Despite the relative error increasing slightly, the proposed method was able to reconstruct the defects even with a certain inaccuracy. The results demonstrated that the ice-coupled ultrasonic FWI method has a high potential to quantitatively inspect complex-shaped structures with an excellent resolution.

It should be noted that a mode conversion in a practical ultrasonic test was not considered in this paper. In the future, an elastic wave equation should be considered in the forward model during the inversion process to obtain more accurate results and a practical experiment will be conducted.

Author Contributions: Conceptualization, M.Y.; Data curation, W.X. (Wenjin Xu); Formal analysis, W.X. (Wenjin Xu) and W.X. (Weiming Xuan); Funding acquisition, M.Y., X.J. and Y.C.; Investigation, W.X. (Wenjin Xu) and W.X. (Weiming Xuan); Methodology, M.Y.; Project administration, X.J. and Y.C.; Software, W.X. (Wenjin Xu); Supervision, M.Y. and X.J.; Validation, W.X. (Wenjin Xu); Writing—original draft, W.X. (Wenjin Xu) and W.X. (Weiming Xuan); Writing—review & editing, M.Y., X.J. and Y.C. All authors have read and agreed to the published version of the manuscript.

Funding: This research was funded by National Key R&D Program of China (No. 2018YFF01010500), National Natural Science Foundation of China (Nos. 51805097, 51975131, 11804059) and Guangdong Innovative and Entrepreneurial Research Team Program (No. 2016ZT06G375).

Institutional Review Board Statement: Not applicable.

Informed Consent Statement: Not applicable.

Conflicts of Interest: The authors declare no conflict of interest.

References

1. Ravi-Kumar, B.V.R. A review on BLISK technology. *Int. J. Innov. Res. Sci. Eng. Technol.* **2013**, *2*, 1353–1358.
2. Pérez, M.; Carou, D.; Rubio, E.M.; Teti, R. Current advances in additive manufacturing. *Procedia CIRP* **2020**, *88*, 439–444. [CrossRef]
3. Brath, A.J.; Simonetti, F. Phased Array Imaging of Complex-Geometry Composite Components. *IEEE Trans. Ultrason. Ferroelectr. Freq. Control.* **2017**, *64*, 1573–1582. [CrossRef] [PubMed]
4. Fletcher, S.; Lowe, M.J.S.; Ratassepp, M.; Brett, C. Detection of Axial Cracks in Pipes Using Focused Guided Waves. *J. Nondestruct. Eval.* **2011**, *31*, 56–64. [CrossRef]
5. Moreau, L.; Drinkwater, B.W.; Wilcox, P.D. Ultrasonic imaging algorithms with limited transmission cycles for rapid nondestructive evaluation. *IEEE Trans. Ultrason. Ferroelectr. Freq. Control* **2009**, *56*, 1932–1944. [CrossRef] [PubMed]
6. Holmes, C.; Drinkwater, B.; Wilcox, P. The post-processing of ultrasonic array data using the total focusing method. *Insight Non Destr. Test. Cond. Monit.* **2004**, *46*, 677–680. [CrossRef]
7. Le Jeune, L.; Robert, S.; Villaverde, E.L.; Prada, C. Plane Wave Imaging for ultrasonic non-destructive testing: Generalization to multimodal imaging. *Ultrasonics* **2016**, *64*, 128–138. [CrossRef] [PubMed]
8. Ji, X.; Yuan, M.; Chen, Y. Cloud Ultrasound System with Flexible Ultrasonic Phased-Array Transducer for Critical Component Monitoring. In Proceedings of the International Symposium on Structural Health Monitoring and Nondestructive Testing 2018, Saarbruecken, Germany, 4–5 October 2018.
9. Jansen, D.; Hutchins, D. Immersion tomography using Rayleigh and Lamb waves. *Ultrasonics* **1992**, *30*, 245–254. [CrossRef]
10. Simonetti, F.; Satow, I.L.; Brath, A.J.; Wells, K.C.; Porter, J.; Hayes, B.; Davis, K. Cryo-Ultrasonic NDE: Ice–Cold Ultrasonic Waves for the Detection of Damage in Complex-Shaped Engineering Components. *IEEE Trans. Ultrason. Ferroelectr. Freq. Control* **2018**, *65*, 638–647. [CrossRef]
11. Simonetti, F.; Fox, M. Experimental methods for ultrasonic testing of complex-shaped parts encased in ice. *NDT E Int.* **2019**, *103*, 1–11. [CrossRef]
12. Tarantola, A. A strategy for nonlinear elastic inversion of seismic reflection data. *Geophysics* **1986**, *51*, 1893–1903. [CrossRef]
13. Pratt, R.G.; Worthington, M.H. Inverse theory applied to multi-source cross-hole tomography. Part 1: Acoustic wave-equation method. *Geophys. Prospect.* **1990**, *38*, 287–310. [CrossRef]
14. Pan, G.S.; Phinney, R.A.; Odom, R.I. Full-waveform inversion of plane-wave seismograms in stratified acoustic media: Theory and feasibility. *Geophysics* **1988**, *53*, 21–31. [CrossRef]

15. Pratt, R.G. Seismic waveform inversion in the frequency domain, Part 1: Theory and verification in a physical scale model. *Geophysics* **1999**, *64*, 888–901. [CrossRef]
16. Pérez-Liva, M.; Herraiz, J.L.; Udías, J.M.; Miller, E.; Cox, B.T.; Treeby, B.E. Time domain reconstruction of sound speed and attenuation in ultrasound computed tomography using full wave inversiona. *J. Acoust. Soc. Am.* **2017**, *141*, 1595–1604. [CrossRef]
17. Bernard, S.; Monteiller, V.; Komatitsch, D.; Lasaygues, P.L. Ultrasonic computed tomography based on full-waveform inversion for bone quantitative imaging. *Phys. Med. Biol.* **2017**, *62*, 7011–7035. [CrossRef]
18. Nguyen, T.D.; Tran, K.T.; Gucunski, N. Detection of Bridge-Deck Delamination Using Full Ultrasonic Waveform Tomography. *J. Infrastruct. Syst.* **2017**, *23*, 04016027. [CrossRef]
19. Nguyen, L.T.; Modrak, R.T. Ultrasonic wavefield inversion and migration in complex heterogeneous structures: 2D numerical imaging and nondestructive testing experiments. *Ultrasonics* **2018**, *82*, 357–370. [CrossRef]
20. Rao, J.; Ratassepp, M.; Fan, Z. Limited-view ultrasonic guided wave tomography using an adaptive regularization method. *J. Appl. Phys.* **2016**, *120*, 194902. [CrossRef]
21. Rao, J.; Ratassepp, M.; Fan, Z. Guided Wave Tomography Based on Full Waveform Inversion. *IEEE Trans. Ultrason. Ferroelectr. Freq. Control* **2016**, *63*, 737–745. [CrossRef] [PubMed]
22. Brossier, R.; Operto, S.; Virieux, J. Seismic imaging of complex onshore structures by 2D elastic frequency-domain full-waveform inversion. *Geophysics* **2009**, *74*, WCC105–WCC118. [CrossRef]
23. Jo, C.H.; Shin, C.; Suh, J. An optimal 9-point, finite-difference, frequency-space, 2-D Scalar wave extrapolator. *Geophysics* **1996**, *61*, 529–537. [CrossRef]
24. Chen, Z.; Cheng, D.; Feng, W.; Wu, T. An optimal 9-point finite difference scheme for the Helmholtz equation with PML. *Int. J. Numer. Anal. Model.* **2013**, *10*, 389–410. [CrossRef]
25. Pratt, R.G. Frequency-domain elastic wave modeling by finite differences: A tool for crosshole seismic imaging. *Geophysics* **1990**, *55*, 626–632. [CrossRef]
26. Belanger, P.; Cawley, P.; Simonetti, F. Guided wave diffraction tomography within the born approximation. *IEEE Trans. Ultrason. Ferroelectr. Freq. Control* **2010**, *57*, 1405–1418. [CrossRef]
27. Norton, S.J. Iterative inverse scattering algorithms: Methods of computing Fréchet derivatives. *J. Acoust. Soc. Am.* **1999**, *106*, 2653–2660. [CrossRef]
28. Guitton, A.; Symes, W.W. Robust inversion of seismic data using the Huber norm. *Geophysics* **2003**, *68*, 1310–1319. [CrossRef]
29. Huthwaite, P.; Simonetti, F. High-resolution guided wave tomography. *Wave Motion* **2013**, *50*, 979–993. [CrossRef]
30. Treeby, B.E.; Cox, B.T. K-Wave: MATLAB toolbox for the simulation and reconstruction of photoacoustic wave fields. *J. Biomed. Opt.* **2010**, *15*, 021314. [CrossRef]
31. Lowe, M.J.S.; Alleyne, D.N.; Cawley, P. The Mode Conversion of a Guided Wave by a Part-Circumferential Notch in a Pipe. *J. Appl. Mech.* **1998**, *65*, 649–656. [CrossRef]
32. Leonard, K.R.; Malyarenko, E.V.; Hinders, M.K. Ultrasonic Lamb wave tomography. *Inverse Probl.* **2002**, *18*, 1795–1808. [CrossRef]

MDPI
St. Alban-Anlage 66
4052 Basel
Switzerland
www.mdpi.com

Applied Sciences Editorial Office
E-mail: applsci@mdpi.com
www.mdpi.com/journal/applsci

www.ingramcontent.com/pod-product-compliance
Lightning Source LLC
Chambersburg PA
CBHW061613220926
51850CB00005B/498
* 9 7 8 3 7 2 5 8 0 7 5 7 4 *